Molecular and Quantitative Animal Genetics

Molecular and Quantitative Animal Genetics

EDITED BY

Hasan Khatib

University of Wisconsin-Madison
Madison, WI, USA

WILEY Blackwell

Published by John Wiley & Sons, Inc., Hoboken, New Jersey
Published simultaneously in Canada

Library of Congress Cataloging-in-Publication Data

Molecular and quantitative animal genetics / edited by Hasan Khatib.
p. ; cm.
Includes bibliographical references and index.
ISBN 978-1-118-67740-7 (paperback)
1. Animal genetics. 2. Intellectual property. I. Khatib, Hasan, editor.
[DNLM: 1. Animals. 2. Genetics. 3. Molecular Biology. QH 432]
QH432.M65 2015
591.3'5–dc23

2014036482

10 9 8 7 6 5 4 3 2 1

Contents

Contributors

Jennifer Minick Bormann
Department of Animal Sciences and Industry
Kansas State University
Manhattan, KS, USA

Samantha Brooks
Department of Animal Science
Cornell University
Ithaca, NY, USA

José A. Carrillo
Department of Animal and Avian Sciences
University of Maryland
College Park, MD, USA

Christopher H. Chandler
Department of Biological Sciences
State University of New York
Oswego, NY, USA

Leigh Anne Clark
Department of Genetics and Biochemistry
College of Agriculture, Forestry and Life Sciences
Clemson University
Clemson, SC, USA

Noelle E. Cockett
Department of Animal, Dairy and Veterinary Sciences
Utah State University
Logan, UT, USA

Mark E. Cook
Animal Sciences Department
University of Wisconsin–Madison
Madison, WI, USA

Elizabeth A. Crate
Department of Animal and Dairy Sciences
Mississippi State University
Mississippi State, MS, USA
Department of Biology
New College of Florida
Sarasota, FL, USA

Jenifer Cruickshank
Department of Biological Sciences
State University of New York
Oswego, NY, USA

Dirk-Jan de Koning
Department of Animal Breeding and Genetics
Swedish University of Agricultural Sciences
Uppsala, Sweden

Şule Doğan
Department of Animal and Dairy Sciences
Mississippi State University
Mississippi State, MS, USA

Ashley Driver
Division of Human Genetics
Cincinnati Children's Hospital Medical Center
Cincinnati, OH, USA

Amin A. Fadl
Animal Sciences Department
University of Wisconsin–Madison
Madison, WI, USA

Michael G. Gonda
Animal Science Department
South Dakota State University
Brookings, SD, USA

Aruna Govindaraju
Department of Animal and Dairy Sciences
Mississippi State University
Mississippi State, MS, USA

Benjamin W. B. Holman
Animal Science and Genetics
Tasmanian Institute of Agriculture
School of Land and Food
Faculty of Science, Engineering and Technology
University of Tasmania
Sandy Bay, Hobart, Australia

Wen Huang
Department of Genetics
North Carolina State University
Raleigh, NC, USA

Brian W. Kirkpatrick
Department of Animal Sciences
University of Wisconsin–Madison
Madison, WI, USA

Congjun Li
Bovine Functional Genomics Laboratory
Agricultural Research Service
United States Department of Agriculture
Beltsville, MD, USA

Jennifer Long
Bureau for Food Security
USAID
Washington, DC, USA

Michael D. MacNeil
Delta G
Miles City, MT, USA

Aduli E. O. Malau-Aduli
Animal Science and Genetics
Tasmanian Institute of Agriculture
School of Land and Food
Faculty of Science, Engineering and Technology
University of Tasmania
Sandy Bay, Hobart;
School of Veterinary and Biomedical Sciences
Faculty of Medicine, Health and Molecular Sciences
James Cook University
Townsville, Queensland, Australia

Erdoğan Memili
Department of Animal and Dairy Sciences
Mississippi State University
Mississippi State, MS, USA

Hayrettin Okut
Department of Animal Science, Biometry and Genetics
University of Yuzuncu Yil
Van, Turkey

Max F. Rothschild
Department of Animal Science
Iowa State University
Ames, IA, USA

Guilherme J. M. Rosa
Department of Animal Science
University of Wisconsin–Madison
Madison, WI, USA

Jiuzhou Song
Department of Animal and Avian Sciences
University of Maryland
College Park, MD, USA

Alison Starr-Moss
Department of Genetics and Biochemistry
College of Agriculture, Forestry and Life Sciences
Clemson University
Clemson, SC, USA

David L. Thomas
Department of Animal Sciences
University of Wisconsin–Madison
Madison, WI, USA

Alison L. Van Eenennaam
Department of Animal Science
University of California
Davis, CA, USA

Michel A. Wattiaux
Department of Animal Science
University of Wisconsin–Madison
Madison, WI, USA

Kent Weigel
Department of Dairy Science
University of Wisconsin-Madison
Madison, WI, USA

Mulumebet Worku
Department of Animal Sciences
North Carolina Agricultural and Technical State University
Greensboro, NC, USA

Chunhua Wu
Department of Animal, Dairy and Veterinary Sciences
Utah State University
Logan, UT, USA

Manuscript Reviewers

Michael Gonda
Lillian Tong
William Lamberson
George Shook
Jon Schefers
Hayrettin Okut
Dave Thomas
Saleh Shahinfar
Mahdi Saatchi
Levent Turkmut
Ron Lewis
Ron Bates
Luca Funtanasi
Carrie Hammer
George Stradin
Phil Sponenberg
Jim Reecy
Jill Maddox
Ralph Noble
Solomon Abegaz
James Koltes
Chris Haley
Yalda Zare
Derek Bickhart
Ehud Lipkin
Shelia Schmutz
Leslie Lyons
Andrew Nok
Masanori Komatsu
Tapas Saha
Pablo Ross
Jianbo Yao
Hasan Khatib
Jenna Kropp
Rick Monson
Sue Lamont
Jordan Sand
Dan Butz
Gary Anderson
Daniel Kevles
June Blalock

Preface

The challenges in animal genetics for today's students and instructors are to cope with the fast growing body of information related to genetics research, to gain basic skills of problem solving, to make use of genetic knowledge in real life applications, and to deal with the ethical and social aspects of genetics research. The aim of this book is to provide students and instructors with up-to-date knowledge to meet these challenges. Recent developments in high throughput technologies (e.g., next-generation sequencing) has had an enormous impact on our understanding of molecular and quantitative genetics. These new technologies require that both molecular and quantitative skills are needed to deal with the challenges of animal genetics research. Genetics is one of the most dynamic topics in biology in terms of flow of new information on a daily basis.

The electronic format of this book (ebook) will allow frequent updates with new genetic information. Importantly, with the ebook, readers can easily download specific chapters, search for any information in the book, resize fonts, and access different web sites through the links, in addition to many other advantages. The 29 chapters of the book were contributed by 34 authors representing 20 different institutions from the USA, Sweden, Turkey, and Australia. Chapters were peer-reviewed by 41 reviewers. The topics of Molecular and Quantitative Animal Genetics have been chosen to help students understand the concepts, principles, and models of animal molecular and quantitative genetics and to learn to apply these concepts for improving the health, production, and well-being of animals.

The changing environment of twenty-first-century teaching and learning is a challenge faced by both students and instructors. What is the role of the instructor in the era of massive internet information available to the students? What is the role of students in the learning process? What is the difference between teaching and learning? These questions and others are discussed in 1 which provides an overview of teaching vs. learning and discusses the fundamental ways by which learning occurs. The first section "Quantitative and Population Genetics" focuses on the statistical and quantitative aspects of animal genetics. Included in this section are inbreeding and inbreeding depression, genomic selection, crossbreeding, basic genetic model for quantitative traits, heritability and selection, heritability and repeatability, and applications of statistics in quantitative traits. Section 2 "Applications of Genetics and Genomics to Livestock and Companion Animal Species" deals with genetic improvement programs for beef cattle, sheep, dairy cattle, and pigs. It also describes the recent research developments of genetics and genomics of the horse, dog, sheep, and goats. The genetics of color and genetic defects in horses, domestication of dogs, the dog genome, current status of the sheep genome, genetics and domestication of goats, and new genetic technologies are discussed in detail in this section.

Bioinformatics tools in animal genetics research, genome-wide association studies, high throughput technologies, mapping and identifying single genes in animal breeding, genetics of coat color, genetics-nutrition interactions in ruminants, and nutritional epigenomics are covered in Section 3 "Molecular Genetics of Production and Economically Important Traits". Section 4 "Genetics of Embryo Development and Fertility" focuses on the genomics of sex determination, chromosome X inactivation and dosage compensation, chromosome X defects, functional genomics of gametes and embryos, transcriptomics and proteomics of embryos, genetics of in vitro-produced embryos, and genetic screening of embryos. The last section "Genetics of Animal Health and Biotechnology" describes the genetic diversity of the major histocompatibility complex proteins and immunoglobulins, genetics of infectious disease susceptibility, applications and challenges of genetic selection for livestock health, genetic advancement and assessment of animal welfare, genetic selection in animal welfare, animal biotechnology, cloning and genetically-engineered animals, and intellectual property rights and animal resources.

The book is designed to serve students at both undergraduate and graduate levels and readers interested in molecular and quantitative genetics.

I wish to acknowledge the authors for their collaboration and the many colleagues who volunteered to review the book chapters (see lists of contributors and reviewers). Finally, I thank my family (Hanan, Karam, Rawi, and Haya) for being patient with me during evening hours.

Hasan Khatib
Madison, Wisconsin

1 Decoding and Encoding the "DNA" of Teaching and Learning in College Classrooms

Michel A. Wattiaux

Department of Animal Science, University of Wisconsin–Madison, WI, USA

Introduction

The success of the human species on Earth has derived in part from its ability to understand, predict, manipulate, govern, and preserve biological lifecycles from the microscopic to the planetary scale. Since the rediscovery of Mendel's work at the beginning of the twentieth century, the mechanisms of inheritance and nature versus nurture have been enduring themes for geneticists. What a living organism (plant, animal, or microbe) exhibits, its recorded performances, and its observed behaviors, depends in part upon the inherited genes, the environment, and their interactions. At the molecular level, the deoxyribonucleic acid (DNA) encodes the genetic instructions. The DNA, organized in chromosomes, is found in the nucleus of cells that make up tissues and organs, which in turn contribute to the proper functions of the organism as a member of a species in constant dynamic equilibrium with other components of an ecosystem. Irrespective of the unique contribution of molecular biology, cellular and organismal physiology, ecology, and evolution to the field of biology as a whole, DNA is the molecule that fundamentally connects these disciplines to each other. Is there such a thing as the "DNA" of teaching and learning? What are the fundamental structures and processes that underpin the education of human beings? Could one look at kindergarten, primary, secondary, and higher education, the household, the workplace, the places of worship, the public squares, and public libraries as places of teaching and learning that form a continuum made of distinct but interacting parts in one's life, and the life of human communities? Could we think of these parts as forming cultural heritages and educational traditions that humans pass on from one generation to the next? If so, what are the roles of institutions of higher education and what are their functions? What educational purpose do they serve when four years on a university campus is approximately 5% of one's lifespan (Figure 1.1)? Is a scientific fact to critical thinking what DNA is to a gene? Could we imagine teaching and learning as the two complementary strands of a DNA molecule where the information that has been acquired by previous generations is stored and preserved, but where mechanisms for change during duplication are essential to allow for adaptation to changing conditions?

In this chapter we have attempted to address what students do when they learn and what teachers do when they teach at a fundamental level. Our goal was to summarize the current literature only to the extent necessary to challenge long-held views about the role of the instructor, the role of the student, and the traditional instructional design of a college classroom. We hope that by analyzing definitions, exploring theories of learning, and reviewing teaching-related institutional reforms, we have created a context and laid the foundation for a deeper understanding of an instructor's role as the designer of a learning environment. The overarching aim is greater fulfillment and reward for both the instructor and the students as they engage in a college classroom intended to equip the latter for a successful career in the twenty-first century.

Teaching and learning: definitions

Teaching and learning are multifaceted and closely related concepts as revealed in the subtleties of their multiple definitions (Table 1.1 and Table 1.2). A close look at these definitions shows the intricacies of the relationships between teaching and learning. For example, although teaching – the act, practice, or profession of a teacher – is to cause to know, to impart knowledge or skills; to teach is also defined as to accustom to some action or attitude (item 1c, Table 1.1) or to guide a study (item 2, in Table 1.1). Not surprisingly, knowledgeable individuals are not the only sources of teaching. One's own experience can be a "teacher" (item 4b, Table 1.1). Similarly, experience may cause learning to occur. To learn – to gain knowledge, understanding of, or skill in – can be completed by the self, by instruction or by experience as illustrated in Table 1.2 items 5a, 5b, and 5c, respectively. The verbs describing teaching and learning are important because they reflect and reveal aspects of the two contrasting, yet complementary, learning theories that will be discussed next: **behaviorism** and **constructivism**. Although the former implies a cause and effect relationship between teaching and learning, the latter describes the relationship in broader associative terms. For example item 5c (Table 1.2) and item 4b (Table 1.1) can be combined to reflect a constructivist approach to learning that emphasizes mental processes: "First year college is a *learning* experience that *taught* us our limitations."

These definitions reveal the diversity of interpretation, and highlight the complexity of the transactions that take place in the classroom. Similarly, the adherence to either learning theory has

Molecular and Quantitative Animal Genetics, First Edition. Edited by Hasan Khatib.

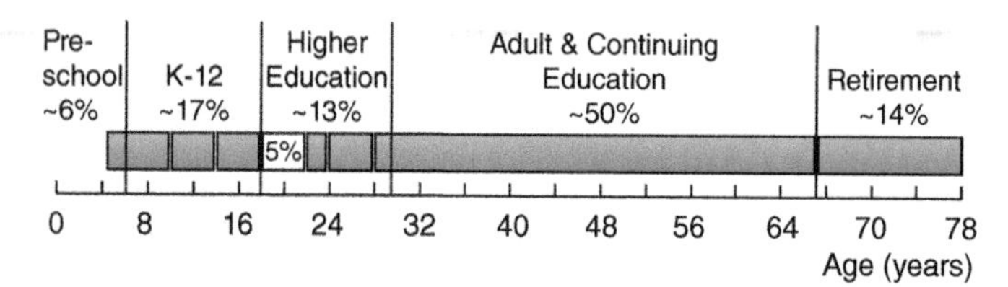

Figure 1.1 Educational stages across the lifespan of a citizen of the United States with an average life expectancy at birth of 78 years.

Table 1.1 Definitions[1] of the verb *to teach* and the noun *teaching*.

Definitions of "to teach"	
1a	To cause to know something <*taught* them a trade>
1b	To cause to know how <is *teaching* me to drive>
1c	To accustom to some action or attitude <*teach* students to think for themselves>
1d	To cause to know the disagreeable consequences of some action <I'll *teach* you to come home late>
2	To guide the studies of
3	To impart the knowledge of <*teach* algebra>
4a	To instruct by precept, example, or experience
4b	To make known and accepted <experience *teaches* us our limitations>
5	To conduct instruction regularly in <*teach* school>
Definitions of "teaching"	
6	Teaching: the act, practice, or profession of a teacher
7	Teaching: something taught (the ideas and beliefs that are taught by a person); *especially* doctrine <the *teachings* of Confucius>

[1]Source: *Merriam-Websters Dictionary* www.merriam-webster.com/.

Table 1.2 Definitions[1] of the verb *to learn* and the noun *learning*.

Definitions of "to learn"	
1a	To gain knowledge or understanding of or skill in by study, instruction, or experience <*learn* a trade>
1b	To memorize <*learn* the lines of a play>
2	To come to be able <*learn* to dance>
3	To come to realize <*learned* that honesty paid>
4[2]	To become informed of; find out <*learned* about the latest findings>
Definitions of "learning"	
	The activity or process of gaining knowledge or skill:
5a	by studying, practicing, <*Learning* ought to be fun>
5b	by being taught <the study guides helped my *learning*>
5c	by experiencing something <First year in college is a *learning* experience>
6	Sum of knowledge or skill acquired by instruction or study <They were people of good education and great *learning*>
7	Modification of a behavioral tendency by experience (as exposure to conditioning)

[1]Source: *Merriam-Webster's Dictionary* www.merriam-webster.com/.
[2]Source: *The Free Dictionary* www.thefreedictionary.com/learn.

profound implications on classroom instructional design and the role played by the instructor and the students therein. For example, unidirectional approaches to teaching have been recognized as providing sub-optimal learning experiences for most students. Quoting the work of educational psychologists (Bloom et al., 1956) who developed a classification of learning objectives to create more holistic forms of education, Kauffman et al. (1971) pleaded with his colleagues in the animal sciences more than 40 years ago to be mindful of their classroom design as they wrote:

> Education experiences in a course should include the recognition of cognitive domain (objectives which deal with ... knowledge and intellectual abilities and skills), the affective domain (objectives which describe changes in interest, motivation attitudes, values and aspirations) and the psychomotor domain (objectives which emphasize manipulation of materials or objects or action which requires neuromuscular coordination).

In addition to these three domains, researchers have recognized the importance of reflection as an essential component of learning (Kolb, 1984). The awareness or analysis of one's own learning or thinking processes, referred to as **metacognition**, has been incorporated in a recently revised version of the taxonomy of learning that identified four knowledge dimensions and characterized them as follows (Anderson and Krathwohl, 2001; Krathwohl, 2002):

- **Factual knowledge**: The basic elements that students must know to be acquainted with a discipline or solve a problem in it;
- **Conceptual knowledge**: The interrelationships among the basic elements within a larger structure that enable them to function together;
- **Procedural knowledge**: How to do something; method of inquiry, and criteria for using skills, algorithms, techniques, and methods;
- **Metacognitive knowledge**: Knowledge of cognition in general, as well as awareness and knowledge of one's own cognition.

Understanding learning

Research in learning

Historically, the data on how people learn have come from the field of educational psychology, which is the study of how humans learn in educational settings. Most of the current recommendations and "best teaching practices" have been derived from more than 100 years of research in child development and efforts to bring learning research into the primary and secondary education ("K-12") classrooms (Bransford et al., 2000). The insights gained from research addressing issues of gender and ethnicity, and specialized fields of investigation targeted at gifted students, students with disabilities, or students from diverse socioeconomic backgrounds have contributed substantially to creating more inclusive and effective learning environments for all. Measures of teaching and learning success have included students' score on standardized tests, reduction in achievement gaps (differential grades of students "at-risk"), and years to graduation and dropout rates. However, there are numerous reasons to caution against simple extrapolation of the K-12 literature to the university classroom. As illustrated in Figure 1.1, college students are in transition from late adolescence ("teenagers") to young adulthood, and thus, in creating learning environments for their students, university instructors may take heed of the emerging literature on adult and continuing education (Cross, 1981; Kazis, 2007) and adult learning theory (Merriam, 2001; 2008). Research in adult and continuing education has focused not only on skills required of individuals interested in promotion within their organization or engaged in career changes but also on meta-objectives such as the understanding that knowledge is neither given nor gotten, but constructed; the ability to assess one's own beliefs; and the realization that learning is a worthy lifelong goal (Taylor and Marienau, 1997). Notably, some of these goals are in alignment with the set of "essential learning outcomes" put forth by the American Association of Colleges and Universities for four-year undergraduate programs in the United States (AAC&U, 2012).

Decisions made by teachers in designing their classroom instructions emerge from their own explicit or implicit adherence to certain views in regards to the nature of teaching and learning. At the risk of oversimplification, we will distinguish in the next sections between two broad theoretical traditions in regards to the fundamental ways by which learning occurs: behaviorism and constructivism. Our attempt is not to describe these theories (which are typically taught in educational psychology courses), but rather the conditions required to make learning happen as implied by the theories.

Behaviorism

The behaviorist views of learning emphasize dualisms: giver and receiver, stimuli and response, right and wrong, reward and punishment. At its roots, behaviorism emerged from early research that succeeded in conditioning (i.e., "training") animals to a certain reflex behavior (a response) from sufficiently repeated pairings with a stimulus, which previously did not elicit the response. This type of research was pioneered by the Russian physiologist Ivan Pavlov (1849–1936) and has led to what is now referred to as **classical conditioning**. Edward Thorndike (1874–1949), another pioneer of this kind of learning formulated the "Law of Effect," which implies that behavior that brings about a satisfying effect (reinforcement) is apt to be performed again, whereas behavior that brings about negative effect (punishment) is apt to be suppressed. Thorndike believed that a neural bond would be established between the stimulus and response. Learning takes place when the bonds are formed into patterns of behavior. However, Burrhus Skinner (1904–1990) considered to be the "grandfather of behaviorism" in educational settings focused on cause-and-effect relationships that could be established by observation, arguing that no scientific measure could address the mental processes that operate in the brain during learning (Roblyer et al., 1997). To Skinner, teaching that brings about learning is a process of arranging effectively "contingencies of reinforcement" or, in other words, arranging situations for the learner in which reinforcement is made contingent upon a correct or desired response. Thus, behaviorists believe that people learn (i.e., change behavior) when stimuli are provided and a voluntary response is either reinforced or punished, a theory referred to by psychologists as **operant conditioning** (Figure 1.2). Teachers and instructional materials are the antecedents/stimuli, whereas the skills that the students demonstrate are the responses/behaviors (Figure 1.2a). Behaviorists concentrate on immediate observable changes in performance (e.g., test scores, athletic level) as indicators of learning and thus they would agree that carefully prepared didactic interactions (e.g., delivery of highly structured lectures), repetition (drills, exercises, practice, etc.) and prompt feedback are important in promoting learning (Figure 1.2b). Skinner believed that high-level capabilities such as critical thinking and creativity could be taught in this way; it was simply a matter of establishing "chains of behavior" through principles of reinforcement.

Behaviorism, however, does not explain all the phenomena observed in learning situations, and as a learning theory it has its limitations. First, it does not account for "internal influences" such as mood, thoughts, and feelings of the learner (i.e., it does not consider internal mental states, emotions, or consciousness).

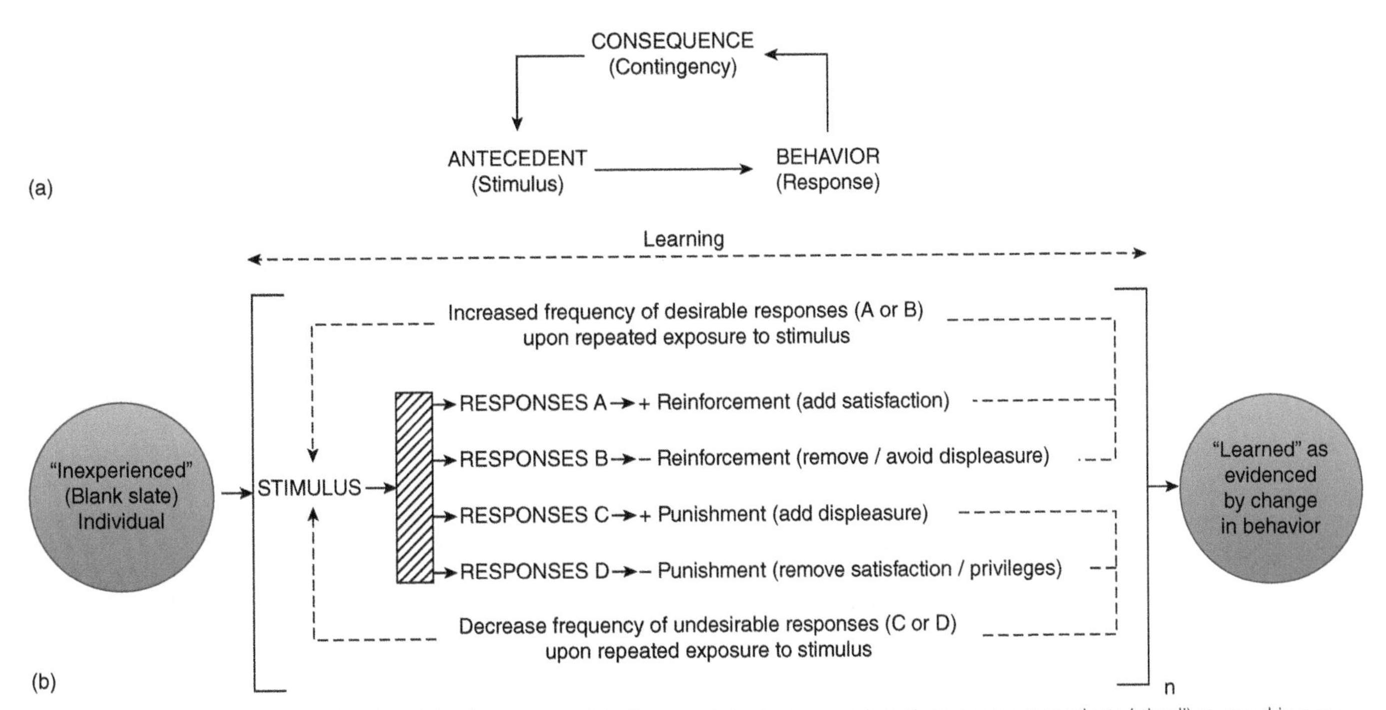

Figure 1.2 Behaviorist approach to learning defined as a measurable change in behavior (response) contingent upon antecedents (stimuli) arranged in a way that alter the frequency of learner-generated trial-and-error responses in which reinforcements are expected to increase the rate of desirable responses and punishments are expected to decrease the rate of undesirable responses.

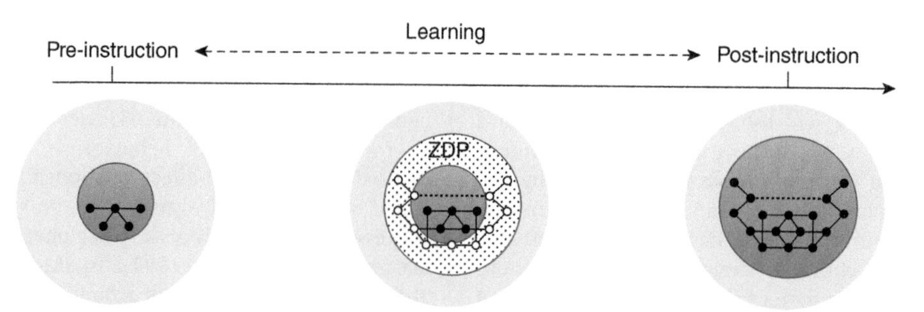

Figure 1.3 Constructivist approach to learning defined as an increasing level of ability in solving complex problem. The zone of proximal development (ZPD) is the area beyond what a student can do alone (blue core circle), but possible with the assistance of a more competent peer or adult with proper temporary scaffolding that helps the learner expand their abilities after internalizing new knowledge and skills (dots), connecting them (lines) into increasingly complex patterns of recognition or abilities.

Second, it does not account for other types of learning that occur without the use of reinforcement and punishment, such as learning based on intrinsic motivation (i.e., learning for the sake of learning). Rewards for attaining a gradually increasing standard of performance may enhance intrinsic motivation in some situations, but undercut it in others, such as when the learner starts with a high level of intrinsic motivation. Third, because the theory assumes that knowledge exists in concrete bits of information, fact and figures independent and separate of the learner, it does not provide a space for the learner to evaluate or reflect as part of the learning process. Thus, in a behaviorist learning environment, a student could confidently assume that the professor knows the answer to any question, and it can only be right or wrong. Finally, behaviorism does not easily account for change in previously established behavioral pattern in response to introduction of new information. In short, excessive reliance on behaviorism in the classroom has led to severe critiques and warnings against the risk of indoctrinating students rather than educating them (Schillo, 1997). Whereas behaviorism is not as dominant today as it was during the middle of the twentieth century among educational psychologists, it has nevertheless provided useful insights for the development of effective classroom instruction techniques (Skinner, 1968), classroom management techniques, and computer-assisted learning, which have been deeply rooted in this approach to learning (Roblyer et al., 1997).

Constructivism

The research in child development by the Russian psychologist Lev Vygotski (1896–1934), the Swiss psychologist Jean Piaget (1896–1980), and many others in the second half of the twentieth century, has provided the foundation for the modern movement in education broadly referred to here as **constructivism**. Toward the end of his life, Piaget challenged the policymakers and educational administrators of his time to rethink the goals of the schools and the educational system as a whole. He asked:

> ... The question is whether schools serve to train children and individuals who are capable of learning what is already known, to repeat what has been gained by prior generations, or to develop creative and innovative minds capable of inventing, discovering, starting at school and then throughout their entire lives.
>
> *(Piaget, undated.)*

The constructivist view of learning emphasizes active mental processes (the very processes deemed unobservable by behaviorists and thus unworthy or scientific experimentation) and the social context in which it takes place. Rather than "acquiring" new knowledge, the learners build their own knowledge because they naturally seek meaning. The learners bring internal motives, distinct levels of aptitudes and consciousness, past experiences, a variety of world-views, preconceptions, and even cultural biases to a learning situation. Learning occurs when one has internalized information and created their own representation of reality by making connection with prior knowledge (or skills) through both individual and social activities. Two inter-related core concepts espoused by constructivists are the **zone of proximal development** (Vygotski, 1978) and **scaffolding** (Wood et al., 1976), which have been merged in the illustration presented in Figure 1.3. The zone of proximal development is "the distance between the current developmental level as determined by independent problem solving and the level of potential development as determined through problem solving under guidance from more competent peers or the guidance of an adult" (Vygotski, 1978). It is in the zone of proximal development that teachers must focus their attention. Scaffolding is a process through which a teacher or a more competent peer helps focus on the elements of the task that are initially beyond the reach of the learner's capacity, but within reach with adequate support. Scaffolds are teaching and learning structures put in place to allow the learner to achieve higher level of performances on their own. However, just as scaffolds in building construction are eventually no longer needed and are removed, instructional scaffolding should eventually become unnecessary as the newly added knowledge or skills have been completely internalized (i.e., have become part of an expanded mental structure or set of skills). Instructional scaffolding is thus the temporary means by which experts (defined here as individuals with higher performance skills than the learner) assist a novice or an apprentice on the path to more experience and greater expertise. For the most part, constructivists see teaching and learning as a highly contextualized process mediated through social interactions.

Thus, constructivism has helped teachers understand that classroom design and implementation should account for diversity in the learners, should rely on a variety of teaching methods and should include formative assessments techniques (CATs) as ongoing processes. Numerous CATs have been tested and are now available (Angelo and Cross, 1993; FLAG, 2012). Furthermore, in a classroom designed according to constructivist principles, the instructor's main role has shifted from a mode of delivery of course content to that of a guide, a facilitator, or a learning coach. However, it would be a misconception to believe that constructivism demands that instructors never "tell" students nor provide "formal" instruction for fear of interfering with their knowledge construction.

Constructivism assumes that all knowledge builds on the learner's previous knowledge, regardless of how one is taught. Thus, even listening to a lecture may involve active attempts to construct new knowledge. However, a more typical constructivist classroom has teams of students working together to form a "knowledge-building" (i.e., learning) community. Such classroom environment could be characterized, "among other things," by activities such as in-class group quiz and problem sets, or out-of-class inquiry-based team projects followed by discussions, written reports, or oral presentations as primary modes of sense-making.

Thus, constructivists would agree that student learning is the focal objective of any instruction. Activities should be student-level appropriate. Constructivist instructors have the ability to perceive students' conceptions of material from the students' perspective. They will see students' "mistakes" as reflecting a particular stage of development. Thus, constructivist instructors do not fall victims to the "expert blind spot" (i.e., the inability to remember what it was like to be a novice learner of the material at-hand; Nathan and Petrosino, 2003). In addition, constructivist instructors recognize that substantial learning occurs in periods of conflict, surprise, over time, and through social interaction (Biggs, 1996; Wood, 1995).

Transferable and life-long learning skills and aptitudes

As illustrated in Figure 1.1, half of one's lifespan is spent in professional settings during which continuing education is becoming increasingly part of the expectation. As indicated earlier, effective teaching and learning in the college classroom should be informed also by the literature on adult and continuing education. In other words, learning at the university should not only be about what is measurable in the short term or what can be done with temporary assistance of a knowledgeable instructor, but should also be about learning to be an adaptable life-long learner, which comes in part with the ability to reflect and exercise metacognition. Wingate (2007) described "learning to learn" in college as first, understanding "learning" and becoming an independent learner, and second, understanding "knowledge" and becoming competent in constructing knowledge within the discipline (Table 1.3). In addition to delivering knowledge, higher education curricula and classrooms should be designed to prepare students with a set of skills and experiences that reflect certain habits of the mind (Wattiaux, 2009). Furthermore, the curricula and the classroom should emphasize transferable skills and aptitudes that are relevant in most professional situations, such as oral and written communication, problem solving, creative thinking, interpersonal skills (e.g., respect of diverse views), leadership skills (e.g., self-motivation, decisiveness, risk-taking), and personal aptitudes (e.g., integrity, reliability, diligence) among many others (Assiter, 1995). Transferable and life-long learning skills are important not only as a means for university graduates to stay current in their professional fields, but also for personal enrichment and enhancement of quality of adult lives (Dunlap and Grabinger, 2003). Authors have grouped life-long learning skills in three categories:

1. Capacity for "self-directed" learning (e.g., identify problems, establish goals, develop an action plan, identify critical resources, capture and use relevant information),
2. Capacity for metacognitive awareness (e.g., the ability to reflect, to strategize, and to self-assess),
3. Capacity toward long-term results (risk taking, intellectual curiosity, seeking deep understanding, gaining intrinsic motivation).

Table 1.3 Components of "Learning to Learn" in higher education as (a) understanding "learning" and (b) understanding "knowledge" of a discipline (reproduced with permission from Wingate, U., 2007. A framework to transition: supporting "learning to learn" in higher education. *Higher Education Quarterly* 61(3):391–405).

(a) Becoming competent in independent learning	(b) Becoming competent in knowledge construction
1. Gaining awareness of conceptions of learning and knowledge in the discipline	
2. Assessing one's current abilities as learner	**2.** Approaching information (lectures, texts) in a focused and critical manner
3. Setting short-term and long-term goals and targets	**3.** Evaluating existing knowledge
4. Planning action for reaching targets	**4.** Synthesizing different sources into a coherent argument
5. Monitoring progress in reaching targets	**5.** Expressing own voice.
6. Evaluating progress/achievements	

Understanding teaching

Research in teaching

In contrast to research on the learner and the learning, the research on the teacher and the teaching has been more limited and has a shorter history, especially in institutions of higher education. Over the course of the twentieth century, higher education has not benefited to the same extent as K12 from the type of research in which a community of educational scientists uses their training and expertise to study teaching effectiveness. Furthermore, with the availability of students' performance on standardized tests as a metric of student learning, researchers have revealed what effective K12 teachers do in their classrooms and the extent of their influence on students' performance (Stronge et al., 2011). In comparison, the quality of undergraduate education in institutions of higher education has suffered for the lack of attention, especially as research accomplishments prevailed as the principal means of recognition and prestige for both individual scientists and their institutions throughout the second half of the twentieth century (Serow, 2000). However, about 20 years ago, the Carnegie foundation for the advancement of teaching released the Boyer's report: "Scholarship Reconsidered: Priorities of the Professoriate" (Boyer, 1990) challenging universities to expand the definition of scholarship beyond the realm of scientific discoveries. The report identified three other types of scholarships that had been core to the professional identity of university professors in the United States since the inception of the land grant system, namely, the **scholarship of teaching**, the **scholarship of outreach**, and the **scholarship of integration**. Since then, at least two distinct areas of research to enhance the quality of teaching in higher education have emerged. The first area focused on changing the institution whereas the second focused on changing the classroom.

Changing the institutional paradigm

Throughout the second half of the twentieth century there were few institutional concerns about the quality of undergraduate education. As institutions were competing to hire the most promising scientists, the assumption that good researchers were good teachers was – and unfortunately, for the most part remains – pervasive. The presence of a disciplinary expert in the classroom was deemed a sufficient guarantee of teaching quality. However, since Boyer's report (Boyer, 1990), research has been conducted to guide and document institutional reform and there is now an increasing body of literature on institutional structure and organization to sustain and advance the educational mission of the university and the instructors within (Dooris, 2002). The following examples illustrate the issues that have been documented through this type of research:

- The relationship between research and teaching (Marsh and Hattie, 2002), and the belief system about teaching at research universities (Kane et al., 2002; Wright, 2005);
- The financial compensation disparity between teaching and research (Fairweather, 2005);
- The faculty evaluation and accountability system (Arreola, 2000);
- The reward system through promotion and tenure (Keele, 2008) and whether multiple forms of scholarship has been increasingly rewarded or not in the last decade (O'Meara, 2005; 2006);
- The current systems of expectations and standards of faculty performances in research universities (Colbeck, 2002; Hardré and Cox, 2009).

Among current concerns is an institutional failure to support adequate teaching-related training for graduate students aspiring to a professional career in academia. At the turn of the twenty-first century, roughly 70% of instructors in the 3200 institutions offering four-year degrees in the United States had received their Ph.D. from one of the roughly 150 "Research and Doctoral" (R&D) institutions (Bob Mathieu, personal communication). Most Ph.D. programs at R&D institutions do not encourage, let alone require, training in undergraduate education and, therefore, most of the current instructors in the United States' higher educational system were utterly unprepared to engage undergraduate students in a semester-long course as they began their academic career. Although some may argue otherwise, there are some signs of change as an increasing number of R&D institutions have begun to offer their graduate students and post-doctoral trainees opportunities to gain knowledge, skills, and meaningful experiences related to undergraduate education (CIRTL, 2012). Furthermore, teaching experience and teaching philosophy have increasingly become points of differentiation in the screening and selection of candidates for faculty positions.

Changing the classroom paradigm

In parallel to institutional reform that began in the 1990s; another important movement emerged as committed educators pointed to a necessary paradigm shift from teaching to learning in the classroom (Barr and Tagg, 1995; Bass, 1999; 2012; Shulman, 1999). Focusing on student learning outcomes rather than the instructor teaching performance has led to what is known as the Scholarship of Teaching and Learning (SoTL). Committed teachers have taken upon themselves to conduct "classroom research" as a way to improve the quality of teaching and learning within their own discipline (Cross and Steadman, 1996; Reagan et al., 2009). Typically, this type of research involves an institutional review board (IRB) approved data collection protocol designed to address pedagogical issues with the intent of contributing to a body of peer-reviewed literature, as illustrated in Crouch and Mazur (2001), Wattiaux (2006), or Wattiaux and Crump (2006). As SoTL gained recognition, confusions arose for lack of a clear definition and differentiation from "good" teaching or "excellent" teaching. Fortunately, Kreber (2002) outlined the distinction among excellence in teaching, expertise in teaching, and the SoTL. In addition, the same author helped bring some consensus to delineate the defining features of SoTL, which accordingly is an activity that, in the context of promoting student learning, meets a series of criteria. For example, it requires high levels of discipline-related expertise, it should break new ground, be replicable, peer-reviewed, and provide significant or impactful insights that can be elaborated upon by others (Kreber 2003).

Although Asmar (2004) described successful attempts to engage the faculty of a research-intensive university to improve their teaching practices, SoTL has met with resistance for a number of reasons. Our current Ph.D. programs rarely engage graduate students in issues related to undergraduate teaching and learning. A lack of knowledge and training is likely associated with a lack of confidence, interest, and motivation for SoTL as a basis to build a successful academic career. Similarly, institutional priorities and reward systems rarely foster faculty engagement in SoTL. According to Boshier (2009) low adoption of SoTL may be related also to continued confusion about the concept, the difficulty to operationalize classroom research in absence of appropriate resources (e.g., lack of expertise and financial support), the over-reliance on peer-review publications as the main criteria to measure scholarship, and the fact that SoTL fits poorly in the twenty-first century modus operandi of universities as businesses delivering education as a commodity to be sold. Notwithstanding the uncertain future of the SoTL in its current form, college science classroom research (the contextualized environment in which the teaching and learning is taking place) has been identified along with brain research (Taylor and Lamoreaux, 2008) and cognitive psychology (which focuses on mental processes including how people think, perceive, remember, speak, and solve problems) as the three components of trans-disciplinary research that will advance our ability to create effective university classroom in the twenty-first century (Wieman, 2012). Furthermore, it is clear that engaging in SoTL should be dependent upon one's institutional context and professional priorities. Not all STEM (science, technology, engineering, and math) scientists should be expected to become educational scientists within their classroom. However, the difference between excellence in teaching and SoTL is not entirely categorical as most features and evaluation criteria lay on continuous scales most indicative of one or the other (Wattiaux et al., 2010). For example, reflective practices (Kane et al., 2004), classroom assessment techniques (FLAG, 2012), and engagement in faculty learning communities are simple steps that may transform one's classroom overtime and place an instructor on the path towards excellence in teaching without necessarily engaging in SoTL.

Implications for classroom design in the twenty-first century

Architecture of an effective classroom

As described so far in this chapter, the implicit or explicit assumptions made about learning, the institutional context, and the inclination of an instructor to engage in teaching improvement initiatives are only some of the factors that influence the instruc-

Table 1.4 Example of an alignment scheme among instructors' activities, students' activities and tests to assess learning using the revised Bloom's Taxonomy (Krathwohl, 2002) as a frame of reference.

Cognitive process dimension[1]	Examples of instructional activities (instructor)	Examples of learning activities (student)	Examples of how students demonstrate learning	Examples of how instructors assess learning
1. Remember: Ability to retrieve knowledge from long-term memory	Lecture; Tell; Show	Read; Review; List; Match	Memorize and recite; Name; Define; Recognize	True-False; Multiple choices
2. Understand: Ability to make sense of oral, written, mathematical or graphical information	Discuss (Q&A); Demonstrate; Illustrate	Give examples; Explain; Solve problems; Infer	Write a summary; Complete worksheet; Compare	Short answers; Mini-essay; Calculation; Comparison
3. Apply: Ability to carry out or use a procedure in a given situation	Train; Coach; Guide	Work on scenarios; Use a procedure	Solve real-world problems	Written report; Oral presentation
4. Analyze: Ability to break material into parts, detect connections and an overall purpose	Provide resources; Model	Compare sources of information, scenarios, or procedures	Organize information; Differentiate associations from causality	Assess a literature review; Evaluate oral answers or reasoning
5. Evaluate: Ability to make judgments based on criteria and standards	Provide resources; Facilitate	Assess; Critique; Check	Provide feedback to peers	Assess student's ability to provide feedback to peers; Portfolio
6. Create: Ability to put elements together to form a coherent and original product	Provide resources; Collaborate	All of the above as needed	All of the above as needed	Assess the design and content of a web page or a research proposal

[1]Listed here as "learning objectives": listed in increasing order of complexity and level of abstraction (from concrete to abstract). Note that the mastery of each simpler category is assumed to be a pre-requisite to mastery of the next more complex one. Level 3 and above are usually referred to as "critical-thinking" skills.

tional design of a university classroom. Interestingly, undergraduate instructors willing to reconsider the architecture of their classroom may act at times as behaviorists ("tell me what to do!") and at times as constructivists ("let me try this!"). The propensity toward one or the other may depend upon professional expectations and personal factors such as intrinsic motivation, level of knowledge, prior experience, prospect of reward, and the context of a particular classroom (e.g., the type of course and the intended learning outcomes).

Regardless of the context, the seven principles for good practice in undergraduate education promoted since the late 1980s (Chickering and Gamson, 1987) remain a solid foundation and a useful guide in designing an impactful undergraduate classroom. These principles are:

1. student-faculty contact,
2. cooperation among students,
3. active learning,
4. prompt feedback,
5. time on task,
6. high expectations, and
7. respect for diverse talents and ways of learning.

Since then, however, books have been written in designing effective courses and curricula (Diamond, 1998; Wiggins and McTighe. 2006). This recent literature emphasizes the importance of proper alignment among three essential components of any course syllabus: the intended learning outcomes, the teaching and learning activities, and the learning assessment (i.e., the grading scheme). It is not difficult to appreciate that this type of alignment serves both the constructivist-leaning classrooms (Biggs, 1996) and the behaviorist-leaning ones. As designers of learning environments, instructors should make these alignments deliberately at the planning stage or as part of the revamping of a course. Table 1.4 was constructed to illustrate what instructors "do" and correspondingly what students "do" at each of the six cognitive processes recently published as a revised Bloom's taxonomy of learning (Krathwohl, 2002). Instructional activities, learning activities, how students demonstrate learning, and how instructors assess learning should vary substantially given the intended learning outcome (remember, understand, apply, analyze, evaluate, and create). Thus, there should be a conscious effort to provide the learner with clearly stated goals, activities that are appropriate for the task, and assessment criteria that reflect the intended goals. In their book *Understanding by Design*, Wiggins and McTighe (2006) coined the phrase "backward design" to describe instructional design as a process that includes the following three sequential steps:

1. Identifying the desired learning outcomes;
2. Determining the acceptable evidences [of learning];
3. Planning the instruction and learning experience.

Although this three-step approach may be counterintuitive, it demands that instructors focus first on the "end-point" (i.e., the goal), then determine how to assess the desired knowledge or skills to be gained (e.g., create the exams, homework, or rubrics for written reports that will be used to assign grades), and finally decide on what tools to use from one's teaching toolbox to provide students with the most appropriate learning experience (i.e., instruction). Note that the backward design is a scalable process that may be applied not only when writing the syllabus

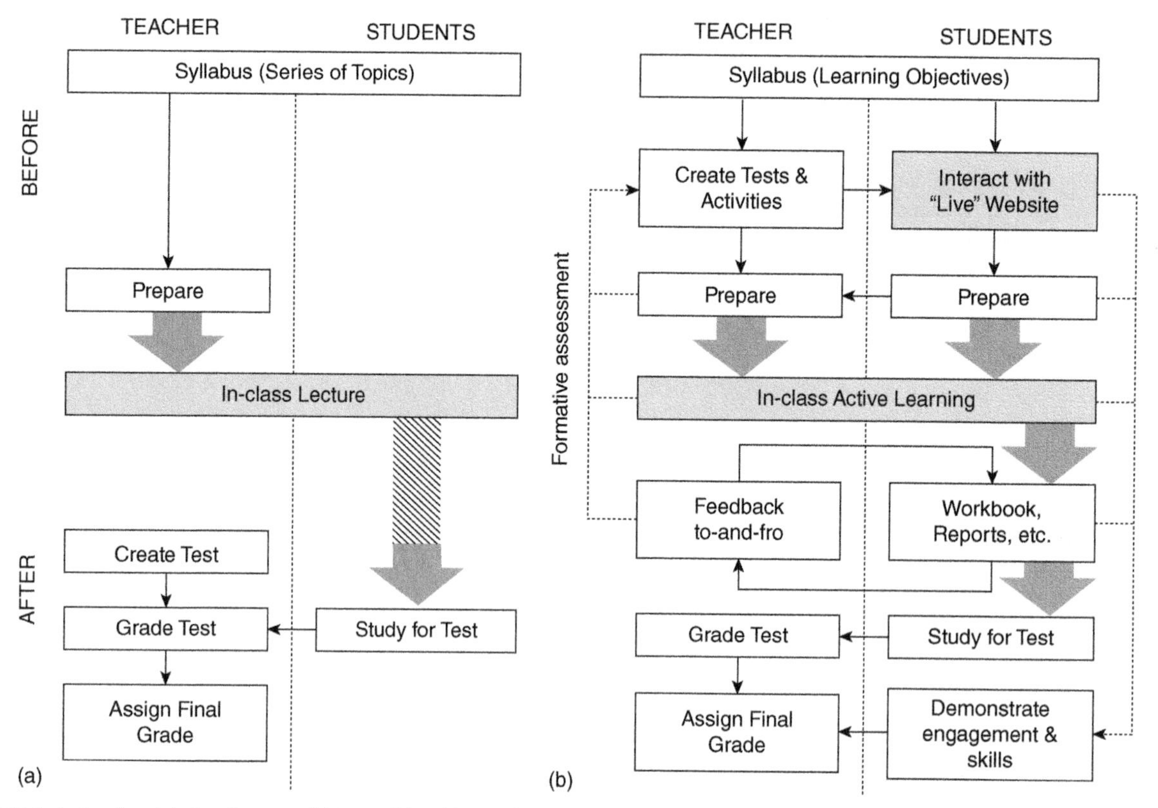

Figure 1.4 Main instructional design features of (a) a traditional lecture-based course and (b) a course designed as an interactive learning environment, highlighting what the teacher and the students do before, during, and after a period of classroom instruction and how grades are assigned.

of a course, but also in planning each unit within a course, and each class period within the unit. Similarly, the concept can be used to improve coordination among courses in a curriculum (Cook et al., 2006).

Instructional features for life-long learning

Separate from the constructivist approach to instructional design discussed so far, there are five instructional features to foster life-long learning as suggested by Dunlap and Grabinger (2003):

- Develop student autonomy, responsibility, and intentionality (e.g., set learning goals, assess current knowledge, set time lines);
- Provide intrinsically motivating learning activities (e.g., relate learning to personal needs and goals, solve problems that students might encounter in their non-school lives, place the learners in authentic decision-making roles to which they may aspire);
- Enculturation into a community of practice (e.g., help students learn the habits of minds, the tools and approaches of the professionals they regards as their role models – a concept referred to as "signature pedagogy": Reagan et al., 2009; Wattiaux, 2009);
- Encourage discourse and collaboration (e.g., engage in debates, authentic team-work, or peer evaluation and review);
- Encourage reflection (e.g., engage students in oral and written self-evaluation of their learning with journals, online posting or portfolios).

As suggested by Onderdonk et al. (2009), adult and continuing education should be essentially problem-centered rather than content-centered. Incorporating instructional features for life-long learning in the design of a college classroom may not only lay the foundation for life-long learning but may also contribute to enhance academic achievement (Masui and De Corte, 2005).

Toward an active learning classroom

A comparative analysis of what instructors do and what students do in two contrasting classroom environments has been illustrated in Figure 1.4. In the traditional lecture-based interactions (Figure 1.4a), an instructor typically spends considerable amounts of time prior to class interacting with the material to find the "best way" to present it so that students can "understand" it. A major assumption of this approach is that there is one way to deliver the lecture so that (almost) all students in the class will grasp the material. Not only this approach disregard the diversity in ways of understanding a body of knowledge but it also implies that the instructor "pitch" the material at a level that is neither to boring nor to challenging for an "average" student in the class. Furthermore, using this approach, instructors are at high risk of missing the "target" because of the so-called expert blind spot discussed previously (Nathan and Petrosino, 2003). In this type of classroom, students are rarely encouraged to engage deeply with the material until a scheduled exam is approaching. Students will be rewarded (positively reinforced) with good grades or punished (negatively reinforced) with poor grades based essentially on a limited number of test scores. Instructors following this model of instruction are likely to be teacher-centered and content-delivery oriented. Also, they are likely to view their role as imparting the content of a discipline rather than supporting students' growth and development (Wingate, 2007).

In contrast, Figure 1.4(b) was constructed to illustrate how this author has attempted to use elements of the constructivist learning paradigm, the backward design process and instructional technology (a course website) in an attempt to create a more student-centered and content-centered learning environment in classes with enrollment of 20–30 students. First, the instructional website serves in part as the repository and mode of delivery of course content that students are expected to view or read as pre-assigned material, and interact with using online quizzes and study guides. The required posting of short comments, questions, or concerns as "blog entries" provides the instructor with an assessment of students' levels of engagement and understanding of the material prior to class. Using students' own "thinking," as a form of formative assessment, the instructor can prepare class activities to engage students in high-order cognitive processes described in Table 1.4. Table 1.5 provides examples of in-class activities that build on students' preparation and thus substitute partially or entirely for PowerPoint ® presentations. Note also in Figure 1.4(b) that after class, students are expected to continue interact with the material and document their learning by completing out-of-class assignments (workbook, project-based activities, etc.). Eventually students' grades are assigned as a combination of level of engagement (demonstrated by out-of-class participation), demonstrated skills gained in completing multi-stage class projects, and test scores reflecting knowledge and understanding of course content. Although this approach to grading students has proven useful, measuring student learning remains a tremendous challenge. The documentation of a three-year transition from lecturing (Figure 1.4a) to using discussion as the primary mode of instruction as described in Figure 1.4(b) in an upper class dairy nutrition course can be found in Wattiaux and Crump (2006).

Table 1.5 Examples of learning activities that may be used to substitute for PowerPoint lectures: What to do in a 50 minute lecture to engage students?

Simple and easy to implement in-class "shared" learning activities	• Pairs of students discuss pre-assigned material and identify an issue for subsequent discussion with the class • Instructor addresses issues raised by students in pre-class web-postings ("discussion board") • Students complete homework or previous year's exam in groups • Students take a group quizzes followed by discussion of quiz items • Instructor uses online material (videos) • Introduce a case-scenario and instruct students to answers multiple choices questions and make recommendations, or predict outcomes based on the information provided • Students make short oral presentations (5-min/student) or larger (group based) PowerPoint presentations of out-of-classroom projects • Mini instructional PowerPoint lecture, only to introduce a topic or synthesize a topic • Students are placed in teams and instructed to draw a figure to illustrate a concept followed by presentation to the whole class • Peer Instruction (Crouch and Mazur, 2001)
More elaborate and complex in-class learning activities	• Wet lab or computer lab to synthesize and/or provide practical applications and illustrations of principles, concepts and theories • Instructor demonstrates a procedure • Field trips • Debates • Case studies
Classroom Assessment Techniques (CATs)	• Students completed a minute paper • Students complete a short survey to evaluate a class project or activity • Students complete an non-graded in-class quiz • Students construct concept maps, analyze figure and tables • Students respond to questions with clickers or an equivalent technique

Final thoughts

Given its intentionality, student learning should be inherently related to teaching in the college classroom. Nevertheless, as anyone who has taught but one course knows – and if students' grades or course evaluations are any indications – teaching and learning may not always be as closely related to one another as one would wish for. It is to be expected that large differences in student learning outcomes occur in response to a given instructional environment. However, research results are becoming increasingly clear. Although the teaching and the learning are rather dichotomous in teacher-centered classrooms, the two are much more tightly correlated when the instructional design incorporates constructivist and life-long learning features.

To end, let's return to, and build on, our introductory metaphor. As Mendel (1822–1884) was attending his experimental plots on plant hybridization in the Augustinian Abbey of Central Europe, another naturalist, the Englishman Charles Darwin (1809–1882) was making history by publishing a book called *On the Origin of Species ...* after traveling around the world. Both men were profoundly intrigued by the mysteries of nature and biological variation. The former gave us the basic laws of inheritance, whereas the latter gave us the theory of evolution. Their work was, however, only the beginning of the journey. As the genome of human and other species have been decoded by the end of the twentieth century, geneticists have moved from reading the letters of the genetic alphabet to reading its words, sentences, paragraphs, and now chapters of genetic information built in the DNA. New fields of investigations (e.g., epigenetics) and applications (e.g., genomics) have emerged. Although advances in our understanding of biological processes in the last 150 years have been mind-boggling, one could wonder about the trajectory of scientific research in teaching and learning. Thorndike Skinner, Piaget, and Vygostki became giants in their fields during their lifetime or posthumously, and collectively they gave us deep insights into basic aspects of learning (and hence teaching). However, so many basic questions remain unanswered and so much variation remains unexplained in our understanding of the college classroom that one wonders whether the field of education is today where the understanding of biological variation was at the time of Mendel and Darwin? One can only contemplate how future research in teaching and learning will transform the college classrooms.

References

AAC&U (Association of American Colleges and Universities). 2012. *The Essential Learning Outcomes*. Online http://www.aacu.org/leap/documents/EssentialOutcomes_Chart.pdf (accessed April 28, 2014).

Anderson, L. and Krathwohl, D. 2001. *Taxonomy for Learning, Teaching and Assessing: A Revision of Bloom's Taxonomy of Educational Objectives*. New York, NY: Longman.

Angelo, T. A. and Cross, K. P. 1993. *Classroom Assessment Techniques: A Handbook for College Teachers*. San Francisco, CA: Jossey Bass.

Arreola, R. 2000. *Developing a Comprehensive Faculty Evaluation System*, 2nd edn. Bolton, MA: Anker.

Asmar, C. 2004. Innovations in scholarship at a student-centered research university: an Australian example. *Innovative Higher Education* 29:49–66.

Assiter, A. (ed.). 1995. *Transferable Skills in Higher Education.*, London: Kogan Page.

Barr, R. B. and Tagg, J. 1995. From teaching to learning – a new paradigm for undergraduate education. *Change* 27:12–25. Online www.ius.edu/ilte/pdf/BarrTagg.pdf (accessed April 28, 2014).

Bass, R. 1999. The scholarship of teaching: What's the problem? *Invention* 1(1), Online: https://my.vanderbilt.edu/sotl/files/2013/08/Bass-Problem1.pdf (accessed May 11, 2014).

Bass, R. 2012. Disrupting ourselves: The problem of learning in higher education. *Educause Review* 47:22–33. Online: www.educause.edu/ero/article/disrupting-ourselves-problem-learning-higher-education (accessed April 28, 2014).

Biggs, J. 1996. Enhancing teaching through constructive alignment. *Higher Education* 32:347–364.

Bloom, B. S., Engelhart, M. D., Furst, E. J., Hill, W. H., and Krathwohl, D. R. 1956. *Taxonomy of Educational Objectives: the Classification of Educational Goals; Handbook I: Cognitive Domain*. New York: Longmans Green.

Boshier, R. 2009. Why is the scholarship of teaching and learning such a hard sell? *Higher Education Research & Development* 28:1–15.

Boyer, E. L. 1990. *Scholarship Reconsidered: Priorities of the Professoriate*. Princeton, NJ: The Carnegie Foundation for the Advancement of Teaching.

Bransford, J. D., Brown, A. L., and Cocking, R. R. (eds). 2000. *How People Learn: Brain, Mind, Experience, and School*. Washington D.C.: National Academy Press.

CIRTL (Center for the Integration of Research, Teaching and Learning). 2012. CIRTL Network. Online: www.cirtl.net/ (accessed April 28, 2014).

Chickering, A. W. and Gamson, Z. 1987. Seven principles of good practice in undergraduate education. *American Association for Higher Education (AAHE) Bulletin* 39: 3–7. Online: http://files.eric.ed.gov/fulltext/ED282491.pdf (accessed May 11, 2014).

Cook, M. D., Wiedenhoeft, M. H., and Polito, T. A. 2006. Using outcomes assessment to change classroom instruction. *Journal of Natural Resources and Life Sciences Education* 35:42–47.

Cross, K. P. 1981. *Adults as Learners*. San Francisco, CA: Jossey-Bass.

Cross, K. P. and Steadman, M. H. 1996. Introduction to Classroom Research. In *Classroom Research, Implementing the Scholarship of Teaching*, pp. 1–28. San Francisco, CA: Jossey-Bass.

Colbeck, C. L. (ed.) 2002. New Directions for Institutional Research, Special issue: *Evaluating Faculty Performance*, pp. 1–108, Hoboken, NJ: Wiley Online Library, Wiley Periodicals, Inc., A Wiley Company.

Crouch, C. H., and Mazur, E. 2001. Peer instruction: Ten years of experience and results. *American Journal of Physics* 69(9):970–977.

Diamond, R. 1998. *Designing and Assessing Courses nd Curricula, A Practical Guide*. Revised edn. San Francisco, CA: Jossey-Bass.

Dooris, M. J. 2002. Institutional research to enhance faculty performance. *New Directions for Institutional Research* 114:85–95.

Dunlap, J. and Grabinger, S. 2003. Preparing students for lifelong learning: A review of instructional features and teaching methodologies. *Performance Improvement Quarterly* 16(2):6–25.

Fairweather, J. 2005. Beyond the rhetoric: trends in the relative value of teaching and research in Faculty salaries. *The Journal of Higher Education* 76:401–422.

FLAG (Field-tested Learning Assessment Guide) for Science, Math, Engineering, and Technology Instructors. 2012. Online: www.flaguide.org/ (accessed April 28, 2014).

Hardré, P. and Cox, M. 2009. Evaluating faculty work: Expectations and standards of faculty performance in research universities. *Research Papers in Education* 24:383–419.

Kane, R., Sandretto, S., and Heath, C. 2002. Telling half the story: A critical review of research on the teaching beliefs and practices of university academics. *Review of Educational Research* 72(2):177–228.

Kane R., Sandretto S., and Heath, C. 2004. An investigation into excellent tertiary teaching: emphasizing reflective practice. *Higher Education* 47:283–310.

Kauffman, R. G., Thompson, J. F., Anderson, D. B., and Smith, R. E. 1971. Improving the effectiveness of teaching animal science. *Journal of Animal Sciences* 32(1):161–164.

Kazis, R. 2007. *Adult Learners in Higher Education Barriers to Success and Strategies to Improve Results* [Washington, D.C.]: U.S. Dept. of Labor, Employment and Training Administration, Office of Policy Development and Research.

Keele, J. 2008. Comparing research and teaching in university promotion criteria. *Higher Education Quarterly* 62(3):237–251.

Krathwohl, D. R. 2002. A revision of Bloom's taxonomy: An overview. *Theory Into Practice* 41(4):212–218.

Kolb, D. A. 1984. *Experiential Learning, Experience as the Source of Learning and Development*. Englewood Cliffs, New Jersey: Prentice Hall.

Kreber, C. 2002. Teaching excellence, teaching expertise, and the scholarship of teaching. *Innovative Higher Education* 27:5–23.

Kreber, C. 2003. The scholarship of teaching: A comparison of conceptions held by experts and regular academic staff. *Higher Education* 46:93–121.

Marsh, H. W., and Hattie, J. 2002. The relation between research productivity and teaching effectiveness; Complementarity, antagonistic, or independent constructs? *Journal of Higher Education* 73(5):603–641.

Masui, C., and De Corte, E. (2005). Learning to reflect and to attribute constructively as basic components of self-regulated learning. *British Journal of Educational Psychology* 75(3):351–372.

Merriam, S. B. 2001. Editor's Notes Special Issue: The New Update on Adult Learning Theory. *New Directions for Adult and Continuing Education* 2001(89):1–2.

Merriam, S. B. 2008. Editor's Notes Special Issue: The New Update on Adult Learning Theory. *New Directions for Adult and Continuing Education* 2008(119):1–4.

Nathan, M. J. and Petrosino, A. J. 2003. Expert blind spot among preservice teachers. *American Educational Research Journal* 40(4): 905–928.

O'Meara, K. A. 2005. Encouraging multiple forms of scholarship in faculty reward systems: Does it make a difference? *Research in Higher Education* 46(5): 479–510.

O'Meara, K. A. 2006. Encouraging multiple forms of scholarship in faculty reward systems: Have academic cultures really changed? *New Directions for Institutional Research* 129(Spring):77–95.

Onderdonk, J. C., Allen D., and Allen D. 2009. Technology and learning: reimagining the textbook. *Journal of Continuing Higher Education* 57:120–124.

Piaget, J. Undated. Video clip retrieved December 31, 2012 from www.youtube.com/watch?v=AyJzvZiCpgo&feature=related (accessed April 28, 2014).

Reagan, N. L. C., Gurung, A. R., and Haynie, A. (ed.). 2009. *Exploring Signature Pedagogies: Approaches to Teaching Disciplinary Habits of Mind*, pp. 1–318. Sterling, VA: Stylus Publishing, LLC.

Roblyer, M. D., Edwards, J., and Havriluk, M. A. 1997. Learning Theories and Integration Models. In *Integrating Educational Technology into Teaching*, pp. 54–79. Saddle River, NJ: Prentice Hall.

Schillo, K. K. 1997. Teaching animal science: Education or indoctrination. *Journal of Animal Science* 75(4):950–953.

Serow, R. C. 2000. Research and teaching at a research university. *Higher Education* 40(4): 449–463.
Shulman, L. S. 1999. Taking learning seriously. *Change* 31:11–17.
Skinner, B. F. 1968. *The Technology of Teaching*. New York: Appleton-Century-Crofts.
Stronge, J. H., Ward, T. J., and Grant, L. W. 2011. What makes good teachers good? A cross-case analysis of the connection between teacher effectiveness and student achievement. *Journal of Teacher Education* 62:339–355.
Taylor, K., and Marienau, C. 1997. Constructive-development theory as a framework for assessment in higher education. *Assessment and Evaluation in Higher Education* 22(2):233–243.
Taylor, K., and Lamoreaux, A. 2008. Teaching with the brain in mind. *New Directions for Adult and Continuing Education* 2008(119):49–59.
Vygotski, L. S. 1978. *Mind and Society: The Development of Higher Mental Processes*. Cambridge, MA: Harvard University Press.
Wattiaux, M. A. 2006. Preparing sophomores for independent learning experiences with a pre-capstone seminar. *NACTA Journal* 50(3):19–25.
Wattiaux, M. A. 2009. Signature pedagogy in agriculture. In N. L. C. Reagan, A. R. Gurung, and A. Haynie (eds), *Exploring Signature Pedagogies, Approaches to Teaching Disciplinary Habits of Mind*, pp. 207–223. Sterling, VA: Stylus Publishing, LLC.
Wattiaux, M. A. and Crump, P. 2006. Students' perception of a discussion-driven classroom environment in an upper-level ruminant nutrition course with small enrollment. *Journal of Dairy Science* 89(1):343–352.
Wattiaux, M. A., Moore, J. A., Rastani, R. R., and Crump, P. M. 2010. Excellence in teaching for promotion and tenure in animal and dairy sciences at doctoral/research universities: A faculty perspective. *Journal of Dairy Science* 93(7):3365–3376.
Wieman, C. 2012. *Taking a scientific approach to science and engineering education*. Presented March 10, 2012, Wisconsin Institutes for Discovery, Madison, WI. Online http://discovery.wisc.edu/home/discovery/recorded-lectures/carl-wieman-32012/ (accessed April 28, 2014).
Wiggins, G., and McTighe, J. 2006. *Understanding by Design*. Upper Saddle River, NJ: Pearson Education, Inc.
Wingate, U. 2007. A framework to transition: supporting "learning to learn" in higher education. *Higher Education Quarterly* 61(3):391–405.
Wood, D., Bruner, J., and Ross, G. 1976. The role of tutoring in problem solving. *Journal of Child Psychology and Child Psychiatry* 17:89–100.
Wood, T. 1995. From alternative epistemologies to practice in education: Rethinking what it means to teach and learn. In L. Steffe and J. Gale (eds), *Constructivism in Education*, pp. 331–339. Hillsdale, NJ: Erlbaum.
Wright, M. 2005. Always at odds?: Congruence in faculty beliefs about teaching at a research university. *The Journal of Higher Education* 76(3):331–353.

Review questions

1. Directions: In the list are examples of simple sentences that describe each of the meaning of the word "To Teach" as defined in Table 1.1. Match each sentence with a meaning as explained in Table 1.1. (Note: that more than one choice may apply – discuss your choices with a partner.)

a. He enjoys **teaching** his students about history (#2).
b. She **taught** English for many years at the high school (#5).
c. The church **teaches** compassion and forgiveness (#4a or #4b).
d. Someone needs to **teach** her right and wrong (#1a).
e. The experience **taught** us that money doesn't mean everything (#1c).
f. Her injury will **teach** her not to be so careless with a knife (#1d).
g. It took patience to **teach** her how to bike (#1b).

2. Directions: In the list are examples of simple sentences that describe each of the meaning of the word "To Learn" as defined in Table 1.2. Match each sentence with a meaning as explained in Table 1.2. (Note: that more than one choice may apply – discuss your choices with a partner.)

a. People **learn** throughout their lives (all definitions).
b. I can't swim yet, but I'm **learning** (#2).
c. She's interested in **learning** French (#1a or #2).
d. We had to **learn** the rules of the game (#1a or #4).
e. I'm trying to **learn** my lines for the play (#1b).
f. She **learned** through a letter that her father had died (#3).
g. I later **learned** that they had never called (#4).

3. Directions: In the list next are behaviors expected to be acquired through operant conditioning. Identify which type of consequence was responsible for the behavior change (i.e., positive/negative reinforcement; positive/negative punishment). Explain briefly.

a. A professor has a policy of exempting students from the final exam if they maintain perfect attendance during the quarter. His students' attendance increases dramatically.
Key: This is an example of **operant conditioning** because attendance is a voluntary behavior and the exemption from the final exam is a **negative reinforcement** because something is taken away that increases the behavior (attendance).

b. When a professor first starts teaching about a concept, she'll praise any answer that is close to the right answer.
Key: This describes the process of **shaping** the **operant behavior** of answering questions, using **positive reinforcement** (praise). In shaping you start by reinforcing anything that is close to the final response. Then you gradually require closer and closer approximations before giving a reinforcer.

c. A student patiently raises her hand, waiting to be called on during a class discussion, after having been rebuffed by the instructors for interrupting others.
Key: This is **operant conditioning**. Because raising one's hand is voluntary. The consequence provides **negative reinforcement** because no rebuff occurs and the behavior of hand-raising increases.

4. Direction: Fill in the table to describe the constructivist and behaviorist attributes of the two models of instructional design illustrated in Figure 1.4. Discuss your findings with a partner.

	Key: *Figure 1.4(a) (twentieth century)*	Key: *Figure 1.4(b) (twenty-first century)*
Constructivist attributes	– Students may construct knowledge during lecture and during their periods of study	– Student / content centered – The interaction between students and the instructor and the materials is more sustained
Behaviorist attributes	Key: – Teacher centered – Teaching and learning are sequential	Key: – More opportunities for repetitions (exposure) to a certain "stimulus"

Section 1

Quantitative and Population Genetics

2

Mating Systems: Inbreeding and Inbreeding Depression

David L. Thomas
Department of Animal Sciences, University of Wisconsin–Madison, WI, USA

Introduction

Replacement males and females are selected for superior estimated breeding values (EBV) in order to make genetic progress in the herd or flock. The rate of genetic progress by selection is determined by accuracy and intensity of selection, generation interval, and genetic variation (Chapter 4). After animals are selected, a decision must be made as to how they will be mated. Matings are often determined by the degree of relationship between the two parents. Males and females can be mated together that are more closely related than the average relationship among individuals in the population (**inbreeding**), or males and females can be mated together that are less closely related than the average relationship among individuals in the population (**outbreeding** and **crossbreeding**). Crossbreeding is discussed in Chapter 3. This chapter will discuss the mating system of inbreeding.

Inbreeding

Inbreeding results from the mating of related individuals. A calf will be inbred if its sire and dam are related. Inbreeding results in an increase in the proportion of loci that are homozygous and, therefore, a decrease in the proportion of loci that are heterozygous. All the effects of inbreeding on performance of animals are due to this increase in homozygosity. Table 2.1 presents the effects of inbreeding on the increase in the proportion of homozygous loci with the most dramatic form of inbreeding – **selfing**. While selfing cannot be accomplished in livestock, it is used here to demonstrate the effects of inbreeding. The intense forms of inbreeding in livestock such as sire–daughter, son–dam, and full-sib matings would also increase the proportion of homozygous loci in successive generations, but at a slower rate than selfing.

Table 2.1 demonstrates that selfing, the most intense form of inbreeding, results in a halving of the proportion of heterozygous loci each generation. By generation 4, approximately 94% of the loci are expected to be homozygous when starting with a population that had 100% heterozygous loci. Starting with all heterozygous loci and after four generations of parent–offspring or full-sib matings, the proportion of homozygous loci is expected to be approximately 59%, and with half-sib matings, the proportion of homozygous loci is expected to be approximately 38%. The increase in proportion of homozygous loci in inbred animals has three primary results – prepotency, increased incidence of genetic defects caused by recessive alleles, and reduced performance for quantitative traits (inbreeding depression).

Presented in Table 2.2 is the increase in expected proportion of homozygous loci in individuals resulting from intensive inbreeding over 10 generations. The values in Table 2.2 are also the inbreeding coefficients (discussed later in this chapter) of the individuals resulting from these matings. Like selfing, the close matings possible in livestock result in a rapid increase in the proportion of homozygous loci.

Prepotency is the ability of a parent to produce offspring that are uniform in appearance and performance and that look and perform like the parent. The uniformity is due to the decreased number of types of gametes that can be produced by a highly homozygous (inbred) animal compared to a highly heterozygous (non-inbred) animal. Consider a trait due to the action of alleles at four loci. As an example, an inbred sire may be homozygous at three of these loci (e.g., AABBCcDD) and can produce only two types of gametes (ABCD and ABcD), whereas a non-inbred sire may be homozygous at only one of these loci (e.g., AABbCcDd) and can produce eight different types of gametes (ABCD, ABCd, ABcD, ABcd, AbCD, AbCd, AbcD, Abcd).

When each sire is mated to a similar group of females, the resulting progeny are going to be more uniform when there are only two possible genotypes contributed by the inbred sire than when there are eight possible genotypes contributed by the non-inbred sire. In addition, if the inbred animal is primarily homozygous for dominant alleles, the offspring are more likely to look and perform like that parent than if the inbred animal is primarily homozygous for recessive alleles or non-inbred and primarily heterozygous.

Prepotency from increased homozygosity is most evident for simply inherited traits and polygenic traits of high heritability. Lowly to moderately heritable polygenic traits, which include most production traits in livestock, are greatly affected by environmental effects, and variable environmental effects among animals tend to overwhelm the increased gametic uniformity of inbred animals for these production traits.

Molecular and Quantitative Animal Genetics, First Edition. Edited by Hasan Khatib.

Table 2.1 Increase in the proportion of homozygous loci in each generation with selfing – the most intense form of inbreeding.

Gen	Genotypes			Proportion of loci:	
				heterozygous	homozygous
0		1.0 Aa		1.0	0
1	.25 AA	.50 Aa	.25 aa	.50	.50
2	.25 AA	.50 (.25 AA + .50 Aa + .25 aa)	.25 aa		
		.125 AA + .25 Aa + .125 aa			
	.375 AA	.25 Aa	.375 aa	.25	.75
3	.375 AA	.25 (.25 AA + .50 Aa + .25 aa)	.375 aa		
		.0625 AA + .125 Aa + .0625 aa			
	.4375 AA	.125 Aa	.4375 aa	.125	.875
4	.4375 AA	.125 (.25 AA + .50 Aa + .25 aa)	.4375 aa		
		.03125 AA + .0625 Aa + .03125 aa			
	.46875 AA	.0625 Aa	.46875 aa	.0625	.9375

Table 2.2 The increase in the expected proportion of homozygous loci over 10 generations of intensive inbreeding with different types of close matings.

Gen	Type of Mating		
	Selfing	Full Brother × Full Sister, Sire × Daughter, or Son × Dam	Half Brother × Half Sister
0	0	0	0
1	0.500	0.250	0.125
2	0.750	0.375	0.219
3	0.875	0.500	0.305
4	0.938	0.594	0.381
5	0.969	0.672	0.449
6	0.984	0.734	0.509
7	0.992	0.785	0.563
8	0.996	0.826	0.611
9	0.998	0.859	0.654
10	0.999	0.886	0.691

Table 2.3 Effects of inbreeding in sheep[a].

Trait	Effect of 1% increase in inbreeding coefficient
Fleece weight	−0.017 kg
Fertility	−1.4 ewes lambing/100 ewes mated
Lamb survival	−2.8 lambs weaned/100 lambs born
Lamb weaning weight	−0.111 kg

[a]Adapted from the review of: Lamberson and Thomas. 1984. *Anim. Breed. Abstr.* 52:287–297. Results from 25 studies with over 25,000 sheep.

Genetic defects

An increased incidence of genetic defects is seen in inbred animals. Simply inherited genetic defects that exist in a population are often due to recessive alleles. Such alleles can exist in a population because, in most individuals, they hide behind a desirable dominant allele in the heterozygous state. It is only when two heterozygous individuals are mated that there is a 25% probability of the recessive allele occurring in the homozygous state and being exposed. If a mutation produces an undesirable dominant allele, the allele is expressed in both the heterozygous and homozygous state and can be easily removed from the population.

Because inbreeding increases the proportion of homozygous loci including recessive homozygotes, which are more likely to be undesirable, there is a greater incidence of genetic defects in inbred compared to non-inbred animals. In Table 2.1, for example, assume that the "a" allele is recessive and results in the theoretical non-lethal defect of "bent tail." In generation 0, all animals are heterozygous with normal tails. In subsequent generations, the proportion of "aa" individuals and the incidence of bent tail increases: generation 1 = 25%, generation 2 = 37.5%, and so on.

People often assume that inbreeding increases the proportion of undesirable recessive **alleles**. This is not true. Note in Table 2.1 that the frequency of the dominant, desirable "A" and the recessive, undesirable "a" alleles, remain constant in each generation at 0.50. In the absence of the forces of selection, migration, mutation, and genetic drift, inbreeding does not change allele frequency over time, but it does change genotypic frequencies by increasing the proportion of both recessive and dominant homozygotes and decreasing the proportion of heterozygotes.

Inbreeding depression

Livestock breeders and animal scientists have known for many years that inbred animals tend to have lower performance for quantitative traits (e.g. milk production, average daily gain, fertility, survival) than do non-inbred animals. Tables 2.3 and 2.4

present the effects for each 1% increase in the inbreeding coefficient on various production traits of sheep and dairy cattle, respectively. The inbreeding coefficient is used to quantify the degree of inbreeding, and its calculation is presented at the end of this chapter. To give some context to the values in Tables 2.3 and 2.4, an animal resulting from the mating of a half-brother to his half-sister will have an inbreeding coefficient of at least 12.5%.

Inbreeding has very large effects on ewe fertility and lamb survival (Table 2.3). A flock of sheep with an average inbreeding coefficient of 10% would be expected to have 14 fewer ewes lamb/100 ewes mated and 28 fewer lambs survive to weaning/100 lambs born than a flock of non-inbred sheep. Likewise, Holstein cows with an average inbreeding coefficient of 10% would be expected to stay in the herd 131 fewer days than non-inbred cows (Table 2.4). These levels of production losses due to inbreeding depression result in very large economic losses to livestock producers.

Inbreeding tends to have its greatest detrimental effects on the fitness traits of reproduction and survival, moderate effects on the production traits (e.g., growth rate and milk yield), and least effects on conformation, product quality (e.g., milk fat percentage), and carcass traits. In general, traits of high heritability suffer the least from inbreeding, and traits of low heritability suffer the most.

Table 2.4 Effects of inbreeding in US Holstein dairy cows[a].

Trait	Effect of 1% increase in inbreeding coefficient
Age at first calving	+0.36 days
Days of productive life	−13.07 days
Lifetime milk	−358.41 kg
Lifetime fat	−13.17 kg
Lifetime protein	−11.41 kg
Lifetime net income	−$24.43
Type traits	little effect

[a]From: Smith, Cassell, and Pearson. 1998. *J. Dairy Sci.* 81:2729–2737. Results from field data from approximately 1.2 million registered Holstein cows.

In general, livestock producers should avoid inbreeding. However, most "purebred" populations of livestock are closed populations, and they necessarily have some level of inbreeding. The inbreeding level is generally higher in those purebred populations with smaller numbers of animals because it is more difficult in small than in large populations to find potential sires and dams that are not related. The Hereford is a popular beef cattle breed in the US, but the mean inbreeding coefficient among the relatively large number of Herefords born in 2001 (approximately 75,000 head) was 9.8% (Cleveland et al., 2005). This inbreeding coefficient is not much less than the inbreeding coefficient expected in animals resulting from the mating of half-sibs (12.5%). The average inbreeding coefficient of animals from beef cattle breeds with smaller populations than the Hereford would be expected to have even higher levels of inbreeding.

Populations that have effective genetic improvement programs also tend to have increased rates of inbreeding accumulation compared to populations where selection is less intense. This arises because performance recording identifies individuals of high genetic merit, and these individuals and their descendants tend to produce the majority of the individuals in subsequent generations. This situation is exacerbated with the reproductive technologies of artificial insemination and super ovulation/embryo transfer where fewer breeding males and females are needed to produce the next generation. Dairy cattle genetic improvement programs in the US are examples of effective genetic improvement programs that have resulted in higher levels of inbreeding than would have been expected with random mating. Figure 2.1 shows the estimated inbreeding coefficients of US Jersey cattle born from 1960 to 2012. Using Jerseys born in 1960 as the base (inbreeding coefficient = 0%), cattle born in 2012 have an average inbreeding coefficient of 7.08%. Since the cattle born in 1960 would have had some level of inbreeding, the actual level of inbreeding of cattle born in 2012 is something larger than 7.08%. Figure 2.1 also presents expected future inbreeding in the Jersey population based upon the pedigrees of

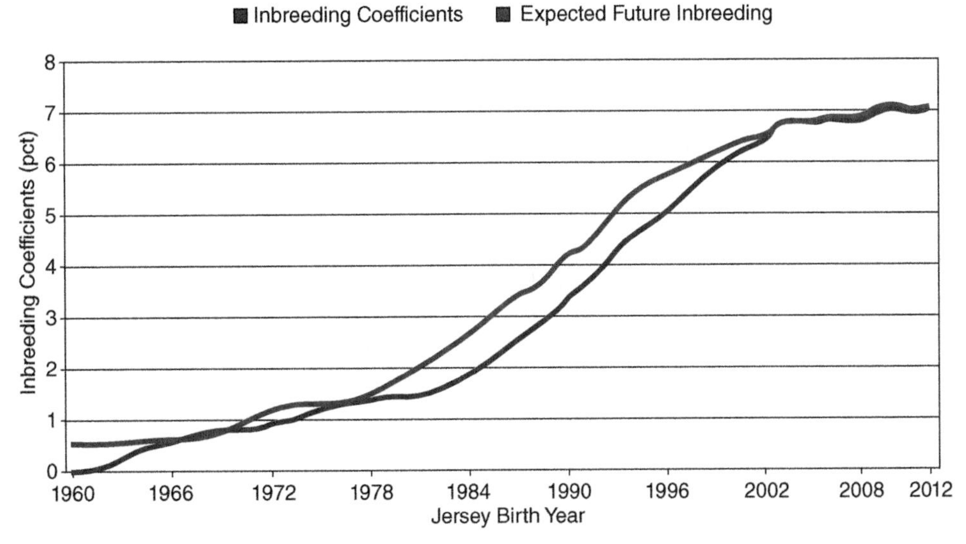

Figure 2.1 Inbreeding coefficients for US Jersey dairy cattle born between 1960 and 2012. From: USDA Animal Improvement Program Laboratory website (www.cdcb.us/eval/summary/inbrd.cfm?), Bovine Inbreeding Trends Menu/Jersey/Inbreeding Trend, Accessed April 28, 2014).

Table 2.5 Outcross Jersey bulls: High Net Merit $ Jersey bulls with a low relationship with current US Jersey cows, December 2012.

Name	Bull ID	Country	Net Merit, $	Percentile rank, NM$	Expected inbreeding of future daughters[a]
ISDK DJ ZUMA	000000302730	Denmark	618	>99	4.7
SUN VALLEY ZUMA POWER-ET	000117660078	USA	586	99	5.5
ISDK DJ HULK	000000302595	Denmark	582	99	4.0
IRISHTOWNS B183 BUNGY JOSHUA-E	000117022993	USA	510	97	5.4
SUNRISE/HACKLINE BUNGY ZIPPER	000067039609	USA	493	96	3.6
SF IMPULS 8916	000114635185	USA	491	93	5.4
JER-Z-BOYZ LEGITIMATE-ET	000117773509	USA	491	93	5.7
VAN DE Q ZIK ZEBULON-ET	000067541066	USA	480	91	5.0
ISDK DJ BROILER	000000302835	Denmark	454	87	3.8
ISDK Q IMPULS	000000301592	Denmark	431	85	3.4
SUNSET CANYON MAXIMUM-ET	000111950696	USA	412	78	5.8
CAVE CREEK KANOO-ET	000114118219	USA	411	77	4.2
FOREST GLEN ARTIST ALEXANDER	000067037158	USA	386	75	5.7
ISAU BROADLIN HATMAN-ET	000020603276	Australia	385	75	5.7
WILSONVIEW BLUEPRINT MADDIX-ET	000067010046	USA	381	75	5.2

[a]The average relationship between the sire and the current cow population is used in the calculation of these inbreeding coefficients.
From: USDA Animal Improvement Program Laboratory website (http://aipl.arsusda.gov/eval/summary/inbrd.cfm, Bovine Inbreeding Trends Menu/Jersey/Outcrosses, Accessed on February 23, 2013).

the current bulls and cows in the population. Until 2002, the expected future level of inbreeding was almost always higher than the current level of inbreeding suggesting that Jersey cattle breeders were not paying a lot of attention to pedigree relationships between bull and cow mates when making mating decisions. In many cases, sons and grandsons of outstanding performance bulls were being mated to daughters and granddaughters of these same outstanding bulls.

However, from 2002 to 2012, inbreeding levels in US Jerseys have increased at a much slower rate than in previous years, and the expected future inbreeding level is almost the same as the current inbreeding level (Figure 2.1). This indicates that breeders are now more aware of the detrimental effects of inbreeding on performance traits and are taking relationships between bulls and cows into consideration when making mating decisions. This has been facilitated by the production and publication of a list of bulls of high genetic merit that are least related to the current cow population by the USDA Animal Improvement Program Laboratory for each of the dairy breeds. This list is routinely updated, and cattle breeders can access it directly online or obtain the information from their artificial insemination bull stud. A current list of such "outcross" Jersey bulls is presented in Table 2.5.

The outcross Jersey bulls listed in Table 2.5 were among the top 25% of active bulls for Net Merit $, an index of cow performance that considers genetic values for production, reproduction, health, and type traits, in December 2012. Use of these bulls is predicted to result in continued improvement in cow breeding value for performance traits while lowering levels of inbreeding in the Jersey population (3.4 to 5.8% inbreeding in daughters of these bulls compared with 7.08% inbreeding in Jersey cattle born in 2012).

Cause of inbreeding depression

The Basic Genetic Model for polygenic traits presented in Chapter 4 is:

$$P - \mu = G + E \qquad (2.1)$$

where:
P = phenotypic performance of an animal,
μ = the mean performance of all animals in the population or contemporary group,
G = the animal's genotypic value; the total effect of all loci affecting the performance trait (+ or − deviation from the mean genotypic value of the population or contemporary group, $\bar{G} = 0$),
E = the environmental effect on the animal's performance (+ or − deviation from the mean environmental effect in the population or contemporary group, $\bar{E} = 0$).

Because genotypic value (G) equals BV + GCV, the formula can be rewritten as:

$$P - \mu = BV + GCV + E \qquad (2.2)$$

where:
BV = the animal's breeding value; the sum of the individual effects of all alleles affecting the trait (+ or − deviation from the mean breeding value of the population or contemporary group, $\overline{BV} = 0$),
GCV = the animal's gene combination value; the sum of the effects of the interaction of alleles within each locus (dominance) and between loci (epistasis); the difference between the genotypic value and the breeding value (+ or − deviation from the mean gene combination value of the population or contemporary group, $\overline{GCV} = 0$).

Selection improves animal performance primarily through an improvement in BV. The choice of mating system, based on pedigree relationship between the parents, will affect animal performance primarily through its effect on GCV. Inbreeding often results in a decrease in GCV, and this decrease in GCV translates into a decrease in performance or inbreeding depression.

The following example will demonstrate how inbreeding can decrease GCV and subsequently decrease genotypic value and

Table 2.6 Theoretical effect of inbreeding on genotypic value (G), breeding value (BV), and gene combination value (GCV).

Genotype	G[a]	BV	GCV (G – BV)
Non-inbred: AaBbCcDd	4(14) = 56 kg	4(7) + 4(−1) = 24 kg	32 kg
Inbred: AAbbCCdd	2(14) + 2(−2) = 24 kg	4(7) + 4(−1) = 24 kg	0 kg

[a]The homozygote is assumed to have a genotypic value equal to the sum of the individual allele effects (e.g., AA = +14 kg, aa = −2 kg), and due to complete dominance, a heterozygote will have the same genotypic value as the dominant homozygote (e.g. AA = Aa = +14 kg.

Table 2.7 Theoretical effect of inbreeding on genotypic value (G), breeding value (BV), and gene combination value (GCV) with different degrees of dominance.

Genotype	G	BV	GCV (G – BV)
No dominance[a]			
Non-inbred: AaBbCcDd	4(6) = 24 kg	4(7) + 4(−1) = 24 kg	0 kg
Inbred: AAbbCCdd	2(14) + 2(−2) = 24 kg	4(7) + 4(−1) = 24 kg	0 kg
Non-inbred – inbred	0 kg	0 kg	0 kg
Partial dominance[b]			
Non-inbred: AaBbCcDd	4(10) = 40 kg	4(7) + 4(−1) = 24 kg	16 kg
Inbred: AAbbCCdd	2(14) + 2(−2) = 24 kg	4(7) + 4(−1) = 24 kg	0 kg
Non-inbred – inbred	16 kg	0 kg	16 kg
Complete dominance[c]			
Non-inbred: AaBbCcDd	4(14) = 56 kg	4(7) + 4(−1) = 24 kg	32 kg
Inbred: AAbbCCdd	2(14) + 2(−2) = 24 kg	4(7) + 4(−1) = 24 kg	0 kg
Non-inbred – inbred	32 kg	0 kg	32 kg
Over-dominance[d]			
Non-inbred: AaBbCcDd	4(18) = 72 kg	4(7) + 4(−1) = 24 kg	48 kg
Inbred: AAbbCCdd	2(14) + 2(−2) = 24 kg	4(7) + 4(−1) = 24 kg	0 kg
Non-inbred – inbred	48 kg	0 kg	48 kg

[a]Each allele at a locus is expressed fully (e.g. AA = 7 + 7 = 14, Aa = 7 + (−1) = 6, aa = (−1) + (−1) = −2).
[b]The heterozygote is less than the dominant homozygote but greater than the average of the two homozygotes in genotypic value (e.g. AA = 14, Aa = 10, aa = −2).
[c]The heterozygote has the same genotypic value as the dominant homozygote (e.g. AA = Aa = 14 kg, aa = y2).
[d]The heterozygote has a greater genotypic value than the dominant homozygote (e.g. AA = 14 kg, Aa = 18, aa = −2).

performance. Assume that weaning weight in beef cattle is due to the action of alleles at four loci where there is complete dominance at each locus. (While we do not know how many loci control weaning weight in beef cattle, it is definitely many times more than four. However, this simple example will demonstrate the effect of inbreeding on GCV.) It is assumed that there is no epistasis among alleles at the different loci. Each dominant allele has an effect of +7 kg and each recessive allele has an effect of −1 kg. Table 2.6 presents the genotypic, breeding, and gene combination values for a possible non-inbred (highly heterozygous) and a possible inbred (highly homozygous) beef animal.

The two individuals in Table 2.6 each have four dominant and four recessive alleles, therefore, their breeding values are the same at 24 kg. However, none of the undesirable recessive alleles in the non-inbred individual can express themselves because all four recessive alleles are in a heterozygous locus with a dominant allele, whereas all the undesirable recessive alleles in the inbred individual are in homozygous loci that allows their expression. Therefore, the non-inbred (more heterozygous) individual has a greater genotypic value due to the greater gene combination value caused by the existence of dominance and the presence of at least one dominant allele at every locus.

Inbreeding depression results from a decrease in gene combination value at loci that are homozygous for undesirable recessive alleles. A precondition for inbreeding depression to occur is the existence of dominance (and/or epistasis) among alleles at some of the loci affecting the trait. When considering dominance as the only cause of inbreeding depression, the degree of dominance determines the amount of inbreeding depression observed. Traits that are influenced by loci where there is no dominance among alleles will show little or no decrease as a result of inbreeding, more decrease will be observed if alleles at the loci exhibit partial dominance, even more decrease will be observed with complete dominance, and the most decrease will be observed with over-dominance. Table 2.7 demonstrates that as the degree of dominance increases, the difference between the non-inbred (heterozygous) and the inbred (homozygous) individual for genotypic value increases as a result of loss of GCV in the inbred individual relative to the non-inbred individual.

Table 2.8 Possible inbred (homozygous) individuals and their genotypic values with complete dominance.

Type of individuals	Genotype	Genotypic value[a]
Non-inbred	AaBbCcDd	4(14) = 56 kg
Inbred	AABBCCDD	4(14) = 56 kg
	AABBCCdd	3(14) + 1(−2) = 40 kg
	AABBccDD	3(14) + 1(−2) = 40 kg
	AABBccdd	2(14) + 2(−2) = 24 kg
	AAbbCCDD	3(14) + 1(−2) = 40 kg
	AAbbCCdd	2(14) + 2(−2) = 24 kg
	AAbbccDD	2(14) + 2(−2) = 24 kg
	AAbbccdd	1(14) + 3(−2) = 8 kg
	aaBBCCDD	3(14) + 1(−2) = 40 kg
	aaBBCCdd	2(14) + 2(−2) = 24 kg
	aaBBccDD	2(14) + 2(−2) = 24 kg
	aaBBccdd	1(14) + 3(−2) = 8 kg
	aabbCCDD	2(14) + 2(−2) = 24 kg
	aabbCCdd	1(14) + 3(−2) = 8 kg
	aabbccDD	1(14) + 3(−2) = 8 kg
	aabbccdd	4(−2) = −8 kg
Inbred average		24 kg

[a]Complete dominance is assumed at each locus with genotypic values at each homozygous dominant locus and heterozygous locus equal to 14 kg and each homozygous recessive locus equal to −2 kg (e.g., AA = Aa = 14 kg, aa = −2 kg).

Of course, the conclusions obtained from Table 2.7 are very much dependent upon the particular inbred individual that was selected to compare against the non-inbred individual that was heterozygous at every locus. For example, an inbred animal that was homozygous for the desirable allele at all four loci (AABBCCDD) would have a genotypic value of 56 kg (4 × 14 kg) and would be equal to or superior to the non-inbred, heterozygous individual under all degrees of dominance except for overdominance. It would appear then that inbreeding would be a mating system to use to produce individuals that were homozygous for desirable alleles and, therefore, would only pass desirable alleles to their offspring. The problem is that in practice, inbreeding results in increased homozygosity for both desirable and undesirable alleles. Table 2.8 presents the possible homozygous genotypes and genotypic values that can result from long-term inbreeding of a population that starts with individuals that are heterozygous at four loci like the non-inbred individual in Table 2.7. The average genotypic value of the 16 possible homozygotes that can result from inbreeding is 24 kg, and all but one are inferior to the non-inbred heterozygote.

One could argue that inbreeding combined with selection to eliminate the individuals with the lowest performance could eventually result in an inbred population of high-performing animals that were homozygous for the desirable alleles. This strategy has been tried with livestock species, and it has failed in most cases. Inbreeding has its greatest negative effects on reproduction and survival traits. Cattle, sheep, and pigs have relatively low reproductive rates when compared to plants and some other animals such as poultry. When inbred lines of these livestock species have been established, the reproductive and survival rates decrease as the level of inbreeding increases to such an extent that there is no or very little selection intensity and, in many cases, the line has become extinct. Inbreeding is not a recommended mating system for mammalian livestock.

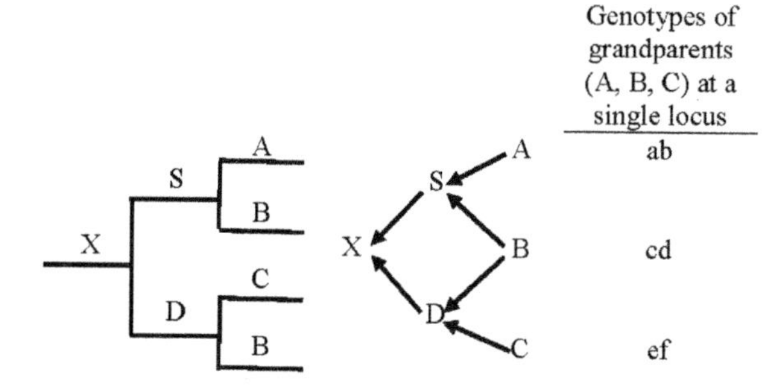

Figure 2.2 Pedigree (*left*) and path diagram (*right*) for individual X resulting from the mating of a half-brother (S) with his half-sister (D), with the alleles present at a single locus in the three grandparents.

However, inbreeding has been used successfully in some populations. Naturally high reproductive rates of many commercial crops (i.e., corn) and poultry have allowed the establishment of inbred lines in these species that are then crossed to provide hybrid commercial varieties. Even in these populations, reproduction will decrease in the early stages of inbreeding, but the initial reproductive rate is high enough that the decrease from inbreeding still allows reasonable selection intensities, and the eventual production of inbreds homozygous for desirable alleles and recovery to reasonable, if not full, reproductive rates.

Quantifying inbreeding

Inbreeding coefficient

The inbreeding coefficient (F) is used to quantify inbreeding. It is defined as the probable proportion of an individual's loci that are homozygous for alleles that are identical by descent. The term "identical by descent" means that the two alleles at a locus are copies of a single allele found in an ancestor. Figure 2.2 presents the pedigree and path diagram for an individual (X) resulting from the mating of a sire (S) and dam (D) that are half-sibs.

Individual X is inbred because his/her parents are related, that is, S and D have the same dam (B). The inbreeding coefficient of X (Fx) is the probability that its loci are homozygous for alleles that are both copies of a single allele in an ancestor common to both its sire and dam. Individual X cannot be homozygous for alleles a, b, e, or f because these alleles can only come from one parent. However, X can be homozygous for allele c or allele d because each of these alleles can come from both the sire and dam. Therefore, Fx is the probability that X has genotype cc or dd. These calculations follow:

Fx = Probability that X has genotype cc or dd = Prob. (X is cc) + Prob. (X is dd)

Prob. (X is cc) = Prob. (S gets c from B) × Prob. (X gets c from S if S gets c from B) × Prob. (D gets c from B) × Prob. (X gets c from D if D gets c from B) = 1/2 × 1/2 × 1/2 × 1/2 = 1/16

Prob. (X is dd) = Prob. (S gets d from B) × Prob. (X gets d from S if S gets d from B) × Prob. (D gets d from B) × Prob. (X gets d from D if D gets d from B) = 1/2 × 1/2 × 1/2 × 1/2 = 1/16

Fx = 1/16 + 1/16 = 1/8 = 0.125

Individual X has an inbreeding coefficient of 12.5%. Individual X is expected to have 12.5% of loci homozygous for alleles that are identical by descent, that is 12.5% of individual X's loci are expected to be homozygous for alleles that are exact copies of a single allele at the same locus in the common ancestor B.

The inbreeding coefficient is an estimate of the "increase" in homozygosity as a result of inbreeding. On average, individuals will have a higher percentage of homozygous loci than indicated by the inbreeding coefficient. For example, if alleles a and e in Figure 2.2 were the same allele, it is possible for individual X to be homozygous for that allele, but the homozygosity would not be for alleles that were identical by descent. Homozygous alleles of this type are referred to as "alike in state." The inbreeding coefficient is a measure of the increase in homozygosity of an individual compared to non-inbred animals in the same population.

Inbreeding coefficient formula

Calculating inbreeding coefficients by directly calculating the probability of an individual's loci being homozygous for alleles that are identical by descent becomes very tedious with even uncomplicated pedigrees. The formula presented in 2.3 for calculating the inbreeding coefficient is more useful:

$$Fx = \sum_{CA=1}^{k} \left(\frac{1}{2}\right)^{n_1+n_2+1} (1+F_{CA}) \tag{2.3}$$

where:

CA = the common ancestor of the sire and dam of individual X,
k = the number of paths connecting the sire and dam of X through a common ancestor,
n_1 = the number of generations from the common ancestor to the sire of X,
n_2 = the number of generations from the common ancestor to the dam of X,
F_{CA} = the inbreeding coefficient of the common ancestor.

In applying this formula to the path diagram in Figure 2.2, we see that there is only one path that connects the sire and dam of X through the common ancestor B. The number of generations from B to S is 1 so $n_1 = 1$. Likewise, the number of generations from B to D is 1 so $n_2 = 1$. Since we do not know the parents of the common ancestor B, F_B is unknown. Without any other information, we assume that B is not inbred, and $F_B = 0$. Substituting these values into the inbreeding formula, we have:

$$Fx = \left(\frac{1}{2}\right)^{n_1+n_2+1} (1+F_{CA}) = \left(\frac{1}{2}\right)^{1+1+1} (1+0) = \left(\frac{1}{2}\right)^{3} (1) = \frac{1}{8} = .125 \tag{2.4}$$

Calculating inbreeding coefficients

The path method is a visual method of calculating an inbreeding coefficient that allows a person to see how alleles flow from a common ancestor, through each parent, and onto the individual. The steps used in the path method for calculation of an inbreeding coefficient are presented below using the pedigree of Hampshire ewe 18420, born January 10, 1979. This ewe was in the Hampshire flock of R.W. Hogg and Sons of Salem, Oregon, which was operated for many years by his sons, Ronald and Glenn Hogg (Figure 2.3). This was one of the most prestigious Hampshire flocks in the US. The flock was dispersed in 1981.

The pedigree of Hampshire ewe 18420 is presented in Figure 2.4. She is the result of the mating of a half-brother with a half-sister so we know that her inbreeding coefficient is at least 0.125. In addition to the common ancestor A (parent of both the sire and dam), individual E is a second common ancestor (a maternal great-grandparent of the sire and a paternal grandparent of the dam), and individual M is a third common ancestor (a maternal great-grandparent of the sire and a paternal great-grandparent of the dam). Therefore, 18420 will have an inbreeding coefficient greater than 0.125.

Figure 2.3 Ronald Hogg with one of his Hampshire rams. The Hampshire flock was in the Hogg family for over 50 years and virtually closed to outside breeding for the last 30 years of its existence. In the last two years of the flock's ownership by the Hogg family (1980 and 1981), the median inbreeding coefficient of lambs was 10.7% and ranged from 1.4 to 29.6%, and the median inbreeding coefficient of ewes was 4.3% and ranged from 0.0 to 28.1%. During these two years, lamb mortality was very high and positively related to level of inbreeding (1.3 lamb deaths per 100 lambs born/0.01 increase in lamb inbreeding coefficient) (Lamberson et al., 1982).

The inbreeding coefficient of Hampshire ewe 18420 will be calculated using the path method following the steps that follow:

1. Draw a path diagram from the pedigree (see Figure 2.5).
   - **a.** All individuals in the pedigree occur only once in the path diagram.
   - **b.** Draw an arrow from each parent to their offspring. The tail of the arrow starts at the parent, and the head of the arrow ends at the offspring.
2. Record each path that connects the parents of the individual for which the inbreeding coefficient is being calculated (Table 2.9).
   - **a.** A path must start with one parent and end with the other parent.
   - **b.** A path cannot move through an individual more than once, that is, an individual can only occur once in a path.
   - **c.** A path can move from an arrow tail to an arrowhead or from an arrowhead to an arrow tail.
   - **d.** A path can change direction (move from an arrow tail to an arrow tail) only once.
   - **e.** A path can never move from an arrowhead to another arrowhead.

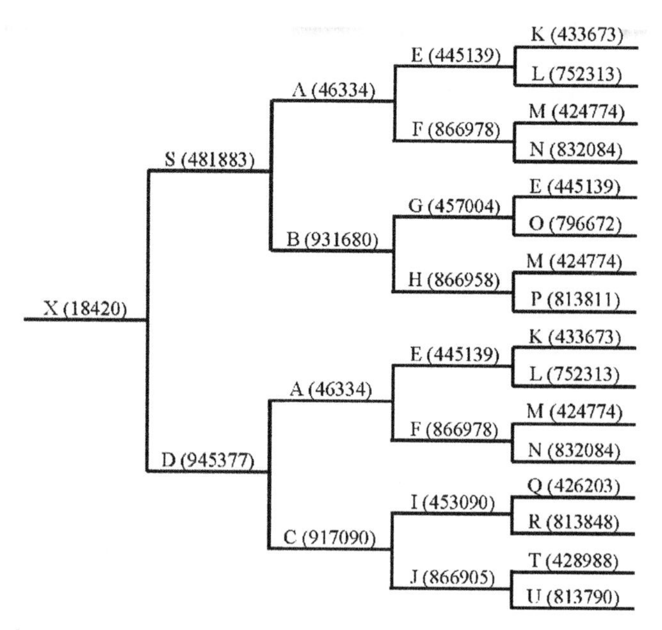

Figure 2.4 Four generation pedigree of Hampshire ewe 18420 from the R.W. Hogg and Sons flock of Salem, Oregon (numbers in parentheses are the registration numbers recorded with the American Hampshire Sheep Association: the letters are arbitrary and used to simplify the identification of each animal).

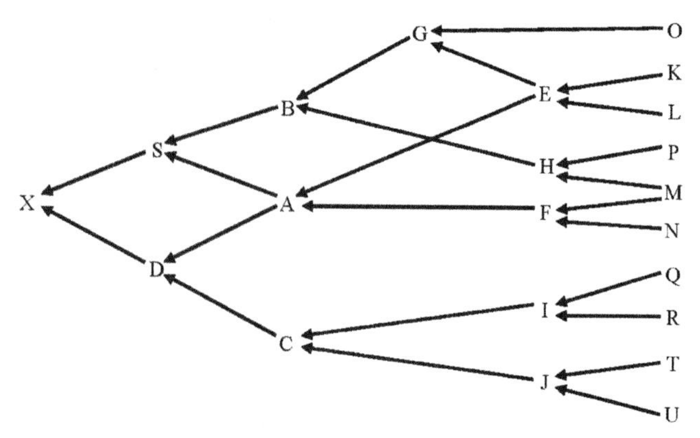

Figure 2.5 Path diagram for individual X (Hampshire ewe 18420).

f. All individuals in the path must be recorded, in Table 2.9, in the order in which they occur, and arrows, moving in the proper direction, must be placed between each pair of individuals.

3. Identify (underline or circle) the common ancestor in each path (Table 2.10).
   a. In paths that change direction, the common ancestor is the individual located where the change of direction takes place.
   b. In paths that do not change direction, the common ancestor is the individual at the tail end of the path.
4. Calculate the inbreeding coefficient of the common ancestor in each path (Table 2.11).
   a. All individuals, including common ancestors, have an inbreeding coefficient greater than zero if their parents are related. If the path diagram indicates that the parents of a common ancestor are related, then a separate exercise must be undertaken to calculate the inbreeding coefficient of the common ancestor.

Table 2.9 For question **2.f.**

Path	F_{CA}	n_1[a]	n_2[b]	$\left(\frac{1}{2}\right)^{n_1+n_2+1}(1+F_{CA})$
$S \leftarrow A \rightarrow D$				
$S \leftarrow B \leftarrow G \leftarrow E \rightarrow A \rightarrow D$				
$S \leftarrow B \leftarrow H \leftarrow M \rightarrow F \rightarrow A \rightarrow D$				
$Fx = \sum_1^3 \left(\frac{1}{2}\right)^{n_1+n_2+1}(1+F_{CA}) =$				

[a]n_1 = the number of arrows in the path between the common ancestor and the sire.
[b]n_2 = the number of arrows in the path between the common ancestor and the dam.

Table 2.10 For question **3.**

Path	F_{CA}	n_1[a]	n_2[b]	$\left(\frac{1}{2}\right)^{n_1+n_2+1}(1+F_{CA})$
$S \leftarrow \underline{A} \rightarrow D$				
$S \leftarrow B \leftarrow G \leftarrow \underline{E} \rightarrow A \rightarrow D$				
$S \leftarrow B \leftarrow H \leftarrow \underline{M} \rightarrow F \rightarrow A \rightarrow D$				
$Fx = \sum_1^3 \left(\frac{1}{2}\right)^{n_1+n_2+1}(1+F_{CA}) =$				

[a]n_1 = the number of arrows in the path between the common ancestor and the sire.
[b]n_2 = the number of arrows in the path between the common ancestor and the dam.

Table 2.11 For question **4.**

Path	F_{CA}	n_1[a]	n_2[b]	$\left(\frac{1}{2}\right)^{n_1+n_2+1}(1+F_{CA})$
$S \leftarrow \underline{A} \rightarrow D$	0			
$S \leftarrow B \leftarrow G \leftarrow \underline{E} \rightarrow A \rightarrow D$	0			
$S \leftarrow B \leftarrow H \leftarrow \underline{M} \rightarrow F \rightarrow A \rightarrow D$	0			
$Fx = \sum_1^3 \left(\frac{1}{2}\right)^{n_1+n_2+1}(1+F_{CA}) =$				

[a]n_1 = the number of arrows in the path between the common ancestor and the sire.
[b]n_2 = the number of arrows in the path between the common ancestor and the dam.

   b. If one or both parents are unknown, assume the inbreeding coefficient of the common ancestor is zero.
5. Record the number of arrows in each path (Table 2.12).
   a. Record n_1 – the number of arrows in the path between the common ancestor and the sire.
   b. Record n_2 – the number of arrows in the path between the common ancestor and the dam.
6. Calculate:

$$\left(\frac{1}{2}\right)^{n_1+n_2+1}(1+F_{CA}) \tag{2.5}$$

for each path, and sum the values across all paths. This sum is the inbreeding coefficient (Table 2.13).

The inbreeding coefficient of Hampshire ewe 18420 is 0.15. Ewe 18420 is expected to have 15% of her loci homozygous for alleles that are identical by descent – 15% of her loci are expected to be homozygous for alleles that are copies of single alleles of common ancestors of her sire and dam. Ewe 18420 is expected to have 15% more of her total loci in the homozygous state than a non-inbred ewe in the same flock.

Notice that common ancestors three or four generations back in the pedigree of 18420 (individuals E and M) contributed relatively little to her inbreeding coefficient. The grandparent of 18420, the common parent of both the sire and dam and the most recent common ancestor in the pedigree, contributed over 84% of the inbreeding coefficient of 18420. Distant common ancestors have relatively little effect on the level of inbreeding of an individual.

The path method is presented here because the path diagram visually shows the flow of alleles from ancestors to subsequent generations and the common ancestors are easily identified by the path diagram. There is a tabular method of calculation of inbreeding coefficients that is much more computationally efficient than the path method demonstrated previously. The tabular method is used for very large pedigrees and when dealing with populations with large numbers of individuals and many pedigrees. The tabular method allows for both the calculation of relationship coefficients among individuals in the population (not discussed here) as well as inbreeding coefficients of individuals. The genetics textbooks listed in the Further Reading section of this chapter present the mechanics of using the tabular method.

Table 2.12 For question **5**.

Path	F_{CA}	n_1[a]	n_2[b]	$\left(\frac{1}{2}\right)^{n_1+n_2+1}(1+F_{CA})$
S ← A → D	0	1	1	
S ← B ← G ← E → A → D	0	3	2	
S ← B ← H ← M → F → A → D	0	3	3	
$Fx=\sum_1^3\left(\frac{1}{2}\right)^{n_1+n_2+1}(1+F_{CA})=$				

[a]n_1 = the number of arrows in the path between the common ancestor and the sire.
[b]n_2 = the number of arrows in the path between the common ancestor and the dam.

Genomics and inbreeding

Inbreeding coefficients derived from pedigree information, as presented previously, estimate the expected or average increase in homozygosity. Due to the random assortment of alleles from parent to progeny, the amount of homozygosity can vary above or below this average. Methods are being developed to characterize an individual's actual homozygosity or heterozygosity using SNP (single nucleotide polymorphism) data. In the future, this genomic data may be able to identify individuals with a high proportion of either heterozygous or homozygous loci within a group of individuals that have the same pedigree-estimated inbreeding coefficient. Selection could then be placed on the more heterozygous individuals or on individuals that are homozygous for the highest proportion of desirable genes, and pedigree-estimated inbreeding coefficients will be of little use. However, that is still in the future.

Summary

Inbreeding results from the mating of related individuals and results in offspring with an increase in the proportion of loci that are homozygous. This increase in homozygosity results in an increase in genetic defects due to deleterious recessive alleles; increased prepotency of the inbred individuals, especially for simply-inherited traits, and a decrease in performance of polygenic production traits. The level of inbreeding in an individual is quantified by the inbreeding coefficient, which is the probable proportion of an individual's loci containing alleles that are "identical by descent." Due to the relatively low reproductive rate of most livestock species and the decrease in reproductive rate and survival that results from inbreeding, it is recommended that inbreeding not be used as a mating system to improve livestock performance.

Table 2.13 For question **6**.

Path	F_{CA}	n_1[a]	n_2[b]	$\left(\frac{1}{2}\right)^{n_1+n_2+1}(1+F_{CA})$
S ← A → D	0	1	1	$\left(\frac{1}{2}\right)^{1+1+1}(1+0)=\frac{1}{8}=.125$
S ← B ← G ← E → A → D	0	3	2	$\left(\frac{1}{2}\right)^{3+2+1}(1+0)=\frac{1}{64}=.015625$
S ← B ← H ← M → F → A → D	0	3	3	$\left(\frac{1}{2}\right)^{3+3+1}(1+0)=\frac{1}{128}=.0078125$
$Fx=\sum_1^3\left(\frac{1}{2}\right)^{n_1+n_2+1}(1+F_{CA})$				$Fx = 0.1484375 \sim 0.15$

[a]n_1 = the number of arrows in the path between the common ancestor and the sire.
[b]n_2 = the number of arrows in the path between the common ancestor and the dam.

Further reading

Bourdon, R. M. 2000. *Understanding Animal Breeding*, 2nd edn. Prentice-Hall, Inc., Upper Saddle River, New Jersey, USA.

Van Vleck, L. D., Pollak, E. J. and Branford Oltenacu, E. A. 1987. *Genetics for the Animal Sciences*. W. H. Freeman and Company, New York, New York, USA.

References

Animal Improvement Program Laboratory, Agricultural Research Service (ARS), United States Department of Agriculture (USDA), Bovine Inbreeding Trends Menu. Online: www.cdcb.us/eval/summary/inbrd.cfm?, (accessed April 28, 2014).

Cleveland, M. A., Blackburn, H. D., Enns, R. M. and Garrick, D. J. 2005. Changes in inbreeding of US Herefords during the twentieth century. *J. Anim. Sci.* 83:992–1001.

Lamberson, W. R. and Thomas, D. L. 1984. Effects of inbreeding in sheep. A review. *Anim. Breed. Abstr.* 52:287–297.

Lamberson, W. R., Thomas, D. L., and Rowe, K. E. 1982. The effects of inbreeding in a flock of Hampshire sheep. *J. Anim. Sci.* 55: 780–786.

Smith, L. A., Cassell, B. G., and Pearson, R. E. 1998. The effects of inbreeding on the lifetime performance of dairy cattle. *J. Dairy Sci.* 81:2729–2737.

Review questions

1. What is the primary effect of inbreeding from which all other characteristics of inbreeding evolve?

2. What component of the Basic Genetic Model for polygenic traits is affected by inbreeding?

3. What is inbreeding depression?

4. Within the pair of traits, which trait is expected to exhibit the greatest amount of inbreeding depression?

a. *trait with high heritability* — *trait with low heritability*
b. *trait affected by loci with complete dominance* — *trait affected by loci with no dominance*
c. *piglet survival* — *loin eye area in pigs*

5. Draw a pedigree and the path diagram for an individual resulting from the mating of two first cousins. Calculate the inbreeding coefficient of this individual. Show your work using the path method.

6. Why have many inbred lines of livestock gone extinct?

7. Show, with a diagram, how an inbred sire and an inbred dam can produce an offspring with an inbreeding coefficient of zero.

8. What is prepotency?

3 Genomic Selection, Inbreeding, and Crossbreeding in Dairy Cattle

Kent Weigel

Department of Dairy Science, University of Wisconsin–Madison, WI, USA

Introduction

Genetic selection for greater productivity and enhanced fitness of dairy cattle has produced impressive results over the past half a century. Any trait that can be measured accurately and inexpensively on tens of thousands of daughters of young dairy bulls that are enrolled in progeny testing programs can be improved through selection. However, because nearly all traits of economic importance in dairy cattle are sex-limited and cannot be measured until the animals begin lactating, the response to selection is limited by long generation intervals. Furthermore, some traits are difficult and expensive to measure on commercial farms, and these traits are not amenable to improvement through progeny testing. Crossbreeding is used widely in beef cattle, swine, poultry, and other species as a means of achieving rapid gains in production and fitness traits, although it did not receive significant attention in dairy cattle until the past decade, despite the fact that dairy producers are familiar with inbreeding and routinely use tools to mitigate its impact. At present, we are witnessing the most remarkable change in dairy cattle breeding since the introduction of artificial insemination. Genomic selection (also known as whole genome selection) based on genotypes for thousands of **single nucleotide polymorphisms** (SNPs) located throughout the genome has transformed breeding programs worldwide. This technology provides estimated breeding values (EBVs) with sufficient accuracy for making selection decisions, even though neither the animal nor its offspring have been measured for phenotypes of interest. Selection among young bulls and heifers based on genomic data greatly reduces generation interval, thereby increasing the rate of genetic progress. Furthermore, genomic selection provides a mechanism for improving traits that are too difficult or expensive to measure routinely on commercial farms.

Genomic selection

Genotyping tools

In the past three years, tens of thousands of North American dairy cattle have been genotyped using the Illumina BovineSNP50 BeadChip (Illumina, Inc., San Diego, CA), and alternative high-density and low-density **genotyping chips** have recently become available (Wiggans et al., 2013). These technologies became possible due to sequencing of the bovine genome and were developed via collaboration between Illumina Inc., USDA-ARS, NAAB, and other commercial and academic partners. A key breakthrough is the ability to carry out thousands of SNP marker tests simultaneously, for a cost of less than ½¢ per marker. These SNP markers represent base differences (A, T, C, or G) within the DNA sequence of a cow or bull – a sequence that consists of approximately 3 billion base pairs distributed over 30 pairs of chromosomes. These SNP markers can be genotyped in an efficient and automated manner, in contrast to the labor-intensive genotyping methods that were used previously. Another key breakthrough is the finding that, once a large number of **genetic markers** become available for an individual animal, it is possible to estimate that animal's breeding value based on associations between marker genotypes and milk yield, SCS, DPR, PL, and other key traits that were observed in other animals of the same breed (Hayes et al., 2009; Meuwissen et al., 2001). Breeding companies use genomic testing routinely to screen young bulls prior to entry in their AI programs, and many progressive cattle breeders genomic test the majority of their cows and heifers in order to identify elite females that received the most favorable combination of genes from their parents. Currently, low-density chips (6K and 9K, with roughly 6900 and 9000 SNPs, respectively) and a medium-density (50K) chip with 54 600 SNPs are the products used most frequently by breeders and AI companies, respectively, and genetic predictions for production, health, and conformation traits can be computed using genotypes from any of these chips. The main difference between the low-density chips and the 50K chip is cost; the low-density chips are more affordable (roughly $40–45) for dairy producers who wish to genotype a large number of cows, heifers, or calves, whereas the 50K chip (roughly $100 to 125) provides more detailed information for pedigree breeders and AI companies who wish to characterize the genetic merit of their animals with greater accuracy. **Genomic testing services** are offered by breed associations, AI studs, and a handful of animal health and biotechnology companies.

Parentage verification and discovery

An immediate application of genomics is **parentage verification**, which could be done previously with blood typing or microsatellite markers, and parentage discovery, which is a new

Molecular and Quantitative Animal Genetics, First Edition. Edited by Hasan Khatib.

benefit that is specific to current SNP genotyping arrays. Why do we need DNA testing to verify or discover parentage in the North American dairy breeds, given that we have nearly a century of calving, parentage, and performance data from the DHI system and over a century of pedigree data from the breed associations? Unfortunately, recording of parentage information is incomplete on many farms, and the error rate in parentage recording varies widely between farms. Some farms have nearly 100% accuracy, whereas others may have more than 50% errors. The actual rate of parentage errors on commercial farms is unknown, but anecdotal evidence suggests that 10–15% errors are likely. Because nearly every AI bull in North America has been genotyped, it is relatively easy to discover the correct sire of a given cow, heifer, or calf if the sire is unknown or incorrectly identified, and the USDA-ARS Animal Improvement Programs Laboratory (in collaboration with the breed associations) routinely finds and corrects parentage errors as part of the national genetic evaluation program for US dairy cattle.

Individual genes with large effects

For more than two decades, dairy cattle geneticists have aggressively pursued the idea of **marker-assisted selection** (Dentine, 1992). We attempted to apply the same principles used for identifying inherited defects to quantitative traits of interest, such as milk production. In other words, we tried to identify individual genes with large effects, so that one-by-one we could try to increase the frequencies of the favorable alleles at these loci, thereby enhancing the phenotype. However, successes were few, and the exact location and mode of action remains unknown for nearly all genes affecting economically important traits in dairy cattle. In reality, genomic testing has confirmed the long-standing hypothesis that traits such as milk production are influenced by a large number of genes scattered throughout the genome, and that the effect of each gene is rather small, as illustrated in Figure 3.1.

How does **whole genome selection** today differ from two decades of attempted marker-assisted selection in dairy cattle? The difference is that we now focus our efforts on selection for desirability of the entire genome, and we do not rely on knowledge regarding individual genes with large effects. While it is true that gains in reliability due to genomic testing are largest for traits for which single genes with large effects had already been discovered, such as fat percentage (DGAT1 gene on chromosome 14; Grisart et al., 2002) and protein percentage (ABCG2 gene on chromosome 6; Cohen-Zinder et al., 2005), we can also achieve significant gains in accuracy of selection in other traits for which no major genes have been identified.

Genotype imputation

To determine if **imputation** (Habier et al., 2009) of medium-density (e.g., 50K) genotypes from subsets of a few hundred or a few thousand equally spaced SNPs was feasible, Weigel et al. (2010a) used a population of Jersey bulls, cows, and heifers that had already been genotyped with the BovineSNP50 BeadChip. This population was divided into a reference panel, comprised of older animals, and a study sample, comprised of younger animals. Genotypes of animals in the study sample were masked (i.e., hidden) for the vast majority of SNPs and subsequently imputed (i.e., filled in) based on patterns of SNPs (commonly known as **haplotypes**) that were present in the **reference population**. Results of the study suggested that a low-density chip with approximately 3000 equally spaced SNPs would be adequate for imputing high-density genotypes from reference animals of the same breed. Next, Weigel et al. (2010b) quantified the impact of imputation errors on the accuracy of genomic predictions for economically important traits. After imputing missing genotypes for young bulls based on haplotypes of animals in the reference population, genomic EBV for milk yield, protein percentage, and daughter pregnancy rate were computed. These were compared with traditional EBV resulting from progeny testing. Results

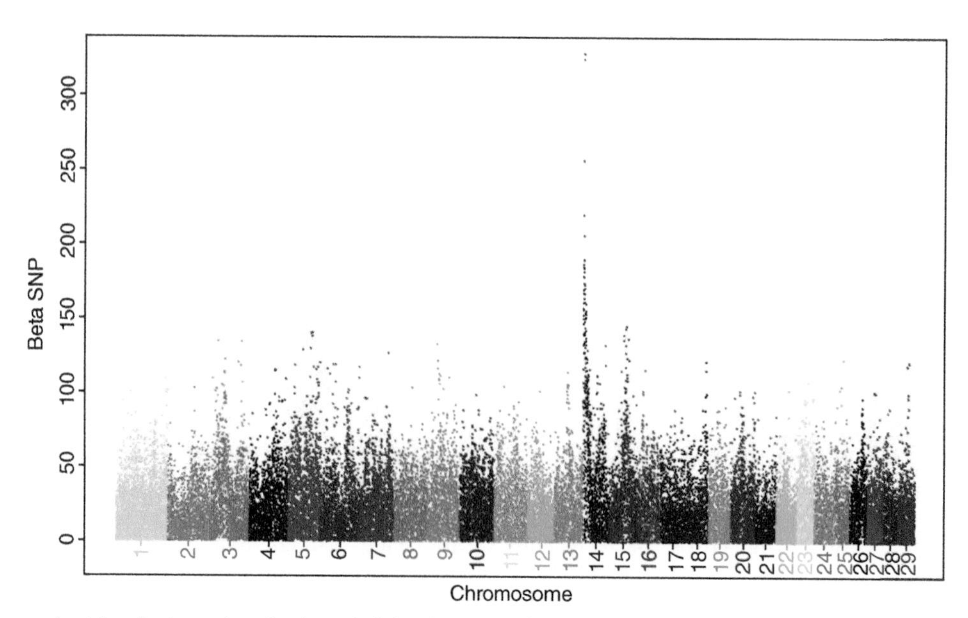

Figure 3.1 Manhattan graph of the absolute value of estimated allele substitution effects for milk yield in US Holstein cattle based on SNP genotypes from the BovineSNP50 BeadChip (source: USDA-ARS Animal Improvement Programs Laboratory).

showed that a low-density genotyping chip with approximately 3000 equally spaced SNPs would provide genomic predictions that are roughly 95% as accurate as predictions from the BovineSNP50 BeadChip, for a small fraction (e.g., one-third) of the price.

Genome-enabled breeding value prediction

Although it sounds mysterious, **genomic selection** is actually rather simple. In the past, all we knew about a young animal's genetic potential was its parent average (PA), which was simply the average predicted transmitting ability (PTA) of its parents, and we had no way to determine whether this young animal got a better than average or poorer than average sample of genes from its parents. The reliability of the PA for a young animal, which is calculated as: 0.25 × (REL of its sire's PTA + REL of its dam's PTA), is typically too low to facilitate an informed selection decision. For calves and heifers with complete pedigree information, REL of the PA typically ranges from 25 to 40%, but REL is much lower in animals with incomplete pedigrees. If we wanted a more accurate prediction of an animal's genetic merit, we had no choice but to wait two years until we could measure the animal's performance, in the case of females, or wait five years until we could measure the performance of the animal's progeny, in the case of males. Now, because the relationships between SNP markers and important functional genes that we observe in an animal's ancestors are maintained for several generations, before recombination breaks down these genetic links, we can glimpse into the crystal ball to see what the future holds for a particular young animal.

As described by VanRaden et al. (2009), genomic PTAs became the official genetic evaluation for US Holsteins and Jersey cattle in January 2009, and official results for Brown Swiss became available in August 2009. The most important animals in this process are the dairy bulls represented in the Cooperative Dairy DNA Repository, which was formed more than 15 years ago, when NAAB member organizations (North American AI studs) began storing semen samples from young bulls entering their progeny testing programs for the purpose of genetic research. For each trait, we can combine a young animal's pedigree with information regarding its SNP genotypes to obtain a genomic PTA of much greater accuracy. For a heifer calf, reliability of the genomic PTA is greater than the information we could obtain by measuring several lactation records on the animal and its daughters. For a young cow, genomic information can be combined with her lactation records to obtain a genomic PTA that is significantly more informative than her traditional PTA. For a bull calf, reliability of the genomic PTA is equivalent to what we could obtain by measuring performance on 25 or 30 of his progeny test daughters, although this figure varies widely by trait. Improvements in accuracy can even be obtained for bulls that have completed progeny testing, although the gain in information for a bull that already has performance data from 80 to 100 daughters is much smaller. At present, typical REL for the genomic PTAs of Jerseys and Brown Swiss (55 to 65% for production traits) is lower than for Holsteins (70 to 80%). This difference is largely due to the fact that fewer progeny tested bulls have been genotyped, and results for these breeds will be improved by combining information from North American sires with that of key populations internationally. Overall, it is clear that genomic evaluations considerably reduce the risk associated with buying young heifers or bulls for the purpose of enhancing genetic improvement.

Applications of genomics to selection of elite breeding stock

What has been the impact of genomics on the AI industry? The AI studs are already in the midst of tremendous change because of this technology (Schefers and Weigel, 2012). Virtually every young bull entering a North American AI center today is tested with the Bovine SNP50 BeadChip, with 20 to 25 young bulls tested per each bull that is chosen, and DNA testing of potential bull dams has become the norm. The success rate (i.e., graduate rate) in progeny testing programs, which is currently about 1 in 10, will increase significantly in the future, because we know prior to entry into the AI stud that each young bull has received a favorable sample of genes from its parents. Progeny testing has been the cornerstone of the dairy cattle breeding industry for nearly a half century, and because genomic testing competes with progeny testing in terms of accuracy, while also dramatically reducing the generation interval, it will have an enormous impact. The AI studs in North America and Europe are now marketing semen from hundreds of young bulls that have genomic PTAs, but no daughters of their own. These genome tested bulls have replaced many older, progeny tested bulls in the marketplace, and some young bulls with superior genomic evaluations are now used for **contract matings** to produce the next generation of young bulls. As AI centers and producers become more comfortable with this technology, we will see a continual decline in progeny testing, because its purpose is the same as that of the BeadChip – to see which young bull received the best sample of genes from its parents. Eventually, some of the genotyping costs may be offset by a lowered investment in progeny testing (e.g., bull housing, semen distribution, incentive payments, etc.).

What is the impact of genomics on pedigree breeders who merchandise breeding stock? Breeders who are selling young bulls to AI have seen a major impact, because the AI studs are genotyping their young bulls, and potentially their dams as well. Based on this initial genomic screening, many bull dams and young bulls are rejected. Conversely, the price paid for young bulls that pass this initial genomic screening is typically higher. Furthermore, because sire analysts now have the ability to distinguish between sets of full brothers that have identical pedigrees, the premium for securing first choice from an embryo transfer flush is much greater. The impact on the female side, whether selling embryos or live heifers, has been similar. Potential buyers of embryos, calves, and heifers want to genotype an animal before the purchase, and genomic information is desired at consignment sales, where buyers pay a premium for young animals with favorable genomic test results.

Applications of genomics to improvement of replacement heifers on commercial farms

What is the impact of genomic testing on commercial producers? First, dairy producers have the option of purchasing semen from hundreds of young bulls that are marketed based on their genomic breeding values. These bulls have high estimated genetic merit, but REL is much lower for young, genome-tested bulls than for progeny tested bulls. Therefore, when buying semen from young, genome-tested bulls, producers should avoid heavy use of one or two top bulls and instead spread their **risk** by using a larger group of bulls and fewer units of semen per bull.

A recent simulation study by Weigel et al. (2012) investigated the potential economic benefits associated with genomic testing of cows, heifers, and calves on commercial dairy farms. Two scenarios were considered: (1) identification of genetically superior females at an early age (e.g., to generate extra income by marketing breeding stock or generate superior herd replacements by using these animals as ET donors), or (2) identification of genetically inferior females at an early age (e.g., to allow early culling of inferior animals that have low potential for future milk yield, thereby reducing the cost of rearing replacement heifers). Genomic testing of all heifer calves was cost-effective in herds in which pedigree information was unavailable or inaccurate. The value of genomic testing was lower in herds that routinely recorded sire identification, and lower yet in herds that had several generations of accurate pedigree data for every animal. Nonetheless, genomic testing of heifer calves was cost-effective in such herds if animals are pre-sorted prior to testing based on their pedigree values (e.g., test only the top 25% of the herd as potential ET donors or the bottom 25% of the herd as potential culls). The value of testing lactating cows with phenotypes of their own was relatively small, particularly if complete pedigree data were available, so investments in genomic testing should focus on calves and yearling heifers. Additional uses of genomic testing results, such as targeting the use of gender-enhanced semen on heifers with above average genomic predictions, or use of genotypic data in mate selection algorithms, may make genomic testing even more valuable in the future.

Crossbreeding

Breed characteristics and complementarity

Today's common dairy breeds can be considered as inbred lines that were formed over a period of several hundred (or more) years due to geographic isolation (e.g., on an island or in a valley), population bottlenecks (e.g., due to disease or transportation to new locations by boat), and selective breeding (e.g., for uniform coat color and other breed characteristics). Selective breeding also occurred, at least at a rudimentary level, for characteristics that were beneficial in a given geographical area, such as heat tolerance and milk, draft, or meat characteristics. As modern tools such as milk recording, type classification, artificial insemination, and embryo transfer became available, more rapid genetic progress could be achieved within breeds, and greater divergence between breeds was possible. It is easy to imagine that one might create a **composite population** that combines the best characteristics of several different breeds, or a **rotational crossbreeding system** that uses the strengths of one breed to offset the weaknesses of another. Studies that compare the performance of different breeds under controlled environmental conditions, as well as studies that quantify the gains or losses associated with various crossbreeding schemes, have been fewer in dairy cattle than in other food animal species. However, the strengths and weakness of common North American dairy breeds are well known. For example, Holsteins tend to be superior for milk yield but inferior for female fertility and calving ability; Jerseys tend to excel in milk composition, female fertility, and calving ability, but lag behind in milk volume; and Brown Swiss tend to excel for hardiness and mobility traits, but lag behind in feed conversion efficiency (milk yield per unit of body size) and female fertility. Complementarity of the breeds is a key consideration in crossbreeding programs, as is the need to select breeds that have numerically large populations with established breeding programs and, hence, accurate breeding value predictions and rapid within-breed genetic progress.

Heterosis and crossbreeding systems

Hybrid vigor, or **heterosis**, occurs when the average performance of crossbred animals exceeds the average performance of the parental breeds. For example, if breed A has average performance of 90 units, breed B has average performance of 110 units, and a first-generation (F_1) cross between breeds A and B yields animals with average performance of 105 units, we would say that we have achieved 5% heterosis. The degree of heterosis observed for a particular trait is specific to each pair of breeds, and it is a function of the genetic distance between these breeds. Crosses between breeds that are closely related to each other, perhaps due to more recent population divergence, will tend to generate less heterosis than crosses between genetically distant breeds. The reason is that allele frequencies at loci affecting economically important traits tend to be similar in breeds that are closely related, and therefore the increase in **heterozygosity** observed in their crossbred offspring is smaller than would be observed in the crossbred offspring of distantly related breeds. In practice, however, the specific level of heterosis for each trait and each pair of breeds is often unknown.

Because **terminal crossbreeding systems** (which produce F_1 animals for slaughter) are not used in dairy cattle, we must also consider the proportion of heterosis that will be maintained in future generations. In a rotational cross, we can consider the percentage of heterosis that is retained relative to the first generation cross. For example, if an F_1 Holstein × Jersey cross generates heifers with 100% heterosis, then breeding these animals back to a Holstein bull will produce backcross animals that are 75% Holstein : 25% Jersey, and only 50% of the heterosis will be retained. At equilibrium, a two-breed rotational cross will retain 67% of the heterosis observed in the F_1 cross, whereas a three-breed rotational cross will retain 86% of the heterosis, and a four-breed cross will retain 93% of the heterosis. Given this information, why wouldn't a dairy producer use as many different breeds as possible in a rotational cross? First, the marginal gain in retained heterosis is smaller with each additional breed. Second, it can be difficult to manage a crossbreeding program with too many different breeds. Third, and most important, is that a rational dairy producer start with most profitable breed, and each additional breed that is added to the rotational cross will be increasingly inferior. Eventually, the genetic inferiority of the breed that is introduced will offset the marginal increase in percentage of retained heterosis associated with using one additional breed.

Experiments and field results

As noted earlier, crossbreeding studies in dairy cattle have been fewer than in many other food animal species, nonetheless there are too many to review in detail. Excellent review articles by McAllister (2002) and Sorensen et al. (2008) describe the challenges and opportunities associated with crossbreeding in dairy cattle production systems. A landmark study by McAllister et al. (1994) involving Ayrshire and Holstein cattle demonstrated the importance of accounting for all costs and expenses incurred throughout the lifetime of each animal when evaluating a crossbreeding system, as opposed to simply comparing groups of purebred and crossbred animals based on specific traits at specific time points, such as first service conception rate or first lactation milk yield. Several experimental studies involving Holstein × Jersey crosses have been carried out, including Bjelland et al. (2011) and Heins et al. (2012), whereas studies involving Holstein × Brown Swiss crosses (Dechow et al., 2007) or studies involving Holstein × Normande, Holstein × Montbeliarde, Holstein × Swedish Red, or Holstein × Norwegian Red crosses (Heins et al., 2006a,b,c) have relied on field data from commercial dairy farms. In general, results have been as expected. Improvement in calving ability, fertility, and longevity is usually observed, particularly among the F_1 crosses, whereas milk production is sacrificed. Nonetheless, it appears that some breeds (e.g., Montbeliarde) can add strength and durability when used as mates for Holstein cows, whereas others (e.g., Jersey, Norwegian Red, Swedish Red) can reduce the incidence of dystocia and stillbirths, and crossbreeding seems to be a viable option for dairy producers who are willing to accept a small decrease in production in order to reduce health problems, labor costs, and management interventions.

Inbreeding and genetic defects

Relationships, inbreeding, and effective population size

A key concern in dairy cattle breeding programs, as well as improvement programs in other food animal species, is achieving a balance between rapid genetic progress and maintenance of **genetic diversity** (Weigel, 2001). The two go hand-in-hand, because **inbreeding** would not be a concern if all males and females in the population were allowed to mate randomly and produce the next generation (i.e., no selection intensity). On the male side of the pedigree, relatively few sires are needed because of the heavy adoption of AI technology more than half a century ago. Roughly 70% of dairy cows and heifers are bred by AI with frozen semen, because it allows widespread (i.e., global) availability of genetically superior bulls, and because it allows farmers to avoid the feed costs and danger associated with keeping mature dairy bulls on the farm. Individual dairy bulls have produced more than 200,000 milk-recorded daughters through the marketing of more than 2 million units of semen. On the female side, **embryo transfer** (ET) and related technologies such as ***in vitro* fertilization** (IVF) are used widely by pedigree breeders who wish to sell young bulls to AI companies, export embryos to foreign customers, and market excess calves and heifers to other farmers. Widespread use of reproductive technologies, such as AI, ET, and IVF, leads to high selection intensity and rapid genetic progress, but genetic diversity is compromised due to selection of fewer parents of the next generation, and because these selected parents (superior AI bulls and elite young cows) are often related to each other.

Inherited defects

Every breed of livestock and, most likely, every individual animal carries one or more defective alleles that can lead to impaired health, fertility, performance, or death if present in the homozygous form. Most of these **genetic defects** are inherited as simple recessives, because dominant genetic defects (particularly if lethal) would be eliminated from the population very quickly. Recessive defects, on the other hand, can linger undetected in a population for many generations. If the frequency of the deleterious allele is low, the likelihood that a male and female that carry the same defect will meet is very small, and even in this case only 25% of their offspring will be affected. However, if the male and female are related through one or more common ancestors, and if one of these common ancestors carries a defective allele, the likelihood of an affected offspring increases significantly. For this reason, offspring of related parents are more likely to suffer from genetic defects, such as bovine leukocyte adhesion deficiency (BLAD), complex vertebral malformation (CVM), progressive degenerative myeloencephalopathy (weaver), syndactylism (mule foot), achondroplasia (bulldog), or other lethal or sub-lethal conditions. It is important to note that not all congenital conditions (i.e., birth defects) are the result of genetic causes, and it is the role of breed associations to determine which conditions are of genetic origin and to label the carriers of known defects. As an example, carriers of the weaver condition in the Brown Swiss breed are denoted with a "W" suffix in their registered names, whereas animals that have been genetically tested as free from weaver are denoted with a "TW" suffix. Once an undesirable recessive allele is identified, genetic testing of young bulls by the AI industry commences, and carrier bulls are typically removed from active service. This leads to a rapid reduction in the risk of producing offspring on commercial farms that are homozygous for the defective allele, and it gradually reduces the frequency of the recessive allele in the female population.

Inbreeding depression for quantitative traits

Accumulation of inbreeding is a natural consequence of intense selection and heavy use of reproductive technologies. Inbreeding occurs whenever an animal's parents are related to each other, and an animal's **inbreeding coefficient** (F) is equal to half of the additive genetic relationship between its parents. For example, if the sire and dam of a particular calf are paternal half-siblings (i.e., they share a common sire), then the additive genetic relationship between its sire and dam is 25%, and the calf's inbreeding coefficient is 12.5%. Why do we care about an animal's inbreeding coefficient? The main reason is that, on average, animals with higher inbreeding coefficients tend to have impaired fertility and fitness relative to their less-inbred counterparts. This phenomenon is known as **inbreeding depression**, and it occurs due to the accumulation of homozygous loci with recessive, sub-lethal effects on fitness and performance. Smith et al. (1998) quantified inbreeding depression in Holstein cattle and estimated that 37 kg milk, 1.2 kg fat, and 1.2 kg protein were sacrificed per lactation for each 1% increase in the inbreeding coefficient. More importantly, 13.1 days of productive life were lost per 1% increase in inbreeding, such that the estimated loss in lifetime net income was $23 per 1% increase in inbreeding. Note that this estimate does not include inbreeding depression for traits such as calf survival, which is measured before the initiation of first lactation. It is important to note that inbreeding depression tends to cause a continuous decline in performance as a function of increasing F (Gulisija et al., 2007), rather than a step function that occurs once F passes a certain threshold (e.g., 6.25%). Furthermore, the concept of threshold values for acceptable versus unacceptable levels of F is not sensible, because the reference point for calculation of inbreeding coefficients (i.e., the point in time at which all animals were considered as unrelated) arbitrarily falls around 1960, when herdbook information was first computerized. Average inbreeding coefficients for US Holstein and Jersey cattle, according to year of birth, are shown in Figure 3.2.

Expected future inbreeding, which measure an animal's relationship to a sample of current animals from the same breed, and inbreeding coefficients are slightly higher for Jerseys than

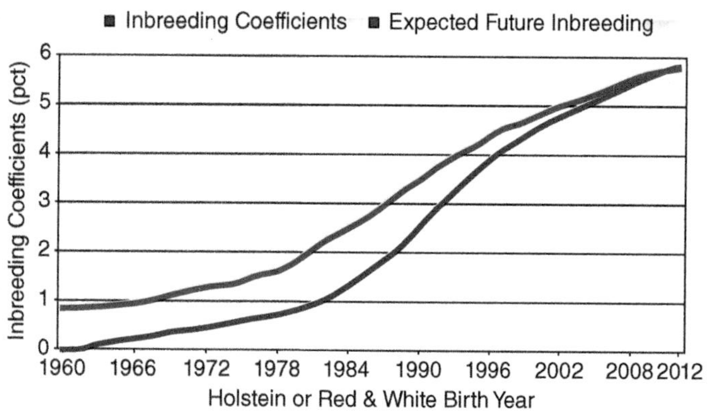

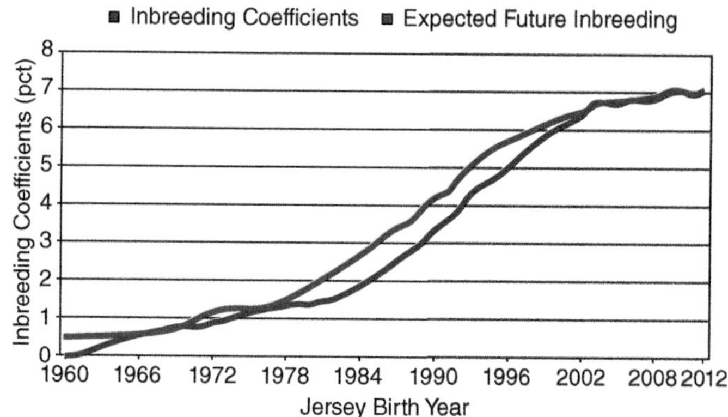

Figure 3.2 Average inbreeding coefficients for US Holstein and Jersey cattle, according to year of birth (source: www.cdcb.us/eval/summary/trend.cfm?).

Holsteins. Prior to 1960 many herds practiced linebreeding, which involved intentional mating of related animals to increase relationships between their offspring and one or more outstanding ancestors. This inbreeding is not considered in current inbreeding calculations, but this omission may not be of great importance, because selection intensity was very low prior to 1960.

Managing inbreeding and genetic diversity

Balancing inbreeding and genetic progress is tricky, and both scientists and dairy producers share concerns about reducing genetic diversity. On the other hand, sacrificing genetic progress in the short-term due to fears of potential and poorly quantified losses in future performance can be difficult to justify. Long-term maintenance of genetic diversity can be facilitated by selection algorithms based on genetic contribution theory (Weigel, 2001; Weigel and Lin, 2002), but implementation of tightly controlled, long-term selection policies in dairy cattle breeding programs is nearly impossible in a free-market economy, because numerous commercial dairy farmers, pedigree breeders, and AI organizations are involved, and most seek to maximize short-term financial gains. Therefore, *ad hoc* breeding policies based on common sense tend to predominate, and both dairy farmers and AI companies simply try to spread the risk by using a variety of different bulls each year. Short-term avoidance of inbreeding (i.e., in the next generation of replacement heifers) is much easier to achieve than long-term maintenance of genetic diversity, due to the availability of computerized **mate selection programs**. These programs, which are offered by breed associations and AI companies, were originally developed for the purpose of finding specific bulls that would correct faults in the physical conformation or appearance of specific cows. By incorporating pedigree and performance information from a cow and each of her potential mates, these programs can compute the expected genetic merit of the hypothetical calf from each pair of potential mates after adjustment for anticipated costs of inbreeding depression (Weigel and Lin, 2002). This approach is now widely used in commercialized mate selection programs, and millions of dairy cows are mated this way each year.

Summary

Crossbreeding has not been used widely in dairy cattle, but recent studies suggest that it may be a viable option for improving fitness traits such as calving ability, health, fertility, and longevity, particularly among dairy producers who seek to minimize time-consuming management interventions and costly veterinary treatments. Mechanisms for controlling inbreeding in herd replacement animals are readily available, yet dairy cattle breeders still struggle to find the optimum balance between rapid genetic progress in the short term and maintenance of adequate genetic diversity in the long term. Information from the bovine genome sequence provides a plethora of opportunities for understanding the genetic mechanisms underlying important traits, as well as enhancing the rate of genetic progress for these traits. Hundreds of thousands of dairy cattle have been genotyped using commercially available SNP chips, and the resulting information has been integrated into genetic evaluation and selection programs in a nearly seamless manner. Dairy producers are readily purchasing semen from young genome-tested bulls that lack progeny information, and the majority of breeding females that are sold at public auctions are genotyped prior to sale. Commercial farms that do not sell breeding stock also utilize genomic testing, due to the development of low-cost, low-density SNP chips coupled with subsequent imputation of medium- or high-density genotypes. Lastly, efforts to establish deeply phenotyped reference populations are underway, and this will allow genomic selection for traits that were previously untouchable, such as the efficiency of feed utilization.

References

Bjelland, D. W., K. A. Weigel, P. C. Hoffman, N. M. Esser, W. K. Coblentz, and T. J. Halbach. 2011. Production, reproduction, health, and growth traits in backcross Holstein × Jersey cows and their Holstein contemporaries. *J. Dairy Sci.* 94:5194–5203.

Cohen-Zinder, M., E. Seroussi, D. M. Larkin, J. J. Loor, A. Everts-van der Wind, J. H. Lee, et al. 2005. Identification of a missense mutation in the bovine ABCG2 gene with a major effect on the QTL on chromosome 6 affecting milk yield and composition in Holstein cattle. *Genome Res.* 15:936–944.

Dechow, C. D., G. W. Rogers, J. B. Cooper, M. I. Phelps, and A. L. Mosholder. 2007. Milk, fat, protein, somatic cell score, and days open among Holstein, Brown Swiss, and their crosses. *J. Dairy Sci.* 90:3542–3549.

Dentine, M. R. 1992. Marker-assisted selection in cattle. *Anim. Biotech.* 3:81–93.

Grisart, B., W. Coppieters, F. Farnir, L. Karim, C. Ford, P. Berzi, et al. 2002. Positional candidate cloning of a QTL in dairy cattle: identification of a missense mutation in the bovine DGAT1 gene with major effect on milk yield and composition. *Genome Res.* 12:222–231.

Gulisija, D., D. Gianola, and K. A. Weigel. 2007. Nonparametric analysis of the impact of inbreeding depression on production in Jersey cows. *J. Dairy Sci.* 90:493–500.

Habier, D., R. L. Fernando, and J. C. M. Dekkers. 2009. Genomic selection using low-density marker panels. *Genetics* 182:343–353.

Hayes, B. J., P. J. Bowman, A. J. Chamberlain, and M. E. Goddard. 2009. Invited review: Genomic selection in dairy cattle: Progress and challenges. *J. Dairy Sci.* 92:433–443.

Heins, B. J., L. B. Hansen, and A. J. Seykora. 2006a. Calving difficulty and stillbirths of pure Holsteins versus crossbreds of Holstein with Normande, Montbéliarde, and Scandinavian Red. *J. Dairy Sci.* 89:2805–2810.

Heins, B. J., L. B. Hansen, and A. J. Seykora. 2006b. Production of pure Holsteins versus crossbreds of Holstein with Normande, Montbéliarde, and Scandinavian Red. *J. Dairy Sci.* 89:2799–2804.

Heins, B. J., L. B. Hansen, and A. J. Seykora. 2006c. Fertility and survival of pure Holsteins versus crossbreds of Holstein with Normande, Montbéliarde, and Scandinavian Red. *J. Dairy Sci.* 89:4944–4951.

Heins, B. J., L. B. Hansen, A. R. Hazel, A. J. Seykora, D. G. Johnson, and J. G. Linn. 2012. *Short communication:* Jersey × Holstein crossbreds compared with pure Holsteins for body weight, body condition score, fertility, and survival during the first three lactations. *J. Dairy Sci.* 95:4130–4135.

McAllister, A. J., A. J. Lee, T. R. Batra, C. Y. Lin, G. L. Roy, J. A. Vesely, et al. 1994. The influence of additive and nonadditive gene action on lifetime yields and profitability of dairy cattle. *J. Dairy Sci.* 77:2400–2414.

McAllister, A. J. 2002. Is crossbreeding the answer to questions of dairy breed utilization? *J. Dairy Sci.* 85:2352–2357.

Meuwissen, T. H. E., B. J. Hayes, and M. E. Goddard. 2001. Prediction of total genetic value using genome-wide dense marker maps. *Genetics* 157:1819–1829.

Schefers, J. M., and K. A. Weigel. 2012. Genomic selection in dairy cattle: Integration of DNA testing into breeding programs. *Animal Frontiers* 2:4–9.

Smith, L. A., B. G. Cassell, and R. E. Pearson. 1998. The effects of inbreeding on the lifetime performance of dairy cattle. *J. Dairy Sci.* 71:1880–1896.

Sorensen, M. K., E. Norberg, J. Pedersen, and L. G. Christensen. 2008. Invited review: Crossbreeding in dairy cattle: A Danish perspective. *J. Dairy Sci.* 91:4116–4128.

VanRaden, P. M., C. P. Van Tassell, G. R. Wiggans, T. S. Sonstegard, R. D. Schnabel, J. F. Taylor, and F. Schenkel. 2009. Reliability of genomic predictions for North American dairy bulls. *J. Dairy Sci.* 92:16–24.

Weigel, K. A. 2001. Controlling inbreeding in modern breeding programs. *J. Dairy Sci.* E177–184.

Weigel, K. A., and S. W. Lin. 2002. Controlling inbreeding by constraining the average relationship between parents of young bulls entering AI progeny test programs. *J. Dairy Sci.* 85:2376–2383.

Weigel, K. A., C. P. Van Tassell, J. R. O'Connell, P. M. VanRaden, and G. R. Wiggans. 2010a. Prediction of unobserved single nucleotide polymorphism genotypes of Jersey cattle using reference panels and population-based imputation algorithms. *J. Dairy Sci.* 93:2229–2238.

Weigel, K. A., G. de los Campos, A. I. Vazquez, G. J. M. Rosa, D. Gianola, and C. P. Van Tassell. 2010b. Accuracy of direct genomic values derived from imputed single nucleotide polymorphism genotypes in Jersey cattle. *J. Dairy Sci.* 93:5423–5435.

Weigel, K. A., P. C. Hoffman, W. Herring, and T. J. Lawlor, Jr. 2012. Potential gains in lifetime net merit from genomic testing of cows, heifers, and calves on commercial dairy farms. *J. Dairy Sci.* 95:2215–2225.

Wiggans, G. R., T. A. Cooper, C. P. Van Tassell, T. S. Sonstegard, and E. B. Simpson. 2013. *Technical note:* Characteristics and use of the Illumina BovineLD and GeneSeek Genomic Profiler low-density chips for genomic evaluation. *J. Dairy Sci.* 96:1258–1263.

Review questions

1. What benefits does the use of genomic selection have over progeny testing? What about over marker-assisted selection?

2. How has genomic information impacted the selection and purchasing of elite breeding stock?

3. How would a commercial dairy producer utilize genomic selection in his or her herd? What are the benefits?

4. What are the benefits and limitations of crossbreeding in dairy cattle breeding?

5. What are some of the negative effects of inbreeding in dairy cattle? Why has there been a large accumulation of inbreeding in dairy cattle populations with these known issues?

4 Basic Genetic Model for Quantitative Traits

Guilherme J. M. Rosa
Department of Animal Science, University of Wisconsin–Madison, WI, USA

Introduction

Simply-inherited traits are determined by a single or a few genes and are generally not affected by environmental factors. Such traits are usually expressed in a few possible phenotypic categories. In contrast, quantitative or complex traits are controlled by a large number of genes and also are affected by environment. As a consequence, quantitative traits are generally expressed in a continuous scale. In addition, there is not a one-to-one relationship between genotypes and phenotypes, which can be established only in terms of probabilistic distributions. This is precisely the focus of this chapter, which introduces a basic genetic model extremely useful to study such complex traits. In the first two sections, the concept of quantitative traits and some discussion on distribution, expectation, and variance of phenotypic traits are introduced. Next, the basic genetic model for quantitative traits is presented, as well as the concepts of additive and non-additive gene action, and environmental effects. The last two sections discuss the definition of heritability and provide an introduction to selection and genetic progress, respectively.

Quantitative traits

Quantitative traits refer to phenotypes (characteristics) that vary in degree, and hence can be measured in some quantitative scale. Many quantitative traits have a continuous distribution of possible values, such as milk yield in dairy goats and fleece weight in sheep. Others, such as litter size in pigs or number of eggs in laying hens, can only be whole numbers as they are counting variables. Lastly, some other quantitative traits are binary, or expressed in a multiple (ordered) categorical scale. Binary traits are those in which individuals can be grouped into two possible outcomes, such as the classification of animals as healthy or ill, or females as pregnant or open. Examples of multiple-category outcomes are body condition score in beef cattle or beef marbling scores. Although some quantitative traits are expressed on a discrete or categorical scale, they can be thought of as the observable outcome of a hypothetical underlying continuous variable (generally called liability), with one or multiple thresholds defining the categories on this liability scale.

The observed distributions of quantitative traits arise because such traits are influenced by many genes, which result in many possible genotypes, and also by environmental effects. For this reason, quantitative traits are often referred to as **complex traits**. Some additional discussion on the distribution of complex traits and on genetic and environmental effects is provided in the next two sections, respectively.

Expected value and variance: the normal distribution

The expected value (or mean) of a variable P (e.g., a phenotypic trait) is the arithmetic average of the phenotypic values of all animals in a population. The expected value of P is usually written as E[P] or μ, where μ is the Greek letter mu.

Suppose $(P_1, P_2, \ldots, P_N)$ represent the phenotypic values of all animals in a population, where N is the total number of animals. The mean of the phenotypic trait P can be calculated as:

$$\mu = \frac{1}{N}(P_1 + P_2 + \ldots + P_N) \quad (4.1)$$

However, the definition of a population in quantitative genetics is a bit "tricky." Although, colloquially, we may call "population" a group of animals of a specific species and breed, statistically speaking such a group of available animals represents a sample of a theoretical population of potential animals that could be generated as different combinations of genetic and environmental effects.

In this case, instead of N phenotypic values related to all potential animals in such a virtual population, we have a sample of n animals with observed phenotypes; that is, $(P_1, P_2, \ldots, P_n)$. Assuming that such a sample of animals is representative of the population, the mean obtained in this sample is an estimate of the population mean. To differentiate the population mean (μ) from a sample mean, the latter will be denoted here by m.

The sample mean (m) is calculated as the arithmetic average of the observed phenotypic values, that is:

$$m = \frac{1}{n}(P_1 + P_2 + \ldots + P_n) \quad (4.2)$$

Molecular and Quantitative Animal Genetics, First Edition. Edited by Hasan Khatib.

Hopefully the sample mean will be a number close to μ, but it will seldom be equal to it. Nonetheless, it will approach the value of μ as the sample size (n) increases.

The mean of a distribution can be interpreted as a central point around which observed values are scattered. It is extremely useful as a summary value of the overall magnitude of the phenotypic values. For example, if we know the average milk yield of a group of cows, simply multiply this average by the number of cows and we have the total milk production in the group.

However, quite often the mean alone is not informative enough as a summary of a distribution. For example, different groups of animals with similar average phenotypic values may have higher or lower variation within groups. Variability is also an important feature of a distribution. For example, when assessing body weight of a group of broilers we should look not only for the average but also for the homogeneity of the flock.

A common measure of variability (or variation) in a distribution is called variance. The population variance of a variable or phenotypic trait P, generally represented as Var[P] or σ^2 (sigma squared), is given by:

$$\sigma^2 = \frac{1}{N}\left[(P_1 - \mu)^2 + (P_2 - \mu)^2 + \ldots + (P_N - \mu)^2\right] \quad (4.3)$$

that is, the average squared deviation from the mean.

As before, if phenotypes of all animals in the population are not known, and the available animals with observation can be thought of as a sample of it, an estimate of σ^2 can be obtained, which is generally represented as s^2. However, in this case, a sample mean (m) must be used in place of μ, which is also unknown, and a slightly different formula is often preferred, in which (n − 1) is used in the denominator instead of n. The sample variance is then calculated as:

$$s^2 = \frac{1}{(n-1)}\left[(P_1 - m)^2 + (P_2 - m)^2 + \ldots + (P_n - m)^2\right] \quad (4.4)$$

Notice that because of the sum of squares in the variance calculation, the unit of variance is the square of the unit of the phenotypic trait. For example, if weights are measured in kg, the variance is expressed as kg^2, and if concentrations are measured as g/l, then the associated variance is expressed as g^2/l^2.

Although variances have some very interesting properties as a measurement of variability, sometimes it is preferable to work on the same scale of the phenotypic trait measurements instead of squared values. In this context, another useful measure of variation is the square root of the variance, called standard deviation. The population standard deviation, denoted by σ, is given by $\sigma = \sqrt{\sigma^2}$, and similarly the sample standard deviation, denoted by s, is given by $s = \sqrt{s^2}$.

As mentioned previously, quantitative traits can be expressed as continuous or discrete outcomes, meaning that the phenotypic values of individuals in a population can present different distributions. For example, in the case of discrete traits, such as categorical or counting data, the distribution of observed values in a population can be represented in a graph indicating the proportion (or probability) of individuals in each class or possible outcome (Figure 4.1).

In the case of continuous traits, such as milk yield and body weight, similar graphical representation can be used if observed phenotypic values are grouped into classes. For example, in Figure 4.2 the observed weaning weight of beef cattle calves were grouped in bins with widths of 10 kg; weaning weight classes are represented on the horizontal axis and the proportion of individuals in each class is represented in the vertical axis.

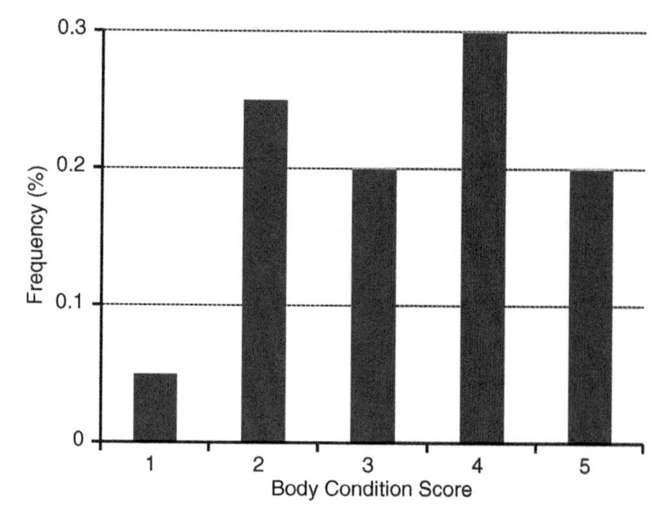

Figure 4.1 Frequency distribution of a group of beef cattle according to their body condition score (BCS). BCS is defined in five categories, varying from BCS = 1 for "very thin" to BCS = 5 for "very fat."

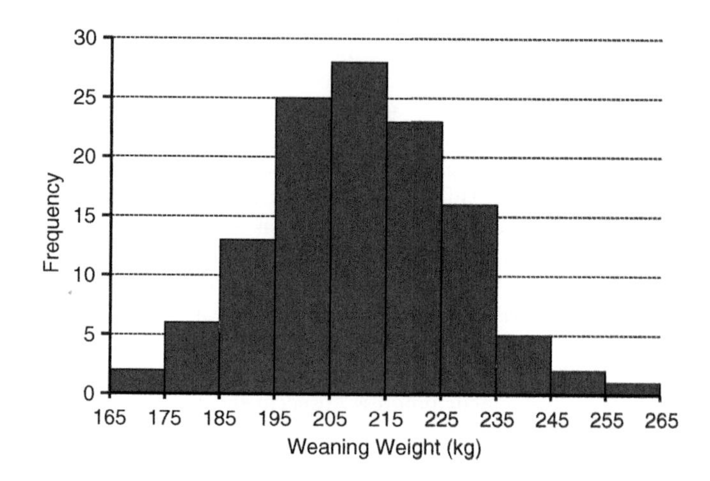

Figure 4.2 Frequency (counting) of beef cattle calves according to their weaning weight, in kg.

Now, if the number of individuals (beef calves in this example) with observed weaning weights is very large and we use bins of narrower width, the observed distribution will be smoother. In the limit, with the number of animals approaching infinity and bin widths tending to zero, the distribution will be perfectly continuous (Figure 4.3) – notice, however, that for this to be true, the gauge device used to measure the phenotypic values (e.g., scale, ruler, or thermometer) must have virtually perfect precision, otherwise observed values will be inevitably grouped by the smallest decimal unit provided by the device.

For many continuous traits, the theoretical distribution of phenotypes is bell-shaped, like the distribution depicted in Figure 4.3, with observed values symmetrically scattered around a population mean. A specific bell-shaped distribution, and probably

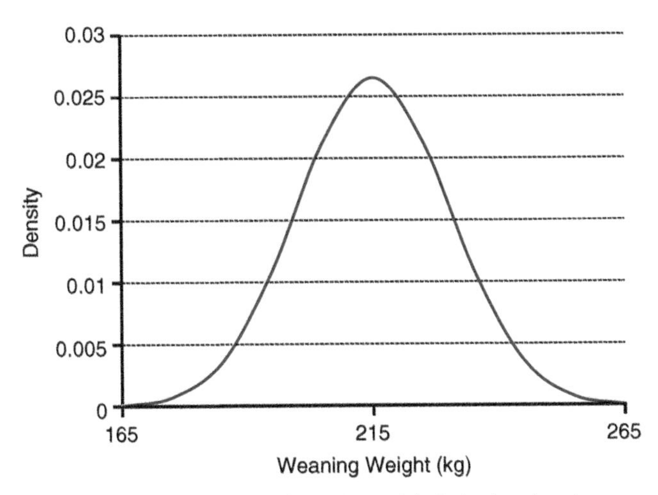

Figure 4.3 Probability density of weaning weight in beef cattle calves, in kg.

the most important one, is known as the **normal distribution** (or Gaussian distribution).

Basic genetic model for quantitative traits

Because of the joint effect of multiple genes and environmental effects on the expression of quantitative traits, differences between the means of genotypes are generally unobservable due to the variability among the environments in which individuals with any particular genotype live. As such, the prediction of genotypes based on phenotypes or vice-versa can be established only in terms of probabilistic distributions.

In this context, a useful and commonly used genetic model to study complex traits can be expressed as $P = \mu + G + E$, where P is the observed phenotype, μ is a population mean for the specific trait, G is the genotypic value of the individual, and E refers to an overall environmental effect affecting the observed phenotype.

The G component of this basic genetic model refers to the joint effects of all alleles an individual carries across all genes affecting the phenotypic trait. Individuals may have different genotypes (allelic combinations) at each locus, and such combinations of different genotypes in each animal provide them with different genetic potential in terms of the expression of the phenotype. The G component relative to each animal in a population is expressed as a deviation from the mean. As such, animals with higher potential (higher genetic merit) than the average have a positive value of G, while those with lower genetic potential have a negative G. For complex traits it is quite often assumed that the values of G in a population have a normal distribution.

The E component of the basic genetic model represents the collective effect of all environmental factors affecting the phenotypic trait of an animal. For example, animals are raised in different herds/farms and are born in different years or seasons, and all this may affect their performance, for better or for worse. Even when you compare animals from the same herd, born in the same year/season, and raised in similar conditions, there will always be environmental effects peculiar to each animal, which boost or reduce their phenotypic performance. All such effects are combined into the E component of the model; animals that benefit from favorable environmental conditions have a positive value of E, whereas those that experience unfavorable conditions have a negative value. Similar to the genetic component G, it is assumed that the E values are normally distributed and centered on zero.

The genetic component G of this model can be partitioned into additive (A) and non-additive (C) genetic effects, where A is also called the **breeding value**, and (C) refers to the **gene combination value**, which encompasses interaction effects between alleles within each gene (i.e., dominance effects) or between alleles in different genes (i.e., epistatic effects). Hence, the basic genetic model can be then expressed as $P = \mu + A + C + E$.

The breeding value (A) of an individual is equal to the sum of additive effects of individual alleles within and across loci, and it is sometimes called **additive genetic deviation** or **additive genetic effect**. Because individual alleles, and therefore independent allele effects, are passed from parent to offspring, the breeding value of an individual is important for predicting its progeny's performance and so it is central to selection of superior animals (Falconer and MacKay, 1996; Lush, 1994).

The gene combination value (C) is the difference between the genetic merit (G) of an animal and its breeding value, that is: $C = G - A$. It is often called **non-additive genetic deviation**. Because the component C involves interactions between alleles (both within and between loci), and only a single allele (as opposed to a pair of alleles) in each locus is transmitted from parents to offspring, non-additive effects are not transmitted in a predictable manner. Hence, while average breeding value in a population can be changed through selection of superior animals, the gene combination value should be explored through specific mating systems. An introduction to selection and genetic improvement of a population in terms of additive genetics effects is provided in Chapter 5. For a discussion on mating systems, such as inbreeding and outbreeding strategies, see for example, Lush (1994), Bourdon (2000), and Crow and Kimura (1970). Additional material on inbreeding depression and heterosis (or hybrid vigor) can be found in Falconer and MacKay (1996) and Lynch and Walsh (1998).

As discussed previously, the breeding value of an individual is equal to the sum of its independent allele effects. Because a parent passes a random sample of half of its alleles to its progeny, an animal's breeding value is twice what is often called **transmitting ability** or **progeny difference** (Hill, 1969; Lush, 1994). The expected breeding value of an offspring is then equal to the average of its parents' breeding values (the same as the sum of its parents' transmitting abilities). However, there will be variability in terms of breeding values within a full-sib family because of the random sampling of parents' alleles that each offspring receives, the so-called **Mendelian Sampling** (Lynch and Walsh, 1998). In the case of full-sibs, as an example, the Mendelian Sampling variance is equal to half the additive genetic variance. This means that the variability expected within a family of full-sibs is equal to half of the variability found in the general population with unrelated individuals.

Heritability and selection

Recall the basic genetic model $P = \mu + G + E$ introduced in the previous section. Assuming that the components G and E are independent from each other, the total phenotypic variance, Var[P], is given as the sum of Var[G] and Var[E], that is $\sigma_P^2 = \sigma_G^2 + \sigma_E^2$, where σ_G^2 and σ_E^2 are the genetic and the environmental variances, respectively. The environmental variance σ_E^2 can be thought of as the variance that would be observed in the distribution of phenotypic values of a population of genetically identical animals (i.e., isogenic population), such as clones. Similarly, the genetic variance σ_G^2 would be the variance observed on phenotypic values of genetically different animals raised in absolutely identical environmental conditions.

From the decomposition of the total phenotypic variance, that is $\sigma_P^2 = \sigma_G^2 + \sigma_E^2$, an important definition is derived, which refers to the **broad sense heritability**. The broad sense heritability is given by:

$$H^2 = \mathrm{Var}[G]/\mathrm{Var}[P] = \sigma_G^2 / \sigma_P^2 \quad (4.5)$$

and it describes the relative importance of the genetic component on the phenotypic variation. The broad sense heritability is a value between 0 and 1, being 0 if there is no genetic variability and so Var[P] = Var[E]; and being close to 1 if environmental variation is negligible, such that the phenotypic variation is the result of solely genetic variability among individuals, that is, Var[P] = Var[G].

Now, if the genetic component is split into additive and non-additive genetic effects, that is $P = \mu + A + C + E$, the overall phenotypic variance can be expressed as: Var[P] = Var[A] + Var[C] + Var[E], with $\mathrm{Var}[A] = \sigma_A^2$ and $\mathrm{Var}[C] = \sigma_C^2$, and remaining terms as before. From this, another definition of heritability arises, which is called **narrow sense heritability**, given by:

$$h^2 = \mathrm{Var}[A]/\mathrm{Var}[P] = \sigma_A^2 / \sigma_P^2 \quad (4.6)$$

The narrow sense heritability measures the contribution of additive genetic effects to the total phenotypic variance. As such, the narrow sense heritability determines the degree of resemblance between relatives, and it is also of utmost importance for the prediction of response to selection, as discussed next.

Predicting rate of genetic change from selection

Since their domestication, artificial selection has greatly changed the shape, size, and production and reproduction performance of livestock and companion animal species. For example, there is an incredible diversity of canine breeds – and between dogs and their wolf ancestor – from differences in overall appearance to behavior and their ability to perform specific tasks. Tremendous genetic changes have been accomplished with livestock species as well, markedly in the last 50 years or so. Although to a lesser degree, the same can be observed in many other companion animal species, such as cats and horses.

The magnitude of genetic progress for a particular trait through artificial selection is highly dependent on the level of heritability of the trait among other factors. Here, we briefly discuss the expected response to selection, with focus on truncation selection of a single normally distributed trait. Further reading for a more comprehensive discussion on the subject can be found in Lush (1994), Bulmer (1985), Falconer and MacKay (1996), and Lynch and Walsh (1998).

The most traditional approach of genetic improvement of livestock is based on selection of animals with the best performance, or **phenotypic selection**. Accordingly, given a group of animals supposedly reared in similar environmental conditions, only those with the highest performance are allowed to breed to produce the next generation.

Let the mean of phenotypic distribution of a trait in a population be μ, as indicated in the basic genetic model discussed previously. Suppose the best performing individuals, that is, those above (or below, if selection favors smaller values) a specified threshold value, are selected and mated to produce the next generation. The difference between the mean of the selected individuals, denoted here by μ_s, and the population mean μ is called **selection differential** (S), that is, $S = \mu_s - \mu$. The selection of the individuals with the highest (or lowest) phenotypic values, however, does not necessary select those with the highest (or lowest) genetic merits. As discussed earlier, a phenotypic value is a combination of genetic and environmental factors. As such, a high phenotypic value (above the selection threshold) can be observed on an individual with unsatisfactory genetic merit but which has been boosted by favorable environmental effects; likewise, a relatively low phenotypic value (below the selection threshold) can be observed on an individual with a superior genetic merit but with unfavorable environmental effects. In this context, although selected individuals do tend to have a higher genetic merit than the overall population, they also tend to have favorable environmental effects. As a consequence, the progeny mean (μ_p) of the selected individuals is not equal to μ_s, and it is actually shrunken towards the original population mean μ. The level of shrinkage is dictated exactly by how much of the phenotypic variation is due to genetic variation, that is, by the heritability h^2.

The **genetic progress** (denoted by R) expected in a single generation of phenotypic selection using the truncation approach is then given by:

$$R = \mu_p - \mu = h^2 \times S \quad (4.7)$$

The formula can be used to calculate expected response to selection, if an estimate of heritability is available.

Lastly, it is worth mentioning that the selection differential is quite often expressed as $S = i \times \sigma_P$, where $\sigma_P = \sqrt{\sigma_P^2}$ is the standard deviation of the phenotypic distribution, and $i = (\mu_s - \mu)/\sigma_p$ is called **selection intensity**, and represents the selection differential in terms of phenotypic standard deviations. Moreover, as R represents the genetic progress expected in a single generation of selection, the genetic improvement per unit of time is then given by $\Delta G = R/L$, where L is the **generation interval**. Hence, the expected genetic progress when phenotypic selection on a single trait is employed is given by:

$$\Delta G = h^2 \times i \times \sigma_P / L \quad (4.8)$$

This equation is a special form of the so-called **breeder's equation** (or **key equation**), for the case of phenotypic selection. In its general form, the breeder's equation is expressed as (Hill, 1969):

$$\Delta G = \frac{\text{accuracy} \times \text{intensity} \times \text{variation}}{\text{generation interval}}, \quad (4.9)$$

meaning that the genetic progress per unit of time is proportional to the accuracy of predicting breeding values, to the selection intensity, and to the genetic variation, and inversely proportional to the generation interval.

Hence, to increase the genetic progress in a population (e.g., breed or line) through selection, animal breeders work to improve the four components of equation (4.9). As the genetic variability is a natural characteristic of a population and cannot be easily changed, genetic progress is generally incremented by improving prediction accuracy (e.g., by using specific statistical techniques to combine different sources of information regarding the animals' genetic merit), by increasing the selection intensity, and by shortening the generation interval, which can be accomplished using molecular genetic techniques (e.g., the use of marker-assisted selection) and biotechnology approaches (e.g., artificial insemination).

Further reading

Darwin, C. 1859. *On the Origin of Species*. John Murray, London.

Wright, S., 1921, Systems of mating. I. The biometric relations between parents and offspring. *Genetics* 6:111–123.

References

Bourdon, R. M. 2000. *Understanding Animal Breeding*, 2nd edn. Prentice Hall, Upper Saddle River, NJ.

Bulmer, M. G., 1985. *The Mathematical Theory of Quantitative Genetics*. Clarendon, Oxford.

Crow, J. and Kimura, M., 1970. *An Introduction to Populations Genetics Theory*. Harper and Row, New York.

Falconer, D. S. and Mackay, T. F. C., 1996. *Introduction to Quantitative Genetics*, 4th edn, Longman, Harlow.

Hill, W. G., 1969. On the theory of artificial selection in finite populations. *Genetical Research* 13:143–163.

Lush, J. L., 1994. *The Genetics of Populations*. Prepared for publication by A. B. Chapman and R. R. Shrode, with an addendum by J. F. Crow. Special Report 94, College of Agriculture, Iowa State University, Ames, IA.

Lynch, M. and Walsh, B., 1998. *Genetic Analysis of Quantitative Traits*. Sinauer Associates, Sunderland, MA.

5 Heritability and Repeatability

Guilherme J. M. Rosa
Department of Animal Science, University of Wisconsin–Madison, WI, USA

Introduction

This chapter revisits the notion of broad sense and narrow sense heritability, and introduces the concept of repeatability. The focus is on concepts and applications using simple terms and circumventing algebraic derivations. In the next section, the definitions of heritability in its broad and narrow senses are discussed. In the section after, a brief introduction to the principles of estimation of genetic and environmental variance components, as well as of heritability is presented. Following that, some applications of heritability are discussed, such as the prediction of breeding values and of response to selection. Lastly, in the final section, the concept of repeatability is introduced, as well as related definitions such as permanent and temporary environmental effects, and producing ability.

Heritability

As discussed in Chapter 4 on "Basic Genetic Model for Quantitative Traits," an observed phenotype (P) can be described as $P = \mu + G + E$, where μ is a population mean for the specific trait, and G and E refer to genetic and environmental components affecting the trait, respectively. Hence, assuming that G and E are independent from each other, the total phenotypic variance, Var[P], is given as the sum of Var[G] and Var[E], that is $\sigma_P^2 = \sigma_G^2 + \sigma_E^2$, where σ_G^2 and σ_E^2 are the genetic (or genotypic) and the environmental variances, respectively.

This partition of the phenotypic variance implies that the total phenotypic variability observed in a population is due to differences between individuals in terms of their genotypic values (G), in addition to an extra variation contributed by environmental factors affecting the expression of the trait. The relative importance of the genetic component on the total phenotypic variation is given by the heritability of the trait, denoted by H^2. Thus, the heritability of a trait is the proportion of phenotypic differences among individuals that is due to genetic differences, that is:

$$H^2 = \frac{\text{Var}[G]}{\text{Var}[P]} = \frac{\sigma_G^2}{\sigma_P^2} \quad (5.1)$$

The coefficient H^2 – which more precisely should be called **heritability in the broad sense** or **degree of genetic determination** – is a value between 0 and 1, which indicates the relative importance of genotypic values in determining phenotypic values. Similarly, the relative importance of environmental effects can be calculated as $(1 - H^2)$.

The broad-sense heritability provides information on the "heredity" of a trait in the sense of how much of its variation is determined by genotypes. However, it is not as useful as an indicator of heredity in terms of how such a trait is transmitted from parents to offspring (Falconer and MacKay, 1996).

Before we introduce another measure of heritability (or heredity) of a trait, recall that the genetic component (genotypic value) of the basic genetic model can be partitioned into additive (A) and non-additive (C) genetic effects. Hence, the phenotypic value of an individual for a specific trait can be further expressed as $P = \mu + A + C + E$, where A is called the **breeding value** and C refers to the **gene combination value**, which encompasses dominance and epistatic effects.

As individual alleles are passed from parent to offspring, the breeding value of an individual is the important component for predicting its progeny's performance (Bourdon, 2000; Falconer and MacKay, 1996; Lush, 1994). The gene combination value, on the other hand, is not transmitted in a predictable manner from parents to offspring, as combinations of alleles within and across loci are broken down during meiosis. Hence, for selection purposes and genetic improvement of livestock and companion animals (and similarly for crops and other agricultural species), the contribution of the additive genetic component to the total phenotypic variance of a trait is of great significance. A measure of the effects of genes on the transmission of phenotypic values from parents to offspring is called **narrow sense heritability**, and is expressed as:

$$h^2 = \frac{\text{Var}[A]}{\text{Var}[P]} = \frac{\sigma_A^2}{\sigma_P^2} \quad (5.2)$$

The narrow-sense heritability (h^2) is also a number between 0 and 1, and is generally smaller than H^2; h^2 will be equal to H^2 only if all the genetic variance is due to additive genetic effects. The heritability (in the narrow sense) of a trait determines the

Molecular and Quantitative Animal Genetics, First Edition. Edited by Hasan Khatib.

degree of resemblance between relatives. If the heritability of a trait is high, genetically-related individuals (such as half-sibs, full-sibs, or parent-offspring) exhibit similar performance for the trait; for example, high performing animals will tend to produce high performing progeny. As such, the heritability (h^2) of a trait is of extreme importance in breeding programs. A discussion on how the concept of resemblance between relatives can be used to estimate heritability is presented in the next section. The subsequent section describes how heritability can be used for prediction of breeding values and of response to selection.

Estimation of heritability and variance components

Given that the total phenotypic variance of a trait in a population and its genetic and environmental variance components are usually unknown (refer to Chapter 4 for a discussion on populations and samples, and on parameters and estimates), the heritability of a trait (either in its narrow or broad sense) is also unknown. However, an estimate of the heritability can be obtained if estimates of the variance components are available. For example, the broad-sense heritability can be estimated as $\hat{H}^2 = \hat{\sigma}^2_G / \hat{\sigma}^2_P$ and the narrow-sense heritability as $\hat{h}^2 = \hat{\sigma}^2_A / \hat{\sigma}^2_P$, where $\hat{\sigma}^2_P$, $\hat{\sigma}^2_G$ and $\hat{\sigma}^2_A$ are estimates of the phenotypic, genotypic, and additive genetic variances, respectively.

The estimation of σ^2_P is relatively simple. Given a sample of individuals with phenotypic measurements, an estimate of the phenotypic variance can be obtained by:

$$\hat{\sigma}^2_P = \frac{1}{(n-1)}\sum_{i=1}^{n}(P_i - m)^2 \qquad (5.3)$$

where n is the sample size (number of animals with phenotypic measurements); P_i represents the phenotypic observation on animal i, with i = 1, 2, ..., n; m is the average phenotypic value; and $\Sigma_{i=1}^{n}$ indicates a summation over index i. Some additional details are provided in Chapter 4.

To infer H^2, it is necessary to estimate σ^2_G (or of σ^2_E). Estimation of these two components of variance, however, is practically difficult, as neither the values G or E of each animal can be observed or estimated directly from phenotypic observations on a single population. However, these two variances can be estimated from experimental populations under some specific circumstances. For example, if one of the variance components could be eliminated, the observed phenotypic variation would provide information regarding the other variance component. Generally, it is impossible to completely remove all the environmental factors contributing to phenotypic variation in a population. Although many environmental factors can be controlled, especially in laboratory settings (such as temperature, water and food availability, light, etc.), there will always be a myriad of other unknown and unmanageable factors affecting the phenotypes of individuals in a population. Experimental elimination of genetic differences, however, can be attained by using individuals from a highly inbred line or by using clones from a single individual.

If a group of such individuals is raised in a specific environmental condition, their phenotypic variance would provide an estimate of the environmental variance existing in such a condition, that is, $\hat{\sigma}^2_E$. Subtracting this variance from the phenotypic variance observed in a group of genetically mixed animals raised in similar settings provides an estimate of the genotypic variance in this population, that is, $\hat{\sigma}^2_G = \hat{\sigma}^2_P - \hat{\sigma}^2_E$.

Although such an experimental approach for estimating genotypic and environmental variances seems straightforward, in practice it has many problems and relies on a number of assumptions. For example, it assumes that the environmental variance is the same for all genotypes. However, it is well known that different inbred lines or their crosses can be more or less sensitive to environmental effects. This means that results from an experiment performed with a specific inbred line (or with clones from a particular individual) cannot always be extrapolated to other lines or clones. Furthermore, estimates obtained with experimental inbred lines may not be representative of the true variance existing in a natural, genetically mixed population.

Improved estimates of genotypic and environmental variances in a mixed population may be obtained with cloned species. In this case, individuals randomly sampled from a population are multiplied by cloning, which are then grown under specific environmental conditions. The phenotypic variances observed in each group of clones (i.e., clones of each individual) provide estimates of the environmental variance, which can then be averaged (and weighted by sample sizes if the number of clones in each group varies) to produce a single estimate. Indeed, there are statistical techniques that automatically average estimates of the environmental variance of all clone groups simultaneously. Such techniques are based on the principle of **analysis of variance**, or ANOVA (Fisher, 1918), which split the total variation into within- and among-group variation. The within-group variability provides information for estimation of the environmental variance, while the between-group variability provides information for the estimation of the genotypic variance. A similar approach, which is often used in human genetics studies, utilizes information on pairs of monozygotic twins. By comparing differences between and within pairs of twins, especially if twin sibs have been raised apart, it is possible to assess the magnitude of environmental and genetic variances.

For the estimation of the narrow-sense heritability, it is necessary to estimate the additive genetic variance. The estimate of the additive genetic variance can be obtained by exploiting information on resemblance between relatives. For example, if data are available on half- or full-sib families, variation within- and among-families can be combined using ANOVA techniques to produce estimates of additive genetic variance and of environmental variance. A comprehensive exposition of heritability estimation can be found in Falconer and MacKay (1996) and Lynch and Walsh (1998). These authors discuss not only ANOVA techniques using groups of relatives but also regression and correlation approaches.

Lastly, an extension of the ANOVA method, which is especially useful in handling unbalanced data of families with different sizes and levels of genetic relationships among individuals (including also complex pedigree structures), refers to the so-called **mixed effects models** or **mixed models** (Henderson, 1975). Mixed models are indeed powerful tools for genetic analyses, and they have been widely used in breeding programs, both in animals and plants. A discussion on mixed model methods however is beyond the scope of this chapter. An introduction to the topic and further reading can be found, for example, in Mrode (2005), Lynch and Walsh (1998), and Rosa (2012).

Prediction of breeding values and of response to selection

The genetic progress resulting from artificial selection on a particular trait is highly dependent on the heritability of the trait. As discussed in Chapter 4, the genetic progress expected in a single generation of phenotypic selection is given by $R = h^2 \times S$, where S is the selection differential. Hence, if an estimate of heritability is available, it is possible to predict the expected response to selection for any given value of S, which is a function of the selection intensity (i) and of the standard deviation of the phenotypic distribution of the trait. Also, using results from a selection experiment, in which the values of R and S are known, an estimate of the heritability of the trait can be calculated as $h^2 = R/S$, which is generally called **realized heritability**. Further reading on the theory of selection for quantitative traits and on empirical results from selection experiments can be found in Bulmer (1985), Falconer and MacKay (1996), Hill (1969), Lush (1994), and Lynch and Walsh (1998).

Another important use of heritability refers to prediction of breeding values, as discussed next. But first, let's reinforce the concept of **breeding value** (and the related concepts of "progeny difference" and "transmitting ability"), and also discuss two other definitions of heritability.

The breeding value (A) of an individual is equal to the sum of additive effects of individual alleles within and across loci, and so it is central for prediction of its progeny's performance (Falconer and MacKay, 1996; Lush, 1994). The expected average performance of a full-sib family is equal to the average breeding value of their parents. Similarly, the expected average performance of a half-sib family originated, for example, by mating a sire to a random sample of dams in a population, is half the breeding value of the sire, or the so-called **transmitting ability** or **progeny difference** (Bourdon, 2000; Hill, 1969; Lush, 1994). Thus, twice the average performance of half-sibs can be used as an estimate of the breeding value of their common parent. The prediction of breeding values (and similarly of transmitting abilities or progeny differences) using progeny information is a commonly used strategy in selection programs involving sex-limited traits, such as selection of sires for higher milk production potential in dairy cattle.

Nonetheless, an estimate of an animal's breeding value can be obtained based on its own performance. The accuracy of such a prediction, however, will depend directly on the heritability of the trait. Another definition or interpretation of the narrow-sense heritability refers to the squared correlation between breeding values and phenotypic values, which consequently corresponds to the regression of breeding values on phenotypic values. This concept is used to estimate heritability by regressing progeny performance (which is an estimate of parents' breeding values) on parents' performance (see, e.g., Falconer and MacKay, 1996 and Lynch and Walsh, 1998). If an estimate of heritability is available, the breeding value of an individual can be estimated by $\hat{A} = \hat{h}^2(P - m)$ as a regression of its phenotypic deviation, given by the difference between its performance and an estimate of the population mean.

Notice that low values of heritability will strongly regress phenotypic deviations towards zero as, for lowly heritable traits, most of the phenotypic variation is due to environmental (and non-additive genetic) effects. In contrast, for highly heritable traits, the phenotypic value of an individual is more highly affected by additive genetic effects. Therefore, phenotypic deviations are not strongly regressed towards zero. Lastly, to be able to compare individuals raised in different conditions (e.g., animals born in different years and seasons or kept in different herds), phenotypic deviations should be calculated using the average performance of groups of animals raised in similar conditions (the so-called **contemporary groups**; Bourdon, 2000), instead of a single mean m. In such cases, predicted breeding values can be expressed as $\hat{A} = \hat{h}^2(P - \overline{CG})$, where $\overline{CG}$ is the average performance of the contemporary group to which an animal belongs.

Repeatability

The concept of repeatability refers to situations in which multiple measurements of the same trait can be recorded (e.g., milk yield in dairy cows, litter size in pigs, and racehorse performance). When more than one measurement is made on each individual for a specific trait, the same ANOVA techniques discussed earlier can be used to partition the total variance into within- and among-individuals variance components. This partitioning leads to a ratio of variances that can be larger than the broad-sense heritability, which is what would be expected under the basic genetic model used here.

This additional source of covariance (or similarity) between repeated records on the same animal refers to environmental factors that affect all such records equally. This means that when studying repeated phenotypic measurements of a specific character, the basic genetic model should be extended to distinguish two groups of environmental effects, as (Falconer and MacKay 1996; Lush, 1994):

$$P = \mu + A + C + E_p + E_t \tag{5.4}$$

where P, μ, A, and C are as previously defined, and E_p and E_t are the so-called **permanent environmental effects** (or **general environmental effects**) and "temporary environmental effects" (or "special environmental effects").

The permanent environmental factors are those which affect all records of an animal and, as such, they contribute to the similarity between repeated measurements. The temporary environmental factors, on the other hand, affect differently each measurement of an animal and so they produce within-animals variability (Falconer and MacKay, 1996).

Using this model, and assuming independence between each of the genetic and environmental factors, the total phenotypic variance can be expressed as:

$$\sigma_P^2 = \sigma_A^2 + \sigma_C^2 + \sigma_{E_p}^2 + \sigma_{E_t}^2 \tag{5.5}$$

where $\sigma_{E_p}^2$ and $\sigma_{E_t}^2$ are the permanent and the temporary environmental variances, respectively. The repeatability (generally denoted by r) is then defined as the ratio:

$$r = \frac{\sigma_A^2 + \sigma_C^2 + \sigma_{E_p}^2}{\sigma_P^2} = \frac{\sigma_G^2 + \sigma_{E_p}^2}{\sigma_P^2} \tag{5.6}$$

It can be shown that the repeatability is also the correlation between pairs of records of a trait of the same animals. In addition, the repeatability is the regression of **producing ability** (or PA) on a single observation of a phenotypic trait. The producing ability is the sum of all permanent effects contributing to a phenotype: $PA = A + C + E_p$, that is, all genetic effects and the permanent environmental effects (Bourdon, 2000). Hence, an estimate of an animal's producing ability based on a single record of its performance can be obtained by: $\widehat{PA} = \hat{r}(P - m)$, where $\hat{r}$ represents an estimate of the repeatability of the trait.

Repeatability and producing ability are therefore extremely important in livestock production systems, where future performance of animals should be predicted based on their previous performance for culling decisions. Similarly, the same concepts are equally important, for example, when purchasing or selling animals based on their performance, or when betting in a horse race.

References

Bourdon, R. M., 2000. *Understanding Animal Breeding*, 2nd edn. Prentice Hall, Upper Saddle River, NJ.

Bulmer, M. G., 1985. *The Mathematical Theory of Quantitative Genetics*. Clarendon, Oxford.

Falconer, D. S. and Mackay, T. F. C., 1996. *Introduction to Quantitative Genetics*, 4th edn. Longman, Harlow.

Fisher, R. A., 1918. The correlation between relatives on the supposition of Mendelian inheritance. *Philosophical Transactions of the Royal Society of Edinburgh* 52:399–433.

Henderson, C. R., 1975. Best linear unbiased estimation and prediction under a selection model. *Biometrics* 31:423–447.

Hill, W. G., 1969. On the theory of artificial selection in finite populations. *Genetical Research* 13:143–163.

Lush, J. L., 1994. *The Genetics of Populations*. Prepared for publication by A. B. Chapman and R. R. Shrode, with an addendum by J. F. Crow. Special Report 94, College of Agriculture, Iowa State University, Ames, IA.

Lynch, M. and Walsh, B., 1998. *Genetic Analysis of Quantitative Traits*. Sinauer Associates, Sunderland, MA.

Mrode, R., 2005. *Linear Models for the Prediction of Animal Breeding Values*. 2nd edn. CAB Int., New York, NY.

Rosa, G. J. M., 2012. Foundations of Animal Breeding. In: *Encyclopedia of Sustainability Science and Technology*. Meyers, R. A. (ed.). Springer, New York, NY.

6 Applications of Statistics in Quantitative Traits

Hayrettin Okut
Department of Animal Science, Biometry and Genetics, University of Yuzuncu Yil, Van, Turkey

Population and sample

The basic idea of statistics is straightforward: One uses information from a **sample** to infer something about a population from which the data was derived. So, what is the population? Simply, a **population** consists of a complete set of individuals, observations, or outcomes that have something in common. For example, a population could be all Angus cattle in Canada, a set of dairy cattle 3 years of age fed a particular diet, or elementary school children in a city in grades 3 and 4.

There should be a distinction between the target population and the study population. The set of all possible study units makes up the **study population**, which is a part of the **target population** of interest. Typically, the population is very large and making observation on all individuals is impractical or impossible. Because of this, researchers typically rely on **sampling** from population to obtain a portion of the population to perform an experiment or observational study. Therefore, a **sample** is a **part** or a **sub-group** of the population of interest (Figure 6.1). **Sampling**, on the other hand, refers to strategies that enable choosing a subgroup from a population. Then a generalization of the results of this subgroup for the population is made from that which the subgroup was drawn from. The sample should be a good representative of the population from which the sample was derived. Therefore, researchers adopt a variety of sampling strategies. The most straightforward is simple random sampling. This is one of the best choices to attain a good representation of the population. Such sampling requires every individual of the population to have an equal chance of being chosen for the sample. With random sampling, the selection of one member must be independent of the selection of every other member. That is, randomly choosing one Angus cow from a farm must not increase or decrease the chance of choosing any other cow. In this sense, it can be said that simple random sampling chooses a sample completely by chance.

Parameter and statistics

Next a distinction between **parameter** and **statistic** should be made. A parameter is a descriptive measure or value used to represent some characteristics of the population. Since usually it is not possible to study the entire population of interest, parameters are generally unknown (and which therefore has to be estimated). A parameter is considered a fixed value that does not vary. A statistic, on the other hand, is a characteristic of the sample. A statistic is used to estimate the value of the parameter. For example, the mean milk yield from a random sample of 150 Jersey cows is used to give information about the overall mean of the Jersey population from where that sample was drawn. Therefore, statistics is used to estimate the corresponding parameter in the population. Contrary to the parameter, the value of a statistic changes from sample to sample, which leads to study of the **sampling distribution of statistic**.

Greek letters are typically used to denote parameters. The symbols for the descriptive terms for population and sample are shown next.

	Mean	Standard deviation	Variance
Parameter of Population	μ	σ	σ^2
Statistics of sample	$\bar{x}$	s	s^2

Descriptive statistics

Types of variables

Variables are descriptive characteristics of some event, unit, or individual that can take on different values or information. Calving ease, heifer pregnancy rate, weaning weight, and body condition score are some example for variables. Variables are called **random variables** because the value or information given for the observation is a numerical or non-numerical event that varies randomly. For example, carcass weight at the age of 3 years is a numerical variable and might take any possible value in an interval from 260 to 350 kg. The possible value of 290.0 kg or 290.0234233 kg depends on the precision of scales or practical use of the number of decimal places to which the values will be reported. The numbers of eggs laid in a month, litter size, number of animals older than 3 years of age are examples of numerical variables as well. Color, sex, calving difficulty, breed, or illness situation are examples of non-numeric variables. The genders of

Molecular and Quantitative Animal Genetics, First Edition. Edited by Hasan Khatib.

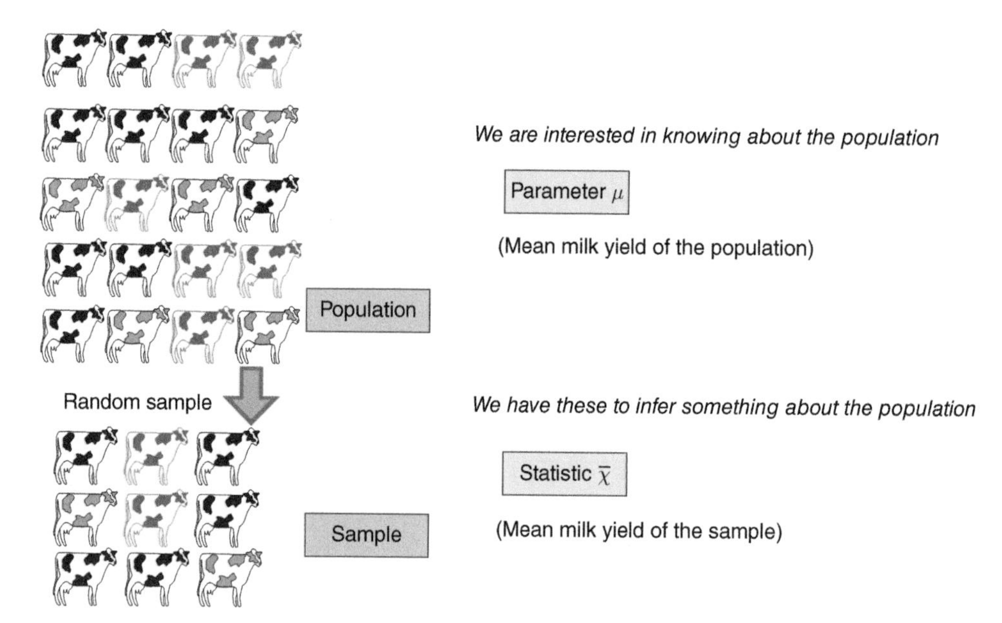

Figure 6.1 Random sampling of a sub-group from a population.

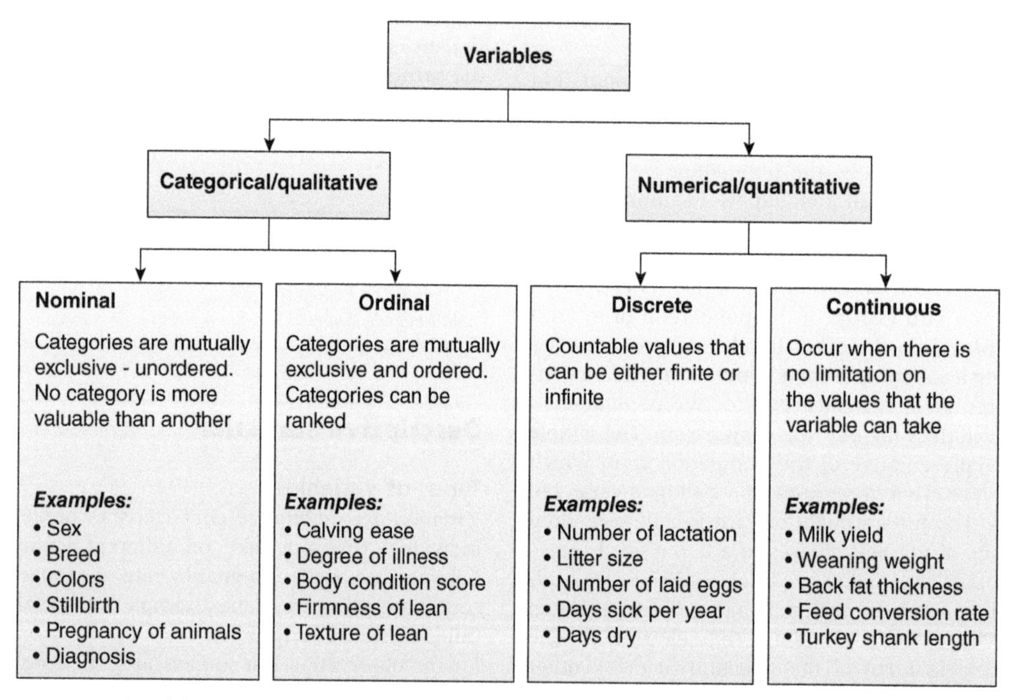

Figure 6.2 Measurement categories of data.

animals, smoke status of an individual, and breed of animal are also examples of non-numeric data. When a statistician receives data, there are some basic statistical analyses that are performed first in order to understand what the variables look like. The type of variables is one of the first steps in deciding which statistical test should be used for statistical inferences.

Statisticians generally conceptualize data as fitting within one of the following two measurement categories (Figure 6.2):

• **Categorical (qualitative) data:** These occur when each individual can only belong to one of a number of distinct categories of the variable (some examples are listed in Figure 6.2).

• **Numerical (quantitative) data:** These occur when the variable takes some numerical value (some examples are listed in Figure 6.2).

Although the distinction between categorical and numerical data is usually clear, in some situations it may become vague. For example, when there is a variable with a large number of ordered categories (e.g., a calving ease can be described in five categories: (1) Normal calving, (2) Calving with little intervention ... and (5) Caesarean section), it may be difficult to distinguish it from a discrete numerical variable. The distinction between discrete and continuous numerical data may be even less clear, although

Table 6.1 The list of some descriptive statistics for qualitative and quantitative variables.

Descriptive Statistics			
Measures of central tendency	**Measures of variability**	**Measures of shape of distribution**	**Measures of relative position**
Frequency percentages	Range	Skewness	Percentiles
Arithmetic mean	Variance	Kurtosis	Deciles
Median	Standard deviation		Quartiles
Mode	Standard error of mean		*z*-values
Geometric mean	Coefficient of variation		
Harmonic mean			

Table 6.2 The body condition score of 487 beef cattle from six different breeding farms.

Body Condition	Score	Frequency	Cumulative Frequency	Percent	Cumulative Percentage
Emaciated	1	12	12	2.46	2.46
Poor	2	34	46	6.98	9.44
Thin	3	46	92	9.48	18.92
Borderline	4	57	149	11.7	30.62
Moderate	5	59	208	12.11	42.73
High moderate	**6**	**62**	**270**	12.73	55.46
Good	**7**	**184**	454	37.78	93.24
Fat	8	23	477	4.72	97.96
Extremely fat	9	10	487	2.05	100.0

in general this will have little impact on the results of most analyses. Age is an example of a variable that is often treated as discrete even though it is truly continuous (Patrie and Sarin, 2000).

Descriptive statistics categorical (qualitative) data

Different measures of descriptive statistics are needed to provide a reasonable explanation for variables, such as: (1) measures of central tendency; (2) measures of variability; (3)measures of the shape of a distribution and (4) measures of relative position (Table 6.1). Only some of the **measure of central tendency** and **measures of variability** will be considered here for categorical and qualitative data structures.

In most cases **categorical data** are classified as **nominal** and **ordinal**. These data are assigned to a specific category and they cannot be manipulated in a meaningful mathematical manner. As an example the **nominal variables** can be converted into numerical scales, such as the gender of animals (Female = 1 and Male = 2), breeds (Angus = 1, Hereford = 2, Limousin = 3, and Simmental = 4), genotype (AA = 2, Aa = 1, and *aa* = 0), a basketball player's jersey number, or a person's social security number. These are used purely for purposes of identification, and cannot be meaningfully manipulated in a mathematical sense. In this sense **nominal scales exemplify the lowest level of measurement.**

Nominal variables are often summarized as **frequency**, **mode**, **proportions**, or **percentages**. The frequency for a certain category is the number of observations in that category. The mode of a distribution with a nominal random variable is the value of the term that occurs the most often. For example if there are 137 total animals with 29 Angus, 34 Hereford, 17 Limousin, and 57 Simmental animals in a herd, the **frequencies** of each breed are 29, 34, 17, and 57, respectively.

The **mode** is Simmental, which occurs most often in that herd. If x is assigned for the number of any breed, say Hereford, and n the total number of animal in herd, here $n = 137$, then the *percentage* (proportion) of Herefords in the herd is:

$$p = \frac{x}{n} * 100 = \frac{34}{137} * 100 = 24.82$$

The two figures x and n are often combined into a statistic, called a proportion, and multiplying proportion by 100 gives the percentage.

Like nominal variables, **ordinal variables** are another type of variable that can be converted into numerical scales, but the values have some inherent order. For example, opinions in questionnaires for **Likert** type scale variables (strongly disagree, disagree, agree, and strongly agree) are ordinal variables. There is an inherent knowledge that strongly disagree is "worse" than disagree. This five-point Likert scale item can be treated numerically, which assumes that the distances between all points on the scale are equal: that is, the distance between 1 and 2 is equal to the distance between 4 and 5. Likewise, the order of the finish in a horse race represents an ordinal scale.

Several types of descriptive statistics can be calculated for these types of variables. The mode is simply the most frequently occurring score. Given the results in Table 6.2, the frequency of good condition with a score of 7 has been counted 184 times (bold face in Table 6.2). Therefore the mode is 7 in this example. The **median** of a set of scores (while ordered) is the middle score; that is, the score that has an equal number frequency both above and below. In other words, if the data (scores) are arranged in order of magnitude starting with the smallest value and ending with the largest value, then the median is the middle value of this ordered set. It is easy to calculate the median if the number of observations, n, is **odd**; it is the $[(n + 1)/2]$th observation in the ordered set. So, for example, if $n = 23$, then the median is the $(23 + 1)/2 = 24/2 = 12$th observation in the ordered set. If n is **even** then, strictly, there is no median. However, usually the median is calculated by taking the mean of the two middle numbers. For example, if the number of observations in $x = \{2, 4, 6, 7, 8\}$ is odd then the median is the middle value and it is 6. The number of observations in, $x = \{2, 4, 6, 7, 8, 13\}$, however, is even, so the median for this data set is $(6+7)/2 = 6.5$. Now, time to return to the example given in Table 6.2 where there is an odd number of observations ($n = 487$) so the middle value is calculated as $(487 + 1)/2 = 488/2 = 244$th observation. This falls into the high moderate (the BCS = 6) group (bold face in Table 6.2, see the cumulative frequency in Table 6.2).

Example: Body condition scores (BCS) are numerical values that reflect fatness or condition of the beef cattle. Scores are subjectively assigned, ranging from 1 (severely emaciated) to 9 (very obese). Body condition scoring is generally done by visual appraisal, but palpation of the animal's condition may be beneficial when it has a thick hair coat (Beef Improvement Federation). Suppose that there are survey results for body condition score of 487 beef cattle from six different breeding farms (Table 6.2). The descriptive statistics concerning this data set are as in Table 6.2.

Descriptive statistics for quantitative variables

As seen earlier, frequency, percentage/proportion, mode, and median are popular descriptive statistics for qualitative variables. As will be shown later on, graphs can also be an excellent tool for summarizing large data sets. The same descriptive statistics and pictorial tools can be used for quantitative variables as well. Because by nature, the sense of qualitative data is more informative, all mathematical operations can be executed with these data. The **most popular measures of central tendency** are the **arithmetic mean**, **median**, and **mode**. You are already familiar with median and mode from our prior explanations; however, reconsider mode and median with an example given next. The arithmetic mean of a set of numbers is the best-known measure of central tendency and it is just their numerical average. The arithmetic mean of a sample of n numbers $x_1, x_2, \ldots, x_n$ is:

$$\bar{x} = \frac{\sum_{1}^{n} x_i}{n} = \frac{x_1 + x_2 + \ldots\ldots + x_n}{n} \qquad (6.1)$$

For example, feed conversion rates of nine bulls are 7.55, 6.89, 7.46, 6.59, 6.13, 7.07, 6.89, 6.75, and 6.95 kg. So the arithmetic mean, median and mode are:

The arithmetic mean:

$$\bar{x} = \frac{\sum_{1}^{n} x_i}{n} = \frac{7.55 + 6.89 + 7.46 + \ldots\ldots + 6.95}{9} = 6.92$$

The median:

Median is the middle value of this ordered data. The ordered data from smallest to the largest values would be:

6.13, 6.59, 6.75, 6.89, **6.89**, 6.95, 7.07, 7.46, 7.55. Then the median is 6.89.

The mode:

Since the mode is the most frequently occurring score, then mode of this data set is 6.89.

Comparing of arithmetic mean, median, and mode

The arithmetic mean is the statistic most commonly used as a measure of central tendency. As given in equation (6.1), the arithmetic mean uses all the data values that are algebraically well-defined and mathematically practicable. But this statistic is sensitive to distortion when raw data sets are comprised of extreme measures. The mode and median are easier to calculate for categorical data and are not affected by extreme values. The main disadvantage of mode and median is that both of these central tendency statistics ignore most of the information in the data set and are not algebraically defined. For example, returning to the feed conversion rate example of nine bulls, consider adding a new bull to the sample with a feed conversion rate 17.8. Now there are 10 bulls with measures of 7.55, 6.89, 7.46, 6.59, 6.13, 7.07, 6.89, 6.75, 6.95, and 17.8.

The central tendency measures will now be calculated as:

The arithmetic mean is

$$\bar{x} = \frac{\sum_{1}^{n} x_i}{n} = \frac{7.55 + 6.89 + 7.46 + \ldots\ldots + 6.95 + 17.8}{10} = 8.208.$$

The median is $(6.89 + 6.95)/2 = 6.92$ and the mode is 6.89. Previously the arithmetic mean was 6.92 for nine bulls and is now calculated to be 8.208 with the extreme value 17.8 added. The median is almost the same; it was 6.89 and now is 6.92. The mode is same as before. To summarize:

- The arithmetic mean uses all data information but is affected by extreme values. It was 6.92 and now is 8.208 (about 19% increased).
- The median uses information of only one or two observations of middle values from ordered data and is not sensitive to extreme values. It was 6.89 and now is 6.92.
- The mode uses only the most frequently occurring score and is not sensitive to extreme values. It was 6.89 and is still the same.

Measures of dispersion

In addition to finding measures of central tendency for a set of observations, the measures of dispersion are calculated to aid in describing the data. The distribution of two distinct populations may differ seriously in their dispersion, although they may have the same mean or median. A proper description of data sets driven from populations should include both the central tendency and dispersion characteristics. The basic question being asked is: How much do the outcomes deviate from the mean? The more bunched up values are around the mean, the better your ability to make accurate predictions in terms of breeding strategy. Hence, there are various methods that can be used to measure the dispersion of a data set, each with its own set of advantages and disadvantages. Only the most used measures of dispersion will be considered here.

Range

The range is one of the simplest measures of variability to calculate. It is calculated as difference between the extreme values in the data set.

$$\text{Range} = X_{max} - X_{min}$$

The range for the example of feed conversion rate of bulls given previously is:

$$\text{Range} = 7.55 - 6.13 = 1.42.$$

As seen from this example, the range is one of the simplest measures of dispersion to calculate. However, it depends only on the extreme values and provides no information about how the remaining data is distributed.

Variance and standard deviation

The **variance** and **standard deviation** are summary measures of the difference of each observation from the mean. These provide a more sensitive description of dispersion for all observations. For these measures each difference between observations and mean is squared first to eliminate negative numbers. The squared differences are then summed and divided by $n - 1$ to find an "average" squared difference. This "average" is termed the **sample variance**. The divisor for sample variance is termed **degrees of freedom**. The use of the divisor $(n-1)$ instead of n is clearly not very important when n is large. Formulas for sample, population variance, and standard deviations are:

$$\text{Variance of sample } s^2 = \frac{\sum_{i=1}^{n}(x_i-\bar{x})^2}{n-1}$$
$$\text{Variance of population } \sigma^2 = \frac{\sum_{i=1}^{n}(x_i-\mu)^2}{n} \tag{6.2}$$

$$\text{Standard deviation of sample } s = \sqrt{\frac{\sum_{i=1}^{n}(x_i-\bar{x})^2}{n-1}}$$
$$\text{Standard deviation of population } \sigma = \sqrt{\frac{\sum_{i=1}^{n}(x_i-\mu)^2}{n}} \tag{6.3}$$

The letters μ, σ,^{2} and σ are parameters and are used for the population mean, variance, and standard deviation, while $\bar{x}$, s^2, and s are statistics and are used for the sample mean, variance, and standard deviation, respectively. Using $(n-1)$ in the calculation of sample variance gives a better **estimate** of a population than if n is the divisor. Here "better" means an **unbiased** estimate for parameters. A statistic used to estimate a population parameter is unbiased if the statistic (e.g., mean of the sample) is equal to the true value of the parameter (e.g., mean of population) being estimated. That is, the sample statistics are not systematically different from population parameters. This is an important consideration for inferential statistics.

Returning to the example of feed conversion with the nine bulls, the sample variance for this data is:

$$s^2 = \frac{(7.55-6.92)^2+(6.89-6.92)^2+(7.46-6.92)^2 + \ldots\ldots\ldots+(6.95-6.92)^2}{9-1} = 0.18445,$$

and the sample standard deviation:

$$\sqrt{s^2} = \sqrt{0.18445} \Rightarrow s = 0.429.$$

The arithmetic mean and standard deviation are two important numerical characters for describing quantitative variables and they are often expressed together as $\bar{x} \pm s$. For this example it is 6.92 ± 0.429 (standard deviation rounded here). Variation in phenotype, genotype, and environment are very important in animal breeding strategy, because it is assumed that phenotypic variance (σ_p^2) = genetic variance (σ_g^2) + non-genetic variance (σ_e^2). That is $\sigma_p^2 = \sigma_g^2 + \sigma_e^2$. In prediction methodology, the non-genetic effects are termed **random environmental effects**. So, since heritability measures the fraction of phenotype variability that can be attributed to genetic variance, the variation structure in quantitative and qualitative traits plays an important role in genetic improvement. If individuals in a population show little variability in terms of genotypes and phenotypes, selection success will be difficult because no individual is better than any other (Bourdon, 2000).

Sometimes the differences of variations between two types of measurement in different units need to be compared. Because different units are used their standard deviations cannot be compared directly. Then the resulting measure is called **coefficient of variation** (CV) that is defined as:

$$CV = \frac{s}{\bar{x}}100. \tag{6.4}$$

For example, 6.13, 6.59, 6.75, 6.89, 6.89, 6.95, 7.07, 7.46, 7.55 is $CV = \frac{0.429}{8.208}100 = 5.23.$

Graphically examining the distribution of the data

An effective and convenient way to visualize the data of any study is through the use of graphs. Graphs increase the understanding of numerical data at a single glance. There are many types of graph, but the basic idea is to provide a sketch that quickly conveys general trends in the data to the reader.

Graphical presentation of categorical data

The pictorial presentation of data given in Table 6.2 is depicted in Figure 6.3. A **pie chart** presented in Figure 6.3(a) is one of the most common ways to portray categorical data. A pie chart consists of a circle that is divided into wedges corresponding to their proportions for body condition scores in the sample. Further, the pie chart in the figure reveals the differences between the sizes of various body condition scores as a decomposition of the total. A line graph in Figure 6.3(b) gives the chance to look at the data from a different perspective. The horizontal axis represents the body condition scores and vertical axis shows the percent that the body condition score occurred. The **relative frequency** represents the **percentage** of the total observations that fall in a given body condition score. For example, the relative frequency for body condition score 7 is 37.78, indicating that 37.78% of the animals have good body condition. The **cumulative percentage curve gives** the accumulation of the previous relative frequencies. The cumulative percentage is **additive**, so it can be said that, for example, over 93.24% of animals in these farms are in good body condition or worse (Figure 6.3b and c).

The bar chart in Figure 6.3 (c) is a common type of graph used to display the frequency (or proportion) for a quick comparison. The body condition scores are represented along the vertical axis and the frequency and the cumulative frequency of each body score are represented on the horizontal axis (these may also be arranged inversely). For instance, the frequency of animals with the moderate body condition score of 5 is 59. The bars should be of equal width and should be separated from one another so as not to imply continuity. Given a particular body condition score, the cumulative frequency of this score is defined as the sum of the frequency of this score and the frequency of all preceding scores. The cumulative frequency of any given body condition score indicates to the reader that the numbers of observations are smaller than or equal to a particular body condition score. For example, the frequency of body condition scores that are in the range of Emaciated (score = 1) to high moderate (score = 6) is 270 (Figure 6.3 and Table 6.2).

Graphical presentation of quantitative data

There are many types of graph options that can be used to portray the pictorial representation of quantitative variables. Some graph types such as stem and leaf displays are best suited to small to moderate data set, while histograms are best suited to larger data

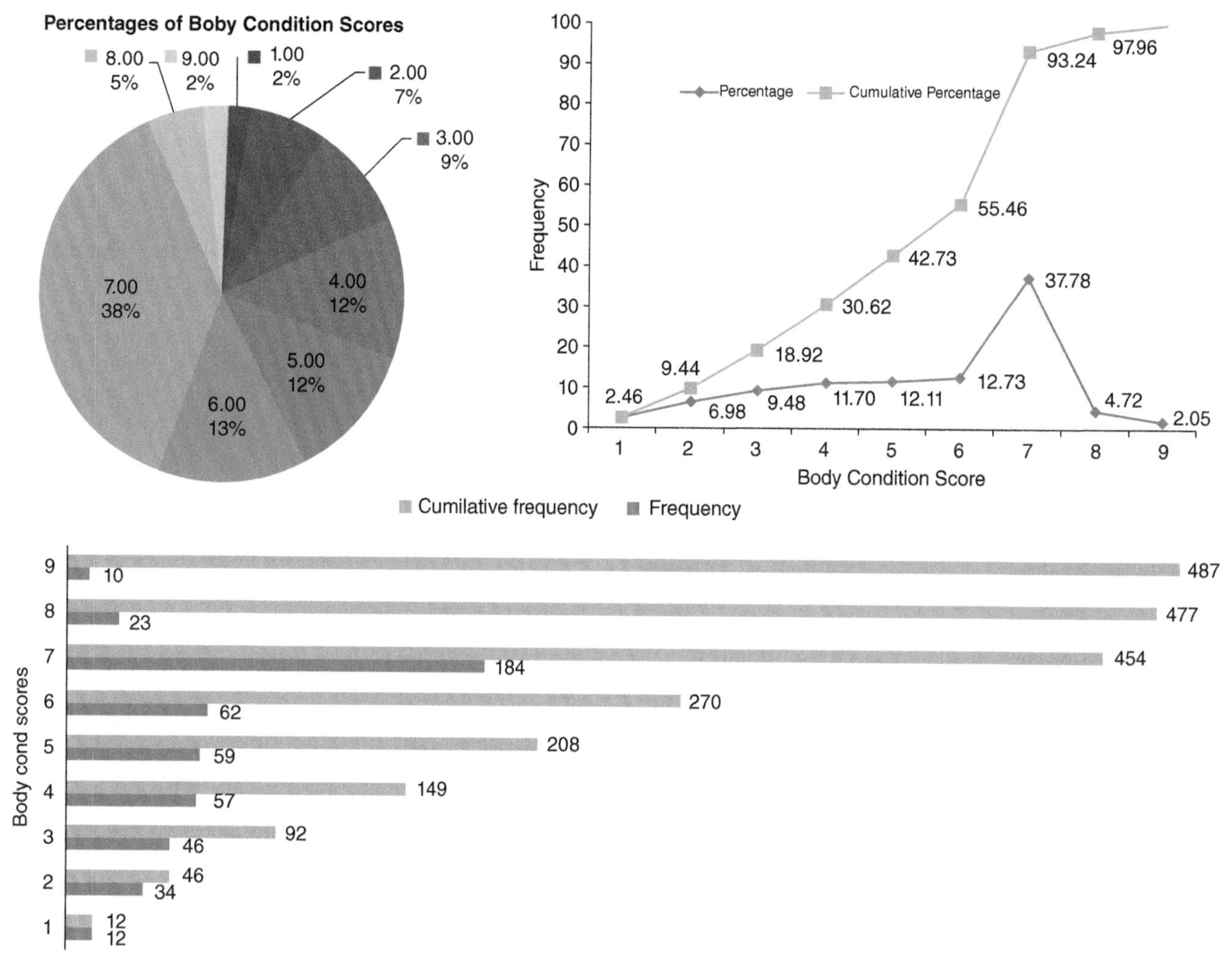

Figure 6.3 Graphical presentations of data given in Table 6.2 by a pie chart (a), line (b) and bar chart.

sets. Box plots are good at depicting differences between distributions of subgroups. Scatterplots are used to show the linear association between two variables. Only histograms and box plots will be considered in detail here while scatterplots will be illustrated later, in the regression and correlation section.

Histograms

Histograms are the most widely used graph for the presentation of quantitative data. A **histogram** is the frequency (or relative frequency) distribution of a set of data. It consists of rectangular bars joining each other, one for each interval. The histogram distribution of simulated data for ribeye area (REA), which is an indicator of muscling, is displayed in Figure 6.4 (ribeye muscle is measured at the 12th rib by using a grid or a ribeye tracing that is measured with a compensating polar planimeter or image analysis system. As ribeye area increases, retail product yield increases). Histograms also give some clues about the type of statistical tests that need to be performed. The parametric tests are the most popular test group uses in data analysis. Several assumptions are needed to perform these test groups. One of them is that the data fit the normal distribution (bell-shaped) or look more or less normal for large data. It is obvious that the distribution of REA across breeds given in Figure 6.4 is more or less bell-shaped (symmetric) but the shapes of distribution for individual breeds are not as symmetric as they are across breeds.

Box and-whiskers plots

The **box and whisker plot** shows the distribution of the data and is particularly useful for comparing individual distributions graphically. There are several steps in constructing a **box plot**. The first relies on a set of five statistics: the median, the 25th percentile (lower or 1st quartile), the 75th percentile (upper or 3rd quartile), the minimum value, and the maximum value. As depicted in Figure 6.5 for the REA data, the horizontal line in the middle of the box is the median (2nd quartile) of the measured values, the upper and lower sides of the box are the upper and lower quartiles, respectively, and two lines extending out from the box, called **whiskers**, denote the minimum and maximum values of lower and upper fences. The length of the central box indicates the spread of the majority of the data between 25 and 75% (the median being 50%) while the length of the whiskers shows how stretched the tails of the distribution are. More instructions as to how to construct a box and whisker plot are provided in Figure 6.5.

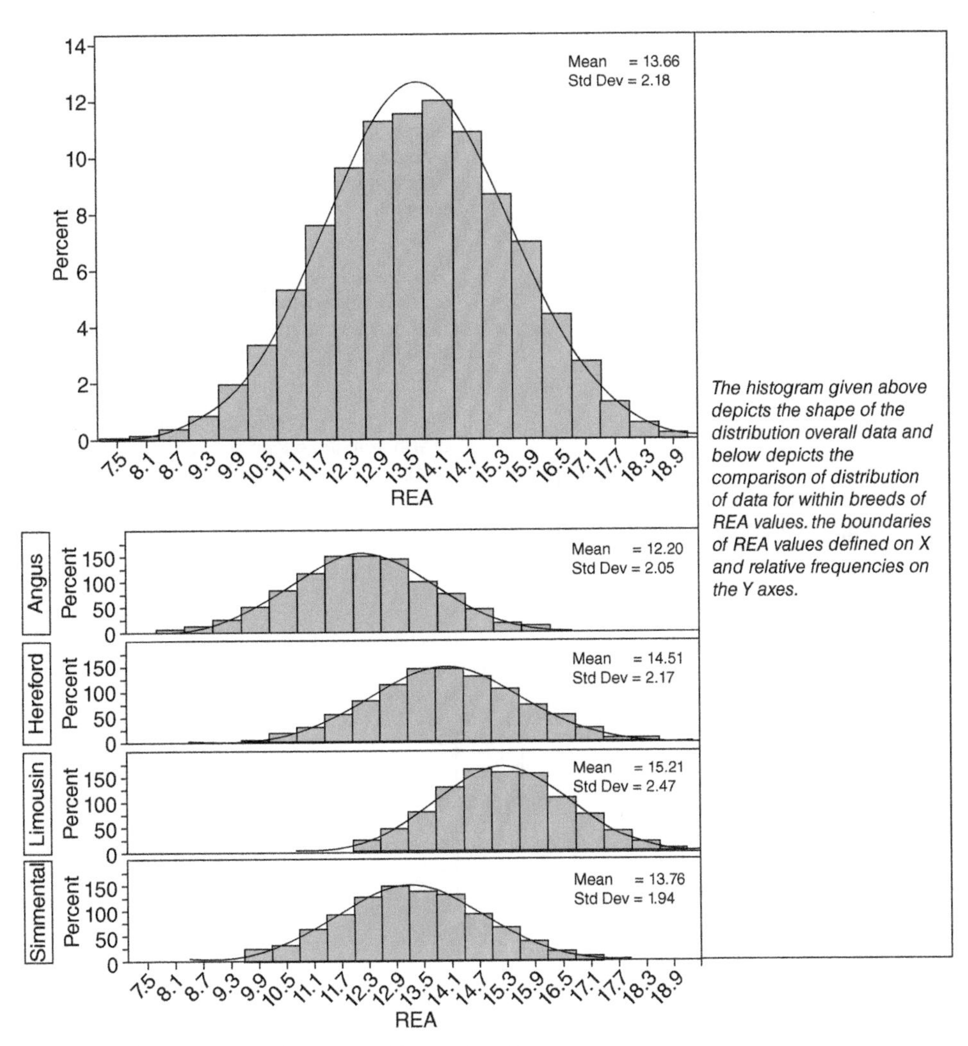

Figure 6.4 Histograms of REA data for across breeds (top) and within breeds (bottom).

Normal distribution

From a statistical perspective, a **distribution** is a theoretical model that describes how a **random variable fluctuates**. In general, introduction to probability theory treats discrete probability distributions and continuous probability distributions separately. Random variables, such as the number of laid eggs, litter size, numbers of lactation, or number of dairy farms, take a finite number of values in any particular range and are called **discrete random variables**. The distribution of discrete variables is termed the **probability function** (pf). For discrete variables, the probability distribution will comprise a measurable probability for each outcome, for example, 0.5 for male and 0.5 for female for given a birth. In contrast to discrete variables, continuous variables are not restricted to integers so there is infinite range of possible outcomes. They are describing an "unbroken" continuum of possible occurrences. The probability distribution of continuous variables is often termed the **probability density function** (pdf). The birth weight, weaning weight, hot carcass weight, back fat thickness, and height are some examples of continuous variables. There are many families of discrete and continuous probability distributions and some of the more common are listed in Figure 6.6. However, the **normal distribution** is the most commonly used in inferential statistical tests for quantitative traits, since the majority of quantitative traits of many biological events assume that they are normally or near-normally distributed. One of the most noticeable properties of the normal distribution is its bell-shaped and symmetrical characteristics, meaning that one half of the observations in the data fall on each side of the middle of the distribution. The mean, median, and mode are all more or less the same at the peak of the maximum frequency of the distribution. The frequencies then gradually decrease at both ends of the curve. Technically, the curve drawn by the normal probability density function is:

$$f(y;\, \mu, \sigma^2) = \frac{1}{\sqrt{2\pi\sigma^2}} e^{-\frac{1}{2}\frac{(y-\mu)^2}{\sigma^2}} \quad for -\infty \leq y \leq +\infty. \qquad (6.5)$$

Where e and π are constants with e denoting the exponential function to the base $e = 2.71828$, $\pi = 3.14$, and y being the value of the random variable. The Greek letter μ is the mean and σ^2 is the variance. Recall that ∞ is the mathematical symbol for **infinity**; thus, the curve goes on forever in both directions. A normal distribution is abbreviated to $y \sim N(\mu,\sigma^2)$. Hence, if $y \sim N(\mu, \sigma^2)$, then y is normally distributed with mean μ and variance σ^2.

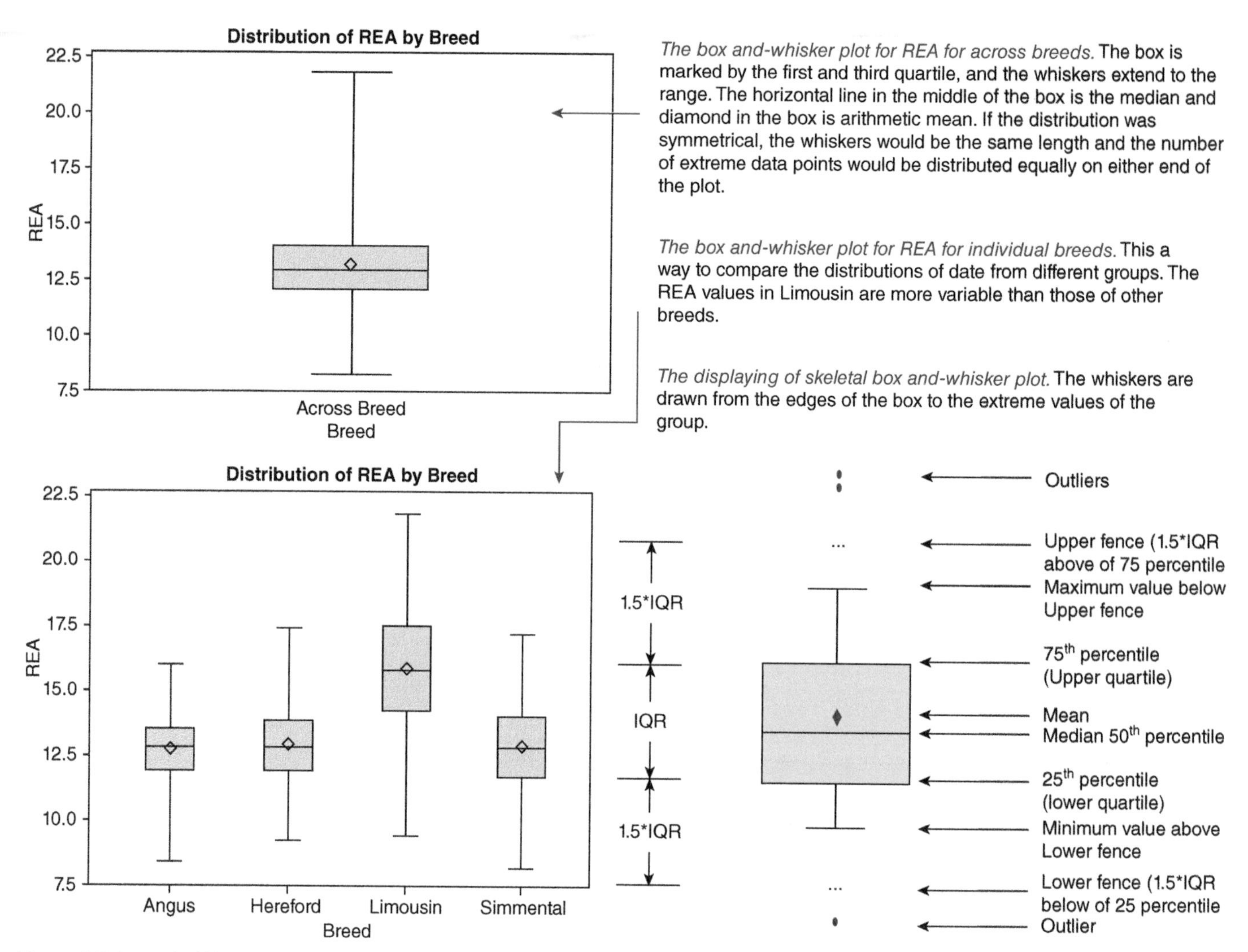

Figure 6.5 Box and whisker plot drawn for ribeye area.

Area under the curve

To calculate probabilities associated with the normal distribution, one must find the area under the normal curve. The probability that the value of a normal random variable is in an interval (y_1, y_2) is:

$$P(y_1 < y < y_2) = \int_{y1}^{y2} \frac{1}{\sqrt{2\pi\sigma^2}} e^{-\frac{1}{2}\frac{(y-\mu)^2}{\sigma^2}} dy \qquad (6.6a)$$

For instance, suppose that the mean and standard deviation of the weight at 60 days of age for a Dorset sheep population are $\mu = 20.5$ kg and $\sigma = 4$ kg. Suppose that the probability of a randomly selected animal's body weight falls between 16.8 and 24.6 kg needs to be calculated. Then, the solution is:

$$P(16.8 < y < 24.6) = \int_{16.8}^{24.6} \frac{1}{\sqrt{2*3.14*16}} * 2.72^{-\frac{(y-20.5)^2}{2*16}} dy.$$

16.8 μ=20.5 24.6

Finding the area under the curve for certain values requires tedious mathematical techniques but is the question that comes to mind. One way to avoid the calculations is to use previously prepared tables of normal distributions to find the probability under the curve. Since each normally distributed variable has its own mean and standard deviation, there are infinite normal distributions and tables. Therefore, data needs to be transformed into the **standardized normal distribution** by calculating the z scores as follows in order to be able to use a single table for the probability under the curve:

$$z = \frac{(x-\mu)}{\sigma} \qquad (6.6b)$$

Any set of normally distributed data can be converted into a standardized form and the desired probabilities can then be determined from a table of standardized normal distribution. However, there is an empirical rule to estimate the spread of data, given the mean and standard deviation, in a normal distribution and this is provided in the Box 6.1 with further visualization in Figure 6.7.

Discrete probability distributions	Continuous probability distributions
Binomial distribution Poisson distribution Geometric distribution Negative binomial distribution Hypergeometric distribution Multinomial Distribution	Uniform distribution Exponential distribution Gamma distribution Normal distribution Standard normal distribution Student's *t* distribution Chi-square (X^2) Distribution F Distribution

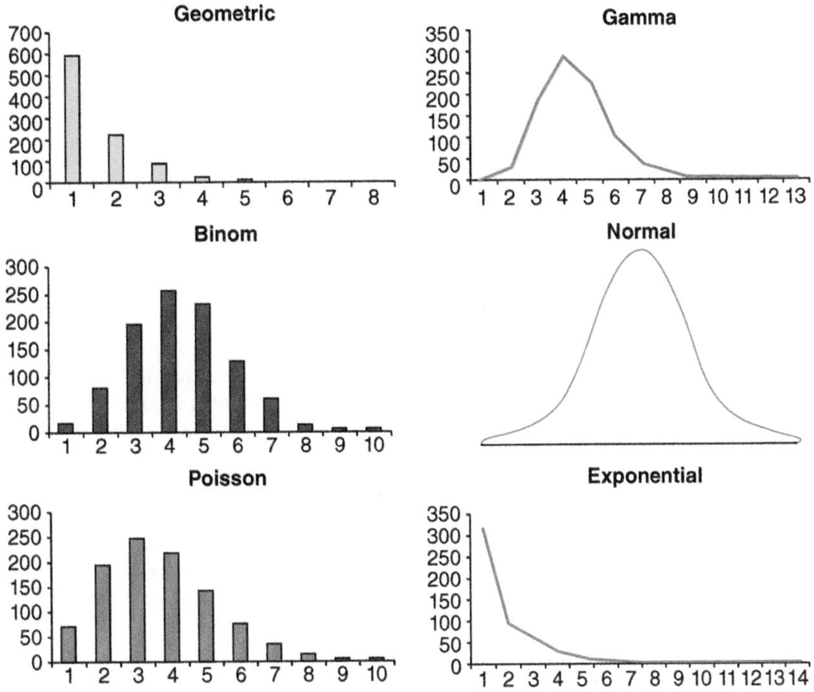

Figure 6.6 Some common probability distributions for discrete and continuous variables.

Box 6.1 The empirical rules for normal distribution.

The empirical rules for normally distributed data

- ✓ Approximately 68% of the measurements will fall within 1 standard deviation of the mean. $P(\mu - \sigma \leq y \leq \mu + \sigma)$.
- ✓ Approximately 95% of the measurements will fall within 2 standard deviations of the mean. $P(\mu - 2\sigma \leq y \leq \mu + 2\sigma)$.
- ✓ And approximately 99.7% of the measurements will fall within 3 standard deviations of the mean. That is $P(\mu - 3\sigma \leq y \leq \mu + 3\sigma)$.

For example, consider the simulated data weight at 60 days of age of the Dorset breed illustrated in Figure 6.7. About 68% of the animals will be between 16.50 and 24.50 kg ($20.50 - 4.00 \leq y \leq 20.5 + 4.00$).

Standard normal distributions

Reconsider the probability density function for a normally distributed random variable:

$$f(y; \mu, \sigma^2) = \frac{1}{\sqrt{2\pi\sigma^2}} e.^{-\frac{1}{2}\frac{(y-\mu)^2}{\sigma^2}} \tag{6.7}$$

Recall e and π are mathematical constants with values of 2.72 and 3.14, respectively. The shape and position of the distribution depends on two parameters μ and σ^2 (or σ). Therefore, each normally distributed variable has its own normal distribution curve, which depends on these two parameters. Figure 6.8 shows several different normally distributed curves from 1000 sheep simulated for 60 days weaning weight (top) and grease fleece weights (bottom). Normal distributions derived from these different data sets, therefore, will have different means and different standard deviations. Figure 6.8 demonstrates how this occurs by comparing different normal curves for 60 days weaning weight of four different breeds. The figure given for 60 days weaning weight and grease fleece weight also demonstrates that normal curves can be drawn in a way that reflects the distribution of data; that is, curves may be tall and narrow, short and wide, or anything in between. As illustrated in Figure 6.8, μ and σ specify the shape and position of a particular normal distribution under consideration. For instance, a decrease in μ (as given for Polypay) corresponds to a shift of the whole curve to the left, and an increase in μ (as given for Targhee) corresponds to a rightward shift. On the other hand, when the value of σ is increased the curve flattens, whereas when σ is decreased, a narrower-type normal distribution is generated.

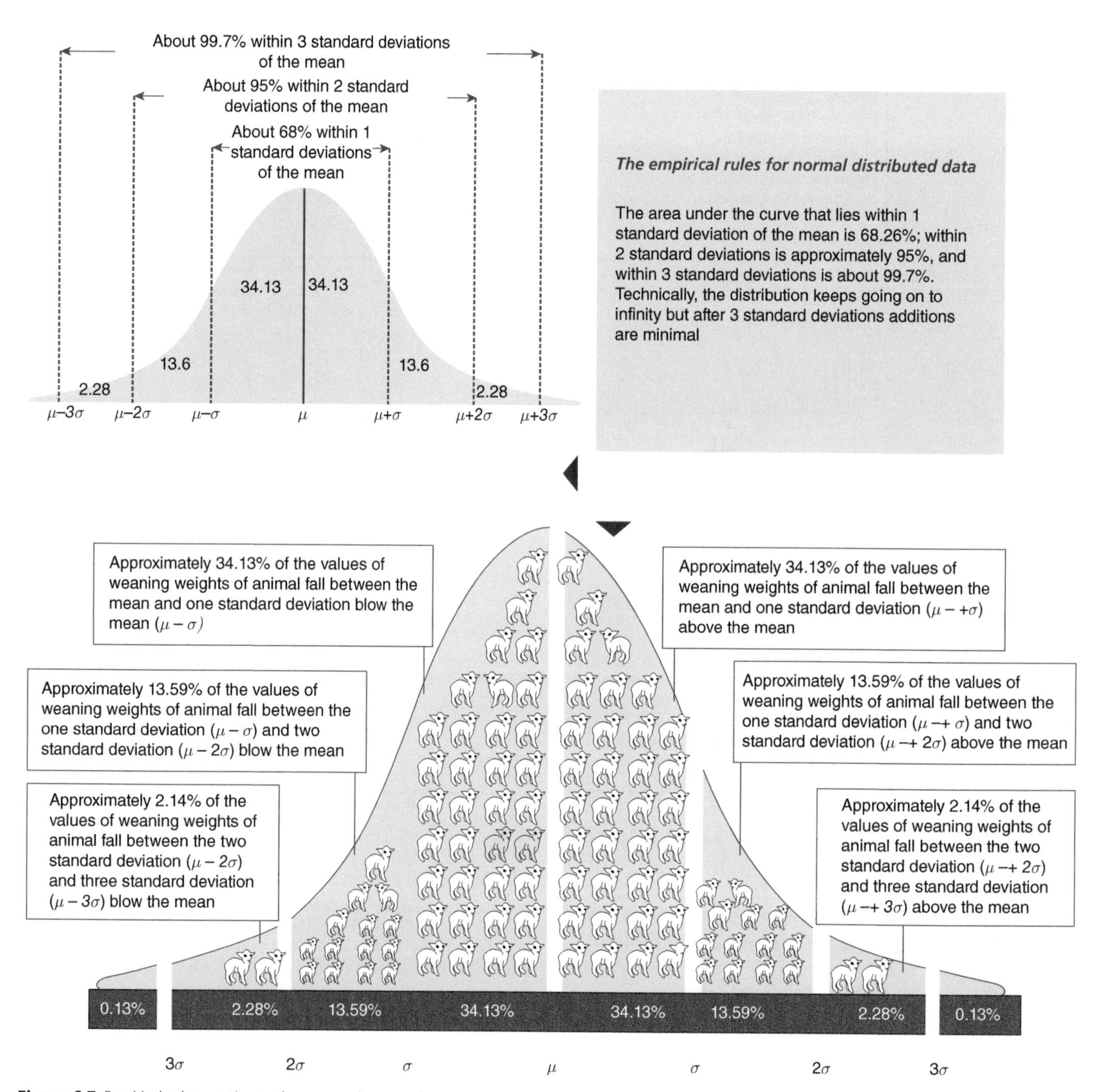

Figure 6.7 Empirical rule to estimate the area under normal curve.

Each normally distributed data set is characterized by its mean, μ, and standard deviation, σ, and there are an infinite number of combinations of μs and σs. Hence, there are an infinite number of normal curves. It would be difficult and tedious to do the calculation for the probability every time there is a new set of parameters for μ and σ. One particular type of normal curve is the **standard normal curve**: A normal curve with $\mu = 0$ and $\sigma = 1$. This is a transformation that makes any normal distribution into a standard normal distribution, using the following formula:

$$z = \frac{(y-\mu)}{\sigma} \sim N(0,1). \tag{6.8}$$

where z is the random variable value on the standard normal distribution, y is the raw value on the original distribution, and μ and σ are mean and standard deviation of the original distribution. While a random y value assumes $y \sim N(\mu, \sigma^2)$, the random z score assumes $z \sim N(0,1)$. After any normally distributed data set with parameter of μ and σ is transformed into z scores, the z scores also are normally distributed with mean $\mu = 0$ and standard deviation $\sigma = 1$. The probability density function of standard normal is called $f(z)$ and is a particular case of the $f(y)$ given earlier:

$$f(z) = \frac{1}{\sqrt{2\pi}} e^{-\frac{1}{2}z^2} \text{ for all values of y.} \tag{6.9}$$

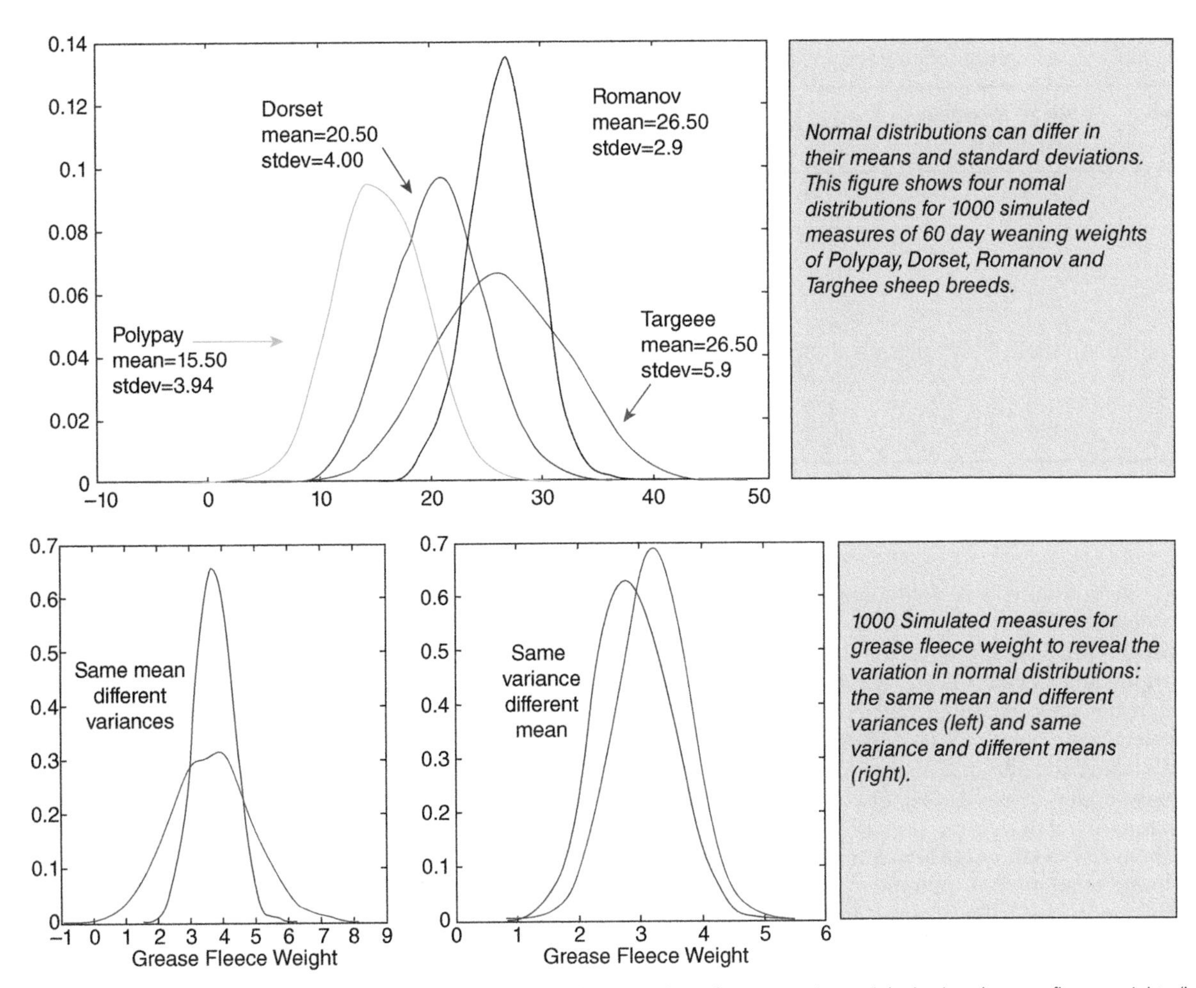

Figure 6.8 Some different normally distributed curves from simulated data for 1000 sheep for 60 weaning weight (top) and grease fleece weights (bottom).

The probability that the value of a standard normal random variable is in an interval (z_1, z_2) is:

$$P(z_1 < z < z_2) = \int_{z1}^{z2} \frac{1}{\sqrt{2\pi}} e^{-\frac{1}{2}\frac{(y-\mu)^2}{}} dz \quad (6.10)$$

To avoid integrating this function, a standard normal table (z scores in Appendix 6.1) was created to find probabilities for this standard normal distribution. To illustrate calculation of the z scores, assume 10 individuals (sheep) are randomly drawn from Dorset and Targhee populations. Recall the mean and standard deviation of these two breeds were $\mu_{Dorset} = 20.5$, $\sigma_{Dorset} = 4.0$, $\mu_{Targhee} = 26\ .5$, $\sigma_{Targhee} = 5.9$. Table 6.3 presents the raw value and corresponding z scores for 10 animals of a 60-day weaning weight from Dorset and Targhee populations (Table 6.3). The z scores in Table 6.3 indicate how many standard deviations fall away from their mean. For example, the third Dorset animal ($y_{Dorset} = 15.19$) had a 60-day weaning weight that is 1.35 standard deviation units below the mean ($\mu_{Dorset} = 20.5$). The tenth animal had a 60-day weaning weight that is 1.71 standard deviation units above the mean. Box 6.2 illustrates rule and steps for calculation of probability of normally distributed data.

Exploring relationships between variables

Various scatterplots for different sets of bivariate (two variables) are illustrated in Figure 6.9. This type of visualization is often employed to explore potential association between variables. A **positive association** between variables, for example, (b) and (c) in Figure 6.9, would be indicated on a scatterplot by an upward trend. The scatterplot illustrated in Figure 6.9(b) explores the association between carcass weight and age at slaughter (days) for a simulated data set. In this scatterplot, you can see a strong positive association: Animals slaughtered at an earlier age generally have lower carcass weight and animals slaughtered in a later age generally have heavier carcass weights. The trend between slaughtered age and carcass weight is fairly linear. One can visualize a summary line going through the center of the data from lower left to upper right. As you move to the right along this line, the points still are closely clustered around the line even though some of the points slightly fan out away from the live. The same interpretation can be done for scatterplot given in Figure 6.9(a). Here, there is a strong negative association between variables. The scatterplots given in Figure 6.9(e) and (f) demonstrate a moderate relationship between variables in a positive and negative association, respectively. The scatterplot illustrated in

Table 6.3 Raw value and corresponding *z* scores for 10 animals from Dorset and Targhee breeds.

Observation	Dorset weaning weight (kg)	z scores for Dorset $z=\frac{(y-20.5)}{4.0}$	Targhee weaning weight (kg)	z scores for Targhee $z=\frac{(y-26.5)}{5.9}$
1	23.93	0.87	29.70	0.54
2	21.80	0.33	31.39	0.83
3	15.19	−1.35	18.83	−1.30
4	21.60	0.28	31.62	0.87
5	18.07	−0.62	18.95	−1.28
6	23.75	0.82	26.93	0.07
7	14.19	−1.60	19.69	−1.16
8	20.07	−0.11	37.15	1.81
9	20.10	−0.10	21.69	−0.82
10	27.23	1.71	29.01	0.42

Figure 6.9(c) demonstrates a perfectly linear association (negative and positive) between variables. Notice that if there is association between variables, almost all the observation points are contained within an imaginary ellipse. The narrower the elliptical profile, the greater the correlation. When there is no association between variables the points scatter widely about the plot with the majority falling roughly in the shape of a circle, see Figure 6.9(d). The scatterplots depicted in Figure 6.9(g) and (h) illustrate curved associations with varying strength. These cases do not mean that there is no relationship between random variables, but rather no linear relation. The scatterplot given in Figure 6.9(i) shows a situation with **outlier** observations. The scatterplots in Figure 6.9(a) and (i) were drawn from the same data set with the exception of including an outlier in Figure 6.9(i), much shows the impact of outliers to the overall pattern.

To explore the relationship between variables and describe the overall **shape** of the relationship, one must pay attention to pattern (linear or curved), the **trend** (positive or negative), and the **strength** (exact, strong, moderate, weak, and none) of the relationship. All of these are determined by **covariation**: a measure of how much two random variables change together.

Covariance

Recall the **covariance** measures the tendency of two variables to increase or decrease together. That is, covariance is a measure of **covariation** of two random variables *X* and *Y* or covariation among relatives for the same trait, say *Y*. For example, consider the covariation between average daily gain and feed intake, if greater values of feed intake mainly correspond with greater values of average daily gain, and smaller value tend to correspond to values similar smaller values the covariance is a positive number. In the opposite occasion, as given in Figure 6.9(a), (f), and (i), when the greater values of one variable primarily correspond to the smaller values of the other, the covariance is then negative. The sign of the covariance therefore shows the tendency of the linear relationship between the variables. The sign of the covariance determines the sign of the correlation and regression coefficients as well. The magnitude of the covariance, however, is not that easy to interpret. Therefore, this magnitude is normalized for practical purposes and the usage and interpretation of the normalized version of the covariance will be reconsidered in the next section.

Variation and covariation among relatives in natural and controlled populations are of vital importance in animal and plant breeding strategies. Some genes only affect a single trait while many genes have an effect on various traits. Because of this, a change in a single gene will have an effect on all those traits. This is calculated using covariance and is termed as the covariance between breeding values. For example, the phenotypic covariance between two traits, say *X* and *Y*, is expressed as $\mathrm{Cov}(P_X, P_Y)$ and calculated as:

$$\mathrm{Cov}\,(P_X, P_Y) = \mathrm{Cov}\,(G_X, G_Y) + \mathrm{Cov}\,(E_X, E_Y) \qquad (6.11a)$$

Here, *P*, *G*, and *E* stand for phenotype, genotype, and environment, respectively. Thus, the phenotypic covariance between two traits can be partitioned into genetic and environmental effects. Consider two metric averages of fiber diameter (AFD) and hogget live weight (HLW), expressed in one individual. The phenotypic value of each character can be described by its own linear model:

$$\begin{aligned} P_{AFD} &= G_{AFD} + E_{AFD} \\ P_{HWL} &= G_{HWL} + E_{HWL}. \end{aligned} \qquad (6.11b)$$

While P_{AFD} expresses the phenotypic value of average fiber diameter, the P_{HLW} expresses the phenotypic value of hogget live weight. The phenotypic covariance between these traits is termed as $\mathrm{Cov}(P_{AFD}, P_{HWL})$, which is expressed as: $\mathrm{Cov}(P_{AFD}, P_{HWL}) = \mathrm{Cov}(G_{AFD}, G_{HWL}) + \mathrm{Cov}(E_{AFD}, E_{HWL})$. In other words, $\mathrm{Cov}(P_{AFD}, P_{HWL})$ measures the degree to which average fiber diameter and hogget live-weight are linearly related. Box 6.3 illustrates the calculation of covariance and Box 6.4 shows the properties of covariance between two traits.

Example of covariance

Listed in the Table 6.4 are simulated data for weaning weight (kg) and grease fleece weights (kg) of 10 sheep. The covariance of these two traits is:

$$\begin{aligned} Cov\,(X,Y) = S_{XY} &= \frac{1}{n-1}\sum_{i=1}^{n}(X_i-\bar{X})(Y_i-\bar{Y}) \\ &= \frac{(66.6-59.51)*(2.2-2.1)+(67.5-59.51)*(2.4-2.1)+\ldots..+(59.2-59.51)*(2.1-2.1)}{10-1} \\ &= 13.15/9 = 1.46111\ \mathrm{kg}^2 \end{aligned}$$

deviations are multiplied together (kg*kg) to give the units kg².

Box 6.2 Examples for calculating the probability of normally distributed data.

Examples and solutions	*Interpretation and pictorial representation of problem*

Example 1

Consider 1000 simulated data for Figure 6.8 and Table 6.3. Use the data and determine the probability that a randomly selected animal from the Dorset population ($\mu = 0.5$ and $\sigma = 4.0$) falls between 15.5 and 23.5 kg.

Solution

1. Convert 15.5 and 23.5 in to standard normal, **z** score

$$z_1 = \frac{(15.5 - 20.5)}{4.00} = -1.25$$

$$z_2 = \frac{(23.5 - 20.5)}{4.00} = 0.75$$

2. Shade the area corresponding to the probability. →

Find the probability corresponding z scores

3. Find the probability using standard normal distribution table given in z scores table at Appendix 6.1. For standard normal score 0.75, we have enclosed the row that represents 0.7 and the column that represents 0.05. Likewise for −1.25, the row represents 1.2 and the column represents 0.05. **Please note** that only positive entries for z are listed in the table, since for a symmetrical distribution with a mean of zero, the area from the mean to +z must be identical to the area from the mean to −z.

- From the z scores table in Appendix 6.1, the probability corresponding to the −1.25 is 0.3944.
- The probability corresponding to the 0.75 is 0.2734.
- The probability that the selected animal lies between 15.5 and 23.5 is total of the shaded area:

$$0.3944 + 0.2724 = 0.6678$$

Interpretation of *z* scores

1. The *z* value represents the number of standard deviations a point is above or below the population mean. The *z*-value for 15.5 is −1.25. This means an animal from this population with 60 day weaning weight has a value that is −1.25 standard deviations below the population mean. An animal with 23.5 has a value that is 0.75 of a standard deviation above the population mean.

Pictorial representation

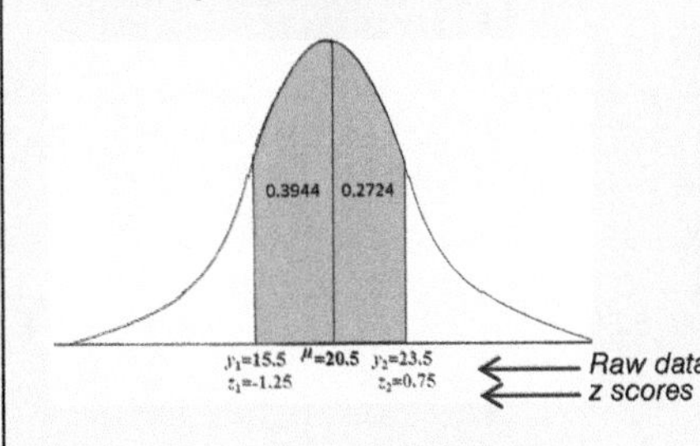

Z	.0	.01	...	.05	...	.08	.09
0.0	.0000	.0040	...	.0199	...	.0319	.0359
0.1	.0398	.0438	...	.0596	...	.0714	.0753
	...	...	...	...	...	...	...
	...	...	...	...	...	...	...
0.5	.1915	.1950	....	.2088	...	.2190	.2224
0.6	.2257	.2291	...	.2422	...	.2517	.2549
0.7	.2580	.2611	...	.2724	...	.2823	.2852
			...		...	...	...
1.2	.3849	.3869	...	.3944			
			...		...	...	...
2.9	.4981	.4982	....	.4984	...	.4986	.4986
3.0	.4987	.4987	...	.4989	...	.4990	.4990

2.

Interpretation of probability

3. In Step 3 we computed the probability associated with $z = 1.25$ and $z = 0.75$ to be 0.3944 and 0.2724, which are the probabilities of a value falling between the mean and 1.25 standard deviations below the mean one 0.75 standard deviations above the mean, respectively. Therefore, the probability any randomly selected animal falls between 15.5 and 23.5 is the sum of 0.3944 and 0.2727 ($P(15.5 \le y \le 23.5)$) is 66.78%.

Example 2

Determine the probability that a randomly selected animal from the Dorset population weighs more than 14.8 kg at the 60-day of age ($P(y > 14.8)$).

Solution

1. Convert 14.8 to standard normal, *z* score

$$z_1 = \frac{(14.8 - 20.5)}{4.00} = -1.425$$

Interpretation of *z* scores

1. The *z* score −1.425 means that an animal from this population that weighs 14.8 kg for a 60-day weaning weight is −1.425 standard deviations lower than the mean of population.

2. Shade the area corresponding to the probability →

Find the probability corresponding *z* score

3. The probability that corresponds to the 1.425 standard score: Because 1.425 lies midway between 1.42 and 1.43 we interpolate halfway between $z = 1.42$ and $z = 1.43$ to get 0.4229 (the average of 0.4222 and 0.4236 that correspond to 1.42 and 1.43 in *z* scores table at Appendix 1).

The resulting probability or area under the curve from the mean to 1.425 standard deviations above it is 0.4436. Then ($P(y > 18.8)$) is 0.4436+0.5=0.94436

Pictorial representation

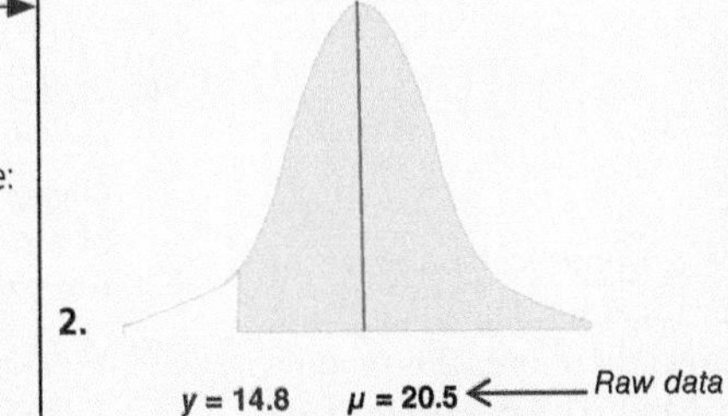

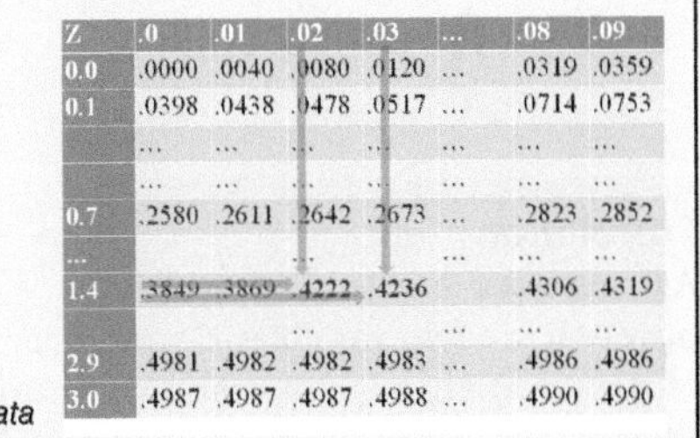

Z	.0	.01	.02	.03	...	.08	.09
0.0	.0000	.0040	.0080	.0120	...	.0319	.0359
0.1	.0398	.0438	.0478	.0517	...	.0714	.0753
	...	...	...	...	...	...	...
	...	...	...	...	...	...	...
0.7	.2580	.2611	.2642	.2673	...	.2823	.2852
...			...		...	...	...
1.4	.3849	.3869	.4222	.4236		.4306	.4319
			...		...	...	...
2.9	.4981	.4982	.4982	.4983	...	.4986	.4986
3.0	.4987	.4987	.4987	.4988	...	.4990	.4990

2.

y = 14.8 μ = 20.5 ← Raw data

z = −1.425 ← z score

Interpretation of probability

3. Thus, the probability that the selected animal weighs more than 14.8 kg is the sum of 0.4436 and 0.5 is 94.436%.

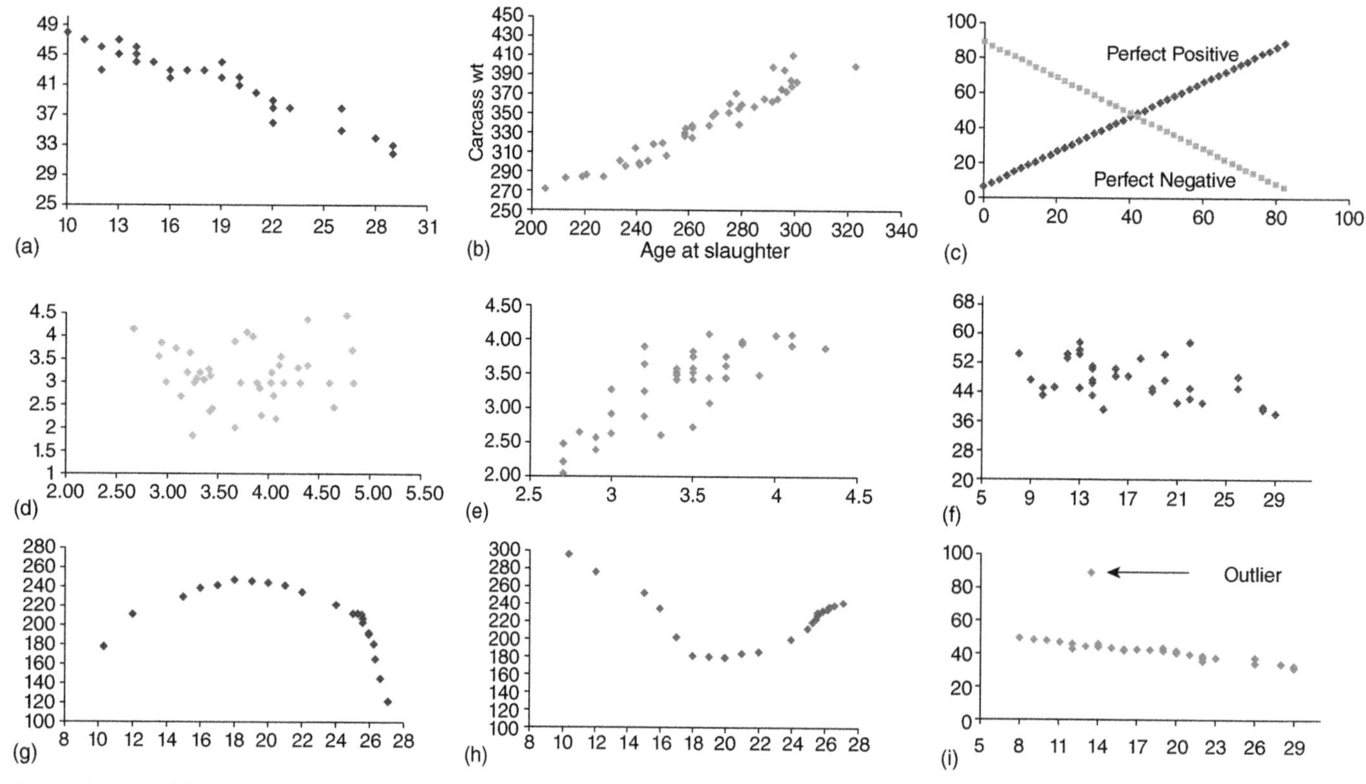

Figure 6.9 Nine different scatterplots with various associations and distributions.

Box 6.3 Calculation of covariance between two traits.

Calculation of covariance

By definition the covariance between two related traits, X and Y, is

Cov(*X*, *Y*) = mean of the product – the product of the mean

Mathematically, the covariance between X and Y for a population and a sample is then:
Covariance for population:

$$Cov(X,Y)=\sigma_{XY}=\frac{1}{n}\sum_{i=1}^{n}(X_i-\mu_x)(Y_i-\mu_y)=\frac{1}{n}\left(\sum_{i=1}^{n}(X_iY_i)-n\mu_x\mu_y\right)$$

Covariance for sample:

$$Cov(X,Y)=S_{XY}=\frac{1}{n-1}\sum_{i=1}^{n}(X_i-\bar{X})(Y_i-\bar{Y})=\frac{1}{n-1}\left(\sum_{i=1}^{n}(X_iY_i)-n\overline{XY}\right)$$

Box 6.4 Some characteristic of covariance between two traits (*X and Y*).

Properties of covariance

1. Covariance can take any positive and negative values, $-\infty \le Cov(X,Y) \le \infty$. In both cases, the absolute value of $Cov(X,Y)$ indicates the strength of the linear relationship between X and Y.
2. The absolute value of a covariance can never be larger than the product of the standard deviations (Cook and Weisberg, 1999),
$$-\sqrt{Var(X)Var(Y)} \le Cov(X,Y) \le \sqrt{Var(X)Var(Y)}$$
3. If $\pm$ $Cov(X,Y) = \{Var(X).Var(Y)\}^{1/2}$ then the relationship between X and Y is perfect, and all pairs of observations must fall on a straight line with negative and positive slope.
4. If $Cov(X,Y) = 0$, then there is no linear relationship between X and Y.
5. The covariance is a symmetric measure of linear dependence: The covariance between X and Y is the same as the covariance between Y and X. That is, $Cov(X,Y) = Cov(Y,X)$.
6. Variances are not related if two effects in the model are independent of each other. Consider the average fiber diameter with a phenotypic value of any individual is $P_{AFD} = G_{AFD} + E_{AFD}$. The variance of a sum of random effects is the sum of these effects plus two times the covariance of each pair of random effects. That is:
7. $Var(P_{AFD}) = Var(G_{AFD} + E_{AFD}) = \sigma_G^2 + \sigma_E^2 + 2Cov(G,E)$. This becomes $\sigma_P^2 = \sigma_G^2 + \sigma_E^2$ when genetic and environmental effects are assumed to be independent. Therefore, variances are not related when random effects in model are independent and this is important for use of covariance in the genetic model.

Correlation

As expressed earlier, the covariance is a numerical measure of how two random variables X and Y vary together. The magnitude of the covariance is not easy to interpret, therefore, the magnitude of covariance needed to be normalized. The **coefficient of correlation** (also known as the **Pearson correlation coefficient**, after Karl Pearson) is the covariance that is standardized to range from -1 to 1. The sample correlation coefficient between X and Y is (see Box 6.5 for more details):

$$r = \frac{Cov(X,Y)}{\sqrt{Var(x)}\sqrt{Var(Y)}} = \frac{S_{XY}}{S_X . S_Y} \quad (6.12)$$

As seen in (6.12), the standardization is done by dividing the quantity of covariance by the product of the standard deviations of X (S_X) and Y (S_Y). The linear correlation coefficient, r, measures the strength of the linear association between the two variables X and Y. If the association between variables tends to go up or down together then a positive correlation should be expected. Otherwise, if low values of one variable are associated with high values of the other, then a negative correlation coefficient should be expected. Therefore, the first thing to do is construct a scatterplot, a graphical display of the data to determine the direction trend between variables. Figure 6.9 illustrates several scatterplots to exhibit the trends between variables. The plots demonstrate that as the correlation coefficient increases in magnitude, the points become more tightly concentrated about a straight line through the data (as given in Figure 6.9a,b, and c). Interpreting correlation using a scatterplot can give only some rough ideas. A more precise way to measure the type and strength of a linear correlation between two variables is to calculate the linear correlation coefficient. An example of correlation between weaning weight and grease fleece weight is given in Box 6.6 and further properties of correlation are given in Box 6.7.

As described for the covariance, the phenotypic correlation is made up of two components: the additive genetic and environmental correlations (the environmental correlation combines both true environmental and non-additive genetic correlation). So, $P_X = G_X + E_X$ and $P_Y = G_Y + E_Y$. The additive genetic correlation ($r_{GX,GY}$) measures the strength of the relationship between the breeding values of two traits. The genetic correlation is important for animal breeding strategy that is used to estimate the selection of parents for one trait that will cause the changing in a second trait in the progeny. The environmental correlation ($r_{EX,\ EY}$), however, measures the degree to which two traits respond to the same environmental variation (Conner and Hartl, 2004). If the full-sibling families are used to estimate the genetic correlation, then the estimate of $r_{GX,GY}$ is termed the **broad-sense genetic correlation**, which includes dominance and the permanent environmental effects. The **causes** of genetic correlation are

Box 6.5 The calculation of correlation coefficient between two traits (two variables).

Calculation of correlation coefficient

As given earlier, any variable drawn from normally distributed population can be standardized as:

$$z_i = \frac{(X_i - \mu_X)}{\sigma}$$

If we standardize the each deviation $\frac{(X_i - \bar{X})}{S_X}$ and $\frac{(Y_i - \bar{Y})}{S_Y}$, and then divide this quantity by degrees of freedom ($n - 1$) gives the correlation coefficient:

$$Cov_{XY} = \left(\sum_{i=1}^{n} \frac{(X_i - \bar{X})}{S_X} \frac{(Y_i - \bar{Y})}{S_Y}\right) \Big/ n-1 \longrightarrow r = \frac{Cov(X,Y)}{S_X S_Y}$$

The correlation coefficient lies between −1 and +1. Here, S_X and S_Y are standard deviations of X and Y traits.

Box 6.6 Calculation of the correlation coefficient between weaning weight and grease fleece weight.

Example

Consider the weaning weight (kg) and grease fleece weights (kg) data given in Table 6.4.

$S_{XY} = 1.4611$, the covariance between weaning weight (X) and grease fleece weight (Y).

$S_X = 5.09877$, the standard deviation of weaning weight.

$S_Y = 0.36128$, standard deviation of grease fleece weight.

Then correlation between weaning weight (kg) and grease fleece weights (kg) is:

$$r_{XY} = \frac{S_{XY}}{S_X S_Y} = \frac{1.4611}{5.09877 * 0.36128} = 0.67299$$

Table 6.4 Weaning weights (kg) and grease fleece weights (kg) simulated for 10 sheep.

Weaning weights (X)	Grease fleece weights (Y)	$X_i - \bar{X}$	$Y_i - \bar{Y}$	$(X_i - \bar{X})*(Y_i - \bar{Y})$
66.6	2.2	7.09	0.1	0.709
67.5	2.4	7.99	0.3	2.397
54	1.7	−5.51	−0.4	2.204
53.3	1.8	−6.21	−0.3	1.863
65.2	2.3	5.69	0.2	1.138
62.6	2.8	3.09	0.7	2.163
55.4	1.9	−4.11	−0.2	0.822
54.1	1.8	−5.41	−0.3	1.623
57.2	2	−2.31	−0.1	0.231
59.2	2.1	−0.31	0	0
Total 595.1	21	0.00	0.00	13.15

Box 6.7 Some properties of coefficient of correlation.

Properties of correlations

1. The correlation is always between $-1 \leq r \leq 1$. An r close 0 indicates negligible linear relationship. A correlation coefficient close to -1 or $+1$ indicates the strong association between traits.
2. r uses the standardized values of the observations, r does not change when units of measurement are changed.
3. Because of symmetry the $r_{XY} = r_{YX}$.
4. Both X and Y may be influenced simultaneously by a third variable, say Z. Thus, a correlation measure of linear association does not necessarily show causation.
5. The relative importance of additive genetic correlation, $r_{GX,GY}$ and $r_{EX,EY}$ in determining the phenotypic correlation between two traits X and Y depend on the heritability of the two traits.

Box 6.8 Partition of phenotypic correlation into genetic and non-genetic correlations.

The phenotypic correlation:

$$r_P = r_{PX,PY} = \frac{Cov(P_X, P_Y)}{\sqrt{Var(P_X)Var(P_Y)}} = \frac{Cov(G_X, G_Y) + Cov(E_X, E_Y)}{\sqrt{Var(P_X)Var(P_Y)}}$$

Additive genetic correlation

$$r_G = r_{GX,GY} = \frac{Cov(G_X, G_Y)}{\sqrt{Var(G_X)Var(G_Y)}}$$

Environmental correlation

$$r_E = r_{EX,EY} = \frac{Cov(E_X, E_Y)}{\sqrt{Var(E_X)Var(E_Y)}}$$

Partitioning of the phenotypic correlation:

$$r_P = h_X h_Y r_A + e_X e_Y r_E.$$

Where h_X and h_Y are the square root of heritabilities for each trait, X and Y, calculated as:

$$h^2 = \frac{\text{Additive genetic variance}}{\text{Phenotypic variance}} = \frac{Var(G)}{Var(P)}$$

Likewise e_X and e_Y are: $e_X = \sqrt{1 - h_X^2}$ *and* $e_Y = \sqrt{1 - h_Y^2}$

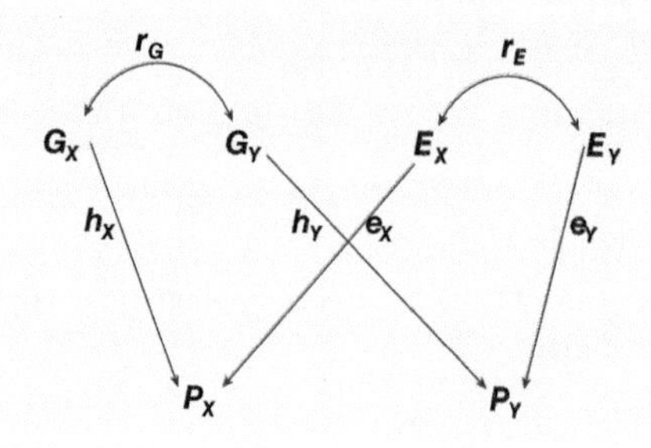

pleiotropic effect and **linkage**. Pleiotropic means two traits are affected by the same gene while linkage means that genes linked on the same chromosome control different traits. The pleiotropic effect is permanent but the linkage effect is temporary. Because some traits are affected by the same gene (pleiotropic), inherently a genetic correlation is expected between the affected for the same gene. It should be noticed that if there is linkage equilibrium in a population, then genetic correlation is only caused by pleiotropic effect (Box 6.7).

Sometimes a variable for each individual has **repeated measurements** and the phenotypic variance can be partitioned into variance between and within individuals (Box 6.8). The strength of the relationship between repeated records of the same individual is termed **repeatability**. This is another useful measure of the correlation between repeated records on the same animals that represents the proportion of the phenotypic variance that is due to permanent effects (genetic effects and permanent environmental effects). This correlation is useful in the prediction of producing ability. If this correlation (repeatability) is high then the animal's next record can be predicted more accurately. Some examples of repeatability estimates in livestock productions are: body measurements in beef cattle (0.8), calving interval in dairy cattle (0.15), egg weight (0.9), number of births in sheep (0.15), and fleece grade (0.60).

Regression

Covariance and correlation quantify the association of two variables but do not predict Y from X and vice versa. Therefore, the decision of which variable is to be called dependent or independent does not matter in correlation or covariance but does matter in regression analysis. With regression analysis, an equation is constructed between dependent and independent variables with the dependent variable being predicted from the independent variable. Therefore, variable Y is a function of variable X, $Y = f(X)$. This relationship is described as the regression of Y on X. The linear relationship between Y and X is basically linear, but is inexact. These two variables can take on very different roles in a simple linear regression.

The linear regression equation will let us predict Y given any amount of X. This is measured by regression coefficient (b_{YX}). Notice that like covariances and correlation coefficients, regressions are not individual values but are population measures. However, because of symmetry, the measures of covariance $Cov(X,Y) = Cov(Y,X)$ and coefficients of correlation $r_{XY} = r_{YX}$ are both symmetric but that is not the case with regression coefficients, $b_{YX} \neq b_{XY}$. Since the relationship is not exact, the statistical model regression equation is (Box 6.9).

$$Y_i = \beta_0 + \beta_1 X_i + e_i \; i = 1, 2, \ldots\ldots, n \tag{6.13}$$

Here, Y_i is the value of the response or dependent variable of the *ith* individual, β_0 and β_1 are the two unknown parameters of the regression equation, X_i is an independent variable, and e_i is the random error (residual) term. β_0 and β_1 in equation (6.13) describe the location and shape of the linear function. The objective in regression analysis is to generate an estimation for these two parameters based upon the information contained in the data set, X_i and Y_i. The X_i is assumed to be measured without error. In other words, the observed values of X are assumed to

Box 6.9 The relation between correlation and regression coefficients.

Mathematical relationship between regression and correlation

Since the relationship between regression and correlation, the regression coefficient (slope) can be converted to the correlation coefficient and vice versa using the following formulas:

$$b_1 = r\left(\frac{S_Y}{S_X}\right) \text{ So, regression coefficient} = \text{correlation coefficient}\left(\frac{\text{Standard deviation of Y}}{\text{Standard deviation X}}\right)$$

The table given here illustrates the regression-correlation coefficients and standard deviation of *ADF* and *DFI*:

b_1	r	S_{ADG}	S_{DFI}
0.02402	0.7757	0.07085	2.28841

Then to convert regression to correlation:

$$0.02402 = r\left(\frac{0.07085}{2.28841}\right) \Rightarrow 0.02402 = r * 0.03096$$

$$\Rightarrow r = \frac{0.02402}{0.03096} = 0.7757$$

To convert correlation to regression

$$b_1 = 0.7757\left(\frac{0.07085}{2.28841}\right) = 0.02402$$

be a set of known constants. The Y_i and X_i are paired observations, that is, both are measured on every individual. The random errors (e_i) are assumed to be normally distributed, which implies that the Y_i are also normally distributed. Further, it is assumed that the expected value (mean) of random errors have mean zero and common variance σ^2. In addition, the random errors are assumed to be mutually independent, that is, the covariance between two error terms is zero ($\text{Cov}(e_i, e_i') = 0$ for $e_i, \neq e_i'$. Since both Y_i and e_i are the random elements of the regression model, the assumptions listed for the random error are valid for dependent random variables *Y*: *Y*s are mutually independent and have common variance. The assumptions for random errors and *Y*s are specified: $e_i \sim NID(0, \sigma^2)$ and $Y_i \sim N(\beta_0 + \beta_1 X_i, \sigma^2)$, where the NID stands for **normally** and **independent distributed property of random error**.

Recall that the regression equation, $Y_i = \beta_0 + \beta_1 X_i + e_i$ is referred to as the **simple linear regression** model. This equation is called simple because it contains only one independent variable that appears only in the first power in the model (e.g., Figure 6.9a, b, and c are linear, g and h in the same figure are nonlinear). Linear regression consists of finding the best-fitting straight line through the data. The best-fitting line is called a **regression line**. Mathematically, the straight-line relationship is represented in a simple linear regression model as:

$$E(Y) = \hat{Y}_i = b_0 + b_1 X_i, \tag{6.14}$$

or equivalently;

$$E(Y_i) = \hat{Y}_i = \bar{Y} + b_1(X_i - \bar{X}) \tag{6.15}$$

where *E* stands for **Expectation** while b_0 and b_1 are estimators of β_0 and β_1. One method to estimate parameters is **ordinary least squares**. Using ordinary least squares, a regression line is selected such that the differences between true and estimated values of *Y* are minimized. The *Y*-values given in equations (6.13–6.15) then comprise this deterministic straight-line relationship plus a random error. Box 6.10 provides more details for estimation of parameters. The random error is:

$$e_i = Y_i - E(Y_i) \Rightarrow e_i = Y_i - \hat{Y}_i \tag{6.16}$$

In other words the error is expressed as:

$$\textit{Error}\ (e_i) = \textit{True value}\ (Y_i) - \textit{Predicted value}\ (\hat{Y}_i)$$

This random error is not an error in the sense of a "mistake," but rather represents variation in *Y* due to factors other than *X* which has not been measured (Pardoe, 2006). For example, consider that average daily gain (*ADG*) of cows can be predicted by using the daily feed intake (*DFI*) measurements given in Figure 6.10. The aim is to determine a linear function that will explain changes in *ADG* as *DFI* changes. This is called the regression of *ADG* on *DFI*. Then the simple linear regression equation between the dependent (*ADG*) and independent (*DFI*) variables is (Box 6.10).

$$ADG_i = \beta_0 + \beta_1 DFI_i + e_i.$$

After fitting the linear regression the expression is $\widehat{ADG_i} = b_0 + b_1 DFI_i \Rightarrow E(ADG_i) = b_0 + b_1 DFI_i$.

With this equation the expected value for *ADG* (the dependent variable) can be found for *i*th cow that correspond to each value of *DFI* for the same cow. The parameters β_0 and β_1 have to be estimated from the data set at hand. Recall that β_0 denotes the intercept, a value of *ADG* when *DFI* = 0, in which β_0 signifies the distance to the baseline at which the regression line generated from *ADF* and *DFI* data cuts the *Y* axis. The parameter β_1 describes the slope of the estimation line and signifies the change in *ADG* for a one-unit change in *DFI*. Here, β_1 represents the degree to which the estimated regression line slopes upwards or downwards. Since β_0 and β_1 are unknown, they are estimated by b_0 and b_1. As illustrated in Figure 6.10, the black diagonal line is the regression line and consists of the predicted *ADG* for each possible value of *DFI*. It is assumed that this straight line represents an appropriate model given that estimates b_0 and b_1 of the parameters β_0 and β_1, respectively, need to be defined. One approach that generates estimators with good properties is to take the line that minimizes the sum of the squared errors. This is called **least-squares estimates**. The vertical lines in Figure 6.10 from the points to the regression line represent the errors of prediction.

Summary

Population and sample

1. Statistics uses information from a **sample** to infer something about a population from which the data were derived.

Box 6.10 Calculation of the regression coefficients with an example.

Calculating of regressions coefficients

We denote the sample estimates of β_0 and β_1 with b_0 and b_1. The least square estimates of β_0 and β_1 are calculated as:

$$\text{the regression of } Y \text{ on } X \Rightarrow b_{Y.X} = b_1 = \frac{Cov(X,Y)}{Var(X)} = \frac{\sum_{i=1}^{n}(X_i - \bar{X})(Y_i - \bar{Y})}{\sum_{i=1}^{n}(X_i - \bar{X})^2}$$

$$\text{the regression of } X \text{ on } Y \Rightarrow b_{X.Y} = b_1 = \frac{Cov(X,Y)}{Var(Y)} = \frac{\sum_{i=1}^{n}(X_i - \bar{X})(Y_i - \bar{Y})}{\sum_{i=1}^{n}(Y_i - \bar{Y})^2}$$

$$\text{and} \qquad b_0 = \bar{Y} - b_1\bar{X}$$

Thus, as stated earlier, unlike calculating of the correlation coefficient, the order of variables in regression is important. It can be seen in equation that $b_{Y.X} \neq b_{X.Y}$.

Example: The average daily gain (ADG) and daily feed intake (DFI) of 17 beef cattle is listed next. The unit of both variables is in kg.

ADG (X)	1.36	1.31	1.22	1.33	1.42	1.33	1.39	1.28	1.32	1.27	1.34	1.42	1.43	1.28	1.34	1.27	1.17
DFI (Y)	22.8	19.3	18.9	21.5	23.1	21.4	21	18.9	22	18.3	21.5	22.8	23.1	20.8	23.3	23.3	14.9

$b_{ADG.DFI} = b_{Y.X} = 0.02402$, $b_0 = \bar{Y} - b_1\bar{X} = 1.3223 - 0.02402 * 20.99412 = 0.81817$. These estimates of the parameters give the regression equation (the regression of *ADG* on *DFI*):

$$ADG_i = 0.81817 + 0.02402DFI_i \text{ or } \hat{Y}_i = 0.81817 + 0.02402X_i.$$

This relationship holds only over the range of the sample DFI-values, that is, from 14.9 to 23.3.

Interpretation: $b_1 = 0.02402$ does have a straightforward practical interpretation. The mean average daily gain is expected to increase by 0.02402 kg per each 1 kg increasing feed intake. Since b_0 is the estimated ADG value to be 0.81817 when DGI = 0, intercept b_0 really only makes sense if the independent variable (here DFI) has a possible value in a real life situation.

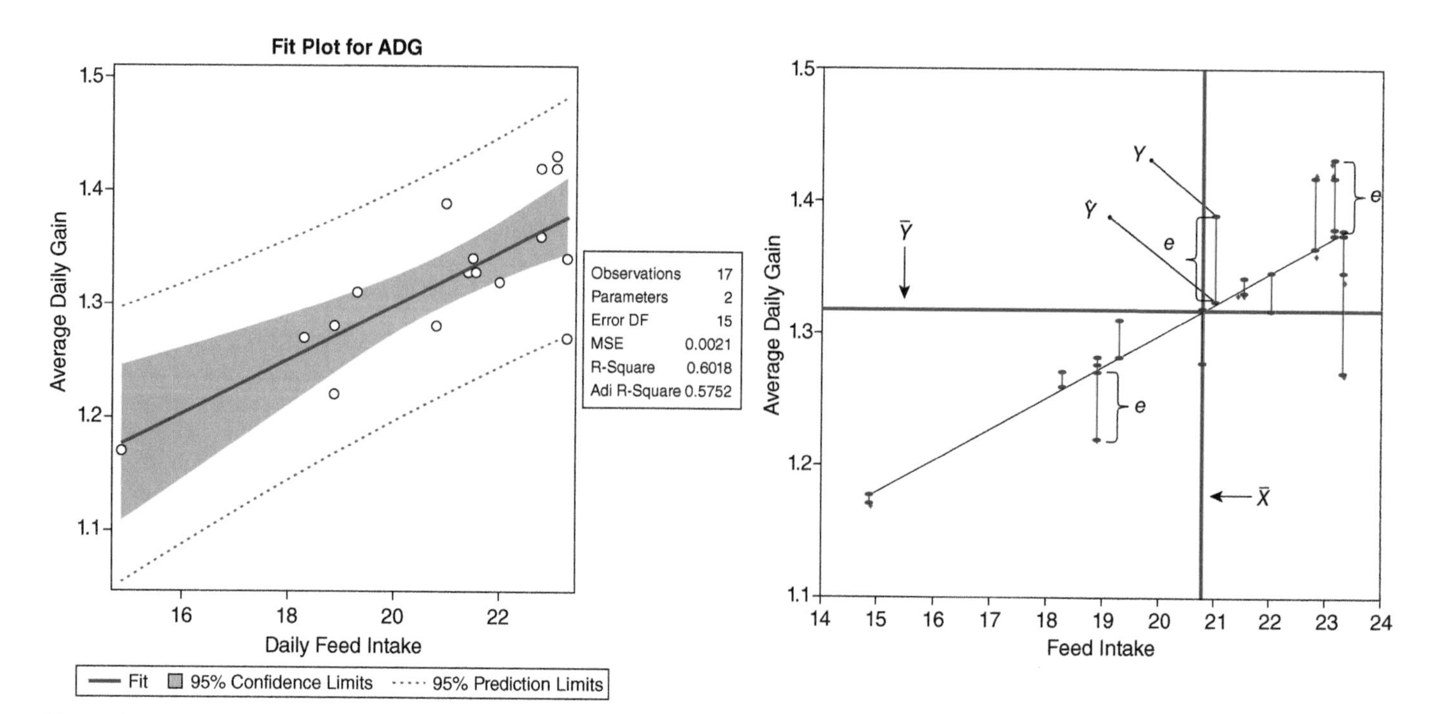

Figure 6.10 Linear regression between average daily gain (*ADG*) and daily feed intake (*DFI*) for 17 cows. Dots represent real observations (X_i, Y_i). The straight regression line $E(y) = \hat{Y}$ shows the expected, or fitted, values of the dependent variable (ADG). The errors e_i are the deviation of the observations from their expected values.

2. The set of all possible study units makes up the **study population**, which is a part of the **target population** of interest. Statistics uses information from a sample to infer something about a population from which the data were derived. By correctly sampling from study population, researchers can analyze the sample and make inferences about population characteristics.
3. Parameter is a characteristic of population and the **statistic** is a characteristic of that sample. the statistic is used to estimate the value of the parameter.
4. The random sampling from a study population allows one to draw valid inferences about the study population characteristics.

Descriptive statistics

1. Descriptive statistics are single values that represent the central tendency and deviation around the central tendency.
2. Data are generally categorized into one of two measurements: **categorical** and **numerical**, which are described by different descriptive statistics.
3. The **percentage**, **mode**, **median**, and **arithmetic** are popular central tendency measures and are used for categorical numerical data.
4. The **arithmetic mean** gives a measure of the center of gravity or balance point for all the data. The **median** represents a score that has an equal number frequency both above and below. The **mode** represents the peak or most numerous values in the data set.
5. The distribution of two distinct populations may differ greatly in their dispersion, although they may have the same central tendency (i.e., mean). The **measure of dispersion** is used in addition to the central tendency to describe the data.
6. Range and **variance/standard deviation** are well known measures of dispersion. The **range** is calculated as difference between the extreme values ($Xmax - Xmin$).
7. The **standard deviation** is calculated as the square root of the **variance** and is usually used in conjunction with the **arithmetic mean**, indicating how closely clustered the observations are to the central tendency of data. The variance of sample is calculated as: $s^2 = \frac{\sum_{i=1}^{n}(x_i - \bar{x})^2}{n-1}$.

Graphically examining the distribution of the data

1. A useful, effective, and most convenient way to reveal the data structure of any study is through the use of graphs.
2. Graphs increase the understanding of numerical data at a single glance.

Normal distribution

1. There are many families of **discrete** and **continuous probability distributions**. The majority of quantitative traits for many biological events assume that they are normally or near normally distributed. A normal distribution is bell-shaped, the mean, median, and mode all overlap at the peak or midpoint. This point has the maximum frequency of the distribution. The frequencies then gradually decrease at both ends of the curve.
2. In normal distributed data:

$$f(y; \mu, \sigma^2) = \frac{1}{\sqrt{2\pi\sigma^2}} e^{-\frac{1}{2}\frac{(y-\mu)^2}{\sigma^2}}$$

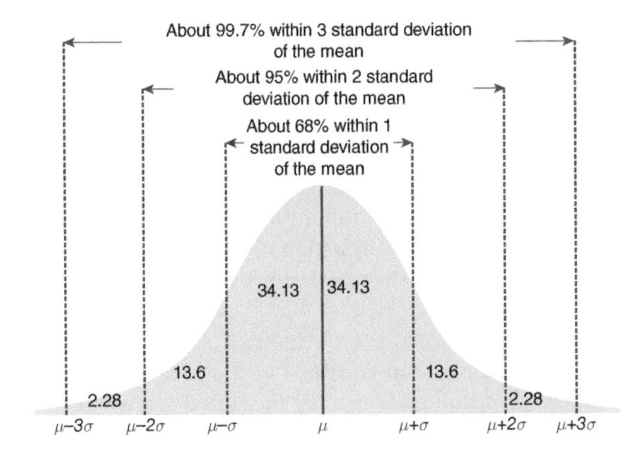

- Approximately 68% of the measurements fall within 1 standard deviation of the mean.
- Approximately 95% of the measurements fall within 2 standard deviations of the mean.
- And approximately 99.7% of the measurements fall within 3 standard deviations of the mean.

3. There are an infinite number of normal distributions because the different combinations of μs and σs.
4. Normal distributed observation is converted to the **standard normal distribution** with a mean and standard deviation of $\mu = 0$ and $\sigma = 1$. A value from any normal distribution is converted into its corresponding value on a standard normal distribution using the $z = \frac{(y-\mu)}{\sigma} \sim N(0,1)$ formula.

Exploring relationships between variables

1. Typically scatterplots are used to illustrate the association between two variables such as feed intake and weaning weight with one variable on the X axis and the other variable on the Y axis.
2. The shape, trend and strength between to traits are measured by the **covariance**. The **sign of covariance** shows the tendency in the linear relationship between the variables. The covariance between to characteristics is calculated as $Cov(X,Y) = S_{XY} = \frac{1}{n-1}\sum_{i=1}^{n}(X_i - \bar{X})(Y_i - \bar{Y})$.
3. If the greater values of one trait correspond with greater values of other trait, and the same holds for smaller values the covariance is a positive number. In contrast, when the greater values of one trait correspond to the smaller values of the other, the covariance is negative. The sign of covariance determines the sign of the correlation and regression coefficients as well.
4. The magnitude of the covariance is not easy to interpret. Therefore, this magnitude is normalized for practical purposes using **correlation**.
5. The coefficient of correlation is just a covariance that is standardized to range from −1 to 1 and is calculated as $r = \frac{Cov(X,Y)}{\sqrt{Var(x)}\sqrt{Var(Y)}} = \frac{S_{XY}}{S_X . S_Y}$.
6. Correlation quantifies the linear association between two variables.
7. The correlation (r) is always between −1 and 1. Large correlation coefficients indicate a strong relationship. An r close 0 indicates a negligible linear relationship. Correlation is completely symmetrical: the correlation between X and Y is the same as the correlation between Y and X, so $r_{XY} = r_{YX}$.

8. Correlation does not equal causation. One variable does not cause the other. Simple correlation means the linear association between two variables.

9. The goals in **regression analysis** are to determine how changes in values of some variables influence the change in values of other variables.

10. For variables such as X and Y, to predict the Y for given any amount of X, variable Y is a function of the variable X ($Y = f(X)$). This is called the **regression of *Y* on *X*** and is measured by the **regression coefficient** (b_{YX}).

11. The regression model is $Y_i = \beta_0 + \beta_1 X_i + e_i$. This is termed a **simple linear regression model**.

12. It is assumed that all X is to be measured without error.

13. The e_i is the random error term and is assumed to be normally distributed, which implies that the $Y$is are also normally and independently distributed with $e_i \sim NID(0, \sigma^2)$, $Y_i \sim N(\beta_0 + \beta_1 X_i, \sigma^2)$.

14. **Simple linear regression** allows to model a straight-line relationship between a dependent Y and independent X variables, as $E(Y_i) = \hat{Y}_i = b_0 + b_1 X_i$, or equivalently; $E(Y_i) = \hat{Y}_i = \bar{Y} + b_1(X_i - \bar{X})$.

15. The random error is expressed as *Error* (e_i) = *True value* (Y_i) − *Predicted value* $(\hat{Y}_i)$.

16. The **least squares** method estimates the regression line and the least square estimates of β_0 and β_1 are calculated as:

$$\text{regression of } Y \text{ on } X \Rightarrow b_{Y.X} = b_1 = \frac{Cov(X, Y)}{Var(X)}$$

$$= \frac{\sum_{i=1}^{n} (X_i - \bar{X})(Y_i - \bar{Y})}{\sum_{i=1}^{n} (X_i - \bar{X})^2} \quad and \quad b_0 = \bar{Y} - b_1 \bar{X}$$

17. Parameters b_0 and b_1 are calculated from the data set with b_0 denoting the intercept, a value of Y when $X = 0$ and b_1 describing the slope of the estimation line and signifying the change in Y for one-unit change in X.

18. The regression method is one way to predict the heritability of traits in plant and animal breeding.

Appendix 6.1

Area under standard normal curve (*z* – scores)

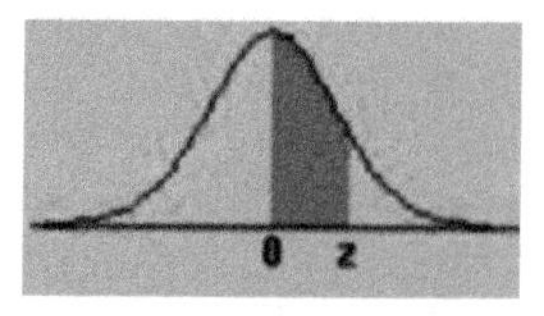

z	.0	.01	.02	.03	.04	.05	.06	.07	.08	.09
0.0	.0000	.0040	.0080	.0120	.0160	.0199	.0239	.0279	.0319	.0359
0.1	.0398	.0438	.0478	.0517	.0557	.0596	.0636	.0675	.0714	.0753
0.2	.0793	.0832	.0871	.0910	.0948	.0987	.1026	.1064	.1103	.1141
0.3	.1179	.1217	.1255	.1293	.1331	.1368	.1406	.1443	.1480	.1517
0.4	.1554	.1591	.1628	.1664	.1700	.1736	.1772	.1808	.1844	.1879
0.5	.1915	.1950	.1985	.2019	.2054	.2088	.2123	.2157	.2190	.2224
0.6	.2257	.2291	.2324	.2357	.2389	.2422	.2454	.2486	.2517	.2549
0.7	.2580	.2611	.2642	.2673	.2704	.2734	.2764	.2794	.2823	.2852
0.8	.2881	.2910	.2939	.2967	.2995	.3023	.3051	.3078	.3106	.3133
0.9	.3159	.3186	.3212	.3238	.3264	.3289	.3315	.3340	.3365	.3389
1.0	.3413	.3438	.3461	.3485	.3508	.3531	.3554	.3577	.3599	.3621
1.1	.3643	.3665	.3686	.3708	.3729	.3749	.3770	.3790	.3810	.3830
1.2	.3849	.3869	.3888	.3907	.3925	.3944	.3962	.3980	.3997	.4015
1.3	.4032	.4049	.4066	.4082	.4099	.4115	.4131	.4147	.4162	.4177
1.4	.4192	.4207	.4222	.4236	.4251	.4265	.4279	.4292	.4306	.4319
1.5	.4332	.4345	.4357	.4370	.4382	.4394	.4406	.4418	.4429	.4441
1.6	.4452	.4463	.4474	.4484	.4495	.4505	.4515	.4525	.4535	.4545
1.7	.4554	.4564	.4573	.4582	.4591	.4599	.4608	.4616	.4625	.4633
1.8	.4641	.4649	.4656	.4664	.4671	.4678	.4686	.4693	.4699	.4706
1.9	.4713	.4719	.4726	.4732	.4738	.4744	.4750	.4756	.4761	.4767
2.0	.4772	.4778	.4783	.4788	.4793	.4798	.4803	.4808	.4812	.4817
2.1	.4821	.4826	.4830	.4834	.4838	.4842	.4846	.4850	.4854	.4857
2.2	.4861	.4864	.4868	.4871	.4875	.4878	.4881	.4884	.4887	.4890
2.3	.4893	.4896	.4898	.4901	.4904	.4906	.4909	.4911	.4913	.4916
2.4	.4918	.4920	.4922	.4925	.4927	.4929	.4931	.4932	.4934	.4936
2.5	.4938	.4940	.4941	.4943	.4945	.4946	.4948	.4949	.4951	.4952
2.6	.4953	.4955	.4956	.4957	.4958	.4960	.4961	.4962	.4963	.4964
2.7	.4965	.4966	.4967	.4968	.4969	.4970	.4971	.4972	.4973	.4974
2.8	.4974	.4975	.4976	.4977	.4977	.4978	.4979	.4979	.4980	.4981
2.9	.4981	.4982	.4982	.4983	.4984	.4984	.4985	.4985	.4986	.4986
3.0	.4987	.4987	.4987	.4988	.4988	.4989	.4989	.4989	.4990	.4990

Further reading

Elston, C. R. and Johnson, D. W. (2008). *Basic Biostatistics for Geneticists and Epidemiologists: A Practical Approach*. John Wiley & Sons, Ltd, Chichester.

McDonald, H. J. (2008). *Handbook of Biological Statistics*. University of Delaware Sparky House Publishing, Baltimore, MA.

Petrie, A and Sarin, C. (2000). *Medical Statistics at a Glance*. Blackwell Science, Inc. Malden, MA.

Swinscow, T. D. V. and Campbell, M. J. (2003). *Statistics at Square One*. 10th edn. BMJ Publishing Group.

References

Bourdon, M. R. (2000). *Understanding Animal Breeding. 2nd Edn*. Prentice Hall. Upper Saddle River, NJ, USA.

Cook, D. R. and Weisberg, S. (1999). *Applied Regression Including Computing and Graphics*. John Wiley & Sons, Inc. New York.

Conner, J. and Hartl, D. (2004). *A Primer of Ecological Genetics*. Sinauer Assoc. Inc. Sunderland, USA.

Pardoe, I. (2006). *Applied Regression Modeling: A Business Approach*. John Wiley & Sons, Inc., Hoboken, NJ.

Review questions

1. Distinguish between the following pairs: target population versus study population, parameter versus statistic, sample versus sampling, covariance versus correlation, and ordinal variable versus nominal variable.

2. Indicate whether each of the following variables is categorical-unordered, categorical-ordered, numerical-amount, or numerical-count. Provide a brief explanation for your answer.

Blood type,
weight,
days in milk,
stage of illness, and
body condition score.

3. Here are meat quality evaluation ratings from a panel study. Construct a pie chart and a bar chart for these data.

9, 7, 5, 3, 5, 4, 7, 6,7, 8, 7, 9, 7, 8, 10, 8, 3, 6, 7, 9, 8, 4, 7, 9, 4, 1, 7, 5, 3, 6, 3, 5, 3, 2, 9, 7, 6

4. Compare the mean, median, and mode of the two sets of data given here and compare the results.

X: 9, 4, 5, 6, 3, 5, 6, 5, 3, 4, 5, 7, 13
Y: 13, 2, 22, 5, 6, 5, 4, 5, 2, 8, 11, 5, 4

5. Indicate three situations when the median and mode would be a better measure of central tendency than the mean.

6. Without doing any calculations, in which sample is the standard deviation smaller? Explain how you reach this conclusion.

Weaning weight: 67.4, 69.3, 73.5, 66.8, 65.8
Grease fleece weight: 1.1, 7.8, 2.4, 5.7, 6.2

7. Use a box and whisker plot to compare the data of 60 day weaning weights for Dorset and Targhee sheep breeds given in Table 6.3.

a. What is the inclusive range?
b. What is the interquartile range?

8. Calculate the measure of dispersion for the data given in Table 6.3 and explain which breed is more variable for 60 day weaning weights.

9. Illustrate a scatterplot for the data given in Table 6.3 and then interpret the plot.

10. The table given here shows 120 mastitis incidents observed from different farms.

Farms	*A*	*B*	*C*	*D*	*E*	*F*	*G*	*H*	*I*	*J*
Number (Frequency) of Mastitis	3	13	–	20	16	14	12	–	9	7
Cumulative frequency of Mastitis	–	–	–	52	–	–	–	–	–	–

a. Fill the blanks in the table.
b. Illustrate the data with pie chart, line, and bar graphs.

11. Suppose that the following figure was drawn for 4800 Angus cattle for normally distributed hot carcass weight (HCW). Use the figure to answer the question that follows:

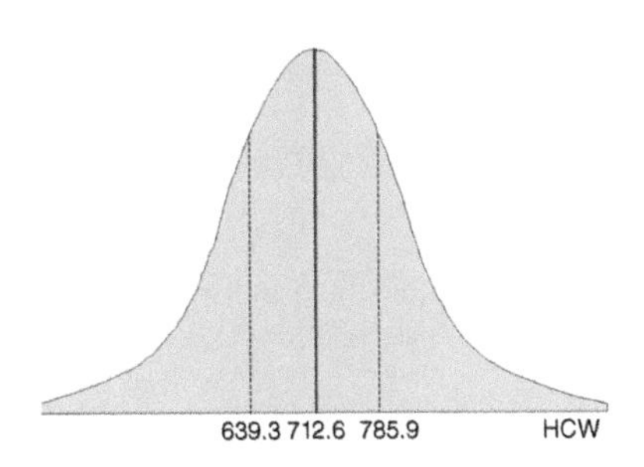

a. What is μ?
b. What is σ?
c. Suppose that the area under the normal curve to the right of $HCW = 756.8$ is 0.7257. Provide two interpretations for this area.
d. Find the z scores for animals with $HCW = 653.4$ and $HCW = 668.8$ and shade the area between these two observations.
e. Find the probability that a randomly selected animal has HCW value between 774.6 and 812.3.
f. Find the number of animals HCW measured between 598.5 and 693.8.

12. Draw a standard normal curve and shade the area to the right of $z = 2.13$. Then find the area of the shaded region.

13. Find the $P(0.36 < z < 1.77)$.

14. Suppose that yearly fourth lactation milk yield for goats has a normal distribution with $\mu = 376.3$ kg and $\sigma = 49.9$ kg.

a. Draw a normal curve based on mean and standard deviation and shade approximately 68% and 95% of the area.
b. Shade the region that represents $P(349.9 \leq X \leq 418.3)$ and find the probability that a randomly chosen animal would fall within this range.

15. Use the data listed here for average daily gain (ADG) and feed conversion rate (FCR) from 15 beef cattle to answer the questions:

ADG	1.71	1.13	1.62	2.02	2.22	1.75	1.89	1.89	1.87	1.44	1.97	2.05	2.27	1.78	1.92
FCR	7.55	5.09	7.46	6.59	6.13	7.07	6.83	6.75	6.95	7.06	6.69	6.29	5.99	7.16	7.46

a. Plot these data. Does it appear that there is a linear relationship between ADG and FCR?
b. Write down the estimated equation for the regression of ADG on FCR.
c. Interpret the intercept and slope.
d. Show the error terms on the graph.
e. Remove the data pair (ADG = 1.13 and FCR = 509) and re-estimate the regression equation. What changes would occur in the b_0 and b_1?
f. Estimate the error term for each data point and calculate the sum of squared error for the regression equation in (b) and (d).
g. What would be the point estimates of ADG when FCR = 4.28 and 9.41.

h. Calculate the covariance and correlation coefficient. Make interpretation for the sign of covariance, correlation and b_1.
i. Convert the correlation coefficient to the slope (b_1) with the formula given in Box 6.9.
j. List all the estimated values of ADG using $E(ADG_i) = \hat{Y}_i = b_0 + b_1 X$ and $E(ADG_i) = \hat{Y}_i = \bar{Y} + b_1(X_i - \bar{X})$.

16. Explain the difference in the purposes of the correlation, covariance, and regression.

17. Interpret the scatterplot given below. What type of associations exists between the five traits?

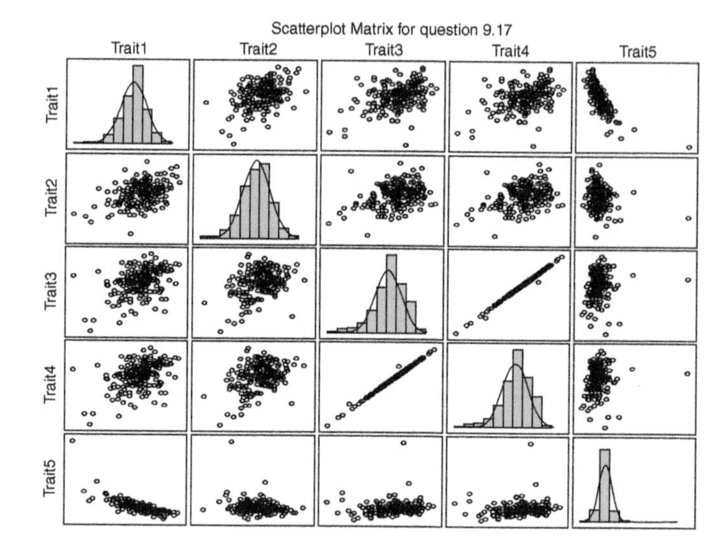

Section 2

Applications of Genetics and Genomics to Livestock and Companion Animal Species

7 Genetic Improvement of Beef Cattle

Michael D. MacNeil
Delta G, Miles City, MT, USA

Introduction

An absolute prerequisite for genetic improvement of any species is defining a quantifiable breeding objective. The breeding objective provides a measureable definition of merit and thus a metric that defines improvement. For the most part, a breeding objective would be composed of several traits, each making some contribution to aggregate merit. The necessity of including multiple traits arises due to the existence of genetic antagonisms among traits. Here, it is assumed that merit can be measured in economic units. Traits having economic value, and thus contributing to aggregate merit, are referred to as **economically relevant traits.**

Genetic improvement results from a process called selection, meaning breeding the "best" individuals or those individuals having the greatest aggregate merit. Selection provides the only opportunity to make improvement in a population that is incremental, cumulative, and permanent. Implementing a selection program depends on a system of genetic evaluation that includes adjustments for extraneous sources of variation. The simplest form of evaluation is ranking individuals based on a single metric; that is, a phenotype for a trait. Predictions of breeding value or genetic merit for individual traits add a level of complexity to the ranking of individuals, but facilitate the fair comparison of many more individuals through systems of national cattle evaluation (NCE). Simultaneous consideration of multiple traits adds another level of complexity in the evaluation, but provides the benefit of facilitating a return to ranking individual candidates for selection on one metric of merit – the aggregate breeding value.

Single trait selection

Response (R) to single trait selection can be estimated by the product of heritability (h^2) and the selection differential (S).

$$R = h^2 S \qquad (7.1)$$

The selection differential measures the deviation of those individuals that become parents from the mean of the population from which they were selected. Heritability is, by definition, the fraction of the parental phenotype that is transmissible to progeny. Thus, response is equivalent to the average breeding value of the parents. If the selection applied to males and females differs, then the selection differential is the mean of the selection differential for males (S_m) and the selection differential for females (S_f). Thus,

$$R = h^2(S_m + S_f)/2 \qquad (7.2)$$

For example: if a group of bulls had an average 365-day weight (corrected for differences in age of dam) of 432 kg (950 pounds) and the bulls chosen for breeding had an average 365-day weight of 550 kg (1100 pounds) then $S_m = 150$ and likewise if the heifers chosen for breeding were on average 20 pounds heavier than the average of the cohort from which they were chosen (i.e., $S_f = 20$), then (given the heritability of 365-day weight ≈ 0.4) the predicted superiority of the progeny relative to the population from which their parents were selected would be:

$$R = \frac{0.4(150+20)}{2} = 34 \qquad (7.3)$$

The selection differential may be partitioned into two components: selection intensity (i) and the phenotypic standard deviation (σ_p). Thus, $i = \frac{S}{\sigma_p}$. Alternatively, i can be estimated from the proportion of animals selected to become parents, assuming the selection criterion is normally distributed and the selection process results in truncation of the normal distribution (Figure 7.1).

Values for i that are applicable to selection from large (technically infinitely large) groups are listed in Table 7.1. Corresponding values for selection from small cohorts are slightly reduced.

Recalling that heritability is the ratio of additive genetic (σ_a^2) and phenotypic (σ_p^2) variance, response to selection can also be estimated by:

$$R = \frac{\sigma_a^2}{\sigma_p^2} i\sigma_p = ih\sigma_a \qquad (7.4)$$

This expression is particularly useful in comparing alternative selection programs. It is noteworthy that response to selection is also, by definition, equivalent to the average **breeding value** of the parents.

Molecular and Quantitative Animal Genetics, First Edition. Edited by Hasan Khatib.

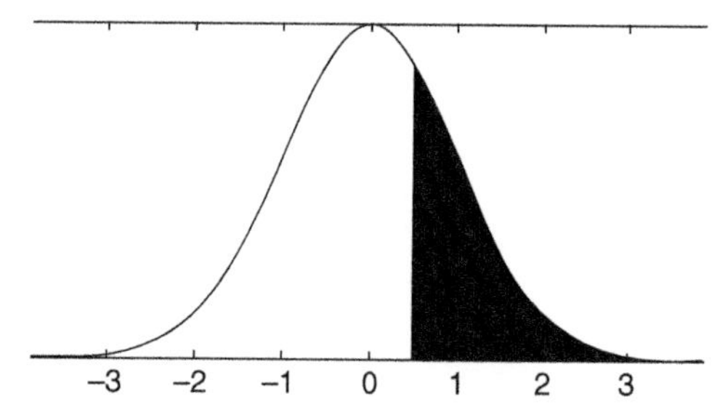

Figure 7.1 Normal distribution truncated at 0.5 SD above the mean of 0. In this example, the proportion selected (shaded region) would be approximately 30% and the selection intensity $i \approx 1.14$ (standard deviation: SD).

Table 7.1 Relationship between the proportion of animals selected (p) and selection intensity (i).

p	i	p	i
1.00	0.00	0.09	1.80
0.90	0.20	0.08	1.85
0.80	0.35	0.07	1.91
0.70	0.50	0.06	1.98
0.60	0.64	0.05	2.06
0.50	0.80	0.04	2.15
0.40	0.97	0.03	2.27
0.30	1.14	0.02	2.42
0.20	1.40	0.01	2.67
0.10	1.76	–	–

To this point, selection response has been measured from parents to offspring or per generation. Considering response relative to time necessitates consideration of generation interval. Generation interval is defined as the average age of the parents at the birth of their selected progeny. Thus, response per year can be predicted as:

$$\frac{R}{yr} = R \bigg/ \left(\frac{L_m + L_f}{2} \right) \tag{7.5}$$

where L_m and L_f are the generational intervals for males and females, respectively. Typically, generation intervals for beef cattle are in the range of 4.5–6 years. Thus, the expected annual progress from selection based on 365-day weight in the example earlier would be 5.7–7.6 lbs. Efficient genetic improvement programs are designed by evaluating the trade-off of selection intensity versus generation interval. For beef cattle, the generation interval for males is typically substantially less than the generation interval for females. The generation interval also can be reduced by culling older cows. However, doing so also causes the herd to be composed of younger and therefore less productive females, and requires allocating more resources to development of replacement heifers.

In the paragraphs that follow, attention is drawn to addressing the question: what is expected to happen to other traits as a consequence of single trait selection? Correlated response (CR) to selection is expected because specific loci affect more than one trait – this phenomenon is called **pleiotropy**. The genetic correlation (r_a) measures the magnitude of the pleiotrophic effect. If selection is applied directly to trait x, the correlated response in trait y can be predicted as:

$$CR_y = ih_x h_y r_a \sigma_{py} \tag{7.6}$$

with the subscripts, x and y = denoting the two traits.

Previously, direct response due to selection on 365-day weight was used as an example, with selection differentials $S_m = 150$ pounds (68 kg) for males and $S_f = 20$ pounds (9 kg) for females. Birth weight has been shown to be genetically correlated with 365-day weight in cattle ($r_a \approx 0.6$) and, like 365-day weight, is fairly highly heritable ($h^2 \approx 0.5$). The selection differential for 365-day weight, which is equivalent to $i\sigma_p$, was 65 pounds (143 kg). Thus, if the phenotype standard deviation of 365-day weight is 100 pounds (45 kg), then the selection intensity would be 0.85 (the average of: 150/100 = 1.5 for males and 20/100 = 0.2 for females). If the phenotypic standard deviation of birth weight is approximately 8 pounds (4 kg), then the expected correlated response in birth weight (y) resulting from direct selection on 365-day weight (x) is predicted to be:

$$CR_y = 0.65 \cdot \sqrt{0.4} \cdot \sqrt{0.5} \cdot 0.6 \cdot 8 \approx 1.4 \text{ lbs}$$

The expected 34-pound (15.5 kg) increase in 365-day weight obtained through direct selection would be expected to also increase birth weight by approximately 1.4 pounds (600 g) as a correlated response.

In presenting examples of predicted direct and correlated responses to selection previously, the desirable increase in growth to a year of age was pitted against the undesirable increase in birth weight, which might cause an increase in calving difficulty and neonatal calf mortality. This is one of the several genetic antagonisms that have potential to compromise production efficiency. Additional genetic antagonisms that could potentially compromise efficient production of beef include the relationships of: (1) birth weight and calving ease versus retail product yield, (2) milk production and cow size versus nutritional requirements of cows, (3) growth versus feed requirements during the post-weaning period, (4) leanness versus age at puberty, and (5) retail product yield versus marbling. Making selection decisions based on simultaneous consideration of multiple traits provides an operational tactic to address these concerns.

By way of a problem to solidify some of the material presented to this point, consider the idea of selecting sires based on marbling with selection intensities $i = 1.76$ and $i = 2.42$ and assess the consequences in terms of retail product percentage. Heritability estimates for marbling score and retail product percentage are both fairly high, 0.52 and 0.50 respectively, indicating selection would be effective in changing either or both of them. The genetic correlation between these two traits has been estimated as being 0.56. Their phenotypic standard deviations are approximately 0.54 and 3.0, respectively.

$$\begin{aligned} \text{Response} = R &= ih^2\sigma_p \\ &= 1.76 \cdot 0.52 \cdot 0.54 = 0.49, \text{ versus} \\ &= 2.42 \cdot 0.52 \cdot 0.54 = 0.68 \end{aligned}$$

marbling score units. Thus, by reducing the fraction of animals selected from the top 10% to the top 2%, response in marbling score would be expected to increase by 10%. The value of this added response depends on the current level of marbling in the population being managed, the break-points in the marbling distribution associated with different premiums and discounts affecting carcass price, and the magnitudes of those premiums and discounts. At what cost, in terms of retail product percentage?

$$\text{Correlated response} = CR_y = ih_x h_y r_a \sigma_{p_y}$$
$$= 1.76 \cdot 0.71 \cdot 0.72 \cdot -0.56 \cdot 3.0 = -1.51, \text{ versus}$$
$$= 2.42 \cdot 0.71 \cdot 0.72 \cdot -0.56 \cdot 3.0 = -2.08$$

percentage points in retail product yield. It would be astute to ask if the more intense selection for marbling is economically justified by the anticipated loss of retail product. Here again, the value of the change in retail product percentage depends on the current level of the population being managed, the break-points in the retail product percentage (or yield grade) distribution associated with different premiums and discounts affecting carcass price, and the magnitudes of those premiums and discounts.

National cattle evaluation

To this point, the description of genetic improvement has been somewhat theoretical and based on simplified conditions. Systems for NCE, the results of which are provided by breed associations in the form of expected progeny differences (EPD), equivalent to one-half of the breeding value, provide a mechanism for making fair comparisons among candidates for selection incorporating a variety of complications. The goal of a NCE is to produce the best possible genetic predictions of breeding values on all animals available as breeding stock for traits of economic importance in commercial beef production (Beef Improvement Federation, 2010). **The foundation of any NCE is in the proper assignment of animals to appropriate contemporary groups.** Information from ancestors, individual performance, performance of collateral relatives, progeny, and genomic testing are all potentially incorporated in a NCE in order to derive the most accurate prediction of an individual's breeding value that is possible. Incorporating information from correlated traits in national cattle evaluation provides opportunities both to increase the accuracy of prediction and reduce selection bias.

In theory, proper formation of contemporary groups is easy; the groups simply identify those individuals with equal opportunity to perform. For example, bull and heifer calves are not equal in their opportunity to perform due to their very different endocrine profiles and are always considered as members of separate contemporary groups. Likewise, individuals offered creep feed have a different opportunity to perform than those without access to a creep and thus these sets of calves should also be placed in separate contemporary groups. However, in reality, there are also challenging decisions that affect formation of contemporary groups. Forming too many contemporary groups compromises the accuracy of the genetic evaluation, while forming too few introduces the possibility of producing biased results. For instance, individuals that receive preferential treatment (e.g., calves fitted for a show) should be placed in their own contemporary group. Also, individuals that experience a significant illness should probably be placed in a contemporary group that is distinct from that of their healthy cohorts.

Table 7.2 Relationship between a candidate for selection and specific relatives, assuming the absence of any inbreeding.

Relative	Relationship	Relative	Relationship
Self	1.00	–	–
Parent	0.50	Grand-parent	0.25
Progeny	0.50	Grand-progeny	0.25
Full-sib	0.50	First cousin	0.125
Half-sib	0.25	Common grand-sire	0.0625

Because organizations performing NCE, such as breed associations, tend to charge for each individual being evaluated, breeders may be tempted to economize by reporting data for only their better performing calves. This temptation is counterproductive on at least two levels. First, the reduced amount of data compromises the accuracy of the NCE for all of the evaluated animals. Second, the selective reporting of better performing individuals creates the opportunity to introduce bias into the NCE. The latter problem, referred to as **selection bias**, can be at least partially overcome by conducting the NCE for multiple traits simultaneously (discussed later). Absent an appropriate multiple trait evaluation that includes all data from the traits upon which the selection decisions were based, selective reporting undervalues the reported calves relative to genetically similar calves from herds that reported all of the data.

Related individuals, by definition, share a fraction of their alleles. Thus, all relatives can contribute to the genetic evaluation of candidates for selection. The degree to which each relative contributes to the evaluation of any individual is proportional to the genetic relationship between them (Table 7.2).

Obviously, it is possible for an individual candidate for selection to have more than one sib and/or progeny. In this case, the value of the sib and progeny information is further increased. As a simple example, the breeding value for an individual can be estimated as (an EBV) based on n progeny by:

$$EBV = \frac{0.5nh^2}{1+(n-1)0.25h^2}\bar{x} \tag{7.7}$$

where, $\bar{x}$ is the average of single records (expressed as deviations from their contemporary group means and adjusted for other known contributing influences) from the n progeny all from different dams. The numerical coefficients in this equation arise from the relationship between parent and offspring. In a NCE, all degrees of relationship inferred from a comprehensive pedigree of the population being evaluated are considered simultaneously.

Including correlated traits in a NCE can be beneficial on at least three levels. First: genomic tests that amalgamate the value of alternative alleles at a great many loci contributing to a phenotype, can be incorporated into an NCE as though the genomically derived predictions of breeding value was a correlated trait. This use of both quantitative data and genomic information together avoids the conflicts that invariably arise when these sources of information are used separately. Second: including information from correlated traits in the same NCE increases the accuracy of the resulting EPD relative to that which would result if the traits were analyzed separately. For instance, intramuscular fat content in the longissimus is genetically correlated with the economically relevant trait of marbling. Including records of intramuscular fat content in the longissimus, obtained using ultrasound, can add appreciable accuracy to a NCE for trait marbling. Third: use of correlated traits in NCE can help to overcome selection bias. For instance, the data for a genetic evaluation of feed intake during the postweaning period is costly to collect. Therefore, seedstock producers are inclined to test only those candidates they believe to be "best," as possibly reflected by their having the heaviest weaning weights amongst their contemporaries. Because feed intake and body weight are highly correlated, if the feed intake NCE were conducted as a single trait evaluation the restricted sample of calves being tested would bias the comparison of the candidates for selection. However, by also including weaning weights in the NCE simultaneously, the bias that results from the selection of calves with recorded feed intake would be taken into account.

In the genetic improvement of beef cattle there is concern for two general categories of traits: those mediated by the genotype of the candidate for selection (called direct effects) and those mediated by the genotype of the candidate's dam (called maternal effects). In case of maternal effects, the genotype of the dam causes her to express a phenotype that environmentally changes expression of the ERT by the candidate for selection. Examples of traits subject to both direct and maternal effects include: (1) birth weight wherein genes of the candidate for selection affect its propensity growth *in utero* and genes of the dam affect blood flow to her uterus and thus the supply of nutrients to the developing fetus; and (2) preweaning gain wherein genes of the candidate for selection affect its propensity growth and genes of the dam affect her milk production and thus the supply of nutrients to the candidate being evaluated.

Commonly available EPD include measures for: birth weight, 205-day weight (direct = d) and 205-day weight (maternal = m, also called milk), 365-day weight, mature weight, calving ease (d), and calving ease (m), marbling score, longissimus muscle (ribeye) area, and subcutaneous fat depth. The suite of EPD consisting of birth weight, 205-day weight (d and m), and 365-day weight often is derived from a single analysis of birth weight, average daily gain (ADG) from birth to weaning, and average daily gain from weaning to yearling. Birth weight and preweaning ADG are typically modeled as having direct and maternal components making this effectively one simultaneous analysis of five traits. The 205-day weight direct EPD (WW_d) would be calculated, after-the-fact, as:

$$WW_d = BW_d + 205G1_d \quad (7.8)$$

where, BW_d is the birth weight direct EPD and $G1_d$ is the EPD for preweaning average daily gain direct. A WWm or "milk" EPD is calculated in a like manner. A value for the "total maternal" EPD (TM) is also derived from this analysis as:

$$TM = \frac{1}{2}WW_d + WW_m \quad (7.9)$$

and reflects the expected performance of the evaluated females' progeny. The 365-day weight EPD (YW) is calculated as:

$$YW = WW_d + 160G2 \quad (7.10)$$

where, $G2$ is the EPD for postweaning average daily gain. Because calving ease is significantly correlated with both birth weight and size of the female at calving, calving ease EPDs are reasonably calculated together with the EPDs for the weight-traits. It is noteworthy that when approached in this way, there is no apparently economic relevance attached to the birth weight EPD because: (1) its contribution to sale weights has been taken into account in the EPD for 205-day and 365-day weights, and (2) all of the information available in the birth weight data is considered in the calving ease analysis negating its further consideration when attempting to reduce dystocia. A practical consequence of this is that the practice of setting a threshold maximum for the birth weight phenotype actually compromises genetic progress. The analyses necessary to obtain EPD for the carcass traits also consider multiple traits simultaneously. The typical assumption is that carcass characteristics of steer progeny define this suite of economically relevant traits. Indicators of carcass merit obtained from candidates for selection using ultrasound and genomic predictions are incorporated into the analyses as correlated traits. A measure of weight should be included in the analysis of longissimus muscle area to increase the accuracy of its genetic prediction due to the large correlation between muscle area and carcass weight (or 365-day weight). As with the calving ease EPD, the EPD calculated for the indicator traits and the phenotypic measurements themselves are of no further use in making selection decisions.

Additional EPD reported by some breed associations include measures for: heifer pregnancy or calving, stayability or longevity, docility, scrotal circumference, maintenance energy requirement, postweaning feed intake, and carcass weight. Fitness traits (e.g., heifer pregnancy, stayability, longevity) typically have relatively high economic value and relatively low heritability suggesting that the improvement in fitness traits may be important, but slow. Scrotal circumference, measured at approximately one year of age in bulls, in and of itself has no economic relevance. However, due to its association with age at puberty in heifers and the relationship between age at puberty and the probability a heifer becomes pregnant, it is typically included as a correlated trait in the genetic evaluation of heifer pregnancy. Yearling weight may also be included in this analysis to account for selection bias resulting from breeders choosing to cull lighter weight heifers and only retain their heavier contemporaries for breeding. Stayability measures the probability that a heifer calving at 2 years of age reaches maturity – typically 5 years of age. Related to stayability is longevity, expressed in units of time, which indicates the age at which a female is culled. The EPD for maintenance energy requirement and postweaning feed intake reflect genetic differences associated with cost of feed for the production system. The maintenance energy EPD is unique among all EPDs currently reported in that there is no data recorded for the maintenance energy requirements of cows directly. Rather this EPD is predicted from the EPD for mature weight, maternal weaning weight, and sometimes body condition score. The postweaning feed intake EPD is derived from the simultaneous analysis of a limited number of postweaning feed intake measurements, and EPD for body weight, postweaning average daily gain, and fat depth. This suite of EPD is frequently transformed to produce an EPD for residual gain or residual feed intake, traits which are indicative of feed efficiency.

The accuracy of any EPD depends on heritability of the trait, correlations with other traits included in the evaluation, number of records, relationships among animals with records, and the distribution of the information across herds. Accuracy of EPD does not affect the ranking of candidates for selection. The best use for accuracy values is in the management of risk because the accuracy reflects the degree to which an EPD might change in the future. For example, in selecting sires to breed heifers where avoiding calving difficulty is a goal, the evaluated sires would first be ranked according to the EPD for direct calving ease. Thereafter, the accuracy of the most favorably ranked sires would be considered (greater accuracy is more favorable) in order to select the sires whose calving ease EPD would be least likely to exceed a specific threshold. For example, consider choosing a Hereford bull for use in breeding heifers. The selection criterion might be stated as identifying a bull with a calving ease EPD that was certainly greater than +5%. All EPD have error associated with their prediction. In this case, the calving ease EPD for Herefords has a possible change of 4% associated with an accuracy of 0.80. Thus, the search would be narrowed to include bulls with an EPD of +9% and an accuracy of 0.80. The threshold level of accuracy also would be reduced for bulls with EPDs greater than +9% and made more stringent as the EPD of bulls being considered approached +5%.

Multiple trait selection

The concept of a selection index as a vehicle to facilitate multiple trait selection was pioneered by Hazel and Lush in the 1940s (Hazel, 1943). More recent developments provide a mechanism to construct a selection index, as a tool for multiple trait selection, based on the EPD directly. Multiple trait selection is predicated on a defined breeding objective wherein aggregate merit is defined by the traits included in it. Often breeders have some implicit objective in mind when making selection decisions. However, the lack of an explicit objective compromises the rate of progress toward their goal.

Implementing a breeding objective is the outcome of a four-step process: First, a mathematical description of the relationships between biological characteristics of a commercial production system and sources of income and expense are explicitly modeled. This goal of this model is to predict profitability of the commercial production system. The tool used (i.e., spreadsheet, programming language, etc.) for this modeling is not terribly important, except that it should facilitate step two. In the second step, the model is exercised to estimate partial derivatives of the profit function with respect to each of the biological traits included in the model. These partial derivatives are the economic values for each of the traits. Those biological traits have non-zero economic values are referred to as economically relevant traits (ERT). The ERT traits may or may not have EPD. For those ERT that do not have EPD, EPD can be predicted for them in a manner similar to that used to predict the maintenance energy EPD from indicator traits. This will require knowledge of the phenotypic variances and heritabilities of the EPD and the ERT, as well as the genetic correlations among them. Finally, the weighted sum of the products of the EPD for the ERT and their economic values (w_i):

$$I = \sum w_i \cdot EPD_i \qquad (7.11)$$

is calculated as the univariate metric of aggregate merit upon which to base selection decisions. A similar process has been used to produce many of the $indexes that are available from the breed associations.

Breed associations have produced $indexes as practical guides to breeders in the weighting of multiple traits as a basis for selection decisions. Depending on the breed, these $indexes address different breeding objectives. For example, several breeds have $indexes for selection of terminal sires – the progeny of which are all to be harvested. As a counterpoint to the terminal sire $indexes, breeds also produce $indexes for use in selection of maternal strains. Traits like cow size, milk production, maternal calving ease, and stayability or longevity are of negligible value in the $indexes for terminal sires. However, these traits are potentially quite important in the $indexes for maternal strains. From an application perspective, it is important to know the ERT that have been included in a $index. Considering an EPD indicative of an ERT that was included in the $index in making a selection decision would likely lead to that ERT being over-emphasized. For traits that are believed to be economically relevant but that are not included in the $index, an EPD would first need to be developed, perhaps even on a within contemporary group basis, and then this EPD could be given an economic weight calculated on a scale that was consistent with the economic weights for the other ERT. If new EPD development seems too onerous, then a pragmatic approach might be to select (say 20%) more animals based in the $index than are actually required for perpetuating the breeding program and to use the additional criteria as independent culling levels to identify the actual replacements. However, this pragmatic approach would lead to some reduction in the response to selection.

Summary

Presented here in the context of beef production, is theory that underlies genetic improvement in terms of direct and correlated responses to selection. Then considerations that impinge on successful systems of national cattle evaluation are discussed, followed by some consideration of the statistics produced as aides to the current evaluation of candidates for selection. Finally, the concept of a profit-motivated metric is put forth as the basis for genetic improvement in beef cattle.

Further reading

Beef Improvement Federation. 2010. *Guidelines for Uniform Beef Improvement Programs*, 9th edn. L. V. Cundiff, L. D. Van Vleck and W. D. Hohenboken (eds). Beef Improvement Federation, Raleigh, NC.

Bourdon, R. M. 2000. *Understanding Animal Breeding*, 2nd edn. Prentice Hall, Inc., Englewood Cliffs, NJ.

Falconer, D. S. and T. F. C. MacKay. 1996. *Introduction to Quantitative Genetics*, 4th edn. Hazel, L. N., G. E. Dickerson, and A. E. Freeman. 1994. The selection index – then, now, and for the future. *J. Dairy Sci.* 77:3236–3251.

Henderson, C. R. 1963. Selection index and expected genetic advance. In: *National Research Council. Statistical genetics and plant breeding: A symposium and workshop*. Committee on Plant Breeding and Genetics, University of North Carolina. State College of Agriculture and Engineering, Raleigh. Pp. 141–163.

Falconer, D. S. and T. F. C. MacKay. 1996. *Introduction to Quantitative Genetics*, 4th edn. Pearson Education Ltd., Essex.

Gregory, K. E., L. V. Cundiff, R. M. Koch, M. E. Dikeman and M. Koohmaraie. 1994. Breed effects, retained heterosis, and estimates of genetic and phenotypic parameters for carcass and meat traits of beef cattle. *J. Anim. Sci.* 72:1174–1183.

MacNeil, M. D. 2005. Breeding objectives for terminal sires for use in US beef production systems. *Proc. Beef Improvement Fed.* 37:82–87.

McClintock, A. E. and E. P. Cunningham. 1974. Selection in dual purpose cattle populations: Defining the breeding objective. *Anim. Prod.* 188:237–247.

Newman, S., C. A. Morris, R. L. Baker, and G. B. Nicoll. 1992. Genetic improvement of beef cattle in New Zealand: breeding objectives. *Livestk. Prod. Sci.* 32:111–130.

Ponzoni, R. W. and S. Newman. 1989. Developing breeding objectives for Australian beef cattle production. *Anim. Prod.* 49:35–47.

Simm, G. 1998. *Genetic Improvement of Cattle and Sheep.* CABI Publishing, Oxfordshire.

References

Beef Improvement Federation. 2010. *Guidelines for Uniform Beef Improvement Programs*, 9th edn. L. V. Cundiff, L. D. Van Vleck and W. D. Hohenboken (eds). Beef Improvement Federation, Raleigh, NC.

Hazel, L. N. 1943. The genetic basis for constructing selection indexes. *Genetics* 28:476–490.

8 Genetic Improvement in Sheep through Selection

David L. Thomas

Department of Animal Sciences, University of Wisconsin–Madison, WI, USA

Products from sheep

Most sheep flocks are raised mainly for meat. Wool also may be an important product, especially under extensive conditions where the environment does not allow high lamb production. Approximately 10% of the world sheep population is composed of hair sheep that do not produce wool, so meat production is definitely of primary importance in these populations. Sheep milk is the most valuable commodity produced by the specialized dairy sheep breeds. Dairy sheep populations are concentrated in southern Europe, North Africa, and the Middle East, and they are producers of all three products (milk, meat, and wool; in descending order of importance).

Selection among breeds

Mason's World Dictionary of Livestock Breeds, Types and Varieties (5th edn, Porter, 2002) has 104 pages devoted to a listing of over 1500 different world sheep breeds. There are over 50 recognized breeds of sheep in the US, and these breeds differ greatly in their genetic value and phenotypic performance for economically important traits.

The US breeds of sheep can be broadly classified into the following groups:

Maternal breeds: moderate body size, high reproductive rate, good mothering ability, used primarily as ewes in farm flocks where there are abundant feed resources (example breeds: Polypay, Dorset).

Range/wool breeds: moderate body size, heavy fleeces of high quality, adapted to arid environments, used primarily as ewes in range flocks in the western US where feed may be of low quality during the grazing season (example breeds: Rambouillet, Targhee).

Hair breeds: small to moderate body size, body covered in hair instead of wool, more tolerant of internal parasites than other breeds, used primarily as ewes in flocks in hotter and more humid environments and as "easy care" ewes in both range and farm flocks (example breeds: Katahdin, Dorper).

Dairy breeds: moderate to large body size, high commercial milk production, used in intensively-managed dairy flocks with abundant feed resources (example breeds: East Friesian, Lacaune).

Terminal/sire breeds: moderate to large body size, fast growth rate, lean, and well-muscled carcasses, used primarily as rams in crossbreeding with ewes of the maternal, range, hair, and dairy breeds to produce high-quality market lambs (example breeds: Suffolk, Hampshire, Texel).

A picture of a representative breed of each of these five classes is presented in Figure 8.1.

Sheep producers make their initial genetic gain by selecting from among available breeds the particular breed or breeds that have a superior genetic value for the traits that are important for efficient sheep production in their environment.

Selection within a breed or population and the Key Equation

Once a breed is selected, further genetic improvement comes from selection within the producer's flock and within the flocks from which replacement animals, primarily rams, are purchased. Progress from selection (change in breeding value per year) in a population is determined by the **Key Equation**:

$$\Delta\frac{BV}{Year} = \frac{Accuracy\ of\ selection \times Selection\ intensity \times Genetic\ variation}{Generation\ Interval\,(years)} = \frac{r_{BV,\widehat{BV}} \times i \times \sigma_{BV}}{L} \tag{8.1}$$

where:

$r_{BV,\widehat{BV}}$ = correlation between true breeding value and estimated breeding value,

i = difference between the average of the selected animals and the average of all animals available for selection in standard deviation units,

σ_{BV} = standard deviation of breeding value, and

L = the number of years it takes to replace one generation with the next.

Adjustment for environmental effects

Almost every trait of a sheep that is of economic importance is influenced by both the genetic value of the animal and the effects

Molecular and Quantitative Animal Genetics, First Edition. Edited by Hasan Khatib.

Figure 8.1 Representative breeds of sheep in the US of the five general classes. From *left* to right, **top row**: maternal – Polypay, range/wool – Targhee, hair – Katahdin. From *left* to *right*, **bottom row**: dairy – East Friesian, Terminal/sire – Hampshire. (Photos courtesy of David L. Thomas. All photographs from Department of Animal Sciences, University of Wisconsin–Madison.)

of the specific environment (non-genetic effects) of the animal (See the Basic Genetic Model in Chapter 4). Since we wish to select replacement animals that have high levels of performance due to high genetic values rather than due to superior environments, phenotypic traits should be adjusted for known environmental effects before selection is practiced. Selection on phenotypes adjusted for environmental effects rather than raw phenotypes improves the progress from selection by increasing the accuracy of selection in the Key Equation.

Raw phenotypic measurements can be adjusted for environmental effects by using statistical models that include both genetic and possible environmental effects. However, such models require very large data sets in order to provide accurate estimates of environmental effects. Standard adjustment factors generated from research are often used to adjust raw phenotypic records for known environmental effects.

In farm flocks, lambs are commonly weaned at 30, 60, or 90 days of age. Lambs in range flocks are more often weaned at 90 or 120 days of age. Non-genetic or environmental factors that are known to affect lamb weaning weight are age of lamb at weaning, sex of lamb, type of birth and rearing of the lamb, and age of dam. Table 8.1 presents multiplicative adjustment factors for adjusting age-corrected lamb pre-weaning and weaning weights to a common sex-type of birth and rearing-age of dam basis. The factors adjust weights of lambs to a ewe lamb, single born and raised, and 3–6 year old dam basis. For example, the ewe age adjustment factors indicate that a lamb born to a 2-year-old ewe is expected to weigh 8% less at weaning than a lamb born to a 4-year-old ewe, largely because younger ewes produce less milk than mature ewes. Therefore, to compare the weaning weights of the two lambs fairly, the lamb from the 2-year-old ewe must have 8% added to its weaning weight.

Lambs are generally weaned in groups when the average age of a group is approximately the target weaning age, for example, 60 days of age. Even within this group of lambs of similar age, there will be lambs both younger and older than 60 days of age, and the older lambs are expected to be heavier than the younger lambs. Before the factors in Table 8.1 are applied, the weaning weights need to be corrected for differences in age at weaning. The following formula provides a 60-day corrected weaning weight:

Table 8.1 Lamb preweaning and weaning weight adjustment factors.

Item: Class	Multiplicative adjustment	Item: Class	Multiplicative adjustment
Ewe age, years:		Type of birth – rearing:	
1	1.14	Single – Single	1.00
2	1.08	Single – Twin	1.17
3–6	1.00	Twin – Single	1.11
>6	1.05	Twin – Twin	1.21
Sex:		≥Triplet – Single	1.19
Ram	.91	≥Triplet – Twin	1.29
Wether	.97	≥Triplet – ≥Triplet	1.36
Ewe	1.00		

Source: *The SID Sheep Production Handbook*. 2003. 2002 edn. Vol. 7. American Sheep Industry Association, Inc. ISBN 0-9742857-0-6. pp. 47 and 52.

$$60\ \textit{day corrected weight} = \left(\frac{\textit{Actual weaning weight} - \textit{birth weight}}{\textit{Age at weaning, days}} \times 60\right) + \textit{birth weight} \quad (8.2)$$

The formula in equation (8.2) is easily modified to accommodate flocks that have target weaning ages younger or older than 60 days by substituting a different target age for 60.

The adjusted weaning weight is obtained by multiplying the age corrected weaning weight from the formula in (8.2) by the appropriate adjustment factors in Table 8.1.

$$60\ day\ adjusted\ weight = 60\ day\ corrected\ weight \times ewe\ age\ adjustment\ factor \times sex\ of\ lamb\ adjustment\ factor \times type\ of\ birth - rearing\ adjustment\ factor \quad (8.3)$$

An example calculation of 60 day adjusted weaning weights for two lambs is presented in Table 8.2.

If a selection decision was to be made between lamb A or lamb B on the basis of the individual's adjusted phenotype for weaning weight, lamb B should be selected. Lamb B had a lower raw weaning weight than lamb A (52 vs 70 lb or 24 vs 32 kg), but this was because lamb A had a more favorable environment (i.e., older when weighed, from a mature dam, and raised as a single) compared to the less favorable environment of lamb B (i.e., younger when weighed, younger dam, and raised as a twin). If both lambs had been weaned at the same age, were of the same type of birth and rearing, and were from the same age of dam, lamb B would be expected to have a heavier weaning weight than lamb A. If these adjusted 60-day weights were the only performance information available on these lambs, lamb B is expected to be genetically superior to lamb A.

Ewe age has an important non-genetic effect on both fleece weight and litter size, and industry adjustment factors also have been developed to adjust for age of ewe when comparing phenotypic records from ewes of different ages for these two traits (Table 8.3).

Phenotypic selection

In the special case of phenotypic selection where individuals are selected on the basis of a single record on themselves, the Key Equation can be simplified to:

Table 8.2 Characteristics of two lambs and calculation of corrected and adjusted 60-day weaning weights (all in lb).

	Lamb A	Lamb B
Birth weight, lb.	12.0	9.5
Weaning age, days	66	54
Weaning weight, lb.	70	52
Dam age, years	4	2
Sex	Ram	Ram
Type of birth – rearing	Twin – Single	Twin – Twin
60 day corrected weaning weight, lb[a]	64.7	56.7
60 day adjusted weaning weight, lb[b]	65.3	67.4

[a]Calculations for 60 day corrected weaning weight:
For Lamb A: (((70 − 12)/66) x 60) + 12 = 64.7 lb.
For Lamb B: (((52 − 9.5)/54) x 60) + 9.5 = 56.7 lb.
[b]Calculations for 60 day adjusted weaning weight using adjustment factors in Table 8.1:
For Lamb A: 64.7 × 1.00 × .91 × 1.11 = 65.3.
For Lamb B: 56.7 × 1.08 × .91 × 1.21 = 67.4.

$$\Delta\frac{BV}{Year} = \frac{Heritability \times Selection\ differential}{Generation\ Interval\ (years)} = \frac{h^2 \times S}{L} \quad (8.4)$$

Heritability (h^2) encompasses the concepts of accuracy of selection and genetic variation from the Key Equation and is an important parameter for determining the amount of progress possible from selection. Table 8.4 presents the estimates of heritability for several economically important sheep traits.

Ewe reproduction (fertility and prolificacy) and lamb survival are lowly heritable (0.05–0.10), early growth traits are lowly to moderately heritable (0.10–0.25), and later weights and carcass, fleece, and dairy traits tend to be moderately to highly heritable (0.25–0.55).

A definition of heritability is the regression of breeding value on phenotypic value, that is, the change in breeding value expected from a 1 unit change in phenotypic value. Therefore, the phenotypic value of an individual for a trait with a very high heritability is a good indicator of the individual's breeding value, whereas the phenotypic value of an individual for a trait with a low heritability is a poor indicator of the individual's breeding value.

In formula 8.4, S = selection differential, which is the difference between the average phenotypic performance of the selected animals and the average phenotypic performance of all animals available for selection.

Estimated breeding values (EBV)

The regression definition of heritability is useful for understanding how heritability is used to estimate breeding value. The simplest estimate of an individual's breeding value is given by the following formula:

Table 8.3 Age of ewe adjustment factors for fleece weight and litter size.[a]

Ewe age, years	Multiplicative adjustment factors	
	Fleece weight lb	Litter size
1	1.10	1.48
2	1.07	1.17
3	1.02	1.05
4	1.00	1.01
5	1.00	1.00
6	1.03	1.00
7	1.03	1.02
8	1.05	1.05
≥9	1.05	1.13

365 day corrected fleece weight = (actual fleece weight/days since last shearing) × 365
365 day adjusted fleece weight = 365 day corrected fleece weight × age of ewe adjustment factor
Adjusted litter size = actual litter size × age of ewe adjustment factor
[a]Source: *The SID Sheep Production Handbook*. 2003. 2002 edn. Vol. 7. American Sheep Industry Association, Inc. ISBN 0-9742857-0-6. pp. 47 and 52.

Table 8.4 Heritability estimates for several sheep traits.[a]

Trait	Heritability
Reproduction:	
Ewe fertility	0.05[b]
Prolificacy[c]	0.10
Scrotal circumference	0.35
Age at puberty	0.25
Lamb survival[d]	0.05
Ewe productivity[e]	0.15
Growth:	
Birth weight	0.15
60-day weight	0.10
90-day weight	0.15
120-day weight	0.20
240-day weight	0.40
Preweaning gain: birth–60 days	0.15
Postweaning gain: 60–120 days	0.25
Carcass:	
Carcass weight	0.35
Weight of trimmed retail cuts	0.45
Percent trimmed retail cuts	0.40
Loin eye area or depth	0.35
12th rib fat thickness	0.30
Dressing percent	0.10
Fleece:	
Grease fleece weight	0.35
Clean fleece weight	0.25
Yield (%)	0.40
Staple length	0.55
Fiber diameter	0.40
Crimp	0.45
Color	0.45
Dairy:	
Milk yield	0.30
Fat yield	0.30
Protein yield	0.30
Fat percentage	0.35
Protein percentage	0.45

[a]From: *The SID Sheep Production Handbook*. 2003. 2002 edn. Vol. 7. American Sheep Industry Association, Inc. ISBN 0-9742857-0-6. pp. 47 and 52.with slight modifications.
[b]May increase to 0.10 in ewe lambs, in ewes lambed in the fall, and in ewes lambed in the spring in flocks with low fertility.
[c]Lambs born per ewe lambing.
[d]May increase to 0.10 in flocks with low lamb survival.
[e]Pounds of lamb weaned per ewe exposed.

$$EBV = h^2(I - CGA) \quad (8.5)$$

where:
EBV = estimated breeding value
h^2 = heritability
I = single phenotypic record on the individual
CGA = contemporary group average (average of single phenotypic records on all animals in the same contemporary group as the individual).

EBVs are positive or negative deviations from the contemporary group average. A sheep with a positive EBV is predicted to have an above average breeding value, whereas a sheep with a negative breeding value is expected to have a below average breeding value. The accuracy of the EBV from a single phenotypic record on the individual is the square root of h^2 or h.

Using multiple sources of information

More accurate EBVs are obtained by combining phenotypic records from multiple sources. Some traits are repeated on an individual. Examples of repeated traits are number of lambs born to a ewe each year and annual fleece weight of a ewe. The average performance of repeated records results in a more accurate EBV than a single record.

Since an individual shares genes with its relatives, phenotypic records on relatives can also provide additional information to increase the accuracy of the EBV of the individual. The general equation for an EBV of an individual for a particular trait is:

$$EBV = b_1(x_1 - CGA_1) + b_2(x_2 - CGA_2) + b_3(x_3 - CGA_3) + \ldots + b_n(x_n - CGA_n) \quad (8.6)$$

where:
EBV = estimated breeding value
b = weighting factor for the information, b = heritability if the **only** source of information is a single record on the individual
x = performance of the individual or relatives
CGA = average of the contemporary group of the individual or relatives.

Calculation of EBVs from multiple sources of information and the accuracies of these EBVs are complicated and generally done using computers with sophisticated statistical genetics software.

Table 8.5 presents the weighting factors used in the calculation of EBVs from some multiple sources of information for the lowly heritable trait of litter size and the highly heritable trait of fleece weight and the accuracies of these EBVs.

Several points can be made from the example presented in Table 8.5:

1. The accuracy of an EBV for a lowly heritable trait is less than the accuracy of an EBV for a highly heritable trait when the same amount and source of information is used.
2. The accuracy of the EBV increases as more information is obtained.
3. Additional information is more important for lowly heritable than highly heritable traits. In this example, the accuracy of the EBV for litter size increased by 75% when going from the least to the most information (0.32 to 0.56), but the EBV for fleece weight increased by only 30% when going from the least to the most information (0.63 to 0.82).

Genetic correlations

If the phenotypes of two traits are influenced by some of the same genes (pleiotropy), the traits are genetically correlated. Genetic correlations can vary from −1.0 to +1.0. A positive genetic correlation between two traits indicates that the genes common to the two traits influence the traits in the same direction, that is, if an individual has genes that increase trait 1, those same genes will tend to increase trait 2. A negative genetic correlation between two traits indicates that the genes common to the two traits influence the traits in the opposite direction, that

Table 8.5 Weighting factors (b) for different sources of information in calculation of EBV for litter size and fleece weight and EBV accuracies.

Trait	Source of information[a]	Weighting factors for calculation of EBV				Accuracy of EBV
		Ind (1)	Ind (3)	5 Half-sibs	5 Progeny	
Litter size	Ind (1)	0.10	–	–	–	0.32
	Ind (3)	–	0.23	–	–	0.48
	Ind (3) + 5 Half-sibs	–	0.23	0.09	–	0.50
	Ind (3) + 5 Half-sibs + 5 Progeny	–	0.21	0.08	0.18	0.56
Fleece weight	Ind (1)	0.40	–	–	–	0.63
	Ind (3)	–	0.60	–	–	0.77
	Ind (3) + 5 Half-sibs	–	0.58	0.15	–	0.78
	Ind (3) + 5 Half-sibs + 5 Progeny	–	0.48	0.12	0.35	0.82

[a]Ind (1) = a single record on the ewe, Ind (3) = the average of three records on the ewe, 5 Half-sibs = the average of single records of 5 half-sisters of the ewe, 5 Progeny = the average of single records of 5 daughters of the ewe.

Table 8.6 Estimates of genetic correlations among important fleece, body weight, and reproduction traits in sheep.[a]

	Fiber diameter	Staple length	Birth weight	Weaning weight	Post weaning weight	Litter size[b]	Number of lambs weaned[c]
Grease fleece weight	0.35	0.45	0.20	0.25	0.35	−0.10	−0.10
Fiber diameter	–	0.20	0.20	0.05	0.20	0.05	0.00
Staple length	–	–	0.05	0.15	0.15	0.00	−.30
Birth weight	–	–	–	0.45	0.30	0.10	−0.05
Weaning weight	–	–	–	–	0.85	0.20	0.05
Postweaning weight	–	–	–	–	–	0.20	0.30
Litter size[b]	–	–	–	–	–	–	0.55

[a]From Safari et al., 2005 but rounded to the nearest 0.05.
[b]Average of pooled estimated genetic correlations with litter size/ewe lambing and litter size/ewe joined.
[c]Average of pooled estimated genetic correlations with number of lambs weaned/ewe lambing and number of lambs weaned/ewe joined.

is, if an individual has genes that increase trait 1, those same genes will tend to decrease trait 2. A genetic correlation of zero, or close to zero, indicates that a different set of genes are influencing the two traits.

Table 8.6 presents estimates of genetic correlations between pairs of economically important sheep traits summarized from the world literature (Safari et al., 2005). Many of these genetic correlations were estimated from fewer than five studies. While they are the best estimates available, there is still considerable uncertainty of the true value of many of these genetic correlations.

Among the fleece traits, grease fleece weight and staple length have the highest genetic correlation (0.45). Longer fleece fibers contribute directly to greater fleece weight so it makes sense that many genes that influence staple length would also influence fleece weight. This is a desirable genetic correlation. Sheep producers desire heavier fleece weights so there is a greater quantity of wool to sell, and processors prefer wool of greater length due to greater versatility of its use. Therefore, direct selection at the farm or ranch for grease fleece weight, which is easy to measure, is expected to result in an indirect increase in staple length.

Grease fleece weight and fiber diameter also are positively correlated (0.35). This again makes sense since individual fleece fibers of greater diameter would contribute directly to increased fleece weight. However, this is an undesirable genetic correlation. Smaller fiber diameters are associated with higher grades of wool. Wools of a higher grade are worth more per pound on the market because they are softer and less scratchy, can be spun into strong yarns with smaller diameters, and are used in high-value garments such as men's and women's suits. Single trait selection on grease fleece weight will result in an increase in fiber diameter and a decrease in the per pound value of the wool. Selection indexes can be developed that can minimize an increase in fiber diameter as fleece weight increases or that can minimize a decrease in fleece weight as fiber diameter decreases.

Even though the genetic correlation between fleece weight and fiber diameter is undesirable and relatively high, it is still much below 1.0. Therefore, individual sheep in a population that have positive EBVs for fleece weight and negative EBVs for fiber diameter can be found. These are the animals, especially rams, that producers interested in maximizing income from wool should be seeking.

The three fleece traits are positively and lowly to moderately correlated with the three body weights (0.05 to 0.35). Since heavy body weights in lambs are desired in order to have more weight of lamb to sell, the positive genetic correlations between the wool traits of fleece weight and staple length and the weight of lambs at weaning and postweaning are very desirable. Direct selection for heavy body weights in lambs is expected to indirectly increase fleece weights and staple lengths. The positive genetic correlation between body weights and fiber diameter is undesirable.

The three fleece traits have low positive (0.05) to moderate negative (−0.30) genetic correlations with the reproductive traits of number of lambs born (litter size) or number of lambs weaned. However, since five of the six correlations had absolute values of 0.10 or less, direct selection for fleece traits is not expected to have large indirect negative effects on these reproductive traits.

The three body weight traits were moderately to highly correlated (0.30 to 0.85) among themselves, and adjacent body weights (birth weight and weaning weight, weaning weight and postweaning weight) had higher genetic correlations than nonadjacent body weights (birth weight and postweaning weight). The genetic correlations between body weights and reproductive traits were either close to zero or low to moderately positive (−0.05 to 0.30) suggesting that direct selection for increased body weight will generally result in an indirect small to moderate increase in reproductive performance.

Selection intensity

As can be seen from the Key Equation, selection intensity is an important component of the amount of progress obtained per year from selection. Selection intensity is directly determined by the proportion of available animals that are selected as replacements, that is, the proportion of available ram lambs and ewe lambs that are retained as breeding animals each year. Selection intensity (i) can be defined as the number of phenotypic standard deviations between the average phenotype of the lambs selected as flock replacements and the average phenotype of the entire population of lambs available for selection and is given by the formula in (8.7):

$$i = \frac{\overline{X}_s - \overline{\overline{X}}}{\sigma_p} \tag{8.7}$$

where:

σ_p is the phenotypic standard deviation,
$\overline{X}_s$ is average phenotype of the selected sheep, and
$\overline{\overline{X}}$ is the average phenotype of all sheep available for selection.

Since $\left(\overline{X}_s - \overline{\overline{X}}\right)$ is the selection differential (S), the selection intensity and the selection differential can be expressed by the two following formulae:

$$i = \frac{S}{\sigma_p} \text{ and } S = i\sigma_p, \text{ respectively.} \tag{8.8}$$

In many sheep flocks, 20–25% of ewes are replaced each year with young ewe lambs. Ewes leave the flock due to death, health issues (e.g., mastitis), old age, or low production. If a flock raises 150 lambs from 100 ewes each year (75 ewe lambs and 75 ram lambs/100 ewes), and 20% of the ewes need to be replaced each year, then 26.7% (20 out of 75) of the ewe lambs need to be selected each year.

Since a ram can service 20–50 ewes with natural mating in a season and can be used for more than one year, the number of ram lambs required for replacements is much smaller than the number of ewe lambs. If a flock requires three rams per 100 ewes, and rams are used for breeding for three years starting when they are 7 months of age, only one ram lamb per 100 ewes in the flock needs to be selected each year. Therefore, only 1.3% (1 out of 75) of the ram lambs need to be selected each year.

It is easy to see that the 1.3% of the ram lambs that are selected will have a higher average phenotypic performance and EBV than the average of the 27% of ewe lambs that are selected if selection is based on performance records. This is why the statement is often made that "the ram is more than half of the flock," and why the majority of the genetic improvement in a flock comes from ram selection. This is shown in Table 8.7 along with the positive effect that an increase in flock reproductive rate has on selection intensity in both rams and ewes.

Across the flock reproductive levels in Table 8.7, ram lambs have approximately 2–3 times greater potential selection intensities than do ewe lambs. Therefore, ram selection can result in more genetic improvement than can ewe selection when considering just the selection intensity component of the Key Equation. It is useful to note that the selection intensity advantage of rams over ewes decreases as the reproductive rate of the flock increases.

Table 8.7 Potential within flock selection intensities (i)[a] in ram lambs and ewe lambs for different flock reproductive rates.

		Number of lambs raised/ewe/year					
Sex	*Item*	0.75	1.00	1.25	1.50	1.75	2.00
Ram lambs	% retained as replacements[b]	2.7	2.0	1.6	1.3	1.1	1.0
	Selection intensity (i)	2.32	2.42	2.52	2.60	2.64	2.67
Ewe lambs	% retained as replacements[c]	53.3	40.0	32.0	26.7	22.9	20.0
	Selection intensity (i)	0.75	0.97	1.12	1.22	1.33	1.40
Ram selection intensity/ewe selection intensity		3.09	2.49	2.25	2.13	1.98	1.91

[a]i = number of phenotypic standard deviations between the average phenotype of the lambs selected as flock replacements and the average phenotype of the entire population of available lambs, assuming truncation selection.
[b]1 ram lamb selected per 100 ewes in the flock each year.
[c]20 ewe lambs selected per 100 ewes in the flock each year.

As we will see later, generation interval also may be less for rams than ewes in many flocks, which also results in more progress from ram selection than ewe selection.

Generation interval (L)

Generation interval is the amount of time required to replace one generation with the next. The shorter the generation interval, the sooner new animals with superior estimated breeding values can join the flock and the greater the rate of increase in average flock breeding value. If replacement animals have an equal probability of being selected from rams and ewes of different ages, the generation interval is the average age of the parents (rams and ewes) when their lambs are born.

Table 8.8 presents the calculation of generation interval for a flock of 100 ewes and three rams. Twenty ewe lambs enter the flock each year and lamb for the first time at 1 year of age. Forty percent of the 20 ewe lambs entering the flock each year have died or been culled from the flock by the time they are 6 years of age, and ewes are culled from the flock after lambing at 6 years of age. One ram lamb enters the flock each year and is used for breeding for three years before being culled. For simplicity of calculations, no death loss or culling is assumed for rams. The 3 rams and 100 ewes raise 150 lambs per year (1.50 lambs/ewe/year).

Table 8.8 Example calculation of generation interval in sires and dams in a flock of 100 ewes.

Sires

Age, yr	Sires, n	Lambs/sire, n	Total lambs, n	Calculation: total lambs × age
1	1	50	50	50
2	1	50	50	100
3	1	50	50	150
Total:			150	300

Generation interval in sires = average age of sires of lambs = 300/150 = 2.00 years.

Dams

Age, yr	Dams, n	Lambs/dam, n	Total lambs, n	Calculation: total lambs × age
1	20	1.13	23	23
2	19	1.43	27	54
3	18	1.59	29	87
4	16	1.65	26	104
5	15	1.68	25	125
6	12	1.68	20	120
Total:			150	513

Generation interval in dams = average age of dams of lambs = 513/150 = 3.42 years.

In this example (Table 8.8), the generation interval is 2 years for rams and 3.42 years for ewes. The values for actual flocks will vary from these values depending upon the age structure of the flock, but when using natural service rams, most flocks will have a longer generation interval for ewes than rams. The generation intervals in this example are 1.7 times greater for ewes than rams, which means that genetic progress per year from selection is expected to be greater from ram selection than from ewe selection due to differences in generation interval if all other components of the Key Equation are the same between ewes and rams.

Predicting progress from selection

As has been pointed out with the previous examples, selection intensity and generation intervals can, and usually do in practice, differ between rams and ewes. Accuracy of selection can also differ between rams and ewes. If rams and ewes are selected based on the performance of their progeny rather than on their own record, rams will almost always have higher accuracies than ewes. That is because rams are capable of producing many more progeny than ewes, and more information means a greater accuracy. However, the one factor of the Key Equation that can be considered the same between rams and ewes is genetic variation.

The Key Equation can be rewritten to accommodate differences between rams and ewes in accuracy of selection, selection intensity, and generation interval by taking the average of the genetic change per generation from ram and ewe selection and dividing it by the average generation interval of rams and ewes:

$$\Delta\frac{BV}{Year} = \frac{\dfrac{(r_{BV_r,\widehat{BV_r}}i_r + r_{BV_e,\widehat{BV_e}}i_e)\sigma_{BV}}{2}}{\dfrac{L_r + L_e}{2}} = \frac{(r_{BV_r,\widehat{BV_r}}i_r + r_{BV_e,\widehat{BV_e}}i_e)\sigma_{BV}}{L_r + L_e}. \quad (8.9)$$

In the special case of selection on a single phenotypic record on the individual in both rams and ewes, the accuracy of selection is the same in both rams and ewes, and the Key Equation can be simplified (with some substitutions) to:

$$\Delta\frac{BV}{Year} = \frac{h^2(i_r + i_e)\sigma_P}{L_r + L_e}. \quad (8.10)$$

An example of where a single phenotypic record on individual lambs may be used in selection is in flocks where ewes and rams are selected on their own 120-day body weight. The estimated heritability (h^2) of 120-day weight is 0.20, and the estimated phenotypic standard deviation is 15 lb (7 kg). The selection intensities and generation intervals in rams and ewes will be assumed to be as in the previous examples in a flock where 150 lambs are raised from each 100 ewes:

selection intensity in rams (i_r) = 2.60
selection intensity in ewes (i_e) = 1.22
generation interval in rams (L_r) = 2 years, and
generation interval in ewes (L_e) = 3.42 years.

$$\Delta\frac{BV}{Year} = \frac{h^2(i_r + i_e)\sigma_P}{L_r + L_e} = \frac{.20(2.60 + 1.22)15}{2 + 3.42} = \frac{11.46}{5.42} = 2.1\ \text{lb./year} \quad (8.11)$$

It is expected that selection of ram lamb and ewe lamb replacements on their 120-day weights will increase the average breeding value for 120-day weight of lambs in the flock by 2.1 lb per year. If environmental effects remain the same across years, the actual average 120-day weight of the lambs is expected to increase by 2.1 lb per year. This is the maximum amount of progress to be expected. In practice, the selection intensities will probably be less than those used here. Some of the ram and ewe lambs with the highest 120-day weights may not be selected if they have low levels of performance for other traits, an undesirable conformation, or a physical defect. Any deviation from truncation selection of the top animals for 120-day weight will reduce selection intensity and decrease expected progress.

The amount and proportion of expected progress from selection due to ram selection compared to ewe selection can be estimated by substituting different selection intensities into the formula in (8.11). If selection is only on the 120-day weights of the ram lambs, and the ewe lambs are selected at random with respect to 120-day weight, the selection intensity for ewes becomes zero. Likewise, if selection is only on the 120-day weights of the ewe lambs, and the ram lambs are selected at random, the selection intensity for rams becomes zero.

Selection on ram lambs only:

$$\Delta\frac{BV}{Year}=\frac{h^2(i_r+i_e)\sigma_P}{L_r+L_e}=\frac{.20(2.60+0)15}{2+3.42}=\frac{7.80}{5.42}=1.4\text{ lb.per year.} \quad (8.12)$$

Proportion of gain due to ram lamb selection = 1.4/2.1 = 0.67.

Selection on ewe lambs only:

$$\Delta\frac{BV}{Year}=\frac{h^2(i_r+i_e)\sigma_P}{L_r+L_e}=\frac{.20(0+1.22)15}{2+3.42}=\frac{3.66}{5.42}=0.7\text{ lb.per year.} \quad (8.13)$$

Proportion of gain due to ewe lamb selection = 0.7/2.1 = 0.33.

In this example, ram lamb selection results in about twice as much genetic progress as does ewe lamb selection due to a greater selection intensity and a shorter generation interval in rams compared to ewes.

National genetic improvement programs

In order to achieve reasonable amounts of genetic progress in a particular breed of sheep within a country, the EBVs for economically important traits must be comparable across flocks so breeders can identify rams in other flocks that can improve their own flocks. By having flocks within a breed tied together genetically (related sheep in several flocks), using sophisticated statistical genetic software, and powerful computer power, many countries have developed national genetic improvement programs that result in the generation of EBVs on individual animals that are valid across flocks.

The program in the US that generates across flock EBVs is the National Sheep Improvement Program (NSIP). The EBVs for sheep enrolled in NSIP are calculated by LambPlan, the national sheep genetic improvement program in Australia. Flock owners in the US collect phenotypic performance information on their sheep throughout the year and submit it electronically to LambPlan at any time. LambPlan combines the new data with the historical data base for that breed and calculates a new set of EBVs every two weeks. Within approximately two weeks of a breeder submitting a set of data to LambPlan, the breeder receives an output with new and/or updated EBVs for every active sheep in his/her flock.

The EBVs generated through NSIP are the most accurate available because they take into account all phenotypic information available for each sheep in the system. The EBV is calculated using phenotypic measurements for the trait and correlated traits on not only the individual but also on every sheep in the data set with a relationship to the individual (e.g., sire, dam, full-sibs, half-sibs, grandparents, aunts, uncles, cousins, progeny). Every additional phenotypic measurement on the individual and its relatives contributes to the EBV and increases its accuracy. However, records of the individual and close relatives have a greater effect on the value of the EBV and result in a greater increase in its accuracy than do the records of more distant relatives.

Following are the traits and their NSIP abbreviations for which EBVs can be obtained. Additional information on each of these EBVs can be obtained from the NSIP website (www.nsip.org/?page_id=1542).

Birth weight (BWT)
Maternal birth weight (MBWT)
Weaning weight (WWT)
Maternal weaning weight (MWWT)
Postweaning weight (PWWT)
Yearling weight (YWT)
Hogget (18-month) weight (HWT)
Yearling grease fleece weight (YGFW)
Yearling fiber diameter (YFD)
Yearling Staple length (YSL)
Fiber diameter coefficient of variation (FDCV)
Fiber curvature (CURV)
Postweaning or yearling fat depth (PFAT, YFAT)
Postweaning or yearling loin muscle depth (PEMD, YEMD)
Number of lambs born (NLB)
Number of lambs weaned (NLW)
Postweaning scrotal circumference (PSC)
Weaning or postweaning fecal egg count (WFEC, PFEC).

In addition to EBVs for these traits, NSIP provides indexes that weight the individual EBVs by their relative economic values and produce a single number (index value) that allows breeders to select for "net merit." Each index is constructed to meet the specific breeding objectives of a particular breed. The following indexes are available:

USA range index = PWWT EBV + (0.26 × MWWT EBV) − (0.26 × YWT EBV) + (1.92 × YGFW EBV) − (0.47 × YFD EBV) + (0.36 × NLB EBV)

USA hair breeds ewe productivity index = (0.246 × WWT EBV) + (2.226 × MWWT EBV) + (0.406 × NLW EBV) − (0.035 × NLB EBV)

USA maternal breeds ewe productivity index = (0.265 × WWT EBV) + (1.200 × MWWT EBV) + (0.406 × NLW EBV) − (0.035 × NLB EBV)

Carcass plus index = (5.06 × PWWT EBV) − (13.36 × PFAT EBV) + (7.83 × PEMD EBV)

LAMBPLAN
Analysis: USA Materials – , 15 June 2013

Sires Animal ID	Inbreeding	Prog:Flks	Bwt Kg	Mbwt Kg	Wwt Kg	Mwwt Kg	Pwwt Kg	Pfat mm	Pemd mm	Yfd u	Ygfw %	NLB %	NLW %	Psc cm	USA Maternal	Sire Dam
620062–2012–L2090X	8.5%	49:2	0.4	1.0	3.6	2.4	9.4					36	38		122.5	620062–2011–L1017X
LAMBSHIRE		Acc:	81	53	79	48	78					42	39			620062–2010–L0026B
620062–2010–L0024B	0.9%	114:4	0.3	1.1	1.1	3.1	2.3					39	36		122.4	620074–2007–PTB730
LAMBSHIRE		Acc:	88	79	85	69	84					67	62			620062–2007–GN1454
620075–2012–E2050B	4.5%	55:1	0.7	1.2	2.3	3.0	4.4					27	28		119.6	620062–2010–L0024B
ELM CREEK		Acc:	81	58	77	51	75					48	44			620075–2008–ECP804
620074–2011–WA1041	7.8%	56:2	0.2	1.1	0.8	2.5	1.7					33	33		119.4	620074–2010–PA0210
WOODHILL		Acc:	78	52	72	45	72					43	39			620074–2006–WH1075
620062–2011–L1170X	1.3%	78:1	0.0	1.0	1.3	2.2	4.0					25	32		118.6	620074–2006–WH1157
LAMBSHIRE		Acc:	85	65	82	61	80					56	53			620062–2006-OPT820
620095–2012–K2024R	3.2%	13:1	0.3	1.0	1.7	2.2	4.3					24	29		117.8	620062–2011–L1011R
CAK		Acc:	71	54	64	48	61					41	37			620092–2009–001046
620074–2010–PA0210	9.6%	60.2	0.5	1.3	1.4	3.2	2.4					31	22		116.9	620074–2007–PA7146
WOODHILL		Acc:	82	63	80	55	82					51	47			620074–2006–OPT820
620017–2009–GV9037	12.5%	22:2	0.8	1.1	2.6	2.7	4.5					27	22		116.6	620017–2002–0GY213
GRANDVIEW		Acc:	78	66	77	60	78					57	53			620017–2005–GV5553
620078–2010–P10190		81:1	0.7	0.5	1.7	1.1	1.8					28	32		116.1	620007–2009–020909
UW-MADISON		Acc:	86	67	85	59	87					56	51			620078–2005–005326
620062–2011–L1017X	0.7%	40:1	0.2	1.0	2.4	1.8	6.3					17	25		115.5	620062–2006–WH1157
LAMBSHIRE		Acc:	82	66	77	59	72					51	48			620074–2009–GN1597
620074–2006–WH1157	0.9%	310:4	0.3	1.4	2.8	2.0	7.3					22	23		115.5	620074–2005–PTX757
WOODHILL		Acc:	95	94	93	89	92					85	82			620074–2003–0WH227
620062–2011–L1094R	3.7%	33:1	−0.1	0.9	0.9	1.6	3.4					27	29		115.4	620074–2009–WR9103
LAMBSHIRE		Acc:	79	60	76	55	73					50	46			620062–2010–L0026B
620074–2010–WR0265	7.7%	42:1	0.1	0.9	0.5	1.7	1.4					21	28		115.3	620074–2007–WHR711
WOODHILL		Acc:	72	54	66	47	62					45	42			620074–2006–WH1075
620078–2012–P12363	5.7%	21:1	1.1	0.8	3.6	1.8	6.3					14	22		115.0	620078–2009–009425
UW-MADISON		Acc:	71	54	67	47	66					43	38			620078–2009–009350
620075–2012–E2041B	3.3%	61:1	0.5	1.2	2.6	3.0	6.1					22	15		114.8	620062–2010–L0024B
ELM CREEK		Acc:	63	56	60	50	57					47	44			620075–2009–ECP925
620061–2011–11B271	1.5%	79:1	0.4	0.9	2.9	2.3	7.5					16	18		114.4	620017–2009–GV9142
UNCOMPAHGRE		Acc:	77	59	75	52	73					45	41			620061–2009–E00267

Sires / Page 1 — 24-Jun-13 — ASBV — LAMBPLAN

Figure 8.2 A list of the top active Polypay sires for the USA Maternal Index (from www.nsip.org).

Lamb 2020 index = (0.32 × WWT EBV) + (0.47 × PWWT EBV) − (0.21 × BWT EBV) − (0.55 × PFAT EBV) + (1.54 × PEMD EBV) − (0.04 × PFEC EBV)

Figure 8.2 is the first page of the listing of the top sires in the Polypay breed ranked on the USA Maternal Index. This one page lists the 16 sires of the Polypay breed in seven different flocks expected to produce ewes with the greatest net merit for lamb production. The sires' EBVs for all other traits available to the Polypay breed are also listed. With information like this, breeders can identify the superior animals in the breed for any particular trait that they wish to improve.

All breeds of livestock participating in well-designed national genetic improvement programs can demonstrate genetic improvement over time, and US breeds of sheep participating in NSIP are no exception. As an example, Table 8.9 presents the EBVs for several traits of US Polypay sheep born in 2002 through 2011. Also presented in Table 8.9 is the linear regression of each EBV and the maternal index value on year, which gives an average change per year in EBV and index value.

On average, Polypay lambs born in each year are genetically superior to the lambs born in previous years. The Polypay is considered a maternal breed with a reputation for high reproductive performance. Polypay lambs born each year have, on average, an EBV that is 0.71 lambs weaned per 100 ewes lambing greater than the average EBV of lambs born in the previous year. If this rate of genetic improvement is maintained in future years, if environmental effects remain constant across years, and if ewe lambs born in 2010 weaned 1.75 lambs per ewe lambing as mature ewes, then expected number of lambs weaned per ewe lambing of ewes born in 2020 is 1.82, of ewes born in 2030 is 1.89, of ewes born in 2040 is 1.96, and so on. In a relatively short period of time, the genetic value of a population and their phenotypic performance can be improved significantly by using modern technology available through a well-designed and implemented national genetic improvement program such as NSIP.

Summary

A sheep producer makes their initial genetic gain by selecting a breed or breeds that have above average performance for traits of importance to their operation. There are over 1500 breeds or races of sheep in the world and over 50 breeds in the US. These breeds are a valuable genetic resource with considerable variation among breeds for reproductive, growth, carcass, wool, and dairy traits.

In order for a breed of sheep to maintain its relevancy, it must continue to improve in its genetic merit or "breeding value" over

Table 8.9 Estimated breeding values and maternal index value for US Polypay sheep born in 2002 through 2011 and enrolled in the National Sheep Improvement Program.

Year	Birth wt., kg	Weaning wt., kg	Postweaning wt., kg	Maternal weaning wt., kg	Lambs born/100 ewes, n	Lambs weaned/100 ewes, n	Maternal index
2002	0.31	1.06	1.98	0.67	−0.70	1.60	103.10
2003	0.30	1.11	2.20	0.75	−0.10	2.70	103.70
2004	0.31	1.22	2.47	0.90	1.30	2.70	104.10
2005	0.25	0.98	2.00	0.76	1.90	2.20	103.40
2006	0.27	1.09	2.30	1.01	−0.20	2.80	104.40
2007	0.30	1.19	2.45	0.90	2.40	4.00	104.60
2008	0.32	1.35	2.89	1.09	4.10	5.50	105.70
2009	0.29	1.31	2.85	1.10	5.20	6.00	105.90
2010	0.35	1.51	3.23	1.25	6.20	7.40	107.00
2011	0.32	1.42	3.09	1.08	5.90	7.90	106.70
$b_{EBV,Year}$	0.00	0.05	0.13	0.05	0.80	0.71	0.43

time through an effective selection program. The factors that determine the change in flock breeding value, over which the sheep breeder has some control, are (1) accuracy of the estimated breeding value, (2) selection intensity, and (3) generation interval. Accuracy of the breeding value is dependent upon the heritability of the trait. In general reproductive traits have the lowest heritabilities, growth and dairy traits are intermediate in heritability, and wool and carcass traits have the largest heritabilities. The accuracy of breeding value for all traits is improved as more phenotypic records are available on an individual and its relatives. National genetic improvement programs, such as the National Sheep Improvement Program, provide breeders with accurate estimates of breeding values. Selection intensities are almost always higher and generation intervals lower in rams than in ewes, resulting in the vast majority of flock genetic gain coming from ram selection.

Further reading

American Sheep Industry Association. 2003. *Sheep Production Handbook*, 2002 edn, Vol. 7. Centennial, Colorado, USA.

Bourdon R. M. 2000. *Understanding Animal Breeding*, 2nd edn. Prentice-Hall, Inc., Upper Saddle River, New Jersey, USA.

Van Vleck, L. D., E. J. Pollak, and E. A. Branford. 1987. *Genetics for the Animal Sciences*. Oltenacu. W.H. Freeman and Company, New York, USA.

References

American Sheep Industry Association. 2003. *Sheep Production Handbook*, 2002 edn, Vol. 7. Centennial, Colorado, USA.

Mason's World Dictionary of Livestock Breeds, Types and Varieties. 2002. Revised by V. Porter. CABI Publishing, Wallingford, UK.

Safari, E., N. M. Fogarty, and A. R. Gilmour. 2005. A review of genetic parameter estimates for wool, growth, meat and reproduction traits in sheep. *Livestock Production Science* 92:271–289.

Review questions

1. List the five broad classifications for sheep breeds and distinguishing characteristics of breeds in each class.

2. Which class of breeds would most likely be used as ewes in intensively managed farm flocks? As ewes in humid or tropical areas? As ewes in more extensive range areas? As terminal sires for the production of high-quality market lambs?

3. Rank the following traits, from lowest to greatest, for their heritability.

120-day weight
Fiber diameter
Prolificacy
Staple length
Ewe fertility
Milk yield
Loin eye area

4. Calculate the adjusted 60 day weaning weights for the following four lambs.

Lamb	Birth wt., lb.	Weaning wt., lb.	Weaning age, days	Sex	Age of dam, years	Type of birth-rearing
A	12	58	60	Ram	2	Twin-Single
B	13	64	67	Ram	4	Twin-Twin
C	12	59	56	Ram	9	Twin – Single
D	10	57	58	Ram	1	Single–Single

5. If the only information you have on four ram lambs is the information in the table in question 4 and your goal is to improve 60 day weaning weight in your lambs, which ram lamb would you select for breeding? Why?

6. Why is the selection intensity usually greater in rams than ewes?

7. Show that Equation (8.10) is equal to: (heritability × selection differential in rams x selection differential in ewes)/(generation interval in rams + generation interval in ewes) when selection is on a single measurement of the rams and ewes.

8. Using the equation in Question 7, calculate the amount of progress per year from selection for fleece weight (selection differential and generation interval in rams is 2.5 lb. and 2 years, respectively, and selection differential and generation interval in ewes is 0.5 lb. and 3.5 years, respectively.

9. What proportion of the total annual progress in Question 8 is due to ram selection? Due to ewe selection?

10. Of the Polypay rams listed in Figure 8.2, which ram would be expected to sire lambs with the heaviest weaning weights?

11. What is the expected difference in average postweaning weight between lambs sired by ram P10190 and ram L1017X from the list of Polypay rams in Figure 8.2?

12. The EBVs for Polypay ram WH1157 in Figure 8.2 have higher accuracies than the EBVs of the other rams. Why?

9 Genetic Improvement Programs for Dairy Cattle

Kent Weigel

Department of Dairy Science, University of Wisconsin–Madison, WI, USA

Introduction

The goals of dairy cattle selection programs are two-fold. First, we seek to create dairy cattle that can efficiently produce large quantities of milk with desirable composition, in terms of human health and manufacturing properties. Second, we seek to create dairy cattle that can resist infectious diseases, metabolic disorders, infertility, and other health problems that lead to high veterinary bills and early culling from the herd. Unlike the centralized, vertically integrated breeding programs that exist in the poultry and swine industries, dairy cattle breeding programs tend to rely on cooperation among commercial dairy producers, government agencies, non-profit organizations, breeding companies, and agricultural universities. Each organization plays a key role in data collection, quality control, statistical analysis, education and outreach, or product development. For more than a century, the basic blocks for genetic progress have been performance records and pedigree information. Widespread use of milk recording programs, coupled with close collaboration among industry partners in developing accurate genetic evaluation methods and facilitating timely selection decisions, have enabled substantial genetic improvement in North American dairy cattle populations. This chapter provides an overview of dairy cattle selection programs that, while largely based on the structure and functionality of the US industry, are applicable to dairy production systems in most developed countries.

Data collection infrastructure

Milk recording

Genetic selection programs in dairy cattle have benefited greatly from the development and widespread utilization of **milk recording programs**, such as the Dairy Herd Improvement (DHI) program in the US, more than a century ago. In many countries, these programs originated from public investment in data collection infrastructure, by institutions such as Cooperative Extension. Organizations such National DHIA (www.dhia.org), AgSource Cooperative Services (www.agsource.com), California DHIA (www.cdhia.org), and their local and regional affiliates set data collection standards, ensure quality control, and manage the day-to-day collection of millions of data points regarding milk volume, milk fat percentage, milk protein percentage, somatic cell count, lactose content, and milk urea nitrogen (MUN). In addition, much of the necessary demographic information regarding individual dairy animals, such as birth dates, calving dates, and so on are gathered through the milk recording program. In the past, these data were recorded manually by DHI testers who traveled to farms on a monthly or bi-monthly basis and, while collecting the milk samples, copied this information from on-farm record-keeping systems. At present, the majority of information regarding birth dates, calving dates, inseminations, culling dates, pedigrees, health problems, and other important events comes directly from on-farm herd management software, such as Dairy-Comp 305 (www.vas.com), DHI-Plus (www.dhiprovo.com), and PCDart (www.drms.org). Milk samples are still collected and sent to milk analysis laboratories for processing, usually on a bi-monthly basis, but many of the milk weights come directly from on-farm milking parlor software programs. In recent years, AM-PM component sampling (i.e., milk sample taken in the morning one month and in the afternoon the following month) have become popular, as producers have sought to reduce data collection costs. Data collection ratings (DCR) are used as indicators of the expected accuracy of production records from different milk recording programs, with larger DCR values indicating greater accuracy. For example, a herd with electronically recorded daily milk weights would have a DCR of 104, whereas a herd with bimonthly supervised recording of one of the three daily milkings would have a DCR of 78 (see http://aipl.arsusda.gov/reference/datarating.htm for details).

After the milk recording data are collected, they move to **dairy records processing centers**, such as AgSource Cooperative Services (www.agsource.com), AgriTech Analytics (www.agritech.com), DHI-Provo (www.dhiprovo.com), or Dairy Records Management Systems (www.drms.org). These organizations carry out several key functions, including compiling the entire milk recording and milk composition data into central databases, and providing management reports and action lists back to their customers to facilitate day-to-day herd management decisions. An increasingly important role for these organizations is to provide data summaries and electronic files to farm consultants, such as nutritionists and veterinarians. Because these processing centers bring data from hundreds or thousands of herds into a central database, they have the ability to provide management benchmarking services to dairy producers and consultants. Lastly, these

Molecular and Quantitative Animal Genetics, First Edition. Edited by Hasan Khatib.

organizations are responsible for sending milk production records and related data to the national genetic evaluation center. Historically, about 50% of dairy herds in the US have routinely participated in milk recording programs and have thereby contributed data to the **national genetic evaluation program**. The remaining herds have captured downstream benefits through the purchase of semen, live cattle, and natural service bulls. Detailed information about milk recording policies, practices, and participation for dairy cattle in the US, including average production levels for specific breeds, can be found at the following link: https://www.cdcb.us/publish/dhi.htm?.

Pedigree information

The task of collecting and maintaining pedigree information has been carried out by **breed associations** since the late 1800s. For nearly a century thereafter, **pedigree records** were kept in hard-covered registry books that noted the name, birthdate, sire, dam, and owner of every animal. With the advent of computers in the 1960s, **herdbooks** were replaced by ancestry databases, and for that reason we often consider animals born during this time period as the base generation in analyses of genetic trends, inbreeding, or related topics. Many herdbooks were open, such that animals with unknown or crossbred ancestry could be recorded and subsequently graded-up to purebred status. Other herdbooks, most notably those of North American Holstein cattle, remained closed for more than a century and as such, individual animals or entire herds could leave the herdbook, but new animals or families (i.e., those that weren't direct descendants of founder animals in the original herdbook) could not enter. For this reason, the vast majority of Holstein cattle in North America are not registered (commonly known as grade Holsteins), even though many have pure Holstein ancestry. This is in contrast to other breeds, such as North American Jersey cattle, for which the vast majority of animals are registered. Other key activities carried out by the breed associations include type classification services (discussed later), certification of milk recording for award programs, computerized mating programs, parentage verification, and maintenance of genetic defect records. These organizations also publish ranking lists of elite sires, cows, heifers, and calves, assist in cattle marketing, and coordinate dairy shows and other youth events. For more detailed information, see links to the Holstein Association USA (www.holsteinusa.com), the American Jersey Cattle Association (www.usjersey.com), the Brown Swiss Association (www.brownswissusa.com), the American Guernsey Association (www.usguernsey.com), the United States Ayrshire Association (www.usayrshire.com), the American Milking Shorthorn Society (www.milkingshorthorn.com), or the Red and White Dairy Cattle Association (www.redandwhitecattle.com).

Type classification

As noted earlier, **type classification services** are provided by the dairy breed associations. Although the exact number, definition, and nomenclature for **conformation** (visual appearance) traits differ among breeds, as well as among countries for a given breed, these traits tend to fall into a few key categories. First, and most important, are udder traits such as udder depth, rear udder height, rear udder width, udder cleft, udder tilt, fore udder attachment, front teat placement, rear teat placement, and teat length. The purpose of measuring these traits is to provide an indication, for genetic selection purposes, of which sire will produce daughters that are able to avoid injuries to the teats and udder, resist mastitis, and milk efficiently (i.e., not too slowly), even while producing large volumes of milk. Second, and also quite important, are mobility traits, such as rear leg set, rear legs – rear view, foot angle, locomotion score, and feet and legs score. The reason for measuring these traits is to get an indication of which sires will produce daughters that will avoid leg injuries, require minimal hoof trimming and veterinary care, and avoid lameness. The remaining traits, many of which have little relationship with health, longevity, or profitability, include stature (or hip height), strength, body depth, rump angle, rump width, and dairy character, and many of these traits have intermediate optima (Caraviello et al., 2003, 2004). For example, cows need a certain amount of size and strength to withstand the rigors of high milk production, but cows that are too large might unnecessarily consume extra feed and therefore incur higher maintenance costs. Likewise, dairy cows should partition energy toward milk production rather than toward the accumulation of body tissue (i.e., fat), but cows that are too thin often experience metabolic health problems and early culling.

Health, fertility, calving ability, and longevity data

For the most part, collection of data regarding **fitness traits**, such as health, fertility, calving ability, and longevity has lagged behind that of production and conformation traits. However, in the early 1990s producers and scientists began to realize that selection for high milk production had brought along undesirable correlated responses in many fitness traits. **Calving ease**, which is measured on a five-point ordinal scale, had been measured for many years and evaluated as a trait of the service sire (i.e., direct calving ease), as noted by Berger (1994). Data regarding culling decisions had been available for decades through DHI milk recording programs, and in 1994 the USDA-ARS Animal Improvement Programs Laboratory (AIPL) began evaluating dairy sires for length of **productive life** (PL) of their daughters (VanRaden and Klaaskate, 1993). This trait was defined as the total number of months in milk between first calving and culling or 84 months of age, whichever occurred first. In the same year, AIPL introduced genetic evaluations for **somatic cell score** (SCS), which is a $\log_2$ transformation of somatic cell count (SCC) and an indirect indicator of mastitis (Shook and Schutz, 1994). Like many other traits, SCC had been measured on commercial dairy farms for many years for management purposes, namely to facilitate identification and culling of cows with poor quality milk, and therefore extensive data were available for genetic evaluation purposes.

Although PL information provided an indirect assessment of fertility (infertile animals have shortened PL), direct selection for female fertility wasn't possible until nearly a decade later, when AIPL introduced genetic evaluations for **daughter pregnancy rate** (DPR) using information from the DHI system regarding days open (VanRaden et al., 2004). Data regarding male fertility, known as service **sire conception rate** (SCR), became available shortly thereafter. Information regarding conception rate of AI breedings is used, along with veterinary pregnancy examination data (when available), to determine the outcome of each insemination (Weigel, 2004a). Lastly, genetic analyses of daughter (maternal) calving ease, as well as **service sire** (direct) **stillbirth rate** (SSB) and **daughter** (maternal) stillbirth rate (DSB) were implemented in 2006 (Cole et al., 2007). Unfortunately, collection and analysis of data regarding other infectious diseases and metabolic disorders, such has clinical mastitis, lameness, ketosis, milk fever, or displaced abomasum, has not yet progressed to the point at which national genetic evaluations are available. A notable exception is in the Nordic countries, where national health recording systems have been in place for decades, and sires are routinely evaluated based on the health of their daughters (e.g., Heringstad et al., 2003; Heringstad, 2010). Research by Zwald et al. (2004a,b) has indicated that selection for health traits is possible in North America using farmer-recorded health event data from herd management software programs, but progress toward routine evaluation of these traits has been complicated by inconsistency in diagnosis and trait definition and incomplete data recording on many commercial dairy farms.

Progeny testing and AI programs

For more than half a century, **progeny testing** has been the cornerstone of genetic improvement in dairy cattle. The reason why progeny testing is so much more important in dairy cattle than other food animal species is simple: virtually all economically important dairy traits are sex-limited and can be measured only in females. This is in contrast to important traits in meat-producing livestock, such as birth weight, average daily gain, or weaning weight, which can be measured in young bulls prior to their usage for **artificial insemination** (AI) or natural service breeding. Progeny testing involves collection of >1000 units of semen per bull from potentially elite young bulls that have been purchased from pedigree breeders based on high **parent average** (PA) for important traits and subsequent distribution of this semen to dozens or hundreds of cooperator herds. In exchange for discounted semen costs and other financial incentives, farmers agree to use this semen randomly and record performance data on the resulting offspring for the purpose of computing genetic evaluations of their sires. Breeding companies, such as ABS Global (www.absglobal.com), Accelerated Genetics (www.accelgen.com), Alta Genetics (www.altagenetics.com), Genex Cooperative (www.crinet.com), Select Sires (www.selectsires.com), Semex (www.semex.com), and Taurus Service (http://www.taurus-service.com), collectively progeny test nearly 1500 young dairy bulls annually, at a cost of roughly $30,000 per bull (Funk, 2006). After progeny testing, when these bulls are roughly 5 years of age, about 1 in 10 bulls are chosen for collection of additional semen and marketing of this semen to domestic and international customers. Because these bulls are used heavily domestically and abroad, they can have a tremendous influence on the future genetics of the population, and accurate evaluation of their genetic merit is critical. Progeny testing programs are undergoing tremendous change at the present time due to the rapid adoption of genomic testing technology, which is discussed in detail later in this chapter. In addition to progeny testing young bulls and marketing semen, AI companies provide several other important services, including computerized mate selection programs and reproductive management (heat detection and timed AI) programs. Collectively, the aforementioned AI companies belong to a trade association, the National Association of Animal Breeders (NAAB, www.naab-css.org) which administers genetic evaluations for calving ability and addresses key animal health and trade issues, as well as its subsidiary Certified Semen Services (CSS), which sets standards for semen preparation and monitors semen collection facilities.

Estimation of breeding values

Pre-adjustment for known environmental effects

In the US, genetic evaluations for dairy cattle are carried out by the AIPL (www.aipl.arsusda.gov), and at present the information is released to the dairy industry three times per year (Wiggans et al., 2011). Virtually all traits of economic importance in dairy cattle are **quantitative traits** – they are measured on a continuous scale and the phenotypes (perhaps after a transformation, such as $\log_2$ function) tend to follow a normal or bell-shaped distribution. Even traits that are observed on a binary scale, such as the incidence of stillbirths, can be expressed on a continuous scale by assuming that the underlying tendency or liability of animals to have the affected phenotype follows a normal distribution. Generally speaking, we assume that each trait follows the simple equation: phenotype = genotype + environment, where **phenotype** represents the observed attribute of an individual animal, **genotype** represents the genetic contribution (predisposition) of the animal for that trait, and **environment** represents the sum of all non-genetic factors that may have affected the animal's performance, such as weather, housing system, feed quality, veterinary care, or behavior of its herdmates. Our objective is to identify and remove all components of the environment, such that we are left with an estimate of the genotype. In a controlled experiment, for example, the effects of environment would be removed by choosing cows of the same breed that are of similar age, stage of lactation, body size, and so on, and then randomly assigning these cows to experimental treatment groups. In contrast, dairy cattle selection programs rely on retrospective analysis of field data from commercial farms, on which the environmental conditions are uncontrolled. Therefore, the effects of the environment must be removed by using statistical adjustments to create **standardized records**. Typically, the observed phenotypes of dairy cows are standardized for known environmental factors that tend to affect performance consistently for all animals, at least within a given breed. These factors include: number of milkings per day, age at calving, and length of lactation, so we commonly express phenotypes for milk production traits as standardized 305-day, 2X, mature-equivalent (ME) records prior to genetic evaluations (Norman et al., 1995; Cole et al., 2009). Other phenotypes, such as scores for physical conformation traits, can also be expressed on an age-adjusted basis.

Contemporary groups

While it is possible to standardize raw phenotypic records for known factors, such as age, milking frequency, or lactation length, these phenotypes are also affected by many environmental factors that cannot be observed. For example, we don't know the exact weather conditions that each cow experienced on every day of her lactation, and even if we could use data from a nearby weather station we still wouldn't have information about the specifications, usage pattern, or effectiveness of the heat abatement devices (e.g., fans, sprinklers, or soakers) that were used on that farm. Furthermore, we don't have information about the farm's vaccination program, nor do we know anything about the ration provided to each animal, such as the type, amount, or quality of specific forages, concentrates, and supplements that were used or the energy density of the total mixed ration. For this reason, we resort to the conception of **contemporary group** or management group when attempting to account for unobserved environmental factors. It is our expectation that all animals within a given contemporary group were exposed to the same environmental conditions throughout their lactations. Therefore, we typically define a contemporary group as all animals that calved in a given herd, in a certain year, and in a particular season of that year. Sometimes we also create separate contemporary groups for primiparous (first lactation) and multiparous (second and later lactation) cows within a given herd-year-season of calving, because these groups of animals are often managed differently. We want contemporary groups to be precise enough to account for short-term environmental conditions that affect specific demographic groups of cows within the herd, but we also want to make sure that contemporary groups are large enough (e.g., at least 5–10 animals) to ensure that average performance of the contemporary group is not influenced too heavily by chance anomalies (e.g., if a few cows get sick). Then, we can express the standardized phenotype of each animal as a deviation from the average standardized phenotype of all cows within that management group – this is commonly known at the cow's **yield deviation**, contemporary group deviation, or management group deviation (VanRaden and Wiggans, 1991).

Animal model genetic evaluations

Once we've accounted for known environmental factors by standardizing the phenotypes and attempted to account for unobserved environmental factors by grouping animals into contemporary groups, we still have a bit more work to do. First, we have to account for non-random mating. Humans tend to select their mates based on commonalities in height, education level, and other factors, rather than select them randomly. Likewise, farmers tend to select the mates for specific cows in a non-random manner to fix faults in their physical conformation or deficiencies in their milk production, reproductive ability, or udder health through **corrective mating**, which can also maximize the benefits from expensive semen by using it to mate the best cows. Therefore, we need to make an adjustment for **merit of mates**. In other words, we have to consider the possibility that a certain bull was mated to better-than-average cows (perhaps because his semen was expensive) or that a given cow was mated to a better-than-average bull (perhaps because the farmer perceived her as valuable). Second, we have to estimate the proportion of each animal's genetic superiority or inferiority that it will pass on to its offspring. Because each parent transmits only one allele at each locus, we know that any portion of an animal's phenotypic superiority or inferiority that was due to dominant (or recessive) combinations of alleles will not be passed to its offspring. Furthermore, because recombination (crossing over) occurs during meiosis, we know that some of the favorable or unfavorable effects of epistasis (i.e., combinations of alleles at different loci) will not be passed to the animal's offspring. Therefore, we seek to isolate the portion of the genetic superiority or inferiority of an animal that is due to alleles that affect the phenotype in an additive manner. That is, we expect that the offspring's performance will reflect the sum of many alleles inherited from the sire and dam, with each of these acting in an additive manner. We estimate the **additive genetic merit** of each cow or bull using a so-called **animal model** (VanRaden and Wiggans, 1991), and we express the result as that animal's **estimated breeding value** (EBV) or **predicted transmitting ability** (PTA). The EBV is an estimate of the animal's total additive genetic merit, and the difference between the EBVs of two animals is an estimate of the difference in their phenotype or performance that is attributable to their genotypes. The PTA is an estimate of the genetic superiority or inferiority that an animal will pass along to its offspring, and the difference between the PTAs of two animals is an estimate of the difference we will observe in the phenotype or performance of their progeny. The PTA is equal to one-half of the EBV, because the PTA represents the value of the **haploid** genome (sperm or egg) that will be transmitted to the offspring, whereas the EBV represents the value of the **diploid** genome of the animal. Each animal's EBV (or PTA) derives from information contributed not only by its own phenotypic records, but also the records of its offspring, parents, grandparents, and other known ancestors and descendants. In a progeny testing system, the animal model is used to calculate differences between sires' daughter groups within each contemporary group, and these differences are subsequently averaged across many contemporary groups. This is possible due to the widespread use of **reference sires** (heavily used AI sires), who have daughters in hundreds of different herds throughout the country. Usage of reference sires in multiple herds, coupled with detailed information regarding the genetic relationships between these sires and their relationships with other animals within the breed, allows us to mathematically separate environmental variation associated with contemporary groups from genetic variation associated with sire progeny groups. Furthermore, usage of exported semen from these key reference sires in hundreds of additional herds around the globe enables an organization called **Interbull** to compute international genetic comparisons (www.interbull.org), which facilitate international semen trade and further enhance genetic progress within the leading dairy breeds.

Four paths of selection

The rate of **genetic progress** can be described in terms of four paths of selection, according to gender of both parent and offspring (Van Tassell and Van Vleck, 1991). The first path, **sires of males**, represents elite males that are selected to be sires of the next generation of young bulls, and this pathway is characterized by high **accuracy** and high **selection intensity**. This group is typically comprised of the top 5% of bulls whose semen is actively marketed to dairy farmers, and these bulls are often referred to as **sires of sons**. The second path, **sires of females**, represents a larger group of males whose semen is used to breed the general population of cows to produce replacement females for commercial farms. These bulls are typically referred to as active AI sires, and this pathway is characterized by high accuracy and relatively high selection intensity. The third path, **dams of males**, represents a group of elite females that are usually rank among the top 1% of the commercial cow population. These cows are mated to bulls from the sires of males group for the purpose of producing bull calves and they are commonly referred to as **bull dams**. In the absence of genomic selection this pathway is characterized by low accuracy, and despite high selection intensity predictions of genetic merit for individual cows can be biased by preferential treatment of these (perceived) high value animals. The fourth path, **dams of females**, represents the large population of females on commercial farms that are primarily used to produce milk, rather than breeding stock. These cows are often referred to as **commercial cows** and are mated to bulls from the sires of females group to initiate lactation and to produce the next generation of **replacement heifers**. In the absence of genomic selection this pathway is characterized by low accuracy, and in the absence of gender-enhanced semen this pathway is characterized by low selection intensity, because nearly every heifer calf that is born must be retained as a future herd replacement.

Selection for increased productivity

Milk yield

Genetic evaluations for 305-day milk, fat, and protein yield, as well as fat and protein percentage, are computed by the USDA-ARS Animal Improvement Programs Laboratory (see http://aipl.arsusda.gov/reference/Form_GE_Yield_1008.pdf for details). Information about the procedures used in more than two dozen other countries can be found on the Interbull website (www-interbull.slu.se/national_ges_info2/framesida-ges.htm). **Heritability**, which is the portion of differences among animals' phenotypes that are attributable to differences in their breeding values, is moderate for most production traits. In routine genetic evaluations for production traits in the US, heritability parameters vary according to the estimated within-herd variance, but on average a heritability value of 30% is used for milk, fat, and protein yield in Holstein cattle. Records from the first five lactations are used, and the genetic **base population** (i.e., the reference group of animals with average EBV and PTA equal to 0) is the population of sire-identified, milk-recorded cows of a given breed that were born in 2005 (this is updated every 5 years). Phenotypic and **genetic trends** for production traits show that remarkable progress has occurred during the past half century, as shown in Figure 9.1. For example, average standardized 305-day milk yield of US Holstein cows increased from 6279 kg in 1960 to 12,222 kg in 2010, a change of 5943 kg. During this time, the average EBV of US Holstein cows increased by 3664 kg, indicating that approximately 62% of the improvement is due to genetic selection. Trends in US Jersey cows were even greater, with phenotypic milk production increasing from 3947 kg in 1960 to 9013 kg in 2010, and average EBV increasing by 3627 kg, which represents 72% of the total increase in production.

To date, selection programs have focused on improvement of total 305-day lactation yield, with little concern about whether this increase is due to higher peak yield (today's high-producing cows may peak at 80–100 kg of milk per day) or greater persistency of lactation (i.e., moderate peak yield but greater ability to sustain a high level of milk production throughout the lactation). Some authors (e.g., Dekkers et al., 1998) have suggested that selection for greater persistency, rather than higher peak yield, would lead to cows that are better able to avoid health problems associated with the stress of high production in early lactation, while also potentially consuming less energy-dense diets (because maximizing feed intake during peak lactation may be less critical). It is important to note that improvements in breeding values due to genetic selection go hand-in-hand with improvements in herd management, such as total mixed rations, sand-bedded free stalls, heat abatement systems, accelerated calf-rearing programs, and recombinant bovine somatotropin. Without these management improvements, the changes in average breeding values of today's dairy cattle would largely represent unrealized potential. To demonstrate this concept, consider the Australian study by Fulkerson et al. (2008, see Figure 9.2). These authors plotted breeding values for milk yield on the *x*-axis and actual milk yield per lactation on the *y*-axis. Our expectation is that a 1-unit increase in breeding value should result in a 1-unit increase in production per lactation. However, in grass-based herds with low levels of supplemental concentrates the slope of the line is much flatter, indicating that the realized gain in milk production phenotype is less than expected. As the level of supplemental concentrate in the diet increases, so does the slope of the line, indicating that the realized increase in milk production phenotype per unit of change in genotype is greater for herds that invest more heavily in increasing the energy density of the diet through use of supplemental concentrates. This relationship between genetic predisposition of the animal and management conditions present in the herd is called **genotype by environment interaction** (G×E).

In an earlier study, Powell and Norman (1984) studied the regression of cows' milk yield phenotypes on their sires' PTA values (they were called predicted difference, PD, at the time), according to herd production level. In the herds with the lowest average milk production (< 5500 kg/cow/lactation) this regression coefficient was 0.75, which was much lower than its expectation of 1.00. Conversely, in the herds with the highest average milk production (≥ 9500 kg/cow/lactation) this regression coefficient was 1.49. Again, this demonstrates that herds with more

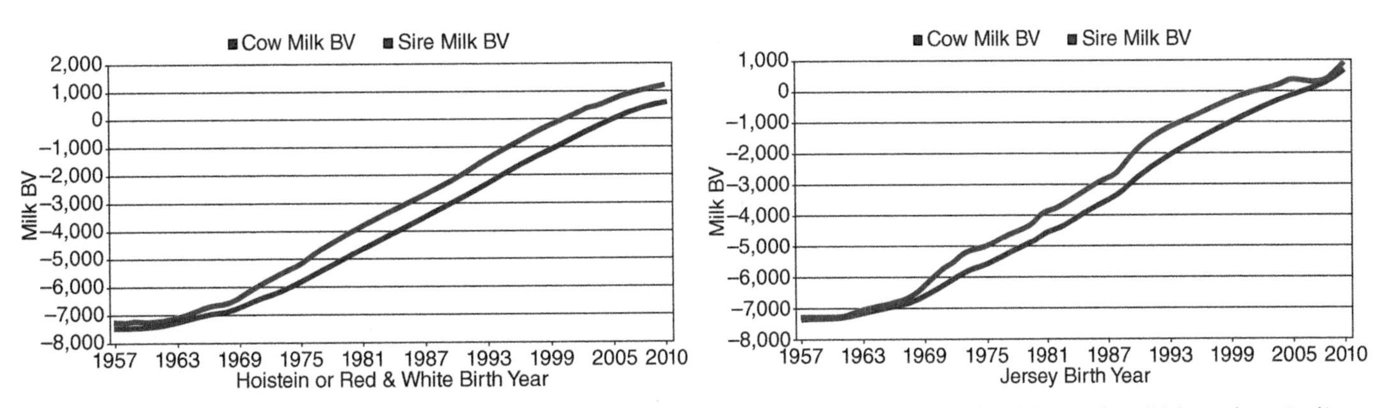

Figure 9.1 Average breeding value for milk yield (milk BV) for US Holstein and Jersey cattle, according to year of birth (source: http://aipl.arsusda.gov/eval/summary/trend.cfm).

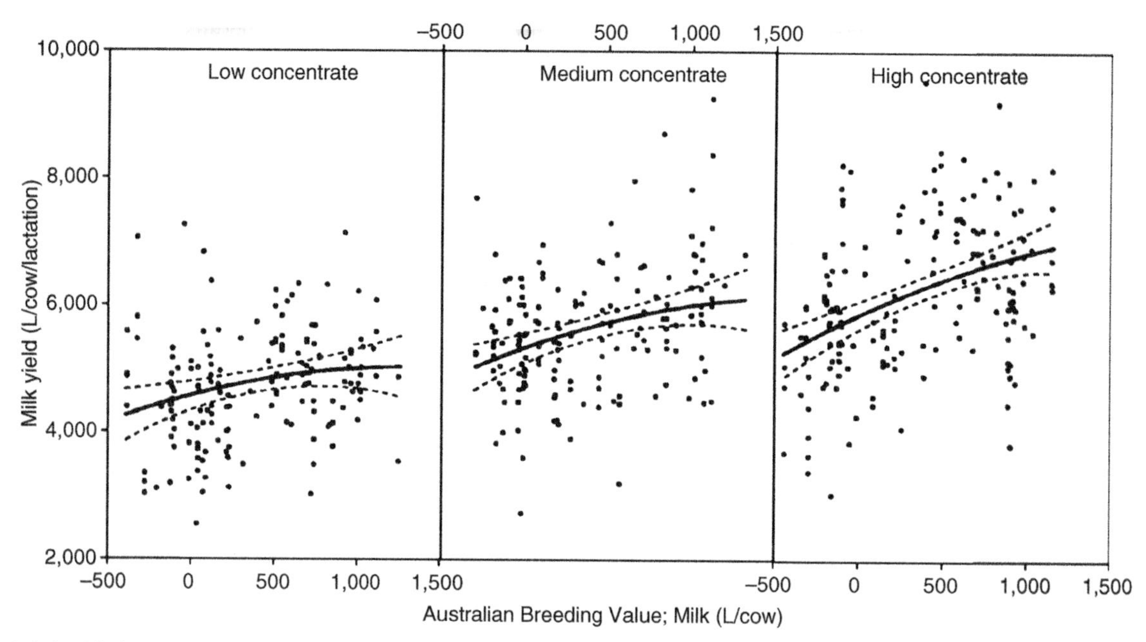

Figure 9.2 Relationship between Australian breeding value for milk yield and actual phenotypic level of milk yield per lactation, for three levels of supplemental concentrate feeding (source: Fulkerson, W. J., T. M. Davison, S. C. Garcia, G. Hough, M. E. Goddard, R. Dobbs, and M. Blockey 2008. Holstein-Friesian dairy cows under a predominantly grazing system: Interaction between genotype and environment. *J. Dairy Sci.* 91:826–839. Copyright (c) Elsevier, 2008).

intensive or effective management will realize greater gains in performance due to genetic selection. In other words, this type of G×E interaction will affect the return on investment that a dairy producer will observe when buying semen from genetically superior bulls, because the full genetic superiority of their offspring will not be realized if environmental conditions are limiting. Another type of G×E interaction occurs when sires actually rank differently based on performance of their daughters under different environmental conditions. For example, do farmers who utilize management-intensive rotational grazing need to select a different group of bulls to use as mating sires, as compared with farmers who feed a total mixed ration in a free stall barn? Research studies (Kearney et al., 2004; Weigel et al., 1999) indicate that the genetic correlation between daughters' performance in different environments is high, usually 0.75 or greater. This means that it is generally not necessary to use different sire families or genetic lines in different management systems, at least within the range of production environments present in leading dairy countries in temperate climates. Estimated genetic correlations between lactation performance in more than two dozen countries in North America, Western Europe, Oceania, the Middle East, and Africa can be found on the INTERBULL website (www-interbull.slu.se/eval/framesida-prod.htm).

Component percentages and cheese yield

In many countries and geographical regions (e.g., Wisconsin), the vast majority of milk is used for making cheese and other manufactured dairy products, rather than for fluid milk consumption. Therefore, increasing the production of **milk solids** (i.e., total fat and protein yield) is of greater interest than increasing milk volume or fat and protein percentage. Increases in total milk solids per cow can be achieved in two ways: (1) increasing milk yield while keeping fat and protein percentages constant; or (2) increasing fat and protein percentages while keeping milk yield constant. In practice, most breeding programs seek to do a combination of both through selection for total volume of milk solids (i.e., increasing fat yield and/or protein yield by direct selection on these traits, with little regard for whether the increase comes from milk volume or component percentages). In this manner, mean fat yield per lactation in US Holstein cows increased from 230 to 453 kg per lactation between 1960 and 2010, whereas mean protein yield increased from 225 to 376 kg per lactation between 1970 and 2010 (note that milk protein content was not measured routinely prior to 1970). To date, most dairy cattle selection programs have considered milk, fat, and protein as generic commodities. Recently, however, several research studies have focused on increasing the concentrations of specific milk fatty acids (e.g., saturated vs unsaturated, monounsaturated vs polyunsaturated) or specific milk proteins (e.g., caseins vs whey proteins) through genetic selection (e.g., Dutch Milk Genomics Initiative, www.milkgenomics.nl/). Fatty acid concentrations can be measured directly in the laboratory using gas chromatography, which is time-consuming and expensive, or they can be approximated using calibration equations and mid-infrared (MIR) spectral data of milk samples collected in DHI milk recording programs (Soyeurt et al., 2008). The primary motivation for changing the fatty acid composition of milk and other dairy products is to improve the nutritional value of milk and other dairy products, and achieving this goal will likely require a partnership between dairy farmers, breeding companies, milk processors, and food retailers. Changing the protein composition of milk through genetic selection is possible using phenotypes measured in the laboratory using high performance liquid chromatography (Wedholm et al., 2006) or capillary zone electrophoresis (Schopen et al., 2009), or perhaps more inexpensively by using phenotypes derived from MIR spectral data and corresponding calibration equations. The primary motivator for changing the ratios of specific caseins and whey proteins will likely be improved cheese-making efficiency in the dairy plant. Cooperation between dairy farmers, breeding companies, and milk processors will be needed to determine appropriate breeding goals and economic incentives.

Maintenance costs and efficiency of milk production

Selection for increased milk yield tends to increase the gross efficiency (GE) of milk production by diluting maintenance costs. Dairy nutritionists refer to this phenomenon in terms of **multiples of maintenance**, because the energy requirements for producing milk (net energy for lactation) in high-producing dairy cows are often 3 to 4 times greater than the energy requirements for routine body functions in the absence of lactation (net energy for maintenance), as noted by VandeHaar and St-Pierre (2006). Increasing the multiples of maintenance through genetic selection and improved management clearly increases farm profitability, because cows producing at 4.5 times maintenance are much more efficient than cows producing at 2.5 times maintenance. Capper et al. (2009) noted that increasing milk production per cow through genetic selection and improved management also confers environmental benefits – from 1944 to 2007 the amounts of manure, methane, and nitrous oxide required to produce 1 billion kg of milk decreased by 24, 43, and 56%, respectively. One concern is that high-producing cows may be prone to more health problems and higher veterinary costs than their low-producing contemporaries, and it is likely that more expensive and energy-dense diets may be required to achieve high levels of production. Furthermore, digestibility of feedstuffs may be depressed at very high levels of production, such that simply selecting for ever higher milk production may lead to smaller marginal gains in efficiency (VandeHaar and St-Pierre, 2006). For this reason, major research projects have been launched to quantify the potential gains that might be achieved by selection for improved feed efficiency in lactating dairy cattle, and to design breeding programs that can achieve this goal through genetic or genomic selection. Selection for **residual feed intake** (RFI), which is the difference between the actual feed energy consumed by a given cow and her expected energy requirements based on age, milk yield, body weight, body weight change, and body condition score, has been considered by researchers in swine and beef cattle for many years. In dairy cattle, selection for lower RFI must be balanced by animal health considerations. In other words, we want to create cows that can utilize feed more efficiently, but we don't want to create cows that lack adequate body condition (i.e., too thin) and therefore have health problems or impaired fertility. Veerkamp (1998) indicated that selection for improved feed efficiency is possible in lactating dairy cattle, given that dry matter intake, live weight, live weight change, and milk yield are moderately heritable traits. However, measurement of the individual cow feed intakes needed to compute RFI is time-consuming and expensive, and it is therefore not feasible to measure the intakes of thousands of daughters of progeny test bulls every year. On the other hand, genomic selection for improved feed efficiency using prediction equations derived from reference populations of cows measured on experimental farms may be feasible in the future.

Net merit index (income portion)

The aforementioned traits comprise the income portion of the **total merit index**. Selection to improve net lifetime profit per animal has been possible using selection indices provided by the USDA-ARS Animal Improvement Programs Laboratory (http://aipl.arsusda.gov/reference/nmcalc.htm) since 1994 (VanRaden et al., 2009). These indices are developed by USDA scientists, in cooperation with researchers at the major Land-Grant Universities, and three alternatives are offered. The flagship index, which is simply called **Net Merit** (NM$), is based on national average prices for fluid milk and its components, and this index represents the most appropriate breeding goal for most US dairy farmers. Relative weights (which sum to 100%) for production traits are 16% for protein yield and 19% for fat yield. Milk volume gets a weight of 0%, because it increases indirectly through its large and positive genetic correlations with fat and protein yield. The total weight on production traits is actually slightly greater than the 35% allocated to fat and protein yield, because length of productive life (which has a relative weight of 22%) is positively correlated with milk production. For dairy producers in cheese-producing regions of the US, who are compensated for producing milk with desirable manufacturing properties, an alternative index called **Cheese Merit** (CM$) is more appropriate. This index has relative weights of 25% for protein yield, 13% for fat yield, and −15% for milk yield, indicating that selection for total milk solids should come from improved fat and protein percentages, rather than through increased milk volume. Lastly, for dairy producers in regions where milk is used primarily for fluid consumption, and where incentives for milk components are negligible, selection should be based on **Fluid Merit** (FM$), which has relative weights of 0% for protein yield, 20% for fat yield, and 19% for milk yield. **Economic weights** for NM$, CM$, and FM$ are updated periodically to reflect trends in the pricing of milk and its components, as well as trends in feed costs. Alternative selection indices are offered by the breed associations, and selection emphasis on conformation traits tends to be somewhat higher for these indices than the Net Merit index. These include the **Total Performance Index** (TPI) by the Holstein Association USA (www.holsteinusa.com/genetic_evaluations/ss_tpi_formula.html) and the **Jersey Performance Index** (JPI) by the American Jersey Cattle Association (www.usjersey.com/Programs/JPIIntro.pdf).

Selection for functional traits

Calving performance

Genetic evaluations for calving ability are administered by the National Association of Animal Breeders (NAAB, Columbia, MO) and consist of service sire (direct) calving ease (SCE), daughter (maternal) calving ease (DCE), service sire (direct) stillbirth rate (SSB), and daughter (maternal) stillbirth rate (DSB) (http://aipl.arsusda.gov/reference/Form_GE_CT_1008.pdf). Calving ease evaluations are computed for the Holstein and Brown Swiss breeds, and calving ease evaluations are published as the expected percentage of difficult births in primiparous dams, whereas stillbirth evaluations are published as the expected percentage of calves dead at birth or by 48 hours of age in both primiparous and multiparous dams (Van Tassell et al., 2003). Heritability estimates are 9% for SCE, 5% for DCE, 3% for SSB, and 7% for DSB, so genetic progress will be slower for calving traits than many other traits, such as milk production. However, tremendous genetic variation exists for these traits. For example, some service sires produce calves that have only 3% difficult births, whereas other service sires produce calves that have 17% difficult births (the average is about 8%). Likewise, daughters of some sires have only 4% stillborn calves, whereas daughters of other sires have 13% stillborn calves (the average is about 12% for primiparous dams and 5% for multiparous dams). Calving difficulty (**dystocia**) and stillbirth rate are highly correlated with each other, as dystocia is a leading cause of stillbirths. Furthermore, calving difficulty and stillbirth rate are correlated with gestation length (Lopez de Maturana et al., 2009), although gestation length is not presently considered in most dairy cattle selection programs.

Male and female fertility

Selection for fertility involves two components: fertility of a straw of semen from a given AI bull, and fertility of the cow into which the semen is deposited (Weigel, 2004a). With regard to the former, virtually every AI stud routinely computes estimates of male fertility based on in-house data from their customers' herds (one dairy records processing center also provides such an evaluation for its customers). As described by Kuhn and Hutchison (2008), the USDA-ARS Animal Improvement Programs Laboratory routinely computes estimates of sire conception rate (SCR) for bulls that have been used in US dairy herds (http://aipl.arsusda.gov/reference/Form_PE_MFertility-SCR_1204.pdf). Insemination data corresponding to lactating cows are used, and the published evaluations reflect a combination of the bull's genetic predisposition for male fertility, the average male fertility of all bulls from the corresponding AI center, and the permanent environmental effects associated with that particular bull (i.e., it is a prediction of his future phenotype, rather than a prediction of the male fertility of his sons). As noted earlier, veterinary pregnancy examination results are typically used to confirm the outcomes of insemination events, in contrast to the practice used routinely in many European countries, where **non-return rate** (NRR) data are used to compute estimates of male and female fertility. With NRR, a cow is assumed to be pregnant if the farmer does not report a subsequent insemination within a predefined time period (usually about 56 days). However, there are many possible reasons for the lack of a subsequent insemination on a given cow, including: an unreported insemination to a natural service (clean-up) bull, a decision by the farmer to forego additional attempts to achieve pregnancy in this cow (i.e., she is marked do not breed), or culling of the cow from the herd. For these reasons, confirmed insemination outcomes are strongly preferred.

With respect to female fertility, the primary trait for selection is daughter pregnancy rate (DPR), which is computed from **days open** (DO) data for each cow (VanRaden et al., 2004). The transformation between DPR and DO is nonlinear, and a 1% increase in DPR corresponds to a decrease of approximately 4% in DO. The DPR evaluation for a particular bull is interpreted as the percentage by which his daughters will exceed, or fall short of, the average 21-day pregnancy rate in a given herd. The **21-day pregnancy rate**, in turn, refers to the ratio of cows that become pregnant in a given 21-day period (i.e., the length of a typical estrus cycle) divided by the number of cows in the herd that were eligible for breeding at the beginning of the 21-day period (i.e., non-pregnant cows that had passed the voluntary waiting period). Although the heritability of DPR is only about 4%, significant genetic variation exists, and the difference in PTA between the best and worst dairy sires for DPR is roughly 7.5%, which corresponds to a difference of about 30 days open in their daughters. In addition to DPR, which is influenced by length of the voluntary waiting period (i.e., the time from calving until the farmer tries to breed the cow, which is usually about 60 days), two additional traits are evaluated from the same data used to compute SCR evaluations for bulls: cow conception rate (CCR) and heifer conception rate (HCR). These traits reflect genetic superiority or inferiority in the probability of achieving conception when lactating cows or yearling heifers are inseminated, respectively (http://aipl.arsusda.gov/reference/Form_GE_FFertility_1008.pdf).

Udder health and mastitis

Mastitis, or inflammation of the mammary gland, is a key concern on commercial dairy farms, because it is the most prevalent and costly disease in the dairy industry. Furthermore, injuries to the teats and udder often lead to premature culling from the herd. For this reason, there is a desire to improve udder health and mastitis resistance through genetic selection. Historically, selection tended to focus on improving the physical conformation of the udder using genetic information for linear type traits. Several studies (e.g., Caraviello et al., 2003, 2004) have shown that udder depth, fore udder attachment, rear udder height and width, udder cleft, teat placement, and teat length are strongly associated with dairy cow survival. Because it is difficult for farmers to interpret genetic information for so many different traits, many of which are highly correlated with each other, the genetic evaluations for udder traits are weighted by their respective economic values and combined into a sub-index called **udder composite** (UDC, www.holsteinusa.com/genetic_evaluations/GenUpdateMain.html). Research has shown that a 1-point increase in the UDC of a dairy sire is associated with an extra 20 days of productive life among his daughters.

In many countries, SCC is routinely measured in milk recording programs as an indicator of udder health. As noted by Shook and Schutz (1994), genetic evaluations for SCS were developed to facilitate indirect selection for mastitis resistance (http://aipl.arsusda.gov/reference/Form_GE_SCS_1008.pdf). Because SCS is computed from SCC using the transformation: $SCS = \log_2(SCC/100000)+3$, a 1-unit increase in SCS means that SCC doubles (e.g., SCS of 3.00 corresponds to 100,000 somatic cells/ml, whereas SCS of 4.00 corresponds to 200,000 cells/ml). The heritability of SCS is about 12%, and the genetic correlation between SCS and the incidence of clinical mastitis is about +0.65, indicating that selection for lower SCS will lead to a reduction in mastitis. Interpreting PTAs for SCS is straightforward – if the PTAs for two bulls differ by 0.60, the difference in average SCS for their daughters is expected to be 0.60 in any given herd environment, and the SCC for daughters of the higher PTA bull is expected to be $2^{0.6} = 1.52$ times the SCC of the lower PTA bull in any given herd environment. Because the heritability of clinical mastitis (about 5%) is less than the heritability of SCC, and because the genetic correlation between SCS and clinical mastitis is high, selection for lower SCS can be an effective approach to reducing susceptibility to mastitis. In addition, SCS is measured objectively and consistently from farm to farm, whereas mastitis treatment decisions are subjective and vary according to the treatment protocols on individual farms. Direct selection against the incidence of clinical mastitis has not been practiced in North America despite research that indicates its feasibility (Zwald et al., 2004ab). In contrast, clinical mastitis has been part of the breeding program in Nordic countries for many years (Andersen-Ranberg et al., 2005; Heringstad et al., 2003).

Mobility

Like udder health, selection for improved mobility has relied largely on selection for improved physical conformation of the feet and legs, with economic weights derived from the genetic correlations between these traits and length of productive life. Linear type traits such as rear legs – side view, rear legs – rear view, and (to a lesser extent) foot angle, are clearly associated with cow survival (Caraviello et al., 2003, 2004). For simplicity of use in selection programs, the PTA values for these traits are typically combined into a sub-index called **foot and leg composite** (FLC, www.holsteinusa.com/genetic_evaluations/GenUpdateMain.html), and research has shown that a 1-point increase in FLC of a dairy sire corresponds to an increase of 10 days of productive life among his daughters. Recently, new traits such as feet and legs score and **locomotion score** have been introduced into selection programs in an attempt to accelerate genetic progress for improved mobility. Selection for clinical **lameness** has not been practiced to date, despite research that suggests its potential (Zwald et al., 2004a,b), largely because lameness data from commercial farms tend to be inconsistent and dairy producers tend to underestimate the severity and incidence of lameness in their herds.

Frame size and body condition

Frame size, as measured by linear type traits such as stature, strength, body depth, and rump width, tends to have little or no association with survival (Caraviello et al., 2003, 2004). Nonetheless, large cows and heifers tend to be favored in cattle shows, and direct selection pressure for these traits has led to increased frame size in North American Holstein cattle. Because space requirements, and hence stall dimensions, tend to be multiplicative functions of the height and length of a cow, farmers frequently face the need to renovate their facilities to accommodate larger cows. Body traits are highly heritable, ranging from about 26% for rump width to 42% for stature, so changes in body size can be achieved very rapidly. In practice, farmers can use a sub-index called **body size composite** (BSC, Tsuruta et al., 2004), which is comprised of the aforementioned linear type traits and for which 1 point corresponds to approximately 10 kg of mature body weight. It should be noted that BSC is penalized in the NM$ index, where a relative weight of −6% reflects the desire to reduce feed costs associated with the maintenance requirements of large cows. **Body condition score** (BCS) has been used by nutrition consultants and veterinarians for many years to assess nutritional management on commercial farms, and recently this trait has gained attention in dairy cattle selection programs. Cows with low BCS values tend to have a greater incidence of early postpartum health problems and a higher risk of culling, so the linear type trait dairy form (which has a large, negative genetic correlation with BCS) receives a slight negative weight (-1%) in the Holstein Association TPI formula, and dairy producers should avoid bulls that sire cows with extreme dairy form or angularity.

Culling and longevity

The most direct measure of a cow's health and ability to survive on a commercial dairy farm is the length of her productive life, from first calving until culling from the herd, with genetic evaluations for PL (http://aipl.arsusda.gov/reference/Form_GE_Longevity_1008.pdf) introduced in 1994 (VanRaden and Klaaskate, 1993; VanRaden and Wiggans, 1995). The most common causes of culling on US dairy farms are infertility and mastitis, followed by low milk production and lameness. The term **voluntary culling** is sometimes used to refer to the removal of healthy animals from the herd due to low production (these animals may be sold to another farm, or they may be sold for slaughter), whereas the term **involuntary culling** is used to refer to the removal of cows with problems such as infertility, mastitis, lameness, injury, other illness, or death. In dairy cattle breeding programs, we seek to reduce mortality rate and the incidence of involuntary culling of sick or injured animals and increase opportunities for voluntary culling of unprofitable animals. The accuracy of genetic evaluations for PL tends to be relatively low for young cows (and young sires), in part because the heritability of PL is only about 8%, but also because these young cows haven't had an adequate opportunity period to express their survival phenotype. Nonetheless, genetic differences between sire families are large, with a range of about 10 months in survival between daughters of the best and worst sires.

Net merit index (expense portion)

Over time, selection indices have been modified to reflect changes in breeding goals (Shook, 2006; VanRaden, 2004). Initially, the vast majority of weight was dedicated to increasing production traits (i.e., higher income), whereas the past two decades have witnessed a shift toward equal or greater emphasis on improvement of fitness traits (i.e., lower expenses). The **selection response** to this change in trait emphasis has been remarkable, as shown for DPR and SCC in Figure 9.3. The recent changes in

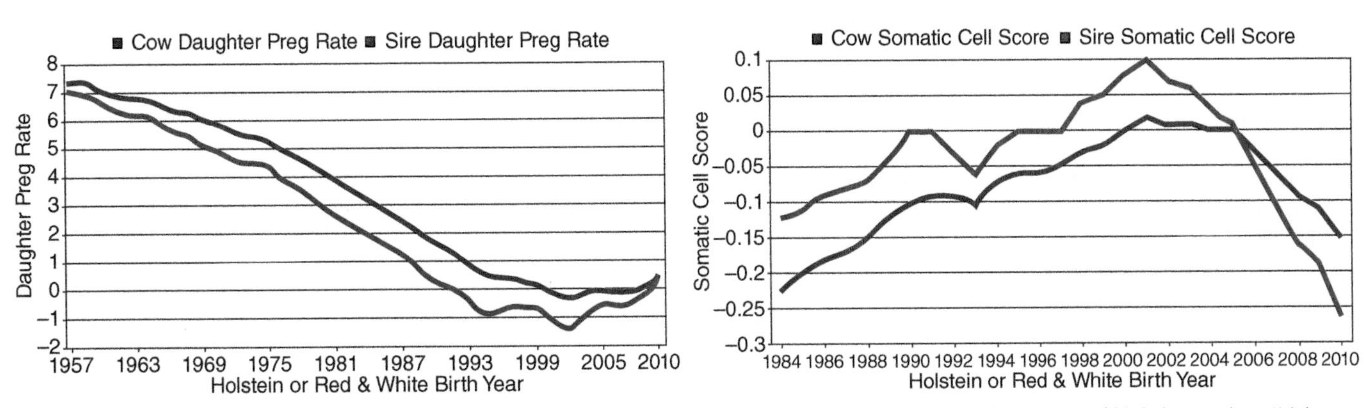

Figure 9.3 Average breeding value for daughter pregnancy rate and somatic cell score in US Holstein cattle, according to year of birth (source: http://aipl.arsusda.gov/eval/summary/trend.cfm).

genetic trends for these traits demonstrate the power of selection and the importance of defining selection goals clearly and updating selection index weights regularly.

At present (http://aipl.arsusda.gov/reference/nmcalc.htm), PL receives the highest relative weight in NM$ among the fitness traits, at 22%, followed by DPR at 11%, and SCS at −10%. Udder composite, feet and legs composite, and body size composite receive relative weights of 7%, 4%, and −6%, respectively, and calving ability (a sub-index comprised of direct and maternal calving ease and stillbirth rate) receives a relative weight of 5%. In general, the breed association indices (JPI, TPI, etc.) tend to put slightly less emphasis on fitness traits than in NM$, whereas type traits receive a slightly higher weight, because some pedigree breeders can capture additional revenue by marketing live cattle or embryos with superior genetic merit for type traits. Detailed information about the selection indices used in leading dairy countries around the world, as well as the economic weights used for production, conformation, health, and fertility traits in each country, can be found in a review paper by Miglior et al. (2005).

Sire selection

Selection index or independent culling levels?

Geneticists typically recommend that dairy producers should choose the bulls to use in their herds using a **selection index**, in which all traits are combined according to their respective economic values, rather than using **independent culling levels**, in which minimum threshold or cut-off values are applied to specific traits. The reason is four-fold. First, selection indices are developed using economic weights that are based on current prices for inputs (e.g., feed costs, veterinary costs, replacement costs) and outputs (e.g., milk price, milk quality premiums, cow salvage values). On the other hand, the cut-off values that are applied by producers who use independent culling levels are arbitrary and are rarely based on scientific evidence. Furthermore, the appropriate cut-off values vary over time due to genetic progress and changes in definition of the genetic base population. Usually the genetic base (i.e., group of animals with average PTA = 0) is updated every 5 years, and culling levels that are appropriate one year may be too lenient or too stringent a year or two later. Second, genetic correlations between traits are considered when calculating the weights used in selection indices (e.g., strong positive correlation between milk yield and protein yield, moderate negative correlation between milk yield and female fertility). However, double-counting of positively correlated traits is common when independent culling levels are used, and undesirable and unexpected correlated responses can occur when negative genetic correlations exist between traits. Third, selection indices work well regardless of the number of traits in the **breeding goal**. Conversely, the effectiveness of selection based on independent culling levels decreases rapidly as the number of traits increases. If selection candidates are forced to surpass the arbitrary cut-off values for a large number of traits, the few animals that meet these criteria will be only marginally superior for any given trait. Fourth, selection indices allow animals that are highly superior for one trait to compensate for slight deficiencies in other traits, and this leads to maximization of the selection response in terms of net economic value.

Reliability and management of risk

Reliability (REL), which is an estimate of precision, refers to the squared correlation between the estimated breeding value and true breeding value of a given animal. Because the true breeding value is unknown, reliability is approximated based on the amount of phenotypic data available for a given animal. The term **daughter equivalents** is often used to quantify the contributions of information from an animal's ancestors, from its own performance, and from its descendants. The most important factor affecting REL is the quantity of phenotypic data available for an animal, its ancestors, and its descendants, but heritability is also a factor. For traits with low heritability, such as female fertility, a large number of progeny (e.g., 200) will be needed to achieve an level of REL that will lead to accurate selection decisions, whereas for traits with high heritability, such as stature, relatively few progeny (e.g., 50) will be needed to achieve a similar level of REL. In practice, the proper way to use REL is to limit the risk associated with imprecision in the EBV of an individual bull. For example, a farmer might choose to buy 300 units of semen from a bull with 95% REL, whereas he or she might choose to buy 30 units of semen from each of ten different bulls with 65% REL. An important concept to consider is the reliability of a **team** of selected bulls (Schefers and Weigel, 2012), where: Team REL = 1 − [(1 − average REL of individual bulls in the team)/(number of bulls in the team)]. For example, if a dairy producer uses a team of 10 bulls, each with 65% REL, the average EBV for this team would have 96.5% reliability.

Gender-enhanced semen

The promise of sexed or **gender-enhanced semen** was finally realized in dairy cattle nearly a decade ago, and in recent years this technology has achieved widespread adoption (Norman et al., 2010). The most cost-effective approach to using gender-enhanced semen is in yearling heifers (Weigel, 2004b), because they have higher conception rates than lactating cows and, on average, higher genetic merit as well. The availability of gender-enhanced semen (along with genomics, which is discussed later) has led to increased interest in genetics and pedigree recording, because unless pedigree records are accurate and complete, a dairy producer cannot determine which heifers are the best candidates for breeding with expensive, gender-enhanced semen. In general, using gender-enhanced semen for the first one or two inseminations per heifer is recommended, because conception rates are compromised with gender-enhanced semen and age at first calving can be delayed if a heifer has multiple failed insemination attempts. Some dairy producers are (for the first time in history) now in a situation in which the number of available replacement heifers exceeds the number that will be needed to maintain the herd, in part due to the availability of gender-enhanced semen, and in part due to modernization of dairy facilities (e.g., adequate stall size, comfortable sand bedding, proper ventilation). Therefore, opportunities for selection among replacement heifers may exist, and this could further accelerate genetic progress.

Summary

Genetic improvement programs in dairy cattle have achieved remarkable progress due to cooperation between milk recording organizations, data processing centers, breed associations, artificial insemination companies, government agencies, and agricultural universities. Enormous databases of pedigree and performance information have been assembled, such that the amount of data that are publicly available for carrying out research projects and making practical breeding decisions vastly exceeds that of any other food animal species. The dairy cattle selection programs discussed herein largely reflect the current situation in North America, but genetic evaluation systems and selection strategies in Europe, Oceania, and other developed dairy-producing regions are quite similar. Rapid genetic progress has been realized for nearly all economically important traits using pedigree-based selection, in conjunction with reproductive technologies such as artificial insemination, embryo transfer, *in vitro* embryo fertilization, and gender-enhanced semen. Selection goals have evolved over time, from increasing milk production, to improving milk solids and physical conformation, to enhancing fertility, longevity, and other fitness traits. As we look to the future, selection programs will be broadened to include production efficiency traits, such as residual feed intake to control feed costs and casein yield to enhance cheese-making potential of the milk, as well as consumer traits, such as alteration of the fatty acid content of milk or modifying the concentrations of certain whey proteins with desirable health properties. At the same time, selection programs should keep one eye on societal trends, such as animal welfare concerns (e.g., selection for polled cattle rather than de-horning) and the desire for organic products (e.g., select for resistance to diseases rather than treatment with antibiotics).

Further reading

Weigel, K. A., P. C. Hoffman, W. Herring, and T. J. Lawlor, Jr. 2012. Potential gains in lifetime net merit from genomic testing of cows, heifers, and calves on commercial dairy farms. *J. Dairy Sci.* 95:2215–2225.

Weigel, K. A., and S. W. Lin. 2002. Controlling inbreeding by constraining the average relationship between parents of young bulls entering AI progeny test programs. *J. Dairy Sci.* 85:2376–2383.

Wiggans, G. R., T. A. Cooper, C. P. Van Tassell, T. S. Sonstegard, and E. B. Simpson. 2013. *Technical note:* Characteristics and use of the Illumina BovineLD and GeneSeek Genomic Profiler low-density chips for genomic evaluation. *J. Dairy Sci.* 96:1258–1263.

Wiggans, G. R., P. M. VanRaden, and T. A. Cooper. 2011. The genomic evaluation system in the United States: Past, present, future. *J. Dairy Sci.* 94:3202–3211.

References

Andersen-Ranberg, I., G. Klemetsdal, B. Heringstad, T. Steine. 2005. Heritabilities, genetic correlations, and genetic change for female fertility and protein yield in Norwegian Dairy Cattle. *J. Dairy Sci.* 88: 348–355.

Berger, P. J. 1994. Genetic prediction for calving ease in the United States: Data, models, and use by the dairy industry. *J. Dairy Sci.* 77:1146–1153.

Capper, J. L., R. A. Cady, and D. E. Bauman. 2009. The environmental impact of dairy production: 1994 compared with 2007. *J. Anim. Sci.* 87:2160–2167.

Caraviello, D. Z., K. A. Weigel, and D. Gianola. 2003. Analysis of the relationship between type traits, inbreeding, and functional survival in Jersey cattle using a Weibull proportional hazards model. *J. Dairy Sci.* 86:2984–2989.

Caraviello, D. Z., K. A. Weigel, and D. Gianola. 2004. Analysis of the relationship between type traits and functional survival in US Holstein cattle using a Weibull proportional hazards model. *J. Dairy Sci.* 87:2677–2686.

Cole, J. B., D. J. Null, and P. M. VanRaden. 2009. Best prediction of yields for long lactations. *J. Dairy Sci.* 92:1796–1810.

Cole, J. B., G. R. Wiggans, and P. M. VanRaden. 2007. Genetic evaluation of stillbirth in United States Holsteins using a sire-maternal grandsire threshold model. *J. Dairy Sci.* 90: 2480–2488.

Dekkers, J. C. M., J. H. ten Haag, and A. Weersink. 1998. Economic aspects of persistency of lactation in dairy cattle. *J. Dairy Sci.* 53:237–252.

Fulkerson, W. J., T. M. Davison, S. C. Garcia, G. Hough, M. E. Goddard, R. Dobbs, and M. Blockey. 2008. Holstein-Friesian dairy cows under a predominantly grazing system: Interaction between genotype and environment. *J. Dairy Sci.* 91:826–839.

Funk, D. A. 2006. Major advances in globalization and consolidation of the artificial insemination industry. *J. Dairy Sci.* 89:1362–1368.

Heringstad, B., R. Rekaya, D. Gianola, G. Klemetsdal, and K. A. Weigel. 2003. Genetic change for clinical mastitis in Norwegian cattle: A threshold model analysis. *J. Dairy Sci.* 86:369–375.

Heringstad, B. 2010. Genetic analysis of fertility-related diseases and disorders in Norwegian Red cows. *J. Dairy Sci.* 93: 2751–2756.

Kearney, J. F., M. M. Schutz, P. J. Boettcher, and K. A. Weigel. 2004. Genotype by environment interaction for grazing versus confinement. I. Production traits. *J. Dairy Sci.* 87: 501–509.

Kuhn, M., and J. L. Hutchison. 2008. Prediction of dairy bull fertility from field data: Use of multiple services and identification and utilization of factors affecting bull fertility. *J. Dairy Sci.* 91:2481–2492.

Lopez de Maturana, E., X. L. Wu, D. Gianola, K. A. Weigel, and G. J. M. Rosa. 2009. Exploring biological relationships between calving traits in primiparous cattle with a Bayesian recursive model. *Genetics* 181:277–287.

Miglior, F., B. L. Muir, and B. J. Van Doormaal. 2005. Selection indices in Holstein cattle of various countries. *J. Dairy Sci.* 88:1255–1263.

Norman, H. D., T. R. Meinert, M. M. Schutz, and J. R. Wright. 1995. Age and seasonal effects on Holstein yield for four regions of the United States over time. *J. Dairy Sci.* 78:1855–1861.

Norman, H. D., J. L. Hutchison, and R. H. Miller. 2010. Use of sexed semen and its effect on conception rate, calf sex, dystocia, and stillbirth of Holsteins in the United States. *J. Dairy Sci.* 93:3880–3890.

Powell, R. L., and H. D. Norman. 1984. Response within herd to sire selection. *J. Dairy Sci.* 67:2021–2027.

Schefers, J. M., and K. A. Weigel. 2012. Genomic selection in dairy cattle: Integration of DNA testing into breeding programs. *Animal Frontiers* 2:4–9.

Schopen, G. C. B., J. M. L. Heck, H. Bovenhuis, M. H. P. W. Visker, H. J. F. van Valenberg, and J. A. M. van Arendonk. 2009. Genetic parameters for major milk proteins in Dutch Holstein-Friesians. *J. Dairy Sci.* 92:1182–1191.

Shook, G. E. 2006. Major advances in determining appropriate selection goals. *J. Dairy Sci.* 89:1349–1361.

Shook, G. E., and M. M. Schutz. 1994. Selection on somatic cell score to improve resistance to mastitis in the United States. *J. Dairy Sci.* 77:648–658.

Soyeurt, H., P. Dardenne, F. Dehareng, C. Bastin, and N. Gengler. 2008. Genetic parameters of saturated and monounsaturated fatty acid content and the ratio of saturated to unsaturated fatty acids in bovine milk. *J. Dairy Sci.* 91:3611–3626.

Tsuruta, S., I. Misztal, and T. J. Lawlor. 2004. Genetic correlations among production, body size, udder, and productive life traits over time in Holsteins. *J. Dairy Sci.* 87:1457–1468.
VandeHaar, M. J., and N. St-Pierre. 2006. Major advances in nutrition: Relevance to the sustainability of the dairy industry. *J. Dairy Sci.* 89:1280–1291.
VanRaden, P. M. 2004. *Invited review:* Selection on net merit to improve lifetime profit. *J. Dairy Sci.* 87:3125–3131.
VanRaden, P. M., and E. J. H. Klaaskate. 1993. Genetic evaluation of length of productive life including predicted longevity of live cows. *J. Dairy Sci.* 76:2758–2764.
VanRaden, P. M., A. H. Sanders, M. E. Tooker, R. H. Miller, H. D. Norman, M. T. Kuhn, and G. R. Wiggans. 2004. Development of a national genetic evaluation for cow fertility. *J. Dairy Sci.* 87:2285–2292.
VanRaden, P. M., C. P. Van Tassell, G. R. Wiggans, T. S. Sonstegard, R. D. Schnabel, J. F. Taylor, and F. Schenkel. 2009. Reliability of genomic predictions for North American dairy bulls. *J. Dairy Sci.* 92:16–24.
VanRaden, P.M., and G.R. Wiggans. 1991. Derivation, calculation, and use of national animal model information. *J. Dairy Sci.* 74:2737–2746.
VanRaden, P. M., and G. R. Wiggans. 1995. Productive life evaluations: Calculation, accuracy, and economic value. *J. Dairy Sci.* 78:631–638.
Van Tassell, C. P., G. R. Wiggans, and I. Misztal. 2003. Implementation of a sire-maternal grandsire model for evaluation of calving ease in the United States. *J. Dairy Sci.* 86:3366–3373.
Van Tassell, C. P., and L. D. Van Vleck. 1991. Estimates of genetic selection differentials and generation intervals for four paths of selection. *J. Dairy Sci.* 74:1078–1086.
Veerkamp, R. F. 1998. Selection for economic efficiency of dairy cattle using information on live weight and feed intake: a review. *J. Dairy Sci.* 81:1109–1119.
Wedholm, A., L. B. Larsen, H. Lindmark-Månsson, A. H. Karlsson, and A. Andrén. 2006. Effect of protein composition on the cheese-making properties of milk from individual dairy cows. *J. Dairy Sci.* 89:3296–3305.
Weigel, K. A. 2004a. Improving the reproductive efficiency of dairy cattle through genetic selection. *J. Dairy Sci.* 87:E86–92.
Weigel, K. A. 2004b. Exploring the impact of sexed semen on dairy cattle improvement programs. *J. Dairy Sci.* 87:E120–130.
Weigel, K. A., T. Kriegl, and A. L. Pohlman. 1999. Genetic analysis of dairy cattle production traits in a management intensive rotational grazing environment. *J. Dairy Sci.* 82: 191–195.
Zwald, N. R., K. A. Weigel, Y. M. Chang, R. D. Welper, and J. S. Clay. 2004a. Genetic selection for health traits using producer-recorded data. I. Incidence rates, heritability estimates, and sire breeding values. *J. Dairy Sci.* 87: 4287–4294.
Zwald, N. R., K. A. Weigel, Y. M. Chang, R. D. Welper, and J. S. Clay. 2004b. Genetic selection for health traits using producer-recorded data. II. Genetic correlations, disease probabilities, and relationships with existing traits. *J. Dairy Sci.* 87:4295–4302.

Review questions

1. How do dairy cattle breeding programs differ from swine and poultry breeding programs?

2. How do the data collection organizations, such as National DHIA, AgSource Cooperative Services, and California DHIA, benefit producers and the dairy cattle genetics as a whole?

3. What traits are utilized to measure fertility and longevity in dairy cattle; how are they calculated?

4. Why is progeny testing important in dairy cattle and how is it performed?

5. What is a contemporary group and why is it used in genetic evaluations?

6. For what reason are breeding values based solely on additive genetic effects and not dominant or epistatic genetic effects?

7. What is the difference between EBV and PTA?

8. What other milk components have dairy producers selected for besides total milk yield; why are these other components important?

9. Which traits are utilized in the calculation of Net Merit (NM$), Cheese Merit (CM$), and Fluid Merit (FM$); how and why do these indices differ?

10. How does SCC differ from SCS and why is SCS used as a trait for udder health instead of SCC or the incidence of clinical mastitis?

11. Give several reasons as to why selection indices, rather than individual traits, should be used in the selection of sires.

10 Genetic and Genomic Improvement of Pigs

Max F. Rothschild
Department of Animal Science, Iowa State University, IA, USA

Introduction

Pork production is an important source of animal protein worldwide. Pork accounts for nearly 43% of all red meat consumed worldwide. It is expected that in the developed world meat consumption will remain steady or increase incrementally in the next 10–20 years, but enormous demand will push meat consumption much higher in developing countries in Asia and Africa. Furthermore, in many of these developing countries livestock production offers families an opportunity for economic survival. Therefore, **pig** production is likely to grow and be an integral part of livestock production both in developed and developing countries except where prohibited for religious reasons. To meet future needs, **genetic** and **genomic** improvement of pigs must increase. This chapter outlines past and present methods to improve the pig genetically for production of meat and puts them in the context of advances in the fields of genomics. Traditional **swine breeding** approaches are presented in earlier textbooks and listed in suggested reading.

Domestication of swine and breed development

Based on current classification, the pig belongs to the order Certartiodactya, which includes even-toe ungulates and whales and dolphins. **Suiformes**, which include pigs and peccaries, are one of four primary suborders. There are two living families within the Suiformes: Tayassuidae (peccaries) and **Suidae** (pigs). The modern Suidae family consists of five or six genera and as many as 18 or 19 recognized species, although this is still not clearly known. The genus *Sus* contains ***Sus scrofa***, the primary ancestor for all domesticated pigs, although other species may have been involved.

The earliest fossil records of ***S. Scrofa*** in Europe date back over 750,000 years ago. *S. scrofa* spread across Europe and Asia and was introduced to other regions of the world. For a more detailed explanation of evolution of the pig see *The Genetics of the Pig* (Ruvinsky et al., 2011).

Domestication of the pig was once believed to have been initiated about 10,000 years ago in a few geographically isolated regions of the Near East and China. However, this model of domestication is now being challenged by information suggesting that the pig was domesticated in several additional regions outside of the Near East and China including Europe and perhaps Japan. One point should be emphasized relative to the modern species that we now use in pork production: although earlier researchers called this present-day species *S. domesticus*, no such species exists and the modern pig is known as *S. scrofa*. The most recent and complete discussion of pig domestication has been presented in *The Genetics of the Pig* (Larson et al., 2011).

The process of domestication of the pig led to the first efforts by humans to improve the pig. Methods used by early humans included selection and systems of mating which are the same approaches used by modern swine breeders. Although records of **animal improvement** date back to about 6000 years ago, most involved ruminants and horses. Animal improvement in the Dark and Middle ages was modest at best.

A resurgence of activities and methods to improve **livestock** occurred in the eighteenth century, and its geographical home was England. Much of this came about due to end of the feudal system and the change to more individual owners who took to learning improved methods of farming and raising livestock. The best known of these master breeders was **Robert Bakewell**, an Englishman born in the early 1700s who began his animal breeding work about 1760. Similar activities began in the Americas as the colonies were being settled. The methods Bakewell espoused were widely imitated and the beginnings of **purebreds** and **breed societies** were constructed.

A **breed** may be defined as a group of animals of the same species and developed by livestock producers to possess similar characteristics, most often phenotypic. Development of a breed was often done by **selection** of animals for similar coat color or other characteristics, and some level of inbreeding was practiced to fix the characteristics. Breeders then formed an association to protect their stock and investment. Early **genetic improvement** in the late 1800s and the first half of the 1900s was in large part due to the formation of breed societies and within breed improvement.

Today there exist over 300 breeds of **swine** worldwide. Many of these breeds are often similar but reside in different countries so are considered by some as different breeds. Information on over 70 breeds of swine, including pictures, can be found at an excellent web site: www.ansi.okstate.edu/breeds/swine/. In addition, the "Breeds of Pigs" chapter in *Genetics of Swine* (Buchanan

Molecular and Quantitative Animal Genetics, First Edition. Edited by Hasan Khatib.

and Stalder, 2011) contains an enormous amount of phenotypic information, references about the breeds and source of origin material. Additional information about breed diversity and pig resources is fully described in Ollivier and Foulley (2011).

Despite the rather large number of breeds available, most of the commercial pork industry uses primarily a limited number of these breeds for pork production as either **purebreds** or in **synthetic (man-made) crosses**. These include breeds such as Landrace and Large White or Yorkshire that are used as dam lines due to their superior **reproductive performance** and **mothering ability**. In addition, breeds such as Pietrain, Hampshire, and Duroc are used as **sire lines** due to their rapid **growth rate** and excellent lean meat production. For specialized niche markets, the Berkshire breed has seen an increase in its use due to its superior **meat quality**. Use of such breeds to produce both **specialized crosses** and synthetic lines is now common and has replaced the older systems of **static** (F1) or **rotational crosses** on large commercial pig operations.

Each year, dire warnings about the loss of **rare breeds** worldwide are given. Certainly many local country breeds are endangered due to their limited economic value and they are often being crossed with more productive breeds. Information on rare pig breeds in the US is available from the **American Livestock Breeds Conservancy** (http://albc-usa.org/) and, for breeds worldwide, from the United Nations' **Food and Agriculture Organization** (FAO) (www.fao.org/). The FAO has adopted a strategy for breed conservation but sadly the financial resources needed to maintain them do not exist. Furthermore, unless they fit some specialized production scheme rare breeds are often overlooked as not viable economically. In the future, genomic evaluation may reveal specific underlying potential genetic contributions that may be worth preserving.

Methods of selection and mating systems

The primary method of genetic improvement is the method of selection, that is, **selection** of the best animals of each generation to be parents of the next generation. Selection, if effective, changes **gene frequency** of desirable alleles and increases their number in the population. Selection, based on phenotype, can be expected to move the mean value of the trait based on the formula:

$$\text{Progeny deviation} = \text{heritability} * (\text{selected parental mean} - \text{population mean}),$$

where **heritability** represents **accuracy** and (selected parental mean – population mean) is the **selection differential**.

This can be seen as:

$$\Delta G = a * i,$$

where "G" is genetic change (improvement), "a" is accuracy, and "i" is **selection intensity**.

ΔG per year is $\Delta G = a * i / t$, where t is **generation interval** of selected parents.

Initially, pig breeders selected animals based on the phenotypic traits of each pig. Breeders might choose animals that were the largest, appeared to be the leanest or had good feet and legs. Today more sophisticated methods are employed including the use of scales to weigh, ultrasound to measure fat and loin eye area and sophisticated scoring methods for anatomical traits. Data on animals and their relatives' records are then used to estimate **breeding values** (see Chapter 5). Some independent culling is performed for deformities (e.g., see Nicholas, 2011) but in general all major breeding organizations collect phenotypes and estimate breeding values to select parents for the next generation based on the best breeding values. Selection based on **markers** or sequence information is discussed later in this chapter.

Once parents are chosen then the use of mating systems is crucial to maximize performance. Such performance depends highly on gene action. Although selection acts to increase gene frequency of the desired alleles, it is effective for traits with a large percentage of additive gene action and hence a higher heritability. **Inbreeding** and **crossbreeding** effects depend greatly on **nonadditive gene effects** (**dominance**, **epistasis**). A review of the relative effects and gene action affecting certain traits is seen in Table 10.1.

Historically, there have been three phases of mating systems employed in swine production. Until about 1950, use of purebreds mating schemes was primarily employed and these often used mild levels of **inbreeding** (mating of relatives). The most popular of these approaches was **line breeding** in herds in which breeders used relatives with a common superior ancestor

Table 10.1 Effects of mating systems on traits of economic importance in swine.

Trait[a]	Heritability	Effect of		Proportion of genetic variance due to different gene action	
		Inbreeding	Crossbreeding	Nonadditive	Additive
Litter size	low	high/unfavorable	high/favorable	high	low
Weaning wt	low	high/unfavorable	high/favorable	high	low
Wt gain to mkt	moderate	moderate/unfavorable	moderate/unfavorable	moderate	moderate
Backfat	high	low/unfavorable	low/favorable	low	high
Carcass quality	high	low/unfavorable	low/favorable	low	high

Source: Modified from Lasley, 1987.
[a]Abbreviations: Wt = weight; mkt = market.

with a desired phenotype. Such approaches are still employed by some purebred breeders. However, such an approach suffers from several problems. First, the "ideal" phenotype may change over time or a better animal may exist outside of a given herd. Second, the ideal individual may carry an undesirable deleterious recessive allele that shows up in some ancestors. After the 1950s, purebred breeders continued to use line breeding and selection but commercial pig producers generally bought purebred males and females and then made static **F1 crosses** to produce commercial females or employed **rotational crosses**. Such approaches of using an F1 female and a purebred sire (of a different breed) maximized (100%) female and pig **heterosis**. Rotational crosses will not maximize piglet heterosis unless a sire breed different from the rotationally produced female is used, and rotational crosses will not maximize female heterosis.

The advent of commercial breeding companies following advances in hybrid corn breeding brought big changes to the pig industry as well. Commercial breeding companies began to use more sophisticated selection methods based on advancing technologies to first measure traits and secondly to estimate breeding values from individual performance and relatives' records. Selection of superior animals (the top 5% males and 20% females) within pure lines occurred primarily at the top or nucleus of the breeding pyramid (see Figure 10.1), and crosses were made then initially to produce synthetic lines. Later the superior individuals in both the pure and synthetic lines were then selected to create the next generation of the elite pure and synthetic lines. For **nucleus herds**, attempts were made to maximize the selection intensity (called i) while minimizing generation interval (called t) such that the ratio of i/t was maximized for the population in question. Also selected were the next level of superior animals (approximately males in the top 20% and females in the top 50%); these were sent to the multiplication farms and were used to produce crossed females and males that were then sold to large commercial production units. Development of males (sire) and dams (lines) and their crosses are based on maximizing total economic output. Therefore, the breed background and traits emphasized are often different to maximize both heterosis and complementarity of strengths from different breeds.

It has been estimated that use of purebreds accounted for over 80% of the commercial pigs produced in 1980. Today, less than 5% of all commercial pigs are produced by purebred schemes and the remaining 95% result from usually static crosses involving synthetic crossbred male and female lines in order to maximize heterosis and total output of the economically important traits at all stages of the pork production.

Traits of economic importance

Traits of economic importance can be defined as those characteristics of the pig that contribute to highest profitability. Such traits have varied over time and hence selection objectives need to be reevaluated as economics and production environments change. Factors affecting pork's efficient production are vitally important as are traits that affect consumer preferences and pork consumption. The most important traits for pork production in the finishing phase are lean growth, feed intake, and pig survival. Arguably, the two most economically important traits for pork production are reproductive traits and **disease resistance**. There are several reproductive traits of interest to the pig industry, the two most important being the number of pigs weaned per sow per year and the other, more overlooked, trait being the reproductive lifetime production of the sow herself. Several research groups have conducted research that has clearly shown genetic variation for these traits. Though consumers are most concerned about the degree of fatness or carcass merit as well as pork quality, pork producers must also pay attention to the ever-growing demand by consumers that the pigs be grown without the use of antibiotics as growth promoters and in facilities that improve animal welfare. Additionally, pork producers must do all of this while becoming more environmentally conscious by having pigs reduce feed wastage, improve feed efficiency, and produce waste that contains less phosphorous.

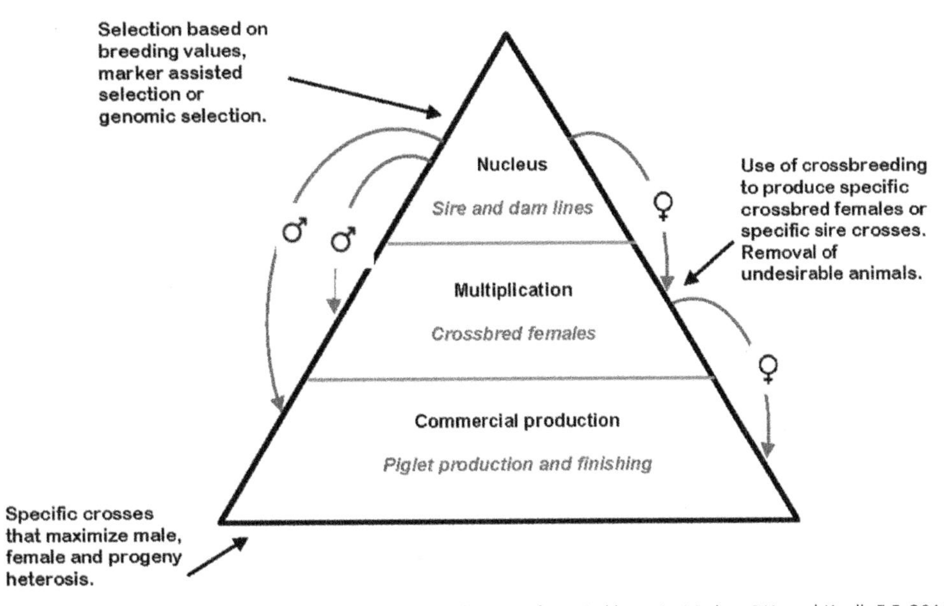

Figure 10.1 Selection and breeding pyramid in modern pig breeding (adapted in part from Dekkers, J., Mather, P.K. and Knoll, E.F. 2011. Genetic improvement of the pig. In Rothschild, M.F. and A. Ruvinsky (eds), *The Genetics of the Pig*. 2nd edn. Ch. 16. CAB International, Wallingford, 507 pp).

Table 10.2 Example of traits and their heritabilities and relative economic values.

Trait	Heritability	Relative Economic Value	Explanation of economic value
Growth rate	Moderate	++++	Faster growth leads to lower facility costs
Feed efficiency	Moderate	+++++	More efficient (less feed) usually leads to slightly slower growth; price and competition for feed determines which trait is more important
Meat/lean %	Moderate	++++	lbs. of lean pork product for most production units
Meat quality	Moderate to high	++	Better quality is desired but not valued except in niche markets
Litter size/ number weaned	Low	+++++	Most important female traits
Sow longevity	Low	++	Sows who produce three litters pay for their replacement costs
Piglet survival	Low	+++	Major economic value depending on total income per pig marketed
Disease resistance	Low	+++	Has major value but no ways to improve at present
Behavior	Low to moderate	+	Value may increase if sows move to pen production

Trait emphasis depends on whether selection is within sire or dam lines. Traits of economic importance in sire lines include growth rate, feed efficiency, carcass fat, meat percentage, meat quality, and structural hardiness. For dam lines, reproductive traits including age at puberty, litter size, number weaned, milking ability, and sow lifetime productivity are usually considered. Some of these traits are summarized in Table 10.2.

In modern pig breeding schemes, selection is often on many traits in order to maximize value. In principle, an index weighting the traits on their relative economic value and their genetic and phenotypic relationships is considered. In fact, breeders must take care that selection for one or more traits does not have a negative influence on others. Some genetic correlations among important traits are listed in Table 10.3.

Other traits may become important if production schemes radically change, and **pleitropic effects** will need to be understood before selection schemes can be initiated. Certainly the pressure to move sows from gestation crates to group housing may affect behavioral and feed intake traits, and their relative economic value.

Development of molecular genetic approaches

Coordinated efforts to better understand the pig genome were initiated in the early 1990s with the development of the international **PiGMaP** gene mapping project as well as projects by the USDA and US agricultural universities. These projects were structured in such a way to include cooperation and collaborations by many different institutions. In the US, the position of USDA Pig Genome Coordinator was created to facilitate collaborative efforts among scientists from both state and private universities as well

Table 10.3 Genetic correlations among several important pork production traits.

Daily gain with	Genetic correlation
Backfat	0.22
Feed per unit gain	−0.50
Loin eye area	−0.10
Reproduction	−0.10
Feed per unit gain with	
Backfat	0.34
Loin eye area	−0.35
Reproduction	−0.20
Backfat with	
Loin eye area	−0.35
Reproduction	0.20

Note: Values are approximate average values assembled from several sources and may differ within specialized lines.

as those from federal labs. These scientists operate cooperatively in a Swine Genome Technical Committee, which has been meeting yearly since 1994. The Committee worked to increase collaborative efforts and share information.

There were three significant **linkage maps** with up to 1000 markers published by the mid-1990s. Since that time, progress in growth of the linkage maps has continued as new gene markers, in particular, **single nucleotide polymorphisms** (**SNPs**) have been continuously identified and mapped, but integration of the linkage maps has been limited because researchers have waited for the pig genome sequence, the ultimate genetic map. Initial **physical maps** were successful, and use of two radiation hybrid (RH) panels allowed significant physical maps of many thousand markers to be developed. These panels were then used to map BAC (bacterial artificial chromosomes) end sequences to form a complete physical map that was used in the pig genome sequencing effort.

QTL, candidate genes, and genetic improvement

In the mid-1990s many **quantitative trait loci** (**QTL**) experiments were undertaken using linkage maps to help determine regions underlying traits of importance to the pig industry (e.g., Andersson-Eklund et al., 1996; Malek et al., 2001). Researchers have identified over 7,000 QTL affecting most traits by using both commercial and exotic pig breeds with various population structures. A new database, PigQTLdb (www.animalgenome.org/QTLdb/), has been constructed combining all the published QTL information into one searchable database and allows the user to search by chromosome, trait, or key words from the publications (Hu et al., 2005). Due to limitations regarding experimental design and classification of phenotypes, QTL associated with immune response traits and disease resistance have been sparse.

An alternative approach to QTL scans undertaken by many researchers was the use of **candidate gene** analyses (Rothschild and Soller, 1997) using biological or mutational candidate genes from other species to investigate a variety of traits. A substantial number of candidate genes have shown significant associations with many traits important to swine production. The first important candidate was the **Estrogen Receptor** (ESR) that was shown to have a significant association for litter size with effects ranging from 0.25 to over 1 pig per allele per gene copy with variations depending on breed background (Rothschild et al., 1996). An **MC4R** mutation has shown a significant association with a reduction in feed intake with less backfat or faster growth, depending on which allele is inherited (Kim et al., 2000). Extensively reviewed meat quality genes (**HAL, RN**) have been reported, and genetic markers identified within these genes allow for genetic testing therefore allowing producers to remove the alleles deleterious to meat quality (Fuji et al., 1991; Milan et al., 2000). Additional genes, including **PRKAG3** and **CAST**, have been shown to be associated with improvements in post mortem pH and tenderness (Ciobanu et al., 2001; 2004). Candidate genes or gene regions (K88, FUT1) have been identified to be associated with differences in **immune response** or disease resistance with FUT1 currently being used to reduce post-weaning diarrhea in commercial pork production (Vogeli et al., 1997). Recently, a polymorphism was identified as showing an association with resistance to K88 *E. coli* (Jørgensen et al., 2003). Additional genes, such as KIT and MC1R, have been used by breeding companies to produce pigs that are white in color, a phenotype that is preferred by commercial meat packing companies (Andersson-Eklund et al., 1996; Kijas et al., 1998). Commercial pig breeding companies initially combined these genetic markers with traditional performance information into **marker-assisted selection** programs to identify and select individuals that have the most genetic potential. Information on commonly used genes and markers is found in Table 10.4.

Table 10.4 Individual genetic markers and genes considered in marker assisted selection in pigs.

Marker or gene/locus name	Traits affected	Reference
RYR1 (Halothane)	Lean growth, porcine stress syndrome, meat quality	Fuji et al., 1991
RN (Rendement Napole)	Meat quality	Milan et al., 2000
ESR (Estrogen Receptor)	Litter size.	Rothschild et al., 1996
MC4R (Melanocortin-4 Receptor)	Lean growth, fatness, feed intake	Kim et al., 2000
IGF2 (Insulin like Growth Factor 2)	Lean growth, litter size	Van Laere et al., 2003
c-KIT Receptor	Coat and skin color	Andersson-Eklund et al., 1996
MC1R (Melanocortin-1 Receptor)	Red/black coat color	Kijas et al., 1998
PRKAG3 (Protein Kinase AMP Activated Gamma3-Regulatory Subunit)	Meat quality	Ciobanu et al., 2001
HMGA1 (High Mobility Group AT-hook1)	Backfat thickness/ growth	Kim et al., 2004
CCKAR (Cholecystokinin type A Receptor)	Feed intake and growth	Houston et al., 2006
CAST (Calpain inhibitor)	Tenderness sand meat quality	Ciobanu et al., 2004
EPOR (Erythropoietin)	Litter size	Vallet et. al., 2005
E. coli receptor F18	*E. coli* diarrhea	Vogeli et al., 1997
E. coli receptor K88	*E. coli* diarrhea	Jørgensen et al., 2003

Source: Dekkers, J., Mather, P.K. and Knoll, E.F. 2011. Genetic improvement of the pig. In Rothschild, M.F. and A. Ruvinsky (eds) *The Genetics of the Pig*, 2nd edn. Ch. 16. CAB International, Wallingford, UK, 507 pp. and others.

Sequencing the pig genome

Sequencing is the unraveling of **DNA** to understand the genetic code. It is equivalent to breaking down books into individual sentences and even specific letters in these sentences and words. The letters in the genetic code (A, T, G, C) are combined into "words," and these words are the genes that control traits or contribute to phenotypes of the animal like rate of growth, level of fat, reproductive performance, and disease susceptibility. Knowing the genetic code requires that one apply modern molecular biology or laboratory methods to break up the code into smaller pieces and then "read" the code.

In 2003, an **international pig genome sequencing consortium** was formed to lead efforts to find funding and to initiate the sequencing of the pig genome. During that same time period, a significant Sino-Danish effort to sequence the pig genome was initiated. This effort produced a 0.6× coverage, and a large number of SNPs were then released to the public domain. Formation of the international consortium also allowed the status of the pig genome to move towards its goal. Using one of the RH panels as a template, a new comparative map was constructed that far exceeded anything to date with an average spacing between comparative anchor loci at 1.15 Mb based on the human genome sequence. Efforts from a group of international swine genome researchers called the International Swine Genome Sequencing Consortium to write a sequencing proposal finally helped to acquire significant funding from the USDA, National Pork Board, Iowa Pork Producers Association, University of Illinois, Iowa State University, North Carolina Pork Council, North Carolina State University, the Wellcome Trust Sanger Institute, UK and a number of research institutions from around the world including those from China, Denmark, France, Japan, Korea, and the UK.

The International Swine Genome Sequencing Consortium continued it activities, and the genome assembly (Sscrofa10.2), which is the template for the Consortium's analysis and annotation efforts and the basis for the pig genome sequence paper, was deposited at National Center for Biotechnology Information (NCBI) in late August 2011 (ftp://ftp.ncbi.nih.gov/genbank/genomes/Eukaryotes/vertebrates_mammals/Sus_scrofa/Sscrofa10.2/). This is the definitive source for the original analyses of the pig genome. The NCBI genome team has released an annotated copy of this draft genome sequence (see ftp://ftp.ncbi.nlm.nih.gov/genomes/Sus_scrofa/GFF/), which is available from the NCBI Genome Browser. Members of the Consortium also have been working with the **Ensembl** genome project team, which recently completed its GeneBuild for Sscrofa10.2 (a gene build is the organization of the sequence). The Ensembl gene models are based on sequence evidence including alignments with **expressed sequences** comprising not only cDNA and EST information in the public databases but also more than

250 Gigabases of RNA-seq data generated by members of the International Swine Genome Consortium. The preliminary Ensembl analysis is available on the Pre-Ensembl site (http://pre.ensembl.org), and the full annotated genome is available on the Ensembl genome browser in Ensembl release 67.

The Swine Genome Sequencing Consortium also released the pig genome sequence data in a timely manner as the project has progressed in accordance with the principles of the Toronto Statement on Pre-publication data sharing (see Toronto International Data Release Workshop Authors. 2009. Prepublication data sharing. *Nature* 461:168–170). Briefly, the Toronto Statement places obligations on the producers of such data sets, including genome sequence data, with respect to prepublication release of the data and confirms the principle that allows the data producers (i.e., the Consortium) to publish the first global analyses of the data set. In 2010, the Consortium published its plans for its first global analyses of the genome sequence and for publication (see Archibald et al., 2010).

The complete genome sequence manuscript, entitled "Pig genomes provide insight into porcine demography, domestication and evolution" (Groenen et al., 2012), describes not only the generation and analysis of a high quality draft reference genome sequence of a single domesticated pig (*Sus scrofa*) but also the analysis of 16 other individual genomes including both 10 European and Asian wild boar (*Sus scrofa*) and six domesticated pigs. Among the major farmed animal species, the pig is unique in as much as the wild ancestors (wild boar from Europe and Asia) from which it was domesticated are still extant. The analyses provided new insights into the demography of wild boar and their subsequent domestication. It includes evidence of a deep phylogenetic split between Asian and European wild boar, which points to their divergence about 1 million years ago. Further insight into the domestication of the pig during the past 10,000 years as well as the more recent development of specific breeds is also discussed.

Researchers have conducted analyses of the pig genome and its gene content in an evolutionary context, and this has revealed accelerated evolution of primarily immune response and olfaction genes. Olfaction, the ability to smell, was clearly important to wild pigs, and this has not been lost in the modern pig, and immune response genes may also have helped to make the pig so flexible in a number of environments. The pig is not only an important agricultural species (pork is the most widely consumed meat globally) but also an important biomedical model. The 1000 Genomes research project is designed to sequence the genomes of 1000 species and has discovered that all people may carry a burden of potential loss of function mutations. Analysis of individual pig genomes has revealed similar mutations. These recently discovered mutations may potentially offer additional reasons to study the pig as a biomedical model. In summary, the pig genome sequencing manuscript represents the efforts of the Consortium, and the authors include scientists from over 50 laboratories in 12 different countries. The Consortium is also associated with a series of companion papers that have been or will be published soon.

Genomic selection

Sequencing the **pig genome** helped to produce millions of SNPs. The SNPs, in turn, were used to create genotyping chips, which have the ability to genotype for thousands of SNPs simultaneously. The first high density pig SNP chip had 62,000 SNPs (Ramos et al., 2009). These were then used to do **genome-wide association studies** (GWAS) for trait discovery and have helped to add to the thousands of QTLs discovered.

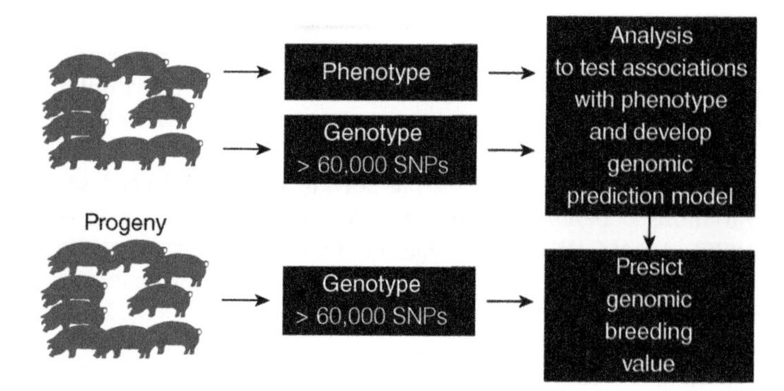

Figure 10.2 Use of SNP chip to do GWAS and genomic selection.

Table 10.5 Single-step genetic evaluation with and without genomic information on 2023 dam line pigs born in and after 2009.

Trait	EBV accuracy	GEBV accuracy	Increase %
Total number born	0.25	0.42	68
Stillborn	0.26	0.43	65
Survival birth – weaning	0.17	0.26	53
Litter weaning weight	0.23	0.35	52
Interval weaning – mate	0.17	0.30	76

Source: Personal communication, S. Forni, unpublished Genus PIC data, 2012.

The large expected outputs from the SNP association trials using the 62K pig SNP chip also offered new challenges and opportunities. Pig breeding companies now have literally hundreds or thousands of markers on which to select. **Genomic selection** (Meuwissen et al., 2001) has evolved rapidly, and estimated effects (Figure 10.2) are used to estimate genomic breeding values (Calus, 2010). Use of genomic selection to select for individual traits or for total economic value is being actively utilized in some genetic improvement programs. The advantage of genomic selection is that it can be devoted to traits that were difficult to improve quickly such as reproduction traits (Table 10.5). Another example is that one can predict the genomic breeding value for litter size on boars without ever progeny testing his daughters but by estimating genomic effects and then just genotyping him (Figure 10.2). Female genomic breeding values also can be predicted before they have their first litter.

Finally, for some traits that have been difficult to measure on the live animal, such as meat quality, once such associations have been estimated this will likely give breeders real opportunities to make noticeable genetic improvement through genomics without having to collect data from progeny. For other traits often overlooked, such as disease resistance, structural and environmental soundness, behavior, and others that until recently had been ignored, genomic selection provides the opportunity for real progress.

Given that the cost of genotyping for 62,000 SNPs is about $100/animal (the price in mid-2012), pig breeding organizations have asked if they need all these markers to get improvement with similar accuracy. One approach that some breeding organizations and other researchers are using is called **imputation**. Genotype imputation is done by genotyping sires and dams of each line for all 62,000 SNPs. Then, by genotyping with a smaller chip (perhaps hundreds to 10,000 SNPS), they can predict the complete genotype (60K SNPs as if from the larger chip) from the genetic and statistical relationships between the fewer markers and the larger marker set. These genomic imputation approaches can save millions of dollars in some breeding schemes and yet obtain similar levels of accuracy.

Additional uses of genomic selection include more effectively measuring genotype by environmental interactions and developing ways to select specific genotypes or "designer genes" for specialized niche markets and products. This will likely raise incomes, at least for those producing specialized products. Finally, one last area of use of the thousands of SNPs is to support the use of DNA to help trace pork products from the "farm to the fork." Such approaches have been developed, and some production and breeding companies are working to develop a traceability approach to brand products.

Databases

Databases play a vital role in providing the tools needed for future genomic discoveries and their uses for pig improvement. Substantial early pig **bioinformatics** efforts were undertaken by the Roslin Institute, Scotland (www.thearkdb.org) and by the US Bioinformatics Coordinator (www.genome.iastate.edu and www.animalgenome.org/). Together these efforts have supported a variety of pig genome efforts and, as well, have displayed the gene maps and other sequence information and tools and information for the pig genome.

Cloning, transgenics, and breeding pigs as biological models

Although **GM** (**genetically modified**) crops are well accepted for many food sources in many parts of the world, GM animals remain at a frontier not yet crossed. **Cloning** of livestock is possible, and it has been pushed as a means to find and multiply superior males for artificial insemination. Some breeders have begun to clone limited numbers of boars. Although helpful to genetic improvement in pigs, at least in the US and Europe, no approval for consumption of such animals has been made by the regulatory agencies.

Use of transgenics falls into a similar place. **Transgenic** pig production has been around for a great deal of time, but only recently have commercially novel and perhaps useful transgenic pigs been produced. A recent example is the transgenic phytase pig, which expresses phytase in its saliva and hence produces less phosphorus in the manure. This has special benefit in areas where pig production is heavy and manure with large amounts of phosphorus has been polluting the environment. Although this transgenic pig is excellent, no approval for the consumption of such pigs was allowed in Canada where it was developed or in the US, and the project was halted in 2011. Another novel transgenic pig is the one that is rich in omega-3 fatty acids. This might have real health advantages for those who consume pork, but this awaits approval for consumption. Given the public's concerns over food safety it is likely such approval is years off in the US, Canada and Europe. However, in some cultures there is a wide range of non-livestock species that are consumed, and therefore it is conceivable that these countries and cultures may be open to transgenics and approve consumption of reasonable transgenic animal products (for an in-depth description of cloning and transgenic animals, see Chapter 28).

There has been continued interest in the pig as a biological model for human biology. A recent survey of US government grants found hundreds of active grants using pigs as models and receiving funding from the National Institutes of Health. Research using pigs as **animal models** of human conditions has covered a vast array of disciplines such as nutrition, digestive physiology, kidney function, heart function, diabetes, obesity, and skin formation and healing. With the growing evidence of the close relationship between pigs and humans, evidenced by the new sequence information as it has become available, the extent of biomedical projects using the pig could be expected to grow in the future. Shortages of human tissues and organs available for transplantation have created interest in xenotransplantation, and the pig is the preferred donor due to its size and comparative physiology. Recent concerns about retroviruses and difficulties producing transgenic pigs meeting the standards required for safe transplantation have slowed the progress in the use of the pig for xenotransplantation and caused some companies to scale back the active research in this area. Use of the genome sequence will help clarify concerns about retroviruses.

Future applications to genetic improvement

Genetic improvement in pigs during the twentieth century was considerable with the development of objective trait measurements, use of breeding values, crossbreeding, and the development of commercial lines of swine that grew fast and efficient and produced more piglets. The close of the century brought the creation of genetic maps, the discovery of important genetic markers and their use in marker assisted selection for traits of economic importance. Such discoveries and their application to the industry have made the pig a major source of lean red meat.

Efforts in the early part of the twenty-first century have been no less amazing. The initial sequencing efforts are now complete and the development of SNP chips and the advent of genomic selection are beginning to revolutionize pig breeding. It is likely that with these new genomic tools some advances in selecting disease resistant pigs will be more successful. It is expected that fine tuning of breeding programs to match specific environments as well as specific niche markets or the production of specialized products will also occur. Unfortunately, lost in this process will likely be the small breeder who led the advancements during the twentieth century.

Pig breeding for biomedical models will likely also accelerate with the development of either transgenic pig models to study human disease or to create animals to supply organs for transplant. Such developments will likely be further advanced using cloning to help create superior individuals in larger number.

Pig breeders, producers, and consumers will all benefit from these advances in genetics and genomics and the resulting genetic improvement. This will require adoption of technology, advanced training by many within the livestock industry, and education of the public of the safety of these advances.

Acknowledgments

Support from the Department of Animal Science and the College of Agriculture and Life Sciences at Iowa State University is gratefully acknowledged. The author also thanks Alan Archibald and Scott Newman for providing information and suggestions for this chapter, Denise Rothschild for editing, and Genus PIC for providing unpublished data.

Further reading

Hudson, G.F.S. and Kennedy, B.W. 1985. Genetic evaluation of swine for growth rate and backfat thickness. *Journal of Animal Science* 61:83–91.

Lasley, J.F. 1987. *Genetics of Livestock Improvement*, 4th edn. Prentice Hall Inc., 482 pp.

Loftus, R. 2005 Traceability of biotech-derived animals: application of DNA technology. *Revue Scientifique et Technique – Office International Des Epizooties* 24:231–242.

Ollivier L. 1998 Genetic Improvement of the Pig. In Rothschild, M.F. and Ruvinsky, A.(eds.), *The Genetics of the Pig*. CAB International, Wallingford, pp. 511–540.

Rothschild, M.F. and Ruvinsky, A. 2011. *The Genetics of the Pig*, 2nd edn. CAB International, Wallingford, 507 pp.

Warwick, E.J. and Legates, L.E. 1990. *Breeding and Improvement of Farm Animals*, 9th edn. McGraw Hill Book Company, 342 pp.

References

Andersson-Eklund, L., Marklund, L., Lundstrom, K., Andersson, K., Hansson, I., Lundheim, N., et al., 1996. Mapping QTLs for morphological and meat quality traits in a wild boar intercross. *Animal Genetics* 27(Suppl. 2):111.

Archibald, A.L., Bolund, L., Churcher, C., Fredholm, M., Groenen, M.A., Harlizius, B., et al., 2010. Swine Genome Sequencing Consortium. Pig genome sequence – analysis and publication strategy. *BMC Genomics* 11:438.

Buchanan, D.S., and Stalder, K. 2011. Breeds of pigs. In *The Genetics of the Pig*, 2nd edn. Rothschild, M.F. and A. Ruvinsky (eds) Chapter 18. CAB International, Wallingford, 507 pp.

Calus, M.P.L. 2010. Genomic breeding value prediction: methods and procedures. *Animal* 4:157–164.

Ciobanu, D.C., Bastiaansen, J.W.M., Lonergan, S.M., Thomsen, H., Dekkers, J.C.M., Plastow, G.S. and Rothschild, M.F. 2004. New alleles in calpastatin gene are associated with meat quality traits in pigs. *Journal Animal Science* 82:2829–2839.

Ciobanu, D., Bastiaansen, J., Malek, M., Helm, J., Woollard, J., Plastow, G. and Rothschild, M. 2001. Evidence for new alleles in the protein kinase adenosine monophosphate-activated 3-subunit gene associated with low glycogen content in pig skeletal muscle and improved meat quality. *Genetics* 159:1151–1162.

Dekkers, J., Mather, P.K. and Knoll, E.F. 2011. Genetic improvement of the pig. In *The Genetics of the Pig*, 2nd edn, Rothschild, M.F. and A. Ruvinsky (eds), Ch. 16. CAB International, Wallingford, 507 pp.

Fuji, J., Otsu, K., Zorzato, F., De Leon, S., Khanna, V.K., Weiler, J.E., et al., 1991. Identification of a mutation in porcine ryanodine receptor associated with malignant hyperthermia. *Science* 253:448–451.

Groenen, M.A.M., Archibald, A.L., Uenishi, H., Tuggle, C.K., Takeuchi, Y., et al., 2012. Pig genomes provide insight into porcine demography and evolution. *Nature* 491:393–398.

Houston, R.D., Haley, C.S., Archibald, A.L., Cameron, N.D., Plastow, G.S. and Rance, K.A. 2006. A polymorphism in the 5′-untranslated region of the porcine cholecystokinin type a receptor gene affects feed intake and growth. *Genetics* 174:1555–1563.

Hu, Z.L, Dracheva, S., Jang, W., Maglott, D., Bastiaansen, J., Rothschild, M.F. and Reecy J.M. 2005. A QTL resource and comparison tool for pigs: PigQTLdb. *Mammalian Genome* 16:792–800.

Jørgensen, C.B., Cirera, S., Anderson, S.I., Archibald, A. L., Raudsepp, T., Chowdhary. B., et al., 2003. Linkage and comparative mapping of the locus controlling susceptibility towards E. COLI F4ab/ac diarrhoea in pigs. *Cytogenetic and Genome Research* 102(1–4):157–162.

Kijas, J.M.H., Wales, R., Tornsten, A., Chardon, P., Moller, M. and Andersson, L. (1998) Melanocortin receptor 1 (MC1R) mutations and coat color in pigs. *Genetics* 150:1177–1185.

Kim, K.S., Larsen N., Short T.H., Plastow G.S. and Rothschild M.F. 2000. A missense variant of the porcine melanocortin-4 receptor (MC4R) gene is associated with fatness, growth, and feed intake traits. *Mammalian Genome* 11:131–135.

Kim, K.S., Thomsen, H., Bastiaansen, J., Nguyen, N.T., Dekkers, J.C., Plastow, G.S. and Rothschild, M.F. 2004. Investigation of obesity candidate genes on porcine fat deposition quantitative trait loci regions. *Obesity Research* 12:1981–1994.

Larson, G., Cucchi, T. and Dobney, K. 2011. Genetic aspects of pig domestication. In *The Genetics of the Pig*, 2nd edn. Rothschild, M.F. and A. Ruvinsky (eds). Ch. 2. CAB International, Wallingford, 507 pp.

Lasley, J.F. 1987. *Genetics of Livestock Improvement*, 4th edn. Prentice Hall. 482 pp.

Malek, M., Dekkers, J.C.M., Lee, H.K., Baas, T.J. and Rothschild, M.F. 2001. A molecular genome scan analysis to identify chromosomal regions influencing economic traits in the pig. I. Growth and body composition. *Mammalian Genome* 12:637–645.

Milan, D., Jeon, J.T., Looft, C., Amarger, V., Thelander, M., Robic, A., et al. 2000. A mutation in PRKAG3 associated with excess glycogen content in pig skeletal muscle. *Science* 288:1248–1251.

Meuwissen T., Hayes B., Goddard M. 2001. Prediction of total genetic value using genome-wide dense marker maps. *Genetics* 157: 1819–1829.

Nicholas, F. 2011. Genetics of morphological traits and inherited disorders. In *The Genetics of the Pig*, 2nd edn. Rothschild, M.F. and A. Ruvinsky (eds.), Ch. 2. CAB International, Wallingford, 507 pp.

Ollivier L. and Foulley, J.L. 2011. Pig genetic resources. In *The Genetics of the Pig*, 2nd edn. Rothschild, M.F. and A. Ruvinsky (eds), Ch. 13.CAB International, Wallingford, 507 pp.

Ramos, A.M., Crooijimans, R.P., Affara, N.A., Amaral, A.J., Archibald, A.L., Beever, J.E., et al., 2009. Design of high density SNP genotyping assay in the using SNPs identified and characterized by next generation sequencing technology. *PLoS One* e6524.

Rothschild, M. F. and Soller, M. 1997. Candidate gene analysis to detect genes controlling traits of economic importance in domestic livestock. *Probe* 8:13–20.

Rothschild, M.F., Jacobson, C., Vaske, D., Tuggle, C., Wang, L., Short, T., et al., 1996. The estrogen receptor locus is associated with a major gene influencing litter size in pigs. *Proceedings of the National Academy of Sciences, USA* 93:201–205.

Ruvinsky, A., M.F. Rothschild, G. Larson and J. Gongora. 2011. Systematics and evolution of the pig. In *The Genetics of the Pig*, 2nd edn.

Rothschild, M.F. and A. Ruvinsky (eds), Ch. 1. CAB International, Wallingford, 507 pp.

Vallet, J.L., Freking, B.A., Leymaster, K.A. and Christenson, R.K. 2005. Allelic variation in the erythropoietin receptor gene is associated with uterine capacity and litter size in swine. *Animal Genetics* 36:97–103.

Van Laere, A.S., Nguyen, M., Braunschweig, M., Nezer, C., Collette, C., Moreau, L., et al., 2003 A regulatory mutation in *IGF2* causes a major QTL effect on muscle growth in the pig. *Nature* 425:832–836.

Vogeli, P., Meijerink, E., Fries, R., Neuenschwander, S., Vorlander, N., Stranzinger, G. and Bertschinger, H.U. 1997. A molecular test for the identification of *E. coli* F18 receptors – a break-through in the battle against porcine oedema disease and post-weaning diarrhoea. *Schweizer Archiv fur Tierheilkunde* 139:479–484.

Review questions

1. Find information on three national or international breeding companies.

a. Examine the commercial female products they are selling. Describe breed composition and what traits have been emphasized.

b. Examine the commercial male products they are selling. Describe breed composition and what traits have been emphasized.

2. Find information on three national or international companies and determine whether the companies use any marker assisted or genomic selection. Describe the programs they advertise.

3. If commercial pig producers are forced to move from gestation stalls to pen housing what types of traits will become more important? Explain.

4. Explain the limitations to selection for single traits.

5. Trait economic value and heritability often affect those traits that are used in selection programs. Read about modern commercial pork production and discuss which traits are most important economically and whether selection for them is effective based on heritability.

6. Explain the difference between inbreeding and crossbreeding and the expected genetic outcomes.

7. What factors might limit the use of marker assisted selection or genomic selection?

8. Consider two breeding programs. The first wishes to produce commodity pork at the least cost. The second wishes to produce high quality tasty pork for high end restaurants. Describe how these programs and their selection objectives might differ and what traits would be most important for each program.

9. Use the Internet to find and describe three specific examples of the use of pig genetics for biomedical research.

10. Go to the pig QTL database (www.animalgenome.org/) and find two examples of trait marker associations. Please describe them.

11 Equine Genetics

Jennifer Minick Bormann
Department of Animal Sciences and Industry, Kansas State University, KS, USA

Color

Introduction

Coat color in horses is a simply-inherited trait that is controlled by many genes. Consult any equine book for a description of the commonly identified horse colors. When describing coat colors, ignore any white markings that may be present. Every horse has the genes for a color. Some of them also have genes to put white markings over the color in certain areas. White patterns and markings will be discussed later. Terminology for horse colors is not consistent with genotype; more than one genotype codes for the same color or one genotype can give rise to more than one color. There are many genotypic combinations that have no color name, so those horses are misclassified as some other color. These factors can make pedigree analysis complicated. All mammals have at least two types of pigment: **eumelanin** (black) and **pheomelanin** (red, yellow, or tan). These pigments are the result of biochemical pathways controlled by genes. A mutation in any of the genes in the pathway results in variation in color. Having a good understanding of gene interaction is critical to learning equine coat colors. Many genes work together to determine the final coat color present on a horse. Throughout this chapter, I will use the notation of Sponenberg (2009). A one or two letter abbreviation will name the gene, and superscripts will be used to designate the alleles of that gene. The superscript + will designate the wild-type. Wild-type is considered to be Przewalski's horse, a primitive breed thought to resemble the ancestral horse. Superscript – means an unknown allele.

Base colors

White

There is a popular saying among old-time horsemen that there is no such thing as a white horse. This is untrue. Most white-appearing horses are gray (discussed later); however, there are several mutations in a gene called KIT that cause white. Depending on the mutation, the horse may be completely white, partially white, or have white spotting similar to sabino (discussed later). If there are non-white areas present, they tend to be on the topline of the horse, or to occur as spots. White horses have pink skin under the white hair and dark eyes. A gray horse that has lightened to appear white will have dark skin. The KIT mutations all appear to be dominant, meaning that only one allele is necessary to produce the phenotype. These mutations are also epistatic to all other color genes. Some of the KIT mutations appear to be lethal when present in the homozygous state.

$Wh^{W}Wh^{W}$ white or die in utero, depending on the mutation
$Wh^{W}Wh^{+}$ white; no other color genes are expressed
$Wh^{+}Wh^{+}$ "normal colored," other color genes are expressed.

There is some evidence that the KIT gene has a high mutation rate, so that new white horses spontaneously appear from dark parents. These horses go on to breed as any other $Wh^{W}Wh^{+}$.

White is rare, but appears in many breeds. There is a color breed made up of white horses that was originally called the American Albino Horse Club. It has since been renamed the American White and American Crème Horse Registry, because it was discovered that these horses were not albinos. In fact, there are no true albino horses.

If a breeder has a goal to produce white horses, it is better to mate a white to a non-white, which gives half white horses. Breeding two whites together may give two-thirds white horses but that is because a quarter of conceived embryos are lost early in development (because some white alleles are lethal). That shows up as a decreased conception rates and more open mares.

Gray

Gray is a common horse color that is progressive in nature. Most gray foals are born a different color and become gray as they shed their foal coat. They will start out a dark gray and gradually lighten in color every year. The rate and extent of lightening is variable among horses, resulting in a wide range of expression of the color. No matter what the shade, all gray horses have dark skin, which distinguishes the lightest grays from white horses, as discussed previously.

All gray horses have a 4.6 kb duplication mutation in the syntaxin 17 (STYX17) locus. It is a dominant mutation, so that any horse with at least one copy of the mutant allele has the gray phenotype. Gray is epistatic to all other color genes except Wh, so the gray allele will mask all other colors except white.

$G^{G}G^{G}$ gray
$G^{G}G^{+}$ gray
$G^{+}G^{+}$ non-gray, other color genes are expressed.

Even though gray is known as a dominant gene, homozygotes tend to fade to white more quickly and tend to end up completely

Molecular and Quantitative Animal Genetics, First Edition. Edited by Hasan Khatib.

white, while heterozygotes are more likely to retain some color. All gray horses are more prone to melanomas than non-grays; however, homozygotes have a higher incidence than heterozygotes. Homozygotes are also more likely to have spots of pink skin than heterozygotes. There is a gene test available to distinguish between homozygous and heterozygous gray horses.

Chestnut, bay, and black

The base color of a non-white, non-gray horse is controlled by two genes called **extension** (E) and **agouti** (A). These genes work together to determine if the horse will be chestnut, bay, or black. A quick note about chestnut: the terms **chestnut** and **sorrel** are both used to describe a reddish-brown horse with non-black points. It is usually a breed- or discipline-specific decision about which term is used, that is, English disciplines and non-stock breeds use the term chestnut. Western disciplines and stock breeds use both terms to describe different shades of reddish-brown, with sorrel being the more common color and chestnut being reserved for the very dark shades of reddish-brown. However, the same gene causes both the light and dark shades. A bay horse has a reddish-brown body with black points, and a black horse is completely black with no brown areas.

Extension is the melanocortin 1 receptor (MC1R) locus. The dominant allele is the wild-type, and causes non-chestnut. The recessive, mutant allele is a single base pair substitution, and causes chestnut. There is a third allele at the extension locus called E^a in some breeds, but its phenotypic effect is identical to E^e. When making crosses and predicting offspring, it is not necessary to distinguish between E^e and E^a.

E^+E^+ non-chestnut, color determined by agouti
E^+E^e non-chestnut, color determined by agouti
E^eE^e chestnut, no matter what is at agouti.

If the horse is non-chestnut, then base color is determined by the agouti signaling protein (ASIP) locus. This gene controls whether black is distributed over the entire body, or restricted to the points.

A^AA^A bay
A^AA^a bay
A^aA^a black.

When both genes are considered together, the possibilities are:

$E^+E^+A^AA^A$ bay
$E^+E^+A^AA^a$ bay
$E^+E^+A^aA^a$ black
$E^+E^eA^AA^A$ bay
$E^+E^eA^AA^a$ bay
$E^+E^eA^aA^a$ black
$E^eE^eA^AA^A$ chestnut
$E^eE^eA^AA^a$ chestnut
$E^eE^eA^aA^a$ chestnut.

There are gene tests available for both extension and agouti. In some competitive disciplines, it is more desirable to have bay or black, rather than chestnut. The extension gene test is widely used on bay and black sires to determine if they carry the E^e allele, and may sire chestnut foals. The common terminology for the extension alleles are black factor/gene (E^+) and red factor/gene (E^e). Stallions that have tested homozygous will be advertised as "homozygous for the black factor," or "does not carry the red gene," and so on.

Other alleles have been hypothesized to exist at extension and agouti. Dominant black (E^D) has been suggested as a potential cause of black phenotype regardless of the agouti genotype, but molecular analysis has not confirmed the existence of this allele. There is possibly a rare allele at the agouti locus called wild bay (A^+), which is thought to be dominant to A^A. This is the wild-type allele and causes the lower legs of the horse to be a mix of red and black. There may also be a black-and-tan allele (A^t) that causes black with brown muzzle, inner upper legs, elbow pocket, and flanks.

Modifiers to base colors

Dun

The dun gene causes dilution of base color, a dorsal stripe, and sometimes zebra stripes on the legs. The manes, tails, and lower legs are unchanged. Dun on a bay base is called dun or zebra dun, and is characterized by yellow or tan body color with black points and dorsal stripe. Dun on a chestnut base results in red dun, which is a light tan or apricot color with chestnut points and dorsal stripe. Dun on a black base is called grullo, and appears as a mousy gray colored body with black points and dorsal stripe. The dorsal stripe is known as the defining characteristic of the dun gene. However, non-dun horses may exhibit light dorsal stripes called countershading. The genetic mechanism of countershading is unknown. Dun is caused by a dominant wild-type allele. The recessive mutation results in non-dun color.

Dn^+Dn^+ dun
Dn^+Dn^{nd} dun
$Dn^{nd}Dn^{nd}$ non-dun.

This gene has not been mapped yet, so there is no direct test. However, it is possible to use markers and pedigree information to determine zygosity of dun horses in some cases.

Cream

Cream is an incompletely dominant gene, so that the heterozygote is intermediate in phenotype to the homozygotes. It is caused by a single nucleotide substitution mutation in the membrane associated transporter protein (MATP) locus. The mutant cream gene C^{Cr} causes lightening of the coat. When only one cream allele is present, red hair is lightened to yellow, and black hair is unchanged. When two cream alleles are present, red and black hair is lightened to off white. Black hair may retain some orangey color. Double dilutes also have pink skin and blue amber eyes. The phenotypic expression of the cream allele depends on the base color of the horse. On a chestnut base, a single copy of cream causes palomino and two copies cause cremello. On a bay base, the single dilute is buckskin and the double dilute is perlino. On a black base, one cream allele causes smoky black and two cream alleles cause smoky cream. Palominos are yellow or golden with a white mane and tail. Buckskins are yellow or golden with black points. Smoky black horses may be entirely dark black, or may be a softer, lighter shade of black. Most breed registries don't have a color category for smoky black, so these horses are usually misclassified as black or brown. Cremellos are off-white with white mane and tail. It can be very difficult to distinguish between cremello, perlino, and smoky cream. Perlinos may retain some orange-appearing pigment in their points, and smoky creams may be slightly orange all over. Shade and sooty (discussed later) will show through one cream allele, to produce all shades of, as well as sooty, palominos, and buckskins. A gene test is available for cream.

Pearl

The pearl phenotype has been described in both stock breeds (originally called Barlink) and Spanish breeds, such as Andalusion and Lusitano. Pearl is a rare recessive mutation and only affects red hair, which is lightened to an apricot color with pale skin.

$Prl^{+}Prl^{+}$ non-pearl
$Prl^{+}Prl^{prl}$ non-pearl
$Prl^{prl}Prl^{prl}$ pearl.

Most registries have no color category for pearl, so these horses are misclassified, usually as having cream or dun dilutions. If a horse has both C^{Cr} and Prl^{prl}, they will appear as a double dilute. A gene test is available for pearl.

Champagne

Champagne is a dominant gene that causes dilution of both red (to gold) and black (to brown) hair, mottled pinkish-tan skin, and green or amber eyes. The coat usually has a metallic sheen. The champagne gene produces different colors depending on the base color of the horse. Champagne on chestnut is called gold champagne, on bay is amber champagne, and on black is champagne. Most breeds have no category for these colors, so they are often misclassified as palomino, buckskin, or brown. The champagne gene has been mapped to a mutation in the solute carrier 36 family A1 (SLC36A1), and has a gene test available.

Silver

A dominant mutation in the pre-melanosomal protein 17 (PMEL17) locus causes the silver (also known as silver dapple) phenotype. This mutation has no effect on red hair, but causes black hair to be lightened to gray-ish or chocolate brown, often with dapples. Black manes and tails are lightened to flaxen or silver. Bays with silver will have unchanged coats but flaxen or silver manes and tails, often causing them to be mistaken for flaxen chestnuts. Silver most commonly occurs in gaited breeds and ponies.

$Z^{Z}Z^{Z}$ silver (only shows up on bay or black based colors)
$Z^{Z}Z^{+}$ silver (only shows up on bay or black based colors)
$Z^{+}Z^{+}$ non-silver.

In Rocky Mountain Horses, silver is associated with vision problems. Because this is not found in other breeds, it is assumed to be caused by a different, linked gene. There is a gene test available for silver.

Roan

Roan is a pattern of white hairs evenly distributed across the body. The head and points of roan horses are unaffected. Roan on chestnut is called red roan, roan on bay is bay roan, and roan on black is called blue roan. Roan is caused by a dominant mutation in the KIT locus (Rn^{Rn}).

$Rn^{Rn}Rn^{Rn}$ roan
$Rn^{Rn}Rn^{+}$ roan
$Rn^{+}Rn^{+}$ non-roan.

In some breeds, such as the Thoroughbred, the term roan has been used to describe chestnut or bay horses that are turning gray due to G^{G}. Some horses can also show scattered white hairs in different parts of the body, variously called white ticking, frosty, roaning, or rabicano. These patterns have unknown genetic mechanism and are not associated with Rn^{Rn}. The roan gene is in the same linkage group as extension, which means that crosses of horses heterozygous for both roan and extension with non-roan chestnut horses produce unexpected ratios of progeny. The roan gene has not been mapped yet, but markers and pedigree can be used to determine roan zygosity.

Other modifications

Mealy

The mealy gene causes lighter colored areas on the eyes, muzzle, elbow pockets, belly, flanks, and inner legs, which can vary in intensity from very pale to barely noticable. The wild-type allele is dominant and causes the mealy phenotype, while the recessive is non-mealy. This gene is abbreviated Pa, from the Spanish pangare.

$Pa^{+}Pa^{+}$ mealy
$Pa^{+}Pa^{np}$ mealy
$Pa^{np}Pa^{np}$ non-mealy (non-pangare).

Some breeds (not stock breeds) have used the term sorrel for a mealy chestnut. This gene is usually not considered when classifying colors in the US.

Shade

Most colors have many variations of shade. The mane and tail of chestnuts can also vary widely in shade. This variation in shade is thought to be polygenic in nature, meaning there is no single gene that determines if a horse will be a light or dark shade within a color. For polygenic traits, offspring tend to resemble their parents on average. However, any one mating may produce offspring with a different phenotype than the parents. Horses may also have black hairs interspersed among the body hairs, contributing to the variation within a color. This is known as sooty or smutty, may be minimal or extensive, and is usually more prominent on the topline of the horse. The genetics of sooty have not been determined but it may be polygenic.

What about brown?

There is no gene in horses that causes the color brown. Horsemen defined the color brown (very dark brown or black body with lighter brown areas around the muzzle, inner upper legs, elbows, and flanks) before the genetics of horse color were known. There are many different genetic mechanisms that cause brown. The most common is probably a bay with dark shade or sooty. The hypothesized black and tan agouti allele would also cause brown. A black horse that shows excessive fading may be categorized as brown. The mealy gene on dark bay or black may cause brown. One copy of the cream gene on a black base color causes smoky black, which may be misclassified as brown.

White markings

White face and leg markings are a polygenic trait, which means the amount of white is controlled by many genes (see Chapter 4 on polygenic/quantitative traits). For unknown reasons, chestnuts tend to be more extensively marked than bays, which have more white than blacks. The amount of face and leg white is usually proportional to each other; if not, it could be that the horse has a white pattern gene.

Paint/pinto patterns

In North American, paint or pinto patterns are classified into three types: tobiano, overo, and tovero. Tobiano and overo are distinct classes, and toveros show characteristics of both patterns. There are actually four different paint/pinto patterns controlled by different gene(s). Any one horse may have more than one pattern. Minimally marked paint/pinto horses may not have enough white to qualify for registry as spotted horses, but will pass on the spotting pattern just as a more extensively marked horse will. In most cases, genetic control of the level of expression (amount of white versus dark) of paint/pinto genes is unknown. There could be polygenic suppressor genes that mask the pattern genes, or the pattern genes may be partially penetrant.

Tobiano

Tobiano is characterized by a dark head, extensive white on the legs, and white that usually crosses the spine. Tobiano spotting is caused by a dominant mutation that is an inversion in the KIT gene. Tobiano is linked to both extension and roan. Horses that are homozygous for tobiano tend to have smudging or roaning of their spots, called paw prints. Zygosity has nothing to do with the amount of white or dark areas present.

$To^T To^T$ Tobiano
$To^T To^+$ Tobiano
$To^+ To^+$ Non-tobiano.

Frame

The frame gene is one type of overo pattern. Frame overo horses usually have dark toplines, dark legs, and extensive white on the head. This gene is associated with lethal white syndrome, in which the foal is born white and apparently normal, but dies soon after from an intestinal abnormality. Frame is caused by a missense mutation in the endothelin receptor b (ENDRB) locus.

$Fr^F Fr^F$ lethal white
$Fr^F Fr^+$ frame pattern
$Fr^+ Fr^+$ non-patterned.

Similarly to tobiano, expression of the gene varies widely, and minimally marked frames may not have enough white to qualify for registry as a patterned horse. A gene test is available, and any potential breeding horse that has overo horses in its pedigree, even if solid colored, should be tested to avoid the production of lethal white foals.

Sabino

Another type of overo is sabino. Sabino is characterized by irregular white spots on the face, and belly, and high white on the legs. Many sabinos have roaning on the edges of the spots or are flecked/speckled. There are several subpatterns of sabino, most of which have unknown genetic mechanisms. A single base pair mutation in the KIT gene is responsible for one type of sabino, called Sabino 1. This is an incompletely dominant gene, with the homozygote completely or almost completely white, and the heterozygote showing the typical sabino markings.

$Sb1^{Sb1} Sb1^{Sb1}$ almost or completely white
$Sb1^{Sb1} Sb1^+$ sabino
$Sb1^+ Sb1^+$ non-sabino.

Minimally marked horses that have $Sb1^{Sb1}$ may be misclassified as non-patterned. There is a gene test available to determine presence and zygosity of the sabino allele. $Sb1^{Sb1}$ is present in many breeds, including Quarter Horses, Paints, Tennessee Walkers, and Shetland ponies.

Some horses that exhibit the sabino pattern test negative for $Sb1^{Sb1}$. There are other unknown genetic mechanisms that result in the sabino pattern in some breeds, such as Arabians, Shires, and Clydesdales.

Splashed white

The splashed white pattern is characterized by head, leg, and belly white with crisp edges and blue eyes. Minimally marked splashed white horses do not have enough white to qualify for registry as a paint/pinto. Some splashed white horses are deaf, but many are not. The genetic mechanism for splashed white has not been completely determined, but there are thought to be three different dominant mutations (Sw1, Sw2, Sw3) that all result in the splashed white phenotype. For example:

$Sw1^S Sw1^S$ splashed white
$Sw1^S Sw1^+$ splashed white
$Sw1^+ Sw1^+$ non-splashed white.

Sw2 and Sw3 may be homozygous lethal. Research is ongoing to further identify the genetic mechanism of splashed white.

Leopard

The leopard pattern is most commonly associated with the Appaloosa breed, but it is also present in other breeds world-wide including the Noriker and Knabstrupper. Leopard is a family of patterns that includes many sub-patterns, including frost, snowflake, mottled, varnish, speckled, blanket, snow cap, and leopard. All patterns are caused by an incompletely dominant mutation in the transient receptor potential cation channel, subfamily M, member 1 (TRPM1) locus.

$Lp^{LP} Lp^{Lp}$ minimally to maximally white, lack leopard spots (snow cap or few spot leopard sub-pattern)
$Lp^{LP} Lp^+$ minimally to maximally white, leopard spots as part of the sub-pattern
$Lp^+ Lp^+$ non-leopard.

There may be several modifier genes that determine the specific sub-pattern present on Lp^{Lp} horses. Leopard has also been associated with a vision impairment called night blindness. Horses that are homozygous for Lp^{Lp} are night blind, while heterozygotes have normal vision.

Table 11.1 is a summary table of the coat color genes and their effects.

Genetic defects

Introduction

Managing genetic defects is a critical component of a horse breeding program. Mechanism of inheritance (dominant or recessive), as well as availability of a gene test, determine how the defect can be managed. Most single gene diseases are autosomal recessive. Dominant disorders usually don't survive in a breed because of natural selection. A lethal dominant disorder cannot be passed on (unless it causes death after reproductive age) because any animal that has the allele will die. If a dominant disease is not lethal but detrimental, it will be naturally eliminated in a population eventually unless it is selected for. Diseases caused by recessive genes can persist in a population for many generations. The disease allele is present in heterozygous form in carrier animals that are perfectly healthy but will pass the allele on to half their offspring. The disease will appear in a quarter of the progeny of two carriers. Many times, the disease won't surface until inbreeding to a common superior ancestor who carried the allele occurs. When a disease becomes associated with a particular ancestor, entire lines of horses are categorized as diseased, when many of them are homozygous normal. Until recently, the only way to know if a horse carried a recessive disease allele was if the horse had produced an affected progeny. It was very difficult to manage a breeding program without knowing the zygosity of unaffected animals. Now, there are many gene tests that can determine whether a phenotypically normal horse is homozygous normal

Table 11.1 Summary of coat color genes and their effects.

Locus	Symbol	Allele	Action	Comment
Agouti	$^{a}A^{+}$	Wild bay	Dom	Causes wild bay
	A^{A}	Bay	Intermed.	Causes bay
	$^{a}A^{t}$	Black and tan	Intermed.	Causes brown
	A^{a}	Black	Rec	Causes black
Cream	C^{+}	Wild-type	Partial dom.	Allows dark color
	C^{cr}	Cream	Partial dom.	Heterozy.: dilutes red to yellow, black unaffected Homozy.: red and black diluted to cream, pink skin and blue eyes
Champagne	Ch^{C}	Champagne	Dom.	Dilutes black to light brown, red to yellow
	Ch^{+}	Wild-type	Rec.	Allows dark color
Dun	Dn^{+}	Wild-type (dun)	Dom.	Causes lineback dun
	Dn^{nd}	Not dun	Rec.	Allows nondun colors
Extension	$^{a}E^{D}$	Dominant black	Dom.	Epistatic to agouti, results in black or brown
	E^{+}	Wild-type	Intermed.	Allows agouti colors
	E^{e}	Chestnut	Rec.	Epistatic to agouti, results in chestnut
Frame	Fr^{F}	Frame	Dom.	Causes frame spotting, lethal in homozygotes
	Fr^{+}	Wild-type	Rec.	Non-spotted
Gray	G^{G}	Gray	Dom.	Causes progressive graying
	G^{+}	Wild-type	Rec.	No graying
Leopard	Lp^{Lp}	Leopard	Dom.	Leopard pattern
	Lp^{+}	Wild-type	Rec.	Non-spotted
Pangare (Mealy)	Pa^{+}	Wild-type	Dom	Causes mealy effect
	Pa^{np}	Non-pangare	Rec	Allows color without mealy effect
Roan	Rn^{Rn}	Roan	Dom.	Cause roaning on base color
	Rn^{+}	Wild-type	Rec.	Allows non-roan colors
Sabino	$Sb1^{Sb1}$	Sabino	Dom.	Causes sabino
	$Sb1^{+}$	Wild-type	Rec.	Non-spotted
Sabino	?	?	?	Unknown genetic mechanism causes sabino in $Sb1^{+}Sb1^{+}$ horses
Shade			Polygenic	Varies relative shade of body color from light to dark
Silver dapple	Z^{Z}	Silver dapple	Dom.	Dilutes black to chocolate brown or flaxen, does not affect red
	Z^{+}	Wild-type	Rec.	Allows non-silver colors
Sooty			Probably polygenic	Causes black hairs in base color
Splashed white	$Sw1^{S}$	Splashed white	Dom.	Causes splashed white
	$Sw1^{+}$	Wild-type	Rec.	Non-spotted
	$Sw2^{S}$	Splashed white	Dom.	Causes splashed white, may be lethal in homozygote
	$Sw2^{+}$	Wild-type	Rec.	Non-spotted
	$Sw3^{S}$	Splashed white	Dom.	Causes splashed white, may be lethal in homozygote
	$Sw3^{+}$	Wild-type	Rec.	Non-spotted
Tobiano	To^{T}	Tobiano	Dom.	Causes tobiano spotting
	To^{+}	Wild-type	Rec.	Non-spotted
White	Wh^{W}	White	Dom.	Causes white, lethal to homozygotes
	Wh^{+}	Wild-type	Rec.	Non white
White markings			Polygenic	White face and leg markings on any base color

[a]Hypothesized allele.

or a carrier for the defect. This allows non-carriers in lines associated with the defect to be identified and used for breeding.

The following list of genetic disorders is not exhaustive, but is a list of those that are most prevalent in the most common US breeds. Breeders should work with their breed association to learn what defects are present in their respective breeds, and what tools are available to manage them. Various breeds also have rules and regulations about testing requirements and registration eligibility pertaining to genetic defects.

Dominant disorders

Hyperkalemic periodic paralysis (HYPP)

Hyperkalemic periodic paralysis is a muscle disorder present in Quarter Horses and related breeds that results from a dominant mutation in the voltage-dependent skeletal muscle sodium channel alpha subunit. The affected allele is called H and the normal is N. HYPP causes episodes of muscle tremors, stiffness, and paralysis, and can result in death from severe episodes. Homozygous horses usually have more severe symptoms than heterozygotes. There is also variable expressivity. Affected horses (both homozygous and heterozygous) have a wide range of severity of symptoms. Some affected horses may never show a symptom, and some may die at a young age. Also, environment, such as diet and medication, may exacerbate or relieve symptoms.

The HYPP allele has been traced back to the influential AQHA stallion Impressive. He was heavily used before the disorder was discovered. A secondary result of the HYPP allele is extreme expression of muscling that is favored in halter competition. Because of this, Impressive descendants that had the defective allele were heavily selected for show ring competition and breeding. This caused the HYPP allele to increase in frequency quickly within the stock breeds. Because of the variable expressivity of the allele, it was a very critical step to have the gene test developed. Now breeders are able to determine the genotype of horses early in life, before symptoms might have appeared. The management of this disorder has been very controversial to the AQHA. Most breeders interested in halter competition used the allele in their breeding programs to produce the desirable muscle type. However, others felt it was unethical to propagate a genetic disorder just to get a certain body type. Beginning with foals born in 2007, any HH horse is no longer eligible for registration. Heterozygous horses are allowed to be registered at this time, but all descendants of Impressive that don't have two homozygous normal parents must have a gene test with the results printed on their registration papers.

Polysaccharide storage myopathy (PSSM)

PSSM is another muscle disorder that is found in many breeds, including draft breeds and Quarter Horses. Horses with PSSM exhibit tying-up symptoms caused by the accumulation of sugars in the muscle. There are two types of PSSM called Type 1 and Type 2. A dominant mutation in the GYS1 gene is responsible for Type 1. The basis for Type 2 may be genetic, but it has not been determined conclusively. There is a gene test for PSSM Type 1. Approximately 6–7% of horses from stock breeds are heterozygous for PSSM Type 1. That percentage is 40% or higher for several draft breeds, such as Belgian, Percheron, and Breton.

Rare defects

Hereditary multiple exostosis (HME) is a rare skeletal disorder that causes benign bone growths shown to be autosomal dominant in a Thoroughbred family. The gene responsible has not been mapped so there is no gene test available currently. Malignant hyperthermia (MH) is an autosomal dominant disorder that causes muscle spasms and death in several breeds, including Quarter Horses. It is analogous to Porcine Stress Syndrome in swine, and can be triggered by stress, exercise, or anesthesia. A mutation in the Ryanodine Receptor 1 gene is been shown to cause MH. There is a gene test available.

Recessive disorders

Severe combined immunodeficiency disease (SCID, CID)

SCID is an autosomal recessive disease found in Arabians that is caused by a deletion in the DNA-protein kinase catalytic subunit gene. This disease is characterized by a lack of immunoglobulins. Most affected foals die by 5 months of age from massive infection. The allele is present in many different bloodlines and is not associated with any particular sire or family. This would indicate that it is a very old mutation and has been present in the breed for many generations. A gene test was developed in 1997 to identify carriers, and the frequency of carriers in the breed has been estimated as 8.4%.

Hereditary equine regional dermal asthenia (HERDA)

This disorder was originally known as hyperelastosis cutis (HC), and is an autosomal recessive missense mutation in the cyclophilin B (PPIB) gene that causes defective collagen. The skin is not properly attached to the underlying tissue and separates after any trauma (such as saddling). This allele was originally traced back to the Quarter Horse stallion Poco Bueno (born 1944). He was an extremely popular sire, and the allele spread throughout the breed in the next several decades. It was later found that his sire, King, also carried the defective allele. Not enough data exists from that era to determine if it goes back further. A gene test was developed in 2007 that allows the identification of carriers.

Glycogen branching enzyme deficiency (GBED)

GBED is an autosomal recessive mutation in the GBE1 gene that causes a defect in the muscle's ability to form glycogen. Affected foals are aborted, stillborn, or die soon after birth. The prevalence of carriers has been estimated to be approximately 10% in Quarter Horses. It is also present in related stock breeds, and traces to the Quarter Horse foundation sires King and Zantanon. There is a gene test available to identify carriers.

Cerebellar abiotrophy (CA)

Cerebellar abiotrophy affects mainly Arabians, but has been reported in other breeds that have Arabian ancestry. This disease is characterized by head tremors and ataxia, and shows variable expressivity. Some affected horses have very severe symptoms, and some show almost no symptoms at all. Many affected horses are euthanized because they cannot be safely handled or ridden. It is caused by a recessive mutation on chromosome 2. The frequency of carriers in the Arabian population has been estimated as 19.7%, although that number may be biased upwards because of selectively testing horses from affected families. The mutation is thought to have occurred before the separation of the breed into different lines (Polish, Spanish, etc.), because it appears in many lines. A gene test is available to identify carriers.

Junctional epidermolysis bullosa (JEB)

JEB is also called red foot disease, and causes blistering of the epithelia and hoof sloughing in foals. Foals die from infection or are euthanized. It is caused by an autosomal recessive mutation in Belgian draft horses that has been mapped to the LAMC2 gene. A form of JEB called epitheliogenesis imperfecta has been identified in draft breeds and American Saddlebreds. This form is caused by a recessive deletion mutation in the LAMA3 gene. The frequency of carriers in the Saddlebred breed is approximately 5%. Gene tests are available for both mutations.

Lavender foal syndrome (LFS)

Another autosomal recessive disorder found in Arabians of Egyptian lines is called lavender foal syndrome. Affected foals are often a pale color with a pink or lavender tinge. They are born alive but are unable to stand and nurse and have neurologic symptoms such as seizures, leg paddling, and arched necks. They are normally euthanized within a few days of birth. Genomic analysis identified the cause as a deletion in the myosin Va gene. It is estimated that approximately 10% of Egyptian Arabians are carriers of LFS. A gene test is available.

Rare defects

A certain type of flexural limb deformity where complete ulnas and fibulas are present has been found in several pony breeds. These bones are improperly attached to the knee or hock. Mating studies showed it to be caused by an autosomal recessive. Another autosomal recessive defect that has been reported in Shetland ponies is lateral patellar luxation. This condition can also be caused by injury. Occipital-atlanto-axial malformation (OAAM) occurs mostly in Arabians, but has also been reported in other breeds. It is a fatal neurologic problem in which there are lesions in the spinal cord and fusion of the cervical vertebrae to the skull. In the Arabian, it is reported to be caused by an autosomal recessive mutation. Equine hemophilia A is similar to human hemophilia and causes failure of blood clotting. Similarly to humans, equine hemophilia is caused by an X-linked recessive gene, and has been reported in Quarter Horses, Thoroughbreds, and Standardbreds.

The genetics of health

All of the disorders discussed previously were caused by single genes. There is a large body of evidence in livestock species that general disease resistance or immune function has a genetic basis. This resistance is polygenic in nature, and can't be predicted by simple gene tests. There are also diseases that appear to have a genetic predisposition, but aren't simply-inherited. Much research into these areas is needed to provide breeders and owners with tools to select against health problems and help to manage those that exist.

Health genetics in the horse is a rapidly changing field. Breeders should stay current with their breed associations to keep abreast of new research as well as rules and regulations pertaining to genetic defects.

Inbreeding and relationship

Inbreeding

As previously discussed in Chapter 2, inbreeding is defined as the mating of relatives. All purebred animals are, to some extent, inbred. Purebreeding is inbreeding. In a practical context, we consider a mating to be inbreeding if the stallion and mare are more closely related than the average of the breed. An animal is inbred if and only if its parents are related. An animal can be highly inbred itself, but if it is mated to an unrelated animal, the offspring has inbreeding coefficient of 0 (see Chapter 2 for a discussion of calculating inbreeding coefficients). This means that inbreeding can be eliminated in one generation if unrelated mates can be found. Particularly in small breeds, it can be very difficult to find unrelated animals. Average levels of inbreeding in several horse breeds have been reported in the literature, and are shown in Table 11.2.

Table 11.2 Inbreeding coefficients in various breeds of horse.

Inbreeding %	Breed	Reference
1.3	Quarter Horse	Tunnell et al., 1983
1.8	Quarter Horse	Tunnell et al., 1983
2.1	Quarter Horse	Tunnell et al., 1983
2.3	Swedish Standardbred	Strom, 1982
2.4	French Selle Francais	Moureaux et al., 1996
2.6	French Thoroughbred	Moureaux et al., 1996
2.6	Quarter Horse	Tunnell et al., 1983
2.9	French Anglo-Arab	Moureaux et al., 1996
3.2	Norwegian Fjord	Bhatnagar et al., 2011
3.6	Finnhorse	Sairanen et al., 2009
5.2	French Trotter	Moureaux et al., 1996
5.8	Norwegian Trotter	Klemetsdal, 1993
6.6	Italian Haflinger	Gandini et al., 1992
7.1	French Arabian	Moureaux et al., 1996
7.4	Standardbred pacers	Cothran et al., 1984
9.0	American Standardbred	MacCluer et al., 1983
9.9	Finnish Standardbred	Sairanen et al., 2009
10.3	Lipizzan	Curik et al., 2003
10.3	Standardbred Trotters	Cothran et al., 1984
13.0	British Thoroughbred	Cunningham et al., 2001

Genetically, inbreeding results in increased homozygosity and decreased heterozygosity throughout the genome. This homozygosity results in increased expression of deleterious recessive alleles. Whenever an autosomal recessive genetic disease crops up, there will be an outcry against inbreeding because it "causes genetic diseases." Inbreeding does not cause disease. Random deleterious recessive mutations occur all the time in individual horses. These mutations are then passed to half of that animal's offspring. The mutation won't be expressed until it is present in the homozygous state. It can't be present in the homozygous state until two carriers are mated. If the mutation occurred in one horse, the only carriers will be descendants of that animal. So it is not until inbreeding of that horse's descendant occurs that the defect will be expressed. Inbreeding does not cause defective alleles to occur (random occurrences of mutation does that), but inbreeding allows those alleles to be expressed in the homozygous state. Another negative effect of inbreeding is depressed performance called inbreeding depression. This is the same logic as used for the expression of deleterious recessive allele applied to polygenic traits. A recessive allele doesn't need to be lethal to be deleterious. Remember, deleterious alleles (lethal or not) tend to be recessive. If an allele with a negative effect was dominant, it would tend to be selected out of the population. Inbreeding causes more expression of recessive alleles that depress performance. Each allele has a small effect, but those effects build up over many genes to affect performance. These effects are most noticeable in lowly heritable traits such as health, survivability, and fertility.

Horses that are inbred are often presumed to have greater prepotency. Prepotency is the characteristic of an animal (usually a sire) to produce offspring that are very much like him and like each other. It's relatively easy to understand how a more homozygous horse would be more prepotent. The more homozygous an animal, the fewer unique gamete that can be produced by that animal, so the fewer different types of progeny that will be seen. As a very simple example, consider two bay stallions: one that is homozygous at both extension and agouti ($E^+E^+A^AA^A$), and one that is heterozygous ($E^+E^eA^AA^a$). Let's assume these stallions are both bred to large numbers of bay, chestnut, and black mares. The homozygous stallion will produce only bay foals, no matter what color the mare. The heterozygous stallion will produce the full spectrum of bay, chestnut, and black foals depending on the genotype of the mares. Therefore, the homozygous bay stallion is considered prepotent for color. However, most important traits in horses (excluding color and single-gene diseases) are polygenic. These would be things like disposition, trainability, speed, quality of movement, jumping ability, cutting ability, and so on. Traits like these are assumed to be controlled by many (maybe hundreds or even thousands) genes, and have the added complication of being affected by the environment. When we talk about polygenic traits, the concept of prepotency is mostly a fallacy. Horses are not inbred to the extent that an animal could be homozygous for hundreds of genes. And even if they were, identical genotypes would have very different phenotypes when environmental effects were factored in. Therefore, the use of inbreeding to increase consistency for anything more than the simplest of traits is not advised.

Why would a breeder want to inbreed? The major reason for inbreeding in our livestock and crop species is to create inbred lines to be used in crossbreeding. Since crossbreeding is not often done in horses in this country (at least not in any systematic, controlled way) there is really no benefit to inbreeding. Inbreeding is used successfully in other species to fix desirable alleles, but for that to succeed, very large populations must be maintained. Inbreeding will fix all alleles, not just desirable ones, so breeders must be prepared to sacrifice large numbers of unfit individuals. Species that have used this strategy successfully are those with very low cost per individual, and very low costs to raise and maintain each individual, such as corn and chickens.

Relationship

Inbreeding is measured by the inbreeding coefficient. Another useful statistic is the coefficient of relationship, which measures the proportion of an animal's genes that are identical to another animal's due to common descent (Chapter 2). This is a value that is often miscalculated and misused in the equine industry. Often horse breeders will use the term "percent blood" to indicate the amount of relatedness between two horses (often a current breeding prospect and a superior ancestor). This value is the numerator of the relationship coefficient, and doesn't appropriately account for inbreeding. If the animals in question are inbred (which is almost always the case when claiming relatedness to a great ancestor) the percent blood actually overestimates the relationship between the horses.

Linebreeding is often used to try and increase the relationship of current horses to superior animals from the past. Linebreeding is just a mild form of inbreeding used to maximize relationship to the superior ancestor while trying to minimize overall inbreeding. To successfully linebreed, very large numbers must be maintained to avoid excessively close matings (i.e., full-sib, father-daughter, etc.). As with all inbreeding, the breeder must cull very strictly to minimize inbreeding depression.

Selection and improvement

Quantitative genetics

In genetic improvement of horses, we are interested in both simply-inherited and polygenic traits. Color and the disease traits discussed previously are examples of economically important traits that are simply-inherited. They are controlled by one or a relatively few number of genes that are known. Simply-inherited traits also tend to be expressed qualitatively or by categories (a horse can be bay or chestnut, but not an intermediate value) and tend to be relatively unaffected by the environment. In contrast, most of the traits that are important to horse breeders are polygenic in nature, meaning they are controlled by many genes. For most of those traits, the number of genes affecting the trait is unknown, but assumed to be in the hundreds or thousands. In most cases, none of the genes are known, so we don't know the animal's genotype for any of them. With a few exceptions, polygenic traits are usually quantitative, or measured on a continuous scale. These would be traits like speed, jumping ability, quality of movement, disposition, trainability, cutting ability, and so on. Because the individual genes aren't identified and we don't know the animal's genotype for them, we have to use advanced statistical methods to predict an animal's collective genetic merit or breeding value for the trait. See Chapters 4–6 for a more complete discussion of polygenic traits and estimating breeding values.

Heritability is an important value that estimates the amount of phenotypic variation that is accounted for by variation in breeding values (Chapter 5). Heritability estimates are indications of the rate of genetic progress that can be made when selecting for traits. There has been extensive work done on estimating heritability for various traits in Europe, but virtually none in the US. Some traits with their heritability estimates are listed in Table 11.3. This list is not exhaustive, but summarizes some results in breeds that are of importance in the US.

Genetic improvement

In order to make genetic progress in a breeding program, the breeder must have a well-defined breeding objective and measureable selection criteria that relate to the objective. Many US breeders have adequate breeding objectives, but industry-wide, there is a shortage of objective, measureable selection criteria. Many sire selection decisions are based on little more than educated guesses about the animal's genetic merit. For example, temperament is very important to most breeders and riders, however, there is no system for measuring and evaluating temperament. One of the oldest rules of animal breeding is that you can't improve what you don't measure. If breeders aren't measuring temperament, jumping ability, speed, barrel racing ability, and so on, those traits can't be improved. Even when breed associations collect data (i.e., times, points earned, placings, etc.) there is no systematic genetic evaluation of the data to account for environmental effects and provide unbiased estimates of genetic merit.

The rate of genetic progress is determined by four factors: accuracy of selection, intensity of selection, genetic variation, and generation interval. Compared to other livestock species, horse

Table 11.3 Heritability estimates in horses.

Trait	h^2	Reference
Racing time (Japanese Thoroughbreds)	0.08–0.25[a]	Oki et al., 1995
Lengths ahead (Australian Thoroughbreds)	0.18–0.98[b]	Williamson and Beilharz, 1998
Finishing position (Australian Thoroughbreds)	0.24–0.93[b]	Williamson and Beilharz, 1998
Earnings (Australian Thoroughbreds)	0.14–1.23[b]	Williamson and Beilharz, 1998
Racing time (American Quarter Horses)	0.00–0.42[c]	Buttram et al., 1988c
Quality of walk (Warmbloods)	0.39	Thoren Hellsten et al., 2006
Quality of trot (Warmbloods)	0.41	Thoren Hellsten et al., 2006
Quality of canter (Warmbloods)	0.40	Thoren Hellsten et al., 2006
Free jumping ability (Warmbloods)	0.50	Thoren Hellsten et al., 2006
Jumping ability (Warmbloods)	0.37	Thoren Hellsten et al., 2006
Rideability (Warmbloods)	0.48	Thoren Hellsten et al., 2006
Cross-country manner (Warmbloods)	0.23	Thoren Hellsten et al., 2006
Racing time (Standardbreds)	0.29	Tolley et al., 1983
Weight (Thoroughbreds)	0.13–0.90[d]	Hintz et al., 1978
Height (Thoroughbreds)	0.33–0.88[d]	Hintz et al., 1978
Cannon bone circumference (Thoroughbreds)	0.12–0.77[d]	Hintz et al., 1978

[a]depending on distance and surface.
[b]depending on model and year.
[c]depending on estimation method.
[d]depending on age of measurement.

breeders are generally disadvantaged on three of them. To maximize progress, accuracy of selection must be high. The most accurate selection is that based on high accuracy, BLUP calculated breeding values (Chapters 4–6). In the US, there is no such genetic evaluation, and selection is based on phenotype and/or pedigree, which is not as accurate. A common breeder technique is to select mares or stallions for breeding because "she has X on her papers," meaning the great-grandfather was an excellent horse. Obviously, this results in relatively low accuracy of selection. Intensity is a function of the percentage of horses selected for breeding. To maximize progress, intensity should be high, meaning the lowest percentage possible (the elite) should be selected. This is another area at which some horse breeders are notoriously bad: "She's too crazy (or lame, or untalented) to ride, so I'll breed her." A short generation interval also leads to greater genetic progress. By the nature of their biology, and by industry structure (riding and competing before breeding), horses in general have a much longer generation interval than other livestock species. All of these factors combine to make genetic progress in the horse industry very slow compared to other livestock species. Another complicating factor is the incredibly wide range of breeding objectives in the horse industry as compared with other livestock species. For example, swine are selected for meat. Individual breeders may put different emphasis on different traits for different lines, but the overall purpose is always for meat. Horse breeders may be selecting for jumpers, or race horses, or cutters, or trail riding horses, or barrel racers, and so on, even within the same breed. Such widely divergent objectives have radically different selection criteria. This has led, in many breeds, to the development of sub-populations, or lines, that are selected specifically for a certain discipline.

The timing of selection decisions in the horse industry has a very negative impact on both accuracy and intensity. Most colts are gelded as weanlings, long they begin performance careers. So the gelding decision is being made with very little information (basically just pedigree), resulting in low accuracy of selection. When it comes time to breed mares, there are relatively few stallions to choose from (because most of the males are gelded) affecting intensity. The conventional wisdom of the conscientious horse breeder to "geld hard," meaning to geld almost everything except the very most elite colts, is actually counterproductive. Gelding all but the most elite is a good plan, but only if you truly know which animals are elite. That is definitely not the case at weaning time. Genetically, the best strategy would be to leave all colts (except those with obvious physical defects) intact until they can be performance tested. However, because of the difficulty in training and competing stallions, that is impractical for the horse industry on a large scale.

Another factor that seriously impedes genetic progress is lack of contemporary grouping. Horses are managed as individuals. Most horses in the US are products of small breeders. Even among large breeders, horses are managed, and more importantly, trained as individuals. Like all other livestock, the normal management factors such as nutrition and health programs contribute to the phenotype. However, for performance events, training is a huge component of overall success. Conventional wisdom is that a great trainer can make a mediocre horse perform well, and a bad trainer can ruin a great one. So when a potential selection candidate has an outstanding performance record, it is impossible to know how much of that is due to great genetics or great training. Progeny records will help sort it out, but even those may be biased. Many good trainers will only bother with horses from certain lines or families. So if progeny of different stallions are preferentially being selected by the good trainers, the progeny performance isn't a fair comparison of the stallions.

Genetic evaluation

In Europe, sophisticated, BLUP based genetic evaluations of Sport Horse breeds have been performed for many years. Sport Horses are those specifically selected for the Olympic events of dressage and jumping. So these breeds (collectively called Warmbloods) have the immediate advantage of a much narrower breeding objective than many of the US breeds. This specific breeding objective and the national genetic evaluation have resulted in a large superiority of European-bred Warmblood breeds in those events. For example, at the recent London Olympics, all of the US dressage and jumper riders were riding Warmbloods or Sport Horses. Every horse that medaled (either team or individual) in dressage or jumping was a Warmblood or Sport Horse. The vast majority of horses that compete in these events at the international level, no matter which country the riders are from, are European bred. Most elite US riders import European horses for competition.

There are some differences between countries and specific breeders but most have the same general premise for genetic evaluation. All stallion prospects are required to pass an inspection by breed officials before being approved for breeding. Stallions that pass this approval are sent to a 70- or 100-day test. This would be analogous to a bull or boar test that used to be popular in the US. The stallions are sent to a testing center where they are trained and ridden by selected professionals for the duration of the test. Each professional works with all the stallions over the test. At the end, the stallions are scored by the riders on things like quality of movement, jumping ability, disposition, trainability, and so on. Because each rider works with all the horses, and each rider scores all the horses, this makes a very nice contemporary group. The environmental variance due to training and management is minimized, so differences expressed are more likely to be genetic differences. The data from the performance test is used for genetic evaluation. Mares don't go through the 70- or 100-day test, however they are inspected and approved. On their inspection, they are scored on conformation traits, as well as quality of movement and jumping ability. Data from mare inspections is also included in the genetic evaluation. To further strengthen the system and increase the amount of data that is used to create genetic predictions, competition records are included in the BLUP evaluation. Most males that don't pass the initial approval go on to be competition horses. In addition, many breeding stallions are able to compete when it's not the breeding season. With the advent of embryo transfer, high quality mares can be competed and bred at the same time. The combination of high quality, properly contemporary grouped stallion tests, mare inspections, and competition records into a BLUP genetic evaluation makes the European system the most advanced horse breeding system in the world. Further information on the European genetic evaluation system can be found at www.biw.kuleuven.be/GENLOG/livgen/research/interstallion_eng.aspx (Interstallion) and www.vit.de/index.php?id=zws-pferd&L=1.

There is nothing in the US that comes even close to the system developed by the Europeans. Over the last couple of decades, many Warmbloods have been imported into the US, and the Warmblood breeding industry in growing. The US Warmblood breeds also have inspections that approve or deny breeding status to stallions; however, the data is not used in any type of genetic evaluation. By requiring approval, there is the hope of guaranteeing some minimum level of quality, but it's definitely not as useful in making genetic progress as BLUP-derived EBVs.

In the 1980s, AQHA sponsored a study to calculate breeding values for racing Quarter Horses. The evaluation was done by graduate students under the supervision of Richard Willham, a respected beef cattle geneticist, and published in a series of articles in the *Journal of Animal Science* (Buttram et al., 1988a,b,c; Wilson et al., 1988). The study showed that a genetic evaluation of racing performance not only was possible, but was an effective way to rank horses. However, the association decided not to implement a breed-wide genetic evaluation program. The board of directors argued that horse breeding is an art, not a science. Perhaps more practically, there was concern about how access to EBV might impact wagering (Wilson, and Willham, personal communication).

Thoroughbred racing is unquestionably the biggest money segment of the horse industry in the US. There are people who make livings doing "pedigree analysis," which is trying to predict winners by subjectively looking at pedigrees. Others do all kinds of statistical gymnastics, calculating various kinds of indexes, speed figures, "dosages," and so on to predict those horses that will be winners. There is extensive study of "nicking," which is trying to find the optimum combination of family lines to produce winners. It is quite interesting the lengths this industry will go to in order to try to predict winners, but they won't do the one thing that is proven to work: breed-wide genetic evaluation using BLUP methodology to predict EBV.

New technologies

Cloning

Cloning is a relatively new technology that has the potential to revolutionize horse breeding. Cloning produces an individual that is genetically identical to the original horse. There will be phenotypic differences due to different environments in which the horses are raised, so performance is not expected to be the same. However, semen produced by a cloned horse will be genetically equivalent to semen produced by the original horse. There has been some cloning of old stallions and mares; proven sires and dams that are too old to breed. That should raise an immediate red flag in the context of genetic improvement. By extending the breeding time of old horses, we are effectively extending the generation interval even more, decreasing genetic progress. The place that cloning is really taking off is in the cloning of superior performance geldings. As mentioned previously, selection decisions (gelding) are made at weaning by most breeders, long before any information about a horse's performance ability is known. Cloning a gelding is a way to "recreate" the testicles that were removed when he was gelded. Semen from the clone of the gelding is genetically equivalent to semen that would have been produced by that gelding if he had stayed a stallion. In a genetic improvement context, this has enormous potential to increase both accuracy and intensity of selection. The geldings that are being chosen for cloning are usually truly superior animals with long careers of sustained excellent performance. This much data on the individual, while not as accurate as progeny data, is still far more accurate than just pedigree data. By cloning superior geldings, intensity is also increasing because there is a larger pool of superior stallions to choose from.

Equine genomics

The equine genome sequence was completed in 2007. The first animal sequenced was a Thoroughbred mare named Twilight. Currently, sequencing of individuals from other breeds is underway. Information from the sequence has been used to develop a 54,000, and more recently a 74,000 SNP chip for the horse. Most of the American researchers involved in the project are part of veterinary schools, and the primary use of the information has been in the study of diseases. The sequence and the SNP chips have been boons to these researchers, enabling them to more quickly find genes that cause simply-inherited diseases, speeding up the time it takes to develop a gene test that can be used by the industry to manage those defects. However, there has been very little work so far to use genomic information to help guide selection decisions for performance traits. In fact, on the Horse Genome Project website (www.uky.edu/Ag/Horsemap/welcome.html), the categories under the link "applications of genome study" are coat color genetics and hereditary disease – single gene defects; no mention of performance traits at all.

Recently, there has been some work on utilizing the genome to predict and select for racing performance in Thoroughbreds. In fact, researchers from Ireland found that a mutation in the myostatin gene (this gene has long been described in other species) can predict a horse's best racing distance. A gene test was quickly commercialized, and press releases touted the discovery of "the speed gene." Since then, several companies have commercialized marker panels that predict some measure of racing performance. While these tests, including the myostatin test, have data that show a statistically significant effect on the measured trait, to date none of them have published the percentage of genetic or phenotypic variation that the test accounts for. This makes it difficult to assess the true value of these tests in identifying horses with superior genetics for speed. In other livestock species, geneticists have shown that using genomic information in combination with BLUP estimated breeding values is the optimum way to realize value from genomic tests in selection programs. In the racing industry in this country, there are no BLUP estimated breeding values, so the optimum method of utilizing genomic information in a selection program is unknown. For example, what weight should the genomic test results be given versus own performance, pedigree or progeny records, or other indicators such as conformation?

There is no question that sequencing the equine genome has been an enormous step forward in the study of equine genetics. However, the application of genomic information into selection program for anything but the most simply-inherited color and disease traits is in its infancy. It is important to remember that the sequence is a tool, not the final answer to all genetic questions. There needs to be a significant commitment by breed associations to research and apply both traditional quantitative genetics and genomics if there is going to be much progress from genomics in selection for performance traits.

Further reading

Color

Adalsteinsson, S. (1974). Inheritance of the palomino color in Icelandic horses. *J. Hered.* 65(1):15–20.

Andersson, L. and K. Sandberg. 1982. A linkage group composed of three coat color genes and three serum protein loci in horses. *J. Hered.* 73(2):91–94.

Archer, S., K. Brown, R. Bellone, D. Bernoco, and E. Bailey. 2007. Investigation of ECA3 as the location of the gene for white patterning (PATN1) in leopard spotted Appaloosas. *Plant and Animal Genomes XV Conference.* P598.

Bellone, R. R., S. A, Brooks, L. Sandmeyer, B. A. Murphy, G. Forsyth, S., E. Bailey, and B. Grahn. 2008. Differential Gene Expression of TRPM1, the Potential Cause of Congenital Stationary Night Blindness and Coat Spotting Patterns (LP) in the Appaloosa Horse (Equus caballus). *Genetics* 179(4):1861–1870.

Brooks, S. and E. Bailey. 2005. Exon skipping in the KIT gene causes a Sabino spotting pattern in horses. *Mammalian Genome* 16(11):893–902.

Brooks, S. A., T. L. Lear, D. L. Adelson, and E. Bailey. 2007. A chromosome inversion near the KIT gene and the Tobiano spotting pattern in horses. *Cytogenet. Genome Res.* 119:225–230.

Bricker, S. J., M. C. T. Penedo, L. V. Millon, and J. D. Murray. 2003. Linkage of the dun coat color locus to microsatellites on horse chromosome 8. *Plant and Animal Genomes XI Conference.* P640.

Brunberg E., L. Andersson, G. Cothran, K. Sandberg, S. Mikko, and G. Lindgren. 2006. A missense mutation in PMEL17 is associated with the Silver coat color in the horse. *BMC Genetics* 7:46.

Cook, D., S. Brooks, R. Bellone, and E. Bailey. 2008. Missense Mutation in Exon 2 of SLC36A1 Responsible for Champagne Dilution in Horses. *PLoS Genetics* 4 (9):e1000195.

Haase B., S. A. Brooks, A. Schlumbaum, P. J. Azor, E. Bailey, F. Alaeddine, et al., 2007. Allelic heterogeneity at the equine KIT locus in dominant white (W) horses. *PLoS Genetics* 3(11):e195.

Haase B., S. A. Brooks, T. Tozaki, D. Burger, P. A. Poncet, S. Rieder, et al., 2009. Seven novel KIT mutations in horses with white coat colour phenotypes. *Animal Genetics* 40:623–629.

Haase, B., R. Jude, S. A. Brooks, and T. Leeb, 2008. An equine chromosome 3 inversion is associated with the tobiano spotting pattern in German horse breeds. *Animal Genetics* 39(3):306–309.

Henner, J., P. A. Poncet, G. Geurin, C. Hagger, G. Stranzinger, and S. Rieder. 2002. Genetic mapping of the (G) locus, responsible for the coat colour phenotype progressive graying with age in horses (*Equus caballus*). *Mammalian Genome* 13:535–537.

Locke, M. M., M. C. T. Penedo, S. J. Brickker, L. V. Millon, and J. D. Murray. 2002. Linkage of the gray coat colour locus to microsatellites on horse chromosome 25. *Animal Genetics* 33(5):329–337.

Locke, M. M., L. S. Ruth, L. V. Millon, M. C. T. Penedo, J. C. Murray, and A. T. Bowling. 2001. The cream dilution gene, responsible for the palomino and buckskin coat colors, maps to horse chromosome 21. *Animal Genetics* 32(6):340–343.

Mariat, D., S. Taourit, and G. Guerin. 2003. A mutation in the MATP gene causes the cream coat colour in the horse. *Genet. Selection Evol.* 35(1):119–133.

Marklund, L., M. J. Moller, K. Sandberg, and L. Andersson. 1996. A missense mutation in the gene for melanocyte-stimulating hormone receptor (MC1R) is associated with the chestnut coat color in horses. *Mammalian Genome* 7(12):895–899.

Marklund, S., M. Moller, K. Sandberg, and L. Andersson. 1999. Close association between sequence polymorphism in the KIT gene and the roan coat color in horses. *Mammalian Genome* 10(3):283–288.

Mau, C., P. A. Poncet, B. Bucher, G. Stranzinger, and S. Rieder. 2004. Genetic mapping of dominant white (W), a homozygous lethal condition in the horse (Equus caballus). *J. Animal Breeding and Genet.* 121(6):374–383.

Metallinos, D. L., A. T. Bowling, and J. Rine. 1998. A missense mutation in the endothelin-B receptor gene is associated with Lethal White Foal Syndrome: an equine version of Hirschsprung Disease. *Mammalian Genome* 9(6):426–31.

Pielberg, G. R., A. Golovko, E. Sundström, I. Curik, J. Lennartsson, M. H Seltenhammer, et al., 2008. A cis-acting regulatory mutation causes premature hair graying and susceptibility to melanoma in the horse. *Nature Genetics* 40(8):1004–1009.

Pielberg, G., S. Mikko, K. Sanberg, and L. Andersson. 2005. Comparative linkage mapping of the Gray coat colour gene in horses. *Animal Genetics* 36:390–395.

Rieder, S., C. Hagger, G. O. Ruff, T. Leeb, and P. A. Poncet. 2008. Genetic analysis of white facial and leg markings in the Swiss Franches-Montagnes Horse Breed. *J. Hered.* 99(2):130–136.

Rieder, S., S. Taourit, D. Mariat, B. Langlois, and G. Guerin. 2001. Mutations in the agouti (ASIP), the extension (MC1R), and the brown (TYRP1) loci and their association to coat color phenotypes in horses (Equus caballus). *Mammalian Genome* 12(6):450–455.

Sandberg, K. and R. K. Juneja. 2009. Close linkage between the albumin and Gc loci in the horse. *Animal Genetics* 9(3): 169–173.

Sandmeyer, L. S., C. B. Breaux, S. Archer, and B. H. Grahn. 2007. Clinical and electroretinographic characteristics of congenital stationary night blindness in the Appaloosa and the association with the leopard complex. *Vet. Ophth.* 6(10):368–375.

Santschi, E. M., A. K. Purdy, S. J. Valberg, P. D. Vrotsos, H. Kaese, and J. R. Mickelson. 1998. Endothelin receptor B polymorphism associated with lethal white foal syndrome in horses. *Mammalian Genome* 9(4):306–309.

Sponenberg, D. P. 1982. The inheritance of leopard spotting in the Noriker horse. *J. Hered.* 5(73):357–359.

Sponenberg, D. P., G. Carr, E. Simak, and K. Schwink. 1990. The inheritance of the leopard complex of spotting patterns in horses. *J. Hered.* 81:323–331.

Sponenberg, D. P., H. T. Harper, and A. L. Harper. 1984. Direct evidence for linkage of roan and extension loci in Belgian horses. *J. Hered.* 75:413–414.

Swinburne, J. E., A. Hopkins, and M. M. Binns. 2002. Assignment of the horse grey coat colour gene to ECA25 using whole genome scanning. *Animal Genetics* 33(5):338–342.

Terry, R. B., S. Archer, S. Brooks, D. Bernoco, and E. Bailey. 2004. Assignment of the appaloosa coat colour gene (LP) to equine chromosome 1. *Animal Genetics* 35(2):134–137.

Thiruvenkadan, A. K., N. Kandasamy, and S. Panneerselvam. 2008. Review: Coat colour inheritance in horses. *Livestock Sci.* 117(2–3): 109–129.

Trommerhausen-Smith., A. 1978. Linkage of tobiano coat spotting and albumin markers in a pony family. *J. Hered.* 69(4):214–216.

Veterinary Genetics Laboratory, School of Veterinary Medicine, University of California, Davis. "Pearl." Online: www.vgl.ucdavis.edu/services/horse/splashedwhite.php. (accessed July 10, 2012).

Veterinary Genetics Laboratory, School of Veterinary Medicine, University of California, Davis. "Splashed-White." Online: www.vgl.ucdavis.edu/services/horse/splashedwhite.php. (accessed July 10, 2012).

Wagner, H. J. and M. Reissmann. 2000. New polymorphism detected in the horse MC1R gene. *Animal Genetics* 31(4):289.

Woolf, C. M. 1989. Multifactorial inheritance of white facial markings in the Arabian horse. *J. Hered.* 80:173–178.

Woolf, C. M. 1990. Multifactorial inheritance of common white markings in the Arabian horse. *J. Hered.* 81(4):240–246.

Woolf, C. M. 1991. Common white facial markings in bay and chestnut Arabian horses and their hybrids. *J. Hered.* 82(2):167–169.

Woolf, C. M. 1992. Common white facial markings in Arabian horses that are homozygous and heterozygous for alleles at the A and E loci. *J. Hered.* 83:73–77.

Diseases

Aleman, M., J. Riehl, B. M. Aldridge, R. A. Lecouteur, J. L. Stott, and I. N. Pessah. 2004. Assocation of a mutation in the Ryanodine Receptor 1 gene with equine malignant hyperthermia. *Muscle Nerve* 30:356–365.

Archer, R. K. 1961. True haemophilia (haemophilia A) in a Thoroughbred foal. *Veterinary Record* 73:338–340.

Bernoco, D. and E. Bailey. 1998. Frequency of the SCID gene among Arabian horses in the USA. *Animal Genetics* 29:41–42.

Brault, L S., C. A. Cooper, T. R. Famula, J. D. Murray, and M. C. T. Penedo. 2010. Mapping of equine cerebellar abiotrophy to ECA2 and identification of a potential causative mutation affecting expression of MUTYH. *Genomics* 97:121–129.

Brooks, S. A., N. Gabreski, D. Miller, A. Brisbin, H. E. Brown, C. Streeter, et al., 2010. Whole-Genome SNP Association in the Horse: Identification of a Deletion in Myosin Va Responsible for Lavender Foal Syndrome. *PLoS Genet.* 6(4): e1000909. doi:10.1371/journal.pgen.1000909.

Gardner, E. J., J. L. Shupe, N. C. Leone, and A. E. Olson. 1975. Hereditary multiple exostoses. *J. Hered.* 66:318–322.

Graves, K. T., P. J. Henney, and R. B. Ennis. 2009. Partial deletion of the LAMA3 gene is responsible for hereditary junctional epidermolysis bullosa in the American Saddlebred Horse. *Animal Genetics* 40(1):35–41.

Henninger, R. W. 1988. Hemophilia A in two related Quarter Horse colts. *JAVMA* 193:91–94.

Hermans, W. A. 1970. A hereditary anomaly in Shetland ponies. *Netherlands J. Vet. Sci.* 3:55–63.

Hermans, W. A., A. W. Kersjes, G. J. W. van der Mey, and K. J. Dik. 1987. Investigation into the heredity of congenital lateral patellar (sub)luxation in the Shetland pony. *Vet. Quart.* 9:1–8.

Hutchins, D. R., E. E. Lepherd, and I. G. Crook. 1967. A case of equine haemophilia. *Australian Vet. J.* 43:83–87.

McCue M. E., S. J. Valberg, M. B. Miller, C. Wade, S. DiMauro, H. O Akmand, J. R. Mickelson. 2008. Glycogen synthase (GYS1) mutation causes a novel skeletal muscle glycogenosis. *Genomics* 91(5):458–466.

Rudolph, J. A., S. J. Spier, G. Byrns, C. V. Rojas, D. Bernoco, and E. P. Hoffman. 1992. Periodic paralysis in Quarter Horses: a sodium channel mutation disseminated by selective breeding. *Nature Genet.* 2:144–147.

Spirito, F., A. Charlesworth, K. Linder, J. P. Ortonne, J. Baird, and G, Meneguzzi. 2002. Animal models for skin blistering conditions: Absence of Laminin 5 causes hereditary junctional mechanobullous disease in the Belgian horse. *J. Investigative Derm.* 119:684–691.

Tryon, R. C., S. D. White, and D. L. Bannasch. 2007. Homozygosity mapping approach identifies a missense mutation in equine cyclophilin B (PPIB) associated with HERDA in the American Quarter Horse. *Genomics* 90(1):93–102.

Valberg, S. and J. Mickelson. Polysaccharide Storage Myopathy (PSSM) in horses. Online: www.cvm.umn.edu/umec/lab/PSSM/home.html (accessed July 12, 2012).

Wagner, M. L., S. J. Valberg, E. G. Ames, M. M. Bauer, J. A. Wiseman, C. T. Penedo, et al., 2006. Allele frequency and likely impact of the glycogen branching enzyme deficiency gene in Quarter Horse and Paint Horse populations. *J. Vet. Int. Med.* 20(5):1207–1211.

Ward, T. L., S. J. Valberg, T. L. Lear, G. Geurin, D. Milenkovic, J. Swinburne, et al., 2003. Genetic mapping of GBE1 and its association with glycogen storage disease IV in American quarter horses. *Cytogenet Genome Res.* 102: 201–206.

Watson, A. G. and I. G. Mayhew. 1986. Familial congenital occipitoatlantoaxial malformation (OAAM) in the Arabian horse. *Spine* 11: 334–339.

Inbreeding and relationship

Gandini, G. C., A. Bagnato, F. Miglior, and G. Pagnacco. 1992. Inbreeding in the Italian Haflinger horse. *J. Anim. Breed. Genet.* 109: 433–443.

Klemetsdal, G. 1993. Demographic parameters and inbreeding in the Norwegian trotter. *Acta Agric Scand* 43:1–8.

Heritability estimates

Oki, H., Y. Sasaki, and R. L. Willman. 1995. Genetic parameter estimates for racing time by restricted maximum likelihood in the thoroughbred horse of Japan. *J. Anim. Breeding Genet.* 112:146–150.

Stallion testing and genetic evaluation

Koenen, E. P. C. and L. I. Aldridge. 2002. Testing and genetic evaluation of sport horses in an international perspective. *Proc. 7th WCGALP, Montpellier*, August 2002.

Koenen, E. P. C., L. I. Aldridge, and J. Philipsson. 2004. An overview of breeding objectives for warmblood sport horses. *Livestock Prod. Sci.* 88:77–84.

Olsson, E., A. Nasholm, E. Strandberg, J. Philipsson. 2008. Use of field records and competition results in genetic evaluation of station performance tested Swedish Warmblood stallions. *Livestock Sci.* 117:287–297.

Genomics

Hill, E. W., J. Gu, S. S. Eivers, R. G. Fonseca, B. A. McGivney, P. Govindarajan, et al., 2010. A sequence polymorphism in MSTN predicts sprinting ability and racing stamina in Thoroughbred horses. *PLoS One*, Jan 20.

Hill, E. W., B. A. McGivney, J. Gu, R. Whiston, D. E. Machugh. 2010. A genome-wide SNP-association study confirms a sequence-variant (g 66493737C>T) in the equine myostatin (MSTN) gene as the most powerful predictor of optimum racing distance for Thoroughbred racehorses. *BMC Genomics* 11:552.

References

Bhatnagar, A. S., C. M. East, and R. K. Splan. 2011. Inbreeding and founder contributions of the Norwegian Fjord horse in North America. *J. Equine Vet. Science* 31:241–242.

Buttram, S. T., R. L. Willham, and D. E. Wilson. 1988a. Genetics of racing performance in the American Quarter Horse: II. Adjustment factors and contemporary groups. *J. Anim. Sci.* 66:2800–2807.

Buttram, S. T., R. L. Willham, D. E. Wilson, and J. C. Heird. 1988b. Genetics of racing performance in the American Quarter Horse: I. Description of the data. *J. Anim. Sci.* 66: 2791–2799.

Buttram, S. T., D. E. Wilson, and R. L. Willham. 1988c. Genetics of racing performance in the American Quarter Horse: III. Estimation of variance components. *J. Anim. Sci.* 66:2808–2816.

Cothran, E. G., J. W. MacCluer, L. R. Weitkamp, D. W. Pfenning, and A. J. Boyce. 1984. Inbreeding and reproductive performance in Standardbred horses. *J. Hered.* 75(3):220–224.

Cunningham, E. P., J. J. Dooley, R. K. Splan and D. G. Bradley. 2001. Microsatellite diversity, pedigree relatedness and the contributions of founder lineages to thoroughbred horses. *Animal Genetics* 32(6):360–364.

Curik, I., P. Zechner, J. Solkner, R Achmann, I. Bodo, P. Dovc, T. Kavar, E. Marti, and G. Brem. 2003. Inbreeding, microsatellite heterozygousity, and morphological traits in Lipizzan horses. *J. Hered.* 94(2):125–132.

Hintz, R. L., H. F. Hintz, and L. D. Van Vleck. 1978. Estimation of heritabilities for weight, height and front cannon bone circumference of Thoroughbreds. *J. Anim. Sci.* 47:1243–1245.

MacCluer, J. W., J. A. Boyce, B. Dyke, L. R. Weitkamp, D. W. Pfenning, and C. J. Parsons. 1983. Inbreeding and pedigree structure in standardbred horses. *J. Hered.* 74:394–399.

Moureaux S., E. Verrier, A. Ricard, and J. C. Meriaux. 1996. Genetic variability within French race and riding horse breeds from genealogical data and blood marker polymorphisms. *Genet. Sel. Evol.* 28:83–102.

Sairanen, J., K. Nivola, T. Katila, A. M. Virtala, and M. Ojala. 2009. Effects of inbreeding and other genetic components on equine fertility. *Animal.* 3(12):1662–1672.

Sponenberg, D. P. 2009. *Equine Color Genetics*. Wiley-Blackwell, Ames, IA.

Strom, H. 1982. Changes in inbreeding and relationship within the Swedish standardbred trotter. *J. Anim. Breeding and Gen.* 99:55–58.

Tolley, E. A., D. R. Notter, and T. J. Marlowe. 1983. Heritability and repeatability of speed for 2- and 3-year-old Standardbred racehorses. *J. Anim. Sci.* 56:1294–1305.

Tunnell, J. A., J. O. Sanders, J. D. Williams, and G. D. Potter. 1983. Pedigree analysis of four decades of Quarter Horse breeding. *J. Anim. Sci.* 57:585–593.

Thoren Hellsten, E., A. Viklund, E. P. C. Koenen, A. Ricard, E. Bruns, and J. Philipsson. 2006. Review of genetic parameters estimated at stallion and young horse performance tests and their correlations with later results in dressage and show jumping competitions. *Livest. Sci.* 103:1–12.

Williamson, S. A. and R. G. Beilharz. 1998. The inheritance of speed, stamina and other racing performance characteristics in the Australian Thoroughbred. *J. Anim. Breed. Genet.* 115:1–16.

Wilson, D. E., R. L. Willham, S. T. Buttram, J. A. Hoekstra, and G. R. Luecke. 1988. Genetics of racing performance in the American Quarter Horse: IV. Evaluation using a reduced animal model with repeated records. *J. Anim. Sci.* 66:2817–2825.

Review questions

1. Cross two white horses. What are the genotypes and phenotypes expected in the progeny?

2. Cross a white horse with a non-white horse. What are the genotypes and phenotypes expected in the progeny?

3. Cross a gray horse that had a non-gray parent to a non-gray horse. What are the genotypes and phenotypes expected in the progeny?

4. Cross a homozygous gray horse with a gray horse that had a non-gray parent. What genotypes and phenotypes are expected in the progeny?

5. Cross two chestnuts. What genotypes and phenotypes are expected in the progeny?

6. Cross two bays. What genotypes and phenotypes are expected in the progeny?

7. Cross two blacks. What genotypes and phenotypes are expected in the progeny?

8. Cross two blacks. Is it possible to get a bay? Or a chestnut?

9. Cross a bay (with a chestnut parent and a black parent) and a chestnut (with two black parents). What genotypes and phenotypes are expected in the progeny?

10. What color are the genotypes $Wh^+Wh^+G^+G^+E^+E^eA^AA^aDn^+Dn^{nd}$ and $Wh^+Wh^+G^+G^+E^+E^eA^aA^aDn^{nd}Dn^{nd}$? If you cross these two horses, what is the probability that you will get a chestnut? Or a grullo?

11. A palomino breeder has a palomino mare and wants to maximize her chances of producing a palomino foal. What color stallion should the breeder choose?

12. How would you handle a genetic defect in your breeding program if the defect was:
  autosomal dominant with a gene test?
  autosomal dominant without a gene test?
  autosomal recessive with a gene test?
  autosomal recessive without a gene test?

13. Explain why linebreeding isn't a very successful strategy in horse breeding.

14. Why hasn't stallion testing and genetic evaluation been implemented in the United States?

12 Genetics and Genomics of the Domestic Dog

Leigh Anne Clark and Alison Starr-Moss

Department of Genetics and Biochemistry, College of Agriculture, Forestry and Life Sciences, Clemson University, SC, USA

Introduction to canine research

Domestication

The domestic dog, *Canis lupus familiaris*, shares its genus with much older species, namely coyotes, jackals, and various wolves. Although it may seem that the vast phenotypic diversity of dogs must be the result of multiple founding ancestors, it is believed that the dog diverged exclusively from the gray wolf, *Canis lupus*. The dog was the first domesticated animal, but the time, location, and number of domestication events remains unclear. Archeological evidence indicates that domestic dogs first appeared 15,000 to 33,000 years ago in Europe and Eastern Siberia, while genetic signatures from dog breeds suggest origins in Europe, the Middle East, and East Asia and point to various times, ranging from 5000 to 100,000 years ago.

Recently, the proposed origins of canine domestication have come under question. Studies of semi-feral African village dogs, which are thought to have a population structure resembling that of early domesticated dogs, have demonstrated comparable levels of diversity to dogs sampled from East Asia. Because the origin of domestication of a species typically harbors the greatest levels of genetic diversity among individuals, this finding suggests that large-scale studies of dogs and wolves around the world are necessary to determine where dogs first originated.

Breeding

A breed of dog is defined as a group of individuals conforming to a single, defined phenotype that is predictable and reproducible through successive generations. In order for a dog to be considered **purebred**, both parents must be registered with a kennel or breed club. Artificial selection for distinct morphological and behavioral traits in dogs has led to more than 400 canine breeds (Figure 12.1). Most breeds are relatively young, having been developed only in the past 250 years.

Not only are breeds genetically isolated because no new genetic information is being introduced from outside the breed population, but many have also experienced significant events that further reduced genetic diversity. Most breeds were established using only a few dogs that possessed certain desirable traits. While limiting the number of founders expedited fixation of the desired traits, a consequence was **genetic drift**, which caused fluctuation of allele frequencies and the loss of rare alleles. **Founder effect** is the loss of genetic variation when a small number of individuals become reproductively separated from a larger group. A few breeds have experienced a sudden dramatic reduction in numbers, known as a **population bottleneck** (Figure 12.2). Recovery of the breed is then dependent on a small number of individuals, resulting in genetic drift and increased genetic homogeneity.

Most breeders utilize linebreeding to ensure uniformity of quality, without the intrinsic risks of close inbreeding. **Linebreeding** is the use of a sire and dam that share a recent common ancestor. Interestingly, studies have shown that linebreeding actually decreases the risk of widespread disease propagation because deleterious alleles are more quickly recognized and purged from the population.

Popular sire effect occurs when a dog with desirable traits is used more frequently for breeding than other males in the population. Artificial insemination and preservation of frozen semen eliminates geographical barriers and allows for desirable studs to be used long after their death. A review of pedigrees of dogs from 10 breeds registered with the United Kingdom Kennel Club revealed popular sires (dogs producing > 100 offspring) in nearly all breeds surveyed, although the proportion varied between breeds. In the Golden retriever, only 5% of all males were used as breeding stock, and 10% of these were popular sires. Because popular sires can cause widespread dissemination of deleterious alleles, some breed clubs have implemented limitations on the number of litters per sire.

A model organism

Animal models are widely utilized for understanding genetic mechanisms and improving treatment of diseases. In the last 15 years, the dog has emerged as a model system for the study of human hereditary diseases. The unique population structure of the dog lends itself to genetic research by allowing scientists to overcome the genetic heterogeneity that hinders the identification of genes in complex human disorders. At the same time, dog varieties are freely able to interbreed to provide diverse genetic backgrounds that more accurately reflect the genetic differences of human populations. Characteristics that also make the dog more amenable to genetic study are large litter sizes, a short gestational period, and accelerated aging/disease progression.

Molecular and Quantitative Animal Genetics, First Edition. Edited by Hasan Khatib.

Figure 12.1 There's a breed for that. Each dog breed was created to possess certain physical and behavioral attributes that enhance their ability to learn and perform a task.

Until recently, the primary model for studying human disease has been the laboratory mouse. Although the mouse is indisputably a fundamental resource, the dog model provides opportunities in areas where the rodent model is deficient. In the US, 40% of households have a dog, and utilization of data from these pet populations eliminates the time and expense necessary for the establishment of research colonies. Owners spend greater than $40 billion per year on veterinary medical care and vaccinations, making healthcare in dogs comparable in many respects to that of humans. Dogs share our environments and are therefore exposed to the same carcinogens and environmental hazards. These factors enhance the probability that abnormalities will be recognized as well as their relevance to human health.

Dogs suffer from approximately 220 naturally-occurring inherited diseases that are analogous to human disorders, including cancers. When modeling the causes and pathogenesis of human disease, environment-gene interactions are better studied in an animal that lives in the same environment. Unlike rodent models in which mutations are generally induced, the dog provides researchers with an opportunity to study naturally-occurring models of human diseases. Because many mouse models are developed from single gene knockouts, they often do not display all of the phenotypes associated with the analogous human disease. Similarity in organ size between large breed dogs and humans makes the dog useful in clinical trials of new drugs, surgical procedures, and gene therapy. Studies in dogs can accelerate drug development because veterinary medicine is governed by fewer regulatory guidelines than human medicine.

Genetic testing

Approximately two-thirds of hereditary diseases in dogs are transmitted in an autosomal recessive fashion. When no phenotype is present in the carrier-state, it is difficult for breeders to eliminate deleterious alleles. Genetic tests enable breeders to make educated decisions to decrease the incidence of disease through the early identification of diseased dogs and carriers of recessive alleles.

Genetic testing is becoming routine in companion animal medicine, with more than 80 DNA tests currently available for mutations associated with diseases and morphological characteristics of the purebred dog. One of the earliest genetic tests available for the dog was for progressive retinal atrophy (PRA), a devastating recessive ocular disorder that causes blindness. As of 2012, 24 mutations causing PRA have been identified in 18 genes. A single disorder that is caused by mutations in different genes is said to have **locus heterogeneity**. Additionally, because independent mutations in the same gene cause PRA, the disease is also said to have **allelic heterogeneity**.

An electronic database of available DNA tests for canine disease can be found online: www.offa.org/dna_alltest.html. When genetic tests are unavailable, breeders may remove the parents and siblings of affected dogs from their breeding program, thereby further reducing genetic diversity (Figure 12.3).

The dog genome

Genomic tools

In order to identify causative mutations and develop genetic tests, it is necessary to assemble a molecular "toolbox" for genetic analyses. The canine genome is organized across 78 chromosomes: 38 autosomal pairs and the sex chromosomes (X and Y). A **karyotype** is a complete characterization of all chromosomes in an organism, including number, size, and morphology. The

Figure 12.2 Population bottleneck in the Leonberger. The Leonberger is a large breed from Germany that was developed in the mid-1800s to be a companion and a protector. During World War I, the Leonberger population was nearly wiped out by violence and starvation. In 1922, a group dedicated to re-establishing the breed identified 25 dogs that possessed the Leonberger phenotype. Only seven of these dogs (five females, two males) were deemed suitable for breeding. Within four years, 350 Leonbergers had been selectively bred, enough to provide foundation stock to new kennels. Today, the Leonberger is an intelligent and loyal family pet that ranks 103rd among registration statistics published by the American Kennel Club.

karyotype was difficult to establish in the dog because the autosomes are all acrocentric (very short *p* arms), small (average 61 Mb), and have uniform banding patterns. These features complicated individual chromosome identification, and researchers did not come to a consensus about chromosome assignment until 1999. The classification of individual chromosomes in the dog was facilitated by localization of genetic markers on each chromosome. These DNA markers included microsatellites, restriction fragment length polymorphisms, and gene sequences.

Microsatellites, also known as short tandem repeats, are sequences comprised of two, three, or four nucleotides recurring in tandem (e.g., $[CAG]_n$). These sequences are found throughout the genome, most often in non-coding regions. Microsatellites are popular for mapping because they are heritable, co-dominant, and **polymorphic** (there are many alleles due to different repeat lengths). Screening sets of polymorphic microsatellites that are spaced regularly across all chromosomes have been developed to facilitate efforts to map disease loci.

Microsatellites were also used to develop the first canine linkage, or meiotic, maps in the late 1990s. A dog/hamster radiation hybrid panel (irradiated dog cells fused to hamster cells) described in 1999 was used to create a physical map comprised of both unique gene sequences and microsatellite markers. In the following decade, several versions of integrated maps (cytogenetic, genetic, and physical) provided higher-resolution and more accurate resources for studies in the dog. By the early 2000s researchers were eager for the ultimate physical map: whole genome sequence.

Whole genome sequence

The first-ever draft sequence of the canine genome was generated in 2001 from a black Standard Poodle. The privately funded genome survey provided 1.5× **coverage** (average number of times each nucleotide is sequenced) and the first resource for

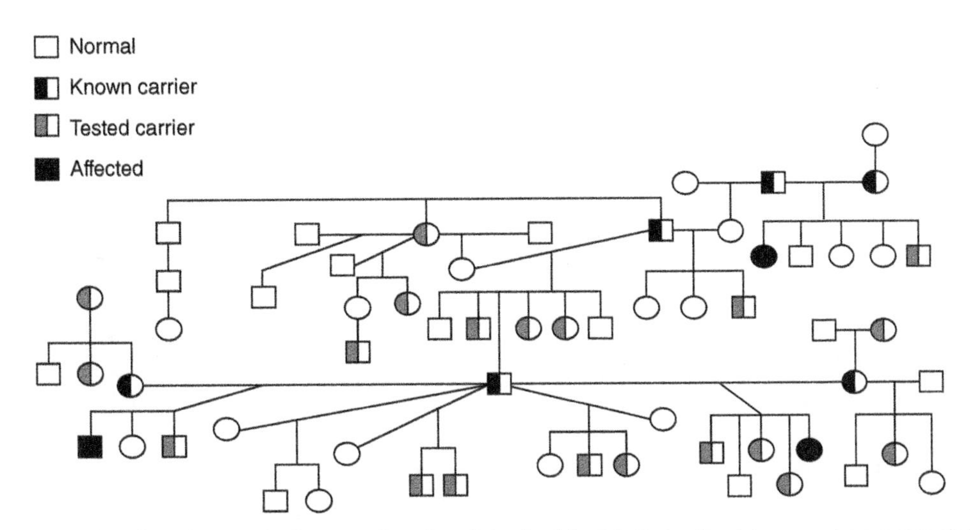

Figure 12.3 Genetic testing identifies phenotypically normal carriers. A family of English Cocker Spaniels segregates a recessively-inherited, fatal renal disease. Prior to development of the genetic test, only obligate carriers (those who had produced affected progeny) could be identified. Development of a genetic test allowed for identification of carriers and enabled continuation of this line with careful breeding.

comparative genomics of the dog. In 2002, the National Institutes of Health (NIH) selected the dog, over several other model and agriculturally important animals, for development of a high quality draft sequence. A 7.5× sequence of a female Boxer was assembled in 2004 at a cost of $30 million (Figure 12.4). Importantly, this dog genome project also generated partial sequences for nine additional breeds, four wolves, and a coyote. These data were used to identify millions of **single nucleotide polymorphisms** (SNPs) for use in linkage studies. A SNP is a difference in the sequence at one nucleotide and is the most common type of genetic variation.

Figure 12.4 The dog behind the genome. A publically available, high-resolution draft sequence of the dog genome was generated using Tasha, a female Boxer. Tasha was selected for her high degree of homozygosity. Previously, the private company Celera Genomics generated a low-resolution draft sequence, using a black Standard Poodle named Shadow. Shadow was selected because he was the personal pet of J. Craig Venter, founder of Celera Genomics.

Genome structure

The dog genome contains 2.4 billion base pairs of DNA. Sequence assembly revealed that the euchromatic portion, that which is gene-rich and genetically active, is 18% smaller in the dog than in the human genome. This is attributed in part to fewer and smaller repeat elements. The human genome shares approximately 650 Mb more ancestral sequence with the dog than the mouse because the murine genome has a higher rate of repeat insertions and large-scale deletions. While humans have over 20,000 protein-coding genes, dogs have about 19,000.

SINEs and **LINEs** are **short** and **long interspersed elements** that are a major source of genome diversity and a driving force in genome evolution. Both repetitive elements use a "copy and paste" mechanism to invade genomic regions, more or less at random. SINEs have an internal promoter for transcription but are dependent upon enzymes produced by autonomous LINEs for reverse transcription (from RNA to cDNA) and integration into the genome. Once integrated, SINEs and LINEs are passed on from generation to generation. An approximately 200 bp element known as SINEC_Cf is highly active in dogs. Greater than 10,000 SINEC_Cf insertion sites in the Boxer and Poodle genomes are bimorphic (present in one breed, but not the other). It is estimated that there are 170,000 SINEC_Cf elements in the dog genome and that half of all genes contain at least one insertion. Several canine phenotypes have been attributed to insertion of a SINEC_Cf element (Figure 12.5). LINEs are much larger, usually 3–4 kb in size, but are generally older and less abundant in the dog genome than in the mouse and human genomes.

Over two million SNPs were identified using the Boxer and Poodle genomes, as well as sequence reads from other breeds. SNP markers are evolutionarily stable and useful in the study of the history of domestication and breed formation. The observed SNP occurrence rate within the Boxer sequence assembly is estimated at 1 per 1600 bp. The SNP rate increases to 1 per 900 bp when the Boxer genome is compared to sequences from other breeds. Some SNPs are more concentrated in one or a few breeds because they are in close proximity to a gene responsible for a selected trait (e.g., short legs). Other SNPs may have occurred after breed formation. Because of their abundance and breed-specificity, SNPs are an effective tool for the discrimination of breeds. Commercially available genetic tests using subsets of informative SNPs have been developed to determine individual ancestry in mixed breed dogs (Figure 12.6).

Selection and inbreeding within domestic dog populations have led to long regions of near complete homozygosity in the canine genome. Sequence assembly from a Boxer showed that

Figure 12.5 SINEC_Cf elements create phenotypic diversity. The merle coat pattern is caused by an insertion at the intron 10/exon 11 boundary of *SILV* (left). Extreme piebald is associated with an insertion upstream of *MITF* (center). The saddle tan phenotype is the result of an insertion in intron 1 of *ASIP* (right).

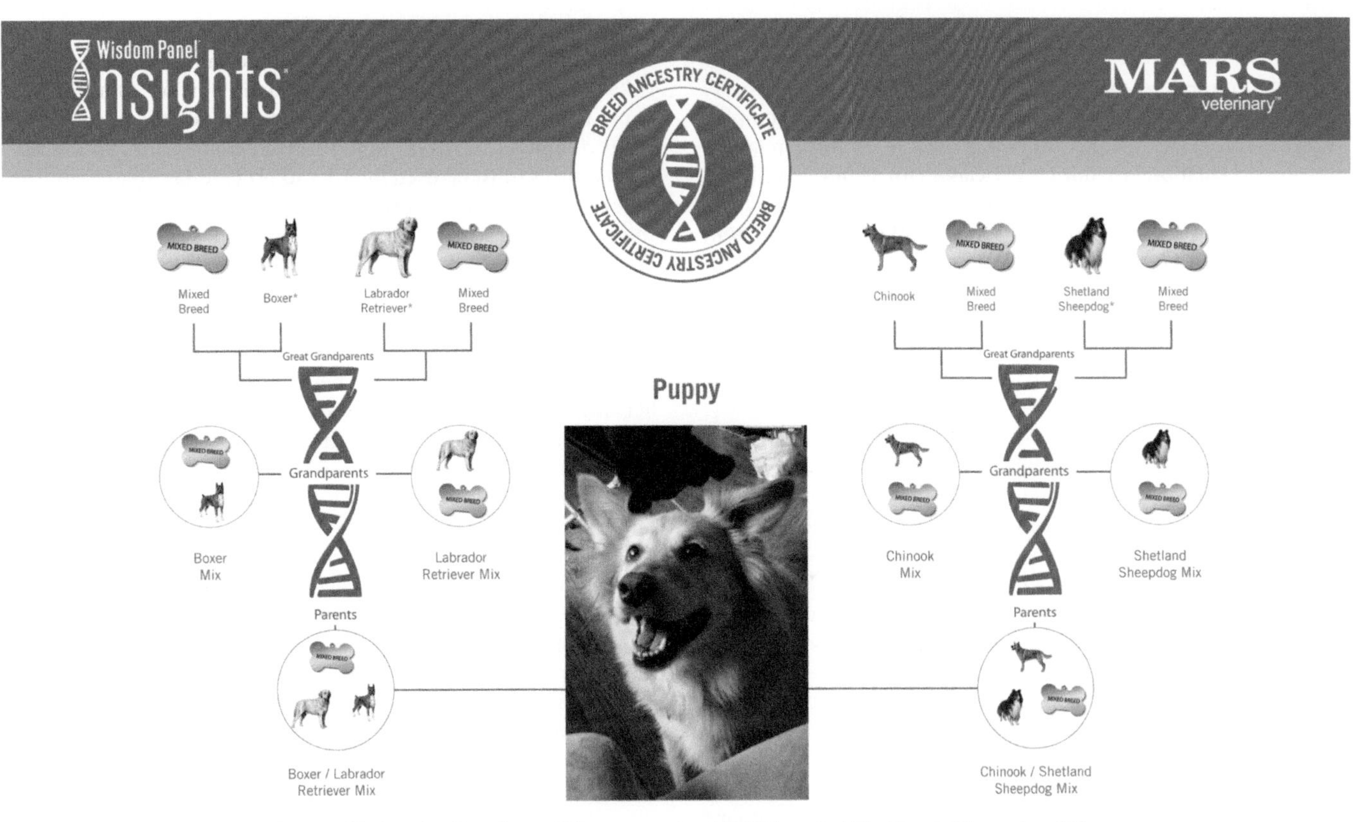

Figure 12.6 Wisdom Panel® Insights™ Mixed Breed Identification Test. A SNP-based test to determine ancestry in a mixed breed dog reveals purebred ancestors. A computer algorithm is used to predict the most likely breeds within three generations. An asterisk (*) denotes a breed detected at a lower confidence.

62% of the genome is homozygous, meaning that both parental chromosomes carry the same combination of alleles, or **haplotype**. Studies in other breeds have revealed that this long-range haplotype structure in the Boxer is typical of most breeds.

Linkage disequilibrium (LD) is the nonrandom association of alleles at different loci along a chromosome. Recombination events in heterozygous individuals disrupt LD by separating adjacent alleles. The average extent of LD in dogs is as much as 100-fold longer than in humans because of long stretches of homozygosity within breeds. LD fluctuates 10-fold between breeds based on their history. Breeds that have larger population sizes tend to have shorter LD than rarer breeds. For example, LD in the popular Golden Retriever is estimated at 0.48 Mb, while in the less common Akita, LD extends for 3.8 Mb. Across breeds, LD is much shorter and comparable to that which is observed in human populations.

Uncovering the genetic basis of phenotypes

Candidate gene approaches

Researchers have begun to unravel the genetic bases of both morphologic and disease phenotypes of domestic dogs. In some cases, analogous traits in other species provide researchers with candidate genes for study. For example, when an undesirable heavy muscling phenotype appeared in the Whippet breed (these dogs were appropriately dubbed "bully" Whippets), a similar phenotype in other mammalian species provided an immediate lead for investigation. Double muscling traits previously described in cattle, sheep, and mice are the result of mutations in ***myostatin***. Evaluation of the canine *myostatin* gene in bully Whippets revealed that a homozygous two bp exonic deletion is responsible for the phenotype.

Many complex disorders of dogs are caused in part by failure of the immune system to distinguish self from non-self. Studies of autoimmune disorders often focus on the dog leukocyte antigen (DLA) locus, part of the major histocompatibility complex in dogs. DLA genes encode proteins central to immune recognition and regulation. In dogs, two classes of genes have been extensively studied: class I and class II. Three class II loci are highly polymorphic and, while the loci are linked, many different haplotypes exist across dog breeds. Significant associations with class II haplotypes have been identified with common autoimmune diseases of dogs, including diabetes mellitus, systemic lupus, and hypothyroidism.

Genome-wide approaches

When no candidate genes can be identified, or too many candidate genes exist, a genome-wide study using linkage-based approaches is necessary. Linkage studies are based on the principle that two loci in close proximity to each other are less often separated by recombination events, and thus are co-inherited. The goal of these studies is to identify a previously mapped genetic marker that is linked, or associated, with the phenotype. When such a marker is identified, its chromosomal location provides a region for candidate gene identification or fine mapping.

Early studies utilized panels of microsatellite markers, which are well suited for **classical linkage analysis**. Multigenerational pedigrees that segregate the phenotype in question are required for classical linkage studies. Figure 12.7 shows a multigenerational pedigree of Scottish Deerhounds with a dominant form of osteosarcoma (bone cancer).

Sixty family members were genotyped for 610 microsatellite markers mapping to the 38 canine autosomes. A statistical calculation known as a **logarithm of the odds** (LOD) score is used to determine if linkage is present between a marker and the phenotype. Generally, a LOD score of 3.0 or above is considered evidence for linkage. In the Scottish Deerhound family, a maximum LOD score of 5.766 was obtained for a marker on chromosome 34. Haplotype analysis identified a 4.5 Mb region harboring the causative locus. Candidate genes within this interval are currently being evaluated for causative mutations for this heritable canine cancer.

Genome wide association studies (GWAS) are a powerful approach for the identification of genetic variants underlying both simple and complex traits. GWAS utilize unrelated cases and controls (eliminating the need to establish multigenerational pedigrees) and a dense set of SNP markers. Rapid generation of comprehensive genetic profiles for GWAS can be accomplished using modern microarray-based technologies that allow for thousands of SNPs to be genotyped simultaneously. To identify an association, allele frequencies are compared between the cases and controls. A SNP allele or genotype that is present more often among cases is said to be "associated" with the phenotype.

In the dog, GWAS for disorders having major loci have been tremendously successful with relatively few affected individuals. Figure 12.8 is a Manhattan plot generated using 58,873 SNP genotypes each for 13 Cavalier King Charles Spaniels: five having a recessive muscle hypertonicity disorder called episodic falling syndrome, one obligate carrier, and seven healthy controls. A strong association with markers on chromosome 7 led to the identification of the causative mutation.

Intense selection for phenotypic attributes within breeds can lead to widespread homozygosity, or **fixation**, of the causative alleles. Most fixed alleles do not have an adverse effect on the health of the dog, although some are associated with medical risks. Across breed, or inter-breed, GWAS are an efficient approach for mapping phenotypes that do not segregate within a breed. Individuals from multiple breeds representing both cases and controls must be used in order to prevent false positives from other breed-specific genetic differences. For example, to identify the locus responsible for chondrodysplasia, or short legs, researchers performed a GWAS using 95 dogs from eight chondrodysplastic breeds (e.g., Bassett Hound, Corgi) and 702 dogs from 64 breeds with longer leg lengths.

Selection for a desired trait can also lead to a **selective sweep**, wherein alleles of proximal genes, some potentially deleterious, become more prevalent or fixed. In the Dalmatian breed, it is hypothesized that selection for the distinctive spotting pattern led to fixation of an adjacent recessive allele causing a defect in urinary metabolism. Mapping of a fixed allele occurring in only one breed poses a challenge. In the Dalmatian, it was necessary to develop an inter-breed backcross with the phenotypically similar Pointer in order to develop a population suitable for mapping studies and to introduce the wild-type urinary metabolism allele to the breed. A single F_1 was backcrossed to a Dalmatian and many generations of backcrossing later, the progeny are indistinguishable from pure Dalmatians and possess the wild-type allele (Figure 12.9). Although this outcrossing was a highly controversial issue for many years, in 2011 the American Kennel Club agreed to allow registration of backcrossed Dalmatians.

Quantitative traits

Unlike the phenotypes discussed to this point in the chapter, which fall into discrete classes (e.g., affected vs unaffected), quantitative traits are continuous and are usually described numerically. Quantitative traits may be attributed to several genes (**polygenic**) or to a combination of one or more genes and environmental factors (**multifactorial**). Genomic regions associated with continuous traits are known as **quantitative trait loci** (QTL).

GWAS are also well suited for the identification of QTL, although larger numbers of samples are usually necessary. Multiple GWAS using both intra-breed and inter-breed populations identified a major QTL on chromosome 15 that is associated with body size in dogs. Deeper analyses of this region using small and giant breed dogs revealed a haplotype that spans **insulin-like growth factor 1** (*IGF1*), a gene encoding an important growth hormone. A common haplotype shared by all small breeds is virtually absent from the giant breeds, suggesting that *IGF1* is a major determinant of small stature in dogs.

In the 1940s, a group producing service dogs for the blind bred for intelligence and temperament. Selection for these two quantitative traits raised the graduation rate of dogs through the training program from nine to ninety percent. Successful selection is correlated with the **heritability** of a trait. Heritability refers to the proportion of differences in phenotypes that can be attributed to genetic control, where 1 indicates complete genetic control and 0 indicates no genetic component. The heritability of quantitative traits is highly variable, and selection for lowly heritable traits is difficult. One study estimated the heritability of fearfulness at 0.44, indicating selection against this trait could greatly diminish the overall occurrence. Other traits including boldness, herding, and sociability, have much lower estimates of heritability, ranging from 0.04–0.27.

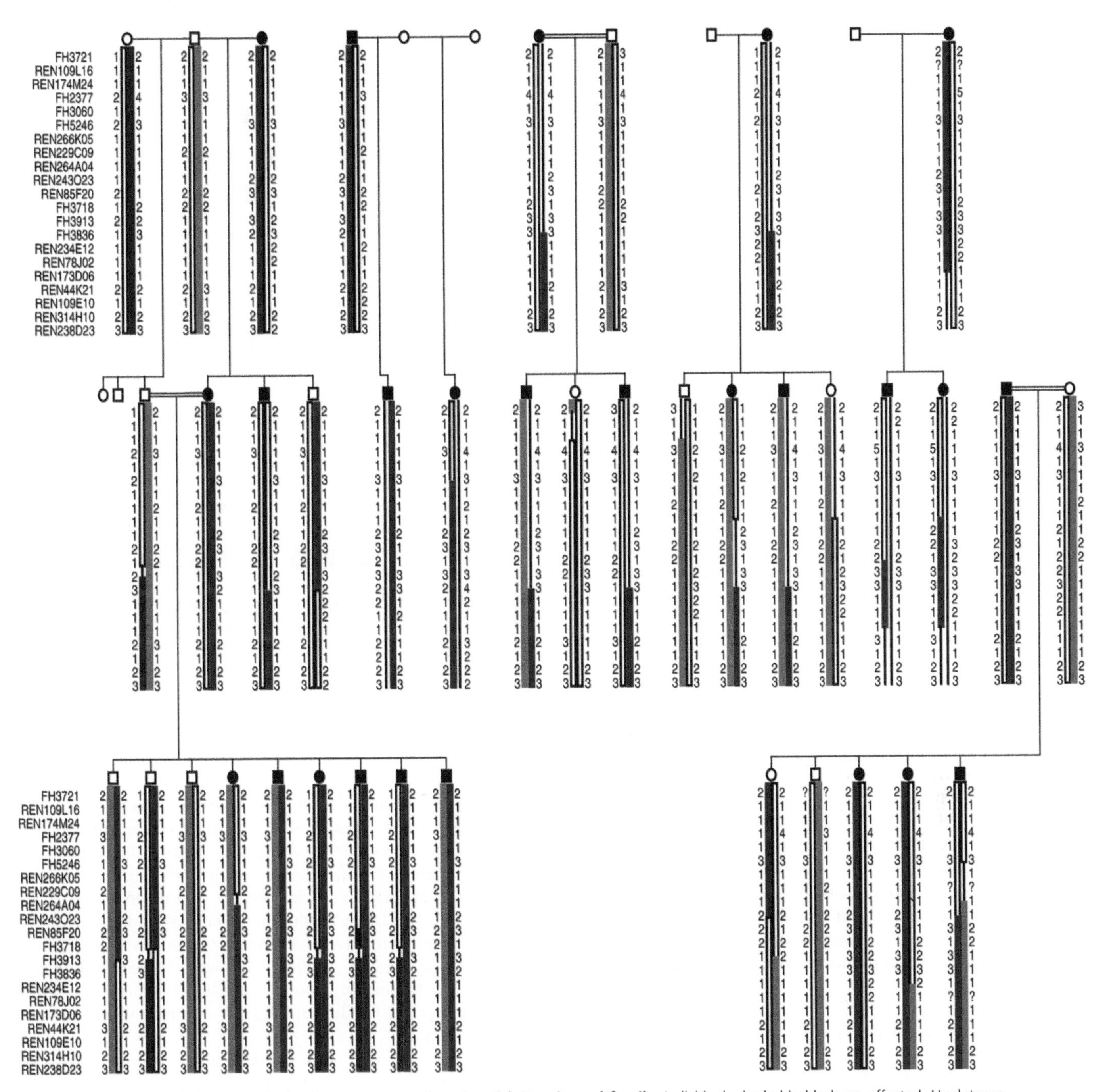

Figure 12.7 Classical linkage analysis of osteosarcoma in a Scottish Deerhound family. Individuals shaded in black are affected. Haplotypes, combinations of alleles that are co-inherited, are depicted below individuals as vertical bars with allele designations to the side. Paternal chromosomes are shown on the left; maternal chromosomes on the right. The haplotype denoted in red segregates with the phenotype. Recombination events delimit the critical interval harboring the causative locus to a 4.5 Mb region between FH3836 and REN44K21. Reprinted with permission from Phillips et al., 2010 *Genomics* 96:220–227.

Future challenges

Behavior

James Watson, the 1962 Nobel Prize winner, stated that the future of research lies in the coming together of biology and psychology. Emerging scientific data support a major role for genetics in human behaviors, such as aggression, nurturing, and sexuality. In dogs, an array of behaviors is observed between breeds, including herding, pointing, retrieving, aggression, and anxiety. Because the underpinnings of behavior are still largely unknown, utilization of the dog model to implicate biological pathways is a growing focus of scientific research.

Certain breeds segregate overt compulsive behaviors, such as tail chasing, blanket sucking, and circling, which are potential models for human obsessive-compulsive disorders (OCD). GWAS for flank sucking and blanket sucking in Doberman Pinschers identified the first major locus associated with an animal

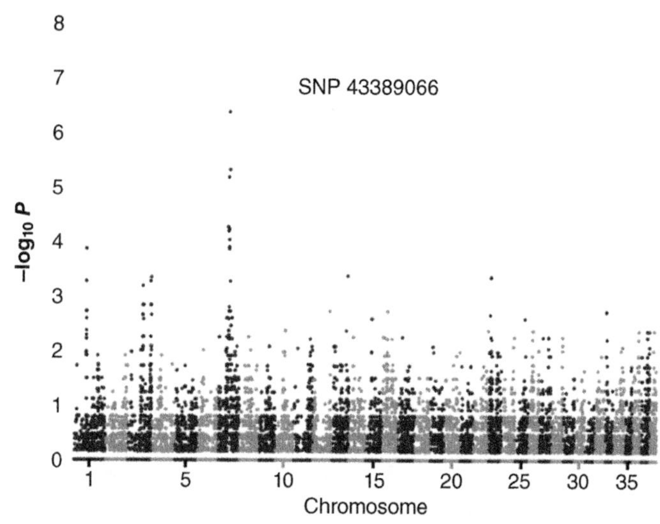

Figure 12.8 GWAS for a neurological disorder in Cavalier King Charles Spaniels. A Manhattan plot (so named because it resembles a skyline) shows the canine chromosomes on the *x*-axis, and the $-\log_{10}$ of the *p*-value on the *y*-axis. Each dot represents the *p*-value obtained for a single SNP, with the highest points indicating the greatest observed associations. These results show a strong association with the phenotype on chromosome 7. Reprinted with permission from Gill et al., 2012 *Neurobiology of Disease*; 45:130–136.

compulsive disorder. A candidate gene in this region is a member of the cadherin superfamily, which has been implicated in learning and memory. This study in dogs supports further investigations of the cadherins in human cases of OCD.

Dogs possess scent detection abilities unmatched by man or machine and thus serve important roles in society. Scent detection dogs are utilized by the military, law enforcement agencies, and the United States Department of Agriculture. Dogs must possess certain behavioral characteristics in order to excel in these roles, including intelligence, trainability, and a desire to please. Other organizations train dogs to be guides for the blind and assistants for disabled individuals. These dogs must possess a calm temperament, intelligence, and motivation. The ability to breed for desirable behavioral traits would lower attrition rates and the expense associated with producing working and service dogs.

Cancer

Another current research emphasis in dogs is cancer. Cancers are caused by abnormal, uncontrolled proliferation of cells in a specific location. This unrestricted growth can be the result of spontaneous genetic changes that give rise to tumor cells. For example, variations in non-coding regions (i.e., regulatory mutations) can change expression of oncogenes and tumor suppressor genes and cause uncontrolled cell growth. The American Cancer Society reports that cancer is the second leading cause of death in Americans and that 1.6 million new cases of cancer will be diagnosed in 2012 (American Cancer Society, 2012). The heterogeneity of human populations has hindered the identification of lowly penetrant hereditary factors. Dogs, on the other hand, have a population substructure better suited for the identification of cancer-susceptibility alleles. The National Cancer Institute estimates that about 6 million dogs in the United States are diagnosed with spontaneous cancer annually. Some cancers are observed primarily within one or a few breeds, indicating a significant hereditary component.

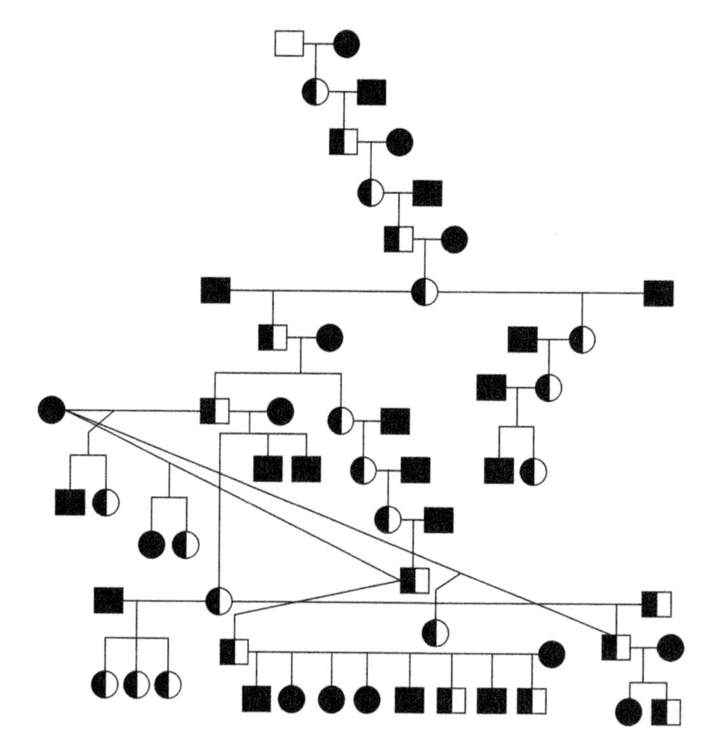

Figure 12.9 Dalmatian/Pointer backcross pedigree. Solid black shading indicates individuals homozygous for the urinary defect; half shading indicates phenotypically normal carriers. In the first generation, a single Pointer (□) possessing the wild type urinary metabolism alleles was crossed with a Dalmatian. In subsequent generations, carriers were backcrossed to Dalmatians in order to maintain breed standards and diversity. Modified and reprinted with permission from *Mammalian Genome* 2006; 17:340–345.

Histiocytic sarcoma is a highly aggressive, lethal cancer of Bernese Mountain Dogs (BMD). In 2012, GWAS identified a locus on canine chromosome 11 homologous to a human tumor suppressor region. A conserved haplotype was identified in 96% of affected BMD. Additional data suggest that histiocytic sarcoma is caused by a regulatory mutation, which has not yet been identified. In people, mutations within this region have been identified in several cancers, including histiocytic sarcoma. For this reason, BMD histiocytic sarcoma is expected to provide a valuable model system for the study of cancer susceptibility in both dogs and people.

Many other canine cancers also have direct human correlates, including osteosarcoma, lung carcinoma, soft tissue sarcoma, breast/mammary cancer, and melanoma. Comparative canine models are being used to carry out genetic studies, as well as clinical trials. Understanding the genetic bases of cancers in the dog could translate to improved detection, diagnosis, and prognosis of human cancers.

Summary

The population substructure of the dog presents challenges and opportunities. Breeders strive to adhere to conformation standards (e.g., morphology, temperament, coat color) and produce healthy puppies. The development of genetic tests is already aiding breeder efforts to eliminate deleterious alleles. A new research focus on the identification of variations associated with behaviors may lead to further tests to guide the selection of reproductive pairs. With increased selection intensity, it will be important to develop breeding strategies that will also preserve genetic diversity within breeds. In addition to improving the health of dogs, there is a desire to utilize this system to elucidate genetic factors in complex phenotypes present in both dogs and humans; such research allows both species to benefit, as opposed to one serving only as a model for the other.

Further reading

Bannasch, D., Safra, N., Young, A., Karmi, N., Schiable, R.S., and Ling G.V. 2008. Mutations in the SLC2A9 gene cause hyperuricosuria and hyperuricemia in the dog. *PLoS Genetics* 4:e1000246.

Boyko, A.R., Boyko, R.H., Boyko, C.M., Parker, H.G., Castelhano, M., Corey, L., et al., 2009. Complex population structure in African village dogs and its implications for inferring dog domestication history. *Proc. Natl. Acad. Sci. USA* 106:13903–13908.

Calboli, F.C., Sampson, J., Fretwell, N., and Balding, D.J. 2008. Population structure and inbreeding from pedigree analysis of purebred dogs. *Genetics* 179:593–601.

Clark, L.A., Wahl, J.M., Rees, C.A., and Murphy, K.E. 2006. Retrotransposon insertion in SILV is responsible for merle patterning of the domestic dog. *Proc. Natl. Acad. Sci. USA* 103:1376–1381.

Demuth, J.P., De Bie, T., Stajich, J.E., Cristianini, N., and Hahn, M.W. 2006. The evolution of mammalian gene families. *PLoS One* 1:e85.

Dewannieux, M., Esnault, C., and Heidmann, T. 2003. LINE-mediated retrotransposition of marked Alu sequences. *Nat. Genet.* 35:41–48.

Dodman, N.H., Karlsson, E.K., Moon-Fanelli, A., Galdzicka, M., Perloski, M., Shuster, et al., 2010. A canine chromosome 7 locus confers compulsive disorder susceptibility. *Mol. Psychiatry* 15:8–10.

Dreger, D.L., and Schmutz, S.M. 2011. A SINE insertion causes the black-and-tan and saddle tan phenotypes in domestic dogs. *J. Hered.* 102 Suppl. 1:S11–18.

Gill, J.L, Tsai, K.L., Krey, C., Noorai, R.E., Vanbellinghen, J.F., Garosi, L.S., et al., 2012. A canine BCAN microdeletion associated with episodic falling syndrome. *Neurobiol. Dis.* 45:130–136.

Goddard, M.E., and Beilharz, R.G. 1982. Genetic and environmental factors affecting the suitability of dogs as guide dogs for the blind. *Theoretical Appl Genet.* 62:97–102.

Karlsson, E.K., Baranowska, I., Wade, C.M., Salmon Hillbertz, N.H., Zody, M.C., Anderson, N., et al., 2007. Efficient mapping of Mendelian traits in dogs through genome-wide association. *Nat. Genet.* 39:1321–1328.

Kennedy, L.J., Davison, L.J., Barnes, A., Short, A.D., Fretwell, N., Jones, C.A., et al., 2006. Identification of susceptibility and protective major histocompatibility complex haplotypes in canine diabetes mellitus. *Tissue Antigens* 68:467–476.

Kennedy, L.J., Quarmby, S., Happ, G.M., Barnes, A., Ramsey, I.K., Dixon, R.M., et al., 2006. Association of canine hypothyroidism with a common major histocompatibility complex DLA class II allele. *Tissue Antigens* 68:82–86.

Kennedy, L.J., Huson, H.J., Leonard, J., Angles, J.M., Fox, L.E., Wojciechowski, J.W., et al., 2006. Association of hypothyroid disease in Doberman Pinscher dogs with a rare major histocompatibility complex DLA class II haplotype. *Tissue Antigens* 67:53–56.

Khanna, C. and Paoloni, M.C. 2006. Cancer biology in dogs. In: *The Dog and its Genome.* E.A. Ostrander, U. Giger, K. Lindblad-Toh (eds). Cold Spring Harbor Laboratory Press, Woodbury, New York, pp. 473–496.

Kirkness, E.F., Bafna, V., Halpern, A.L., Levy, S., Remington, K., Rusch, D.B., et al., 2003. The dog genome: survey sequencing and comparative analysis. *Science* 26:1898–1903.

Lark, K.G., and Chase, K. 2012. Complex traits in the dog. In: *The Genetics of the Dog.* E.A. Ostrander and A. Ruvinsky (eds). CABI, Oxford, pp. 435–457.

Leonberger Club of America. 2012. *Leonberger Owner's Guide.* Leonberger University.

Leroy, G., and Baumung, R. 2011. Mating practices and the dissemination of genetic disorders in domestic animals, based on the example of dog breeding. *Anim. Genet.* 42:66–74.

Lindblad-Toh, K., Wade, C.M., Mikkelsen, T.S., Karlsson, E.K., Jaffe, D.B., Kamal, M., et al., 2005. Genome sequence, comparative analysis and haplotype structure of the domestic dog. *Nature* 438:803–819.

Mellersh, C. 2012. DNA testing and domestic dogs. *Mamm. Genome* 23:109–123.

Miyadera, K., Acland, G.M., Aguirre, G.D. 2012. Genetic and phenotypic variations of inherited retinal diseases in dogs: the power of within- and across-breed studies. *Mamm. Genome* 23:40–61.

Moody, J.A., Clark, L.A., Murphy, K.E. 2006. Working dogs: History and Applications. In: *The Dog and its Genome.* E.A. Ostrander, U. Giger, K. Lindblad-Toh (eds). Cold Spring Harbor Laboratory Press, Woodbury, New York, pp. 1–18.

Mosher, D.S., Quignon, P., Bustamante, C.D., Sutter, N.B., Mellersh, C.S., Parker, H.G., and Ostrander, E.A. 2007. A mutation in the myostatin gene increases muscle mass and enhances racing performance in heterozygote dogs. *PLoS Genetics* 3:e79.

National Cancer Institute. 2012. Comparative Oncology Program Disease Information. Online: https://ccrod.cancer.gov/confluence/display/CCRCOPWeb/Disease+Information (accessed September 13, 2012).

Pang, J.F., Kluetsch, C., Zou, X.J., Zhang, A.B., Luo, L.Y., Angleby, H., et al., 2009. mtDNA data indicate a single origin for dogs south of Yangtze River, less than 16,300 years ago, from numerous wolves. *Mol. Biol. Evol.* 26:2849–2864.

Parker, H.G., Shearin, A.L., and Ostrander, E.A. 2010. Man's best friend becomes biology's best in show: genome analyses in the domestic dog. *Annu. Rev. Genet.* 44:309–336.

Parker, H.G., VonHoldt, B.M., Quignon, P., Margulies, E.H., Shao, S., Mosher, D.S., et al., 2009. An expressed fgf4 retrogene is associated with breed-defining chondrodyspasia in domestic dogs. *Science* 325:995–998.

Rowell, J.L., McCarthy, D.O., and Alvarez, C.E. 2011. Dog models of naturally occurring cancer. *Trends Mol. Med.* 17:380–388.

Safra, N., Schaible, R.H., and Bannasch, D.L. 2006. Linkage analysis with an interbreed backcross maps Dalmation hyperuricosuria to CFA03. *Mamm. Genome* 17:340–345.

Saetre, P., Strandberg, E., Sundgren, P.E., Pettersson, U., Jazin, E., Bergström, T.F. 2006. The genetic contribution to canine personality. *Genes Brain Behav.* 5:240–248.

Savolainen, P., Zhang, Y.P., Luo, J., Lundeberg, J., and Leitner, T. 2002. Genetic evidence for an East Asian origin of domestic dogs. *Science* 298:1610–1613.

Schmutz, S.M., and Schmutz, J.K. 1998. Heritability estimates of behaviors associated with hunting in dogs. *J. Hered.* 89:233–237.

Shearin, A.L., and Ostrander, E.A. 2010. Leading the way: canine models of genomics and disease. *Dis. Model Mech.* 3:27–34.

Shearin, A.L., Hedan, B., Cadieu, E., Erich, S.A., Schmidt, E.V., Faden, D.L., et al., 2012. The MTAP-CDKN2A locus confers susceptibility to a naturally occurring canine cancer. *Cancer Epidemiol. Biomarkers Prev.* 21:1019–1027.

Sutter, N.B., Bustamante, C.D., Chase, K., Gray, M.M., Zhao, K., Zhu, L., et al., 2007. A single IGF1 allele is a major determinant of small size in dogs. *Science* 316:1284.

Sutter, N.B., Eberle, M.A., Parker, H.G., Pullar, B.J., Kirkness, E.F., Kruglyak, L., and Ostrander, E.A., 2004. Extensive and breed-specific linkage disequilibrium in *Canis familiaris. Genome Res.* 14:2388–2396.

Tsai, K.L., Clark, L.A., and Murphy, K.E. 2007. Understanding hereditary diseases using the dog and human as companion model systems. *Mamm. Genome* 18:444–451.

Vignaux F., Hitte, C., Priat, C., Chuat, J.C., Andre, C., and Galibert, F. 1999. Construction and optimization of a dog whole-genome radiation radiation hybrid panel. *Mamm. Gen.* 10:888–894.

Vilà, C., Savolainen, P., Maldonado, J.E., Amorim, I.R., Rice, J.E., Honeycutt, R.L., et al., 1997. Multiple and ancient origins of the domestic dog. *Science* 276:1687–1689.

VonHoldt, B.M., Pollinger, J.P., Lohmueller, K.E., Han, E., Parker, H.G., Quignon, P., et al., 2010. Genome-wide SNP and haplotype analyses reveal a rich history underlying dog domestication. *Nature* 464:898–902.

Wade, C.M. 2006. The dog genome: sequence, evolution, and haplotype structure. In: *The Dog and its Genome*. E.A. Ostrander, U. Giger, K. Lindblad-Toh (eds). Cold Spring Harbor Laboratory Press, Woodbury, New York, pp. 209–219.

Wang, W., and Kirkness, E.F. 2005. Short interspersed elements (SINEs) are a major source of canine genomic diversity. *Genome Res.* 15: 1798–1808.

Wayne, R.K., and Ostrander, E.A. 2007. Lessons learned from the dog genome. *Trends Genet.* 23:557–567.

Wayne, R.K., and vonHoldt, B.M. 2012. Evolutionary genomics of dog domestication. *Mamm. Genome* 23:3–18.

Wilbe, M., Jokinen, P., Hermanrud, C., Kennedy, L.J., Strandberg, E., Hansson-Hamlin, H., et al., 2009. MHC class II polymorphism is associated with a canine SLE-related disease complex. *Immunogenetics* 61:557–564.

References

American Cancer Society. 2012. *Cancer Facts and Figures*. Online: www.cancer.org/acs/groups/content/@epidemiologysurveilance/documents/document/acspc-031941.pdf (accessed September 13, 2012).

Phillips, J.C., Lembcke, L., and Chamberlin, T. 2010. A novel locus for canine osteosarcoma (OSA1) maps to CFA34, the canine orthologue of human 3q26. *Genomics* 96:220–227.

Review questions

1. Consider characteristics of a purebred dog you know. Which traits are fixed in the breed? Which occur across breeds? If a trait segregates within the breed, can you postulate the mode of inheritance?

2. Why are two-thirds of canine hereditary diseases transmitted in an autosomal recessive fashion, while dominant disorders are more prevalent among humans?

3. Using GWAS, only five affected individuals were necessary to localize the recessive muscle hypertonicity disorder in Cavalier King Charles Spaniels. Would you expect the mapping of a dominant disorder to require a larger or smaller population? What about a complex disorder?

4. You are interested in mapping the trait for curly tails in dogs using a GWAS. You have resources to generate whole genome SNP genotypes for 200 dogs. Describe the population you would assemble for this study.

5. You are interested in the genetics of two behavioral traits in dogs: tug-of-war (heritability 0.19) and fearfulness (heritability 0.44). A QTL would be easier to identify for which trait?

13 The Sheep Genome

Noelle E. Cockett and Chunhua Wu
Department of Animal, Dairy and Veterinary Sciences, Utah State University, Logan, UT, USA

Investment in sheep genome research

There are more than 1 billion sheep belonging to over 1300 breeds spread across the world (Scherf, 2000), with vast amounts of genetic variability within and across sheep breeds in response to genetic selection for meat, milk, and wool production. Sheep significantly contribute to the food production of third world countries, primarily because of this species' small size, which requires less feed for maintenance and growth, a short generation interval, and multiple production outputs of meat, fiber, and milk from a single animal over its lifetime or within a small flock. However, world-wide consumer demand for lamb meat and wool products have declined over the last 40 years, while demand for beef, pork, and chicken has increased. This shift in consumption has led to declining numbers of sheep over the last decade in developed countries including the USA, New Zealand, Australia, France, and the UK (FAOSTAT, February, 2013). There are emerging niche markets for sheep in the USA, including organic lamb, dairy products such as milk and cheese, and specialized wool products such as rugs, blankets, and felting.

There are numerous ways that a better understanding of the sheep genome will enhance production efficiency and quality. For example, the identification of genetic markers for resistance to internal parasites in sheep is the subject of ongoing research by several international groups (Marshall et al., 2011; Matika et al., 2011; Silva et al., 2012).

In addition to the world-wide production of food and fiber, sheep have served an important role in biomedical research. The size, anatomy, physiology, temperament, and low maintenance requirements of sheep make them a suitable model for studying multiple mammalian biological functions, including embryology, fetal development, immunology, endocrinology, and reproduction. Historically, sheep have been used to explore several human disorders and health issues (reviewed by Greep, 1970) ranging from chronic renal failure (Bernstine, 1970) to the life span of red blood cells (Carter et al., 1965). More recently, sheep are well-established animal models for human osteoarthritis (Cake et al., 2012), ligament regeneration (Kon et al., 2012), asthma (Van der Velden and Snibson, 2011; Zosky and Sly, 2007), fetal renal disease (Springer et al., 2012) and hemophilia A (Porada et al., 2010).

A highly publicized experiment resulted in "Dolly" the sheep, which was the first animal produced by transfer of the nuclei from adult cells was (Wilmut et al., 1997). Subsequent cloning studies in sheep have furthered our understanding of oocyte, embryo, and fetal development, as well as genetic reprogramming (Galli and Lazzari, 2008; Vajta and Gjerris, 2006; Wilmut et al., 2009).

Thus, an understanding of the genetic makeup of sheep will enhance searches for the genes and genetic regions that underlie production traits in sheep, as well as contribute to biomedical research directed towards humans that use sheep as a large-animal model.

Overview of the sheep genome

Sheep (2n = 54) have 26 pairs of autosomes, including three pairs of submetacentric chromosomes (OAR1, OAR2, and OAR3), 23 pairs of acrocentric chromosomes, and a pair of acrocentric sex chromosomes. The three submetracentric autosomes are likely the result of Robertsonian translocations that occurred during the divergence of caprine and ovine species. Although numerous sheep karyotypes have been published, a standard G-band karyotype for sheep was agreed upon at the ninth North American Colloquium om Domestic Animal Cytogenetics and Gene Mapping in 1995 (Anasari et al., 1999).

Genomic resources in sheep

Researchers worldwide have contributed to the development of resources and tools that can be used for genetic studies in sheep. Although not as extensive as resources in humans, mice, and cattle, an array of important tools are now available for research in sheep genomics. These resources are described next.

In the last 10 years, prioritization of efforts to develop genomic resources for sheep have been undertaken by the International Sheep Genomics Consortium (ISGC), which includes active participation of scientists from Australia, Canada, China, France, New Zealand, UK, and USA. In addition, the ISGC has collectively developed and distributed resources that advance research in sheep genomics. Significant leveraging of funds, expertise, and

Molecular and Quantitative Animal Genetics, First Edition. Edited by Hasan Khatib.

Table 13.1 Useful URLs.

Australian Sheep Gene Mapping:	http://rubens.its.unimelb.edu .au/~jillm/jill.htm
BACPAC Resources Center:	http://bacpac.chori.org/library. php?id=162
CSIRO Sheep Genome:	www.livestockgenomics.csiro.au/ sheep/
Food and Agriculture Organization (FAO) of the United Nations:	http://faostat3.fao.org/home/ index.html#HOME
Illumina Ovine SNP50 BeadChip:	www.illumina.com/documents// products/datasheets/datasheet _ovinesnp50.pdf
International Sheep Genomics Consortium:	www.sheephapmap.org/
NCBI Sheep Genome Resources:	www.ncbi.nlm.nih.gov/projects/ genome/guide/sheep/
Sheep Genome Assembly:	www.livestockgenomics.csiro.au/ sheep/oar3.1.php

efforts has been a hallmark of ISGC, resulting in a highly effective pipeline for the development of resources necessary for exploration of the sheep genome. In addition, the collaborative nature of the ISGC has ensured that sheep genomics resources are developed in a well-coordinated and non-redundant manner. Bimonthly ISGC conference calls provide a regular venue for updates and discussions on "next steps." These calls are documented and minutes are circulated to all ISGC members for comment. In addition, members of the research team meet face-to-face once a year at the Plant and Animal Genome meeting in San Diego, CA for coordination of results and presentation of results to the greater scientific community.

All projects undertaken by the ISGC are conducted within the public domain. Regular updates from the ISGC are available through several established sheep genomics websites (Table 13.1).

Physical map

The first large-scale, international effort to increase knowledge of the sheep genome focused on the physical mapping of genes and genetic markers to specific locations along sheep chromosomes. The ovine physical map has been primarily developed using *in situ* hybridization (e.g., Di Meo et al., 2007) and somatic cell hybrid panels (Burkin et al., 1997; Saidi-Mehtar et al., 1979). The most current review of the cytogenetic map summarized a total of 566 loci assigned to specific chromosome regions (Goldammer et al., 2009c).

Linkage maps

The ovine linkage map lagged somewhat behind the physical map, and prior to 1994, included only 17 markers assigned to 7 syntenic groups (Broad et al., 1997). In 1994, a linkage map containing 19 linkage groups containing 52 markers, including microsatellites and candidate gene restriction fragment length polymorphisms (RFLPs), was published by Crawford et al. (1994). These assignments were the result of a genome scan initiated to map the Booroola fecundity gene (Montgomery et al., 1993) using 12 pedigrees segregating for the Booroola gene. However, the creation of the International Mapping Flock (IMF) by AgResearch, New Zealand allowed loci to be much more efficiently assigned to the linkage map and since the pedigree's formation in 1995 (Crawford et al., 1995), the number of loci on the ovine linkage map have increased exponentially. The IMF includes 127 animals in a three-generation pedigree of nine full sib families descending from a single male and generated by multi-ovulation and embryo transfer. The first published map using the IMF contained 246 markers with all 26 sheep autosomes having at least two assignments (Crawford et al., 1995). Another big advancement to the ovine linkage map occurred in 2001 (Maddox et al., 2001) with the publication of a map containing 1093 loci. One unusual finding reported in this paper is that the sheep autosomal male map is about 21% longer than the autosomal female map, with about 50% of the increase occurring at the centromeric and telomeric ends of the chromosomes. In other livestock species, the female map is longer. The cause of this elongation is not known.

Loci have continued to be added to the ovine linkage map on a regular basis (Figure 13.1), with the most recent count of loci being 2528 on version vSM5 as of January 15, 2012 (Jill Maddox, personal communication). A large increase in coverage will occur when approximately 50,000 single nucleotide polymorphisms (SNP) on a medium-density SNP array are added to the linkage map (Jill Maddox, personal communication). The planned analysis of SNP genotypes will include data from the IMF as well as five other two-generation and three-generation half-sib sheep families, totaling around 12,000 animals with genotypes.

Radiation hybrid maps

Another resource that is now available for assigning genes and genetic markers to locations within the sheep genome is an ovine radiation hybrid (RH) panel. This panel, referred to as USUo5000RH, was developed through a collaborative project between Utah State University and Texas A&M University. Ninety clones with retention frequencies between 15–40% have been selected for inclusion in the 5000 rad RH panel (Wu et al., 2007). The panel has been typed by Utah State University and the Research Institute for the Biology of Farm Animals in Dummerstorf, Germany, with over 500 microsatellites, 500 expressed sequence tags or ESTs, and 100 BAC-end sequences (Goldammer et al., 2009a,b; Wu et al., 2008, 2009) and the resulting framework maps have been orientated with respect to the multiple ovine linkage maps that exist. The USUo5000RH panel has also been typed with the ovine medium-density SNP array and at least 4000 of the SNPs will be orientated on the ovine RH map (Wu, unpublished data).

A second ovine radiation hybrid panel has been constructed by INRA. This 12,000 rad panel contains 90 clones with an average retention of 31.8% and a resolution of 15 kb/cR (Maddox and Cockett, 2007). Sixty-seven markers have been mapped to a 23 Mb region on OAR18 using the INRA panel. Because a high radiation dose creates relatively small DNA fragments resulting in enhanced mapping resolution, this 12,000-rad panel will be useful for physical contiguous development and fine mapping of genetic regions. However, development of a whole-genome map using this panel will be difficult.

SNP arrays

The availability of a high density SNP array for the sheep has been a significant milestone for researchers investigating the sheep genome. The Illumina Ovine SNP50 BeadChip was developed by the International Sheep Genomics Consortium and released to the public in January, 2009. Researchers can now obtain genotypes for over 50,000 SNPs for hundreds of animals in a single analysis. The identification of SNPs on the BeadChip was done through a large, international sequencing effort. The first source of sequence data (9.7 Gbp) was generated from six sheep (Romney, Texel, Merino, Dorset, Rambouillet, and Suffolk) using funding from the International Science Linkage Program (Australia) and Ovita (New Zealand). The second source of sequence data used for SNP mining (3 Gbp) was generated from a pool of DNA comprised of 60 genetically divergent animals. The sequences then were compared to identify single nucleotide differences (i.e., SNPs) that differed in at least 5% of the sequenced animals. This approach assured that the SNPs were actual genetic differences rather than sequencing anomalies.

To date, at least 11,000 sheep have been genotyped with the SNP50 chip and analyses of the genotypes are ongoing in a myriad of research projects across the world. For example, SNP genotypes of 2810 sheep from 74 breeds have been combined with genotype data from seven species of wild sheep and nine outgroup species such as bighorn sheep and Mouflons as part of the world-Dwide ovine HapMap project. Results from the HapMap analysis indicate that domestic breeds diverged from their wild ancestors about 11,000 years ago and modern breeds started to differentiate around 200 years ago (Kijas et al., 2012a). Although not unexpected based on a previous study using a low density chip containing 1532 SNPs (Kijas et al., 2009), analysis of the SNP50 data also indicated that American breeds are most closely related to Europe and Middle East breeds than to Asian or African breeds. The SNP50 chip has also been used to identify population substructure in wild bighorn and thinhorn sheep (Miller et al., 2011) as well as in domestic sheep breeds such as the Gulf Coast Native (Kijas et al. 2012b).

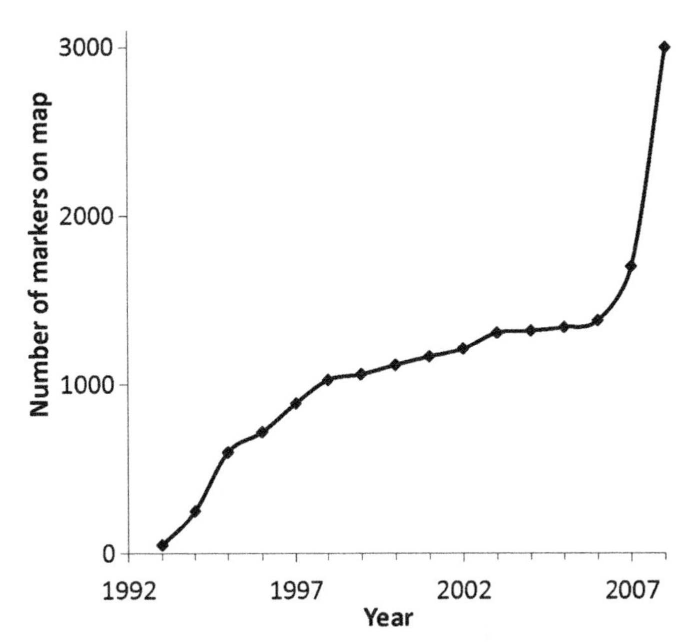

Figure 13.1 Assignments on the sheep linkage map over time. (Courtesy J. Maddox, 2011)

BAC library and end sequences

Two ovine bacterial artificial chromosome (BAC) libraries were constructed in 1999 (Gill et al., 1999; Vaiman et al., 1999). These two libraries are remarkably similar, with 90,000 and 60,000 clones respectively, an average clone size of 123 and 103 kb respectively, and covering three and two genome equivalents respectively. DNA from each of the libraries has been pooled, allowing PCR-based screening for sequences of interest.

Another ovine BAC library, called CHORI-243, was constructed in 2003 by Pieter de Jong's group (Nefedov et al. 2003) using DNA from a Texel ram contributed by the USDA/ARS group. This library consists of 202,752 clones with an average insert size of 184 kb, and has 10-fold genome coverage. It is organized in two segments; bacterial colonies for each segment have been arrayed on filters which can be used for screening the library. The library, filter arrays, and individual BAC clones are available through the BACPAC Resources Center (Table 13.1).

In 2005, each of the BAC clones in the CHORI-243 BAC library was end-sequenced, resulting in 376,493 sequences from 193,073 BAC clones. The average BAC end-sequence length was 687 bp, with a range of 64–1044 bp. Over 99% of the clones provided at least 100 bp sequence. A total of 258,650,691 bp sequence (approximately 6% of the genome) was produced from this project. These sequences were critical for initiating the construction of the ovine genome assembly (see next).

Whole genome reference sequence

A key resource for advancing research on the sheep genome is the ovine reference genome sequence (International Sheep Genomics Consortium 2010), which is the linear order of DNA nucleotides found in the sheep genome. The assembled reference sequence contains approximately 2,472,000,000 of ordered nucleotides (2.47 Gigabases) assigned to specific chromosomes and covers approximately 92% of the total ovine genome (Jiang, 2011; Jiang et al., 2012). Sequence data for the whole genome reference sequence were generated at two sequencing facilities (Beijing Genomics Institute and the Roslin Institute) from DNA of a Texel ewe and a Texel ram, respectively. The first step in assembling the reference sequence involved assembly of 75X reads from the Texel ewe into contigs, scaffolds, and super-scaffolds based on overlaps among the sequence reads. Once that was completed, sequences from both animals were used for filling gaps. Information from the sheep linkage and radiation hybrid maps has been used to refine the assembly of the reference sequence. In order to define the expressed portion of the genome, mRNA-seq data was collected by BGI on seven tissue samples (heart, liver, ovary, kidney, brain, lung, and white fat) of the Texel ewe. This information is being used for annotation of genes within the ovine genome, along with mRNA-seq data submitted by other researchers. In addition to the reference assembly, about 5 million SNPs were identified in separate analyses of the male and female Texel sequences. Many of these SNPs will be used in the construction of an ovine high-density (HD) SNP chip containing over 700,000 SNPs; this HD array should be released in early 2013.

During this process a number of issues with the reference sequence assembly have been identified. These difficulties in assembling the genome are in part due to the assembly software that was used but also because of the variation that exists between and within the sequences of the two Texel animals. More than 1000 small tandem duplications in the current assembly are probably incorrect and the size of large gaps may be over-estimated. In addition, regions of high GC are under-represented in the assembly, with regions >60% GC being particularly underrepresented in the assembly.

Although several other problematic regions in the reference genome have been identified, the ground work has been laid for a detailed assessment of the strategy for improving the genome reference sequence particularly in filling gaps, connecting scaffolds and super-scaffolds, and annotating genes and genetic elements.

Whole genome assembly

Another important genomic resource that is now available for sheep is the whole genome reference assembly (Dalrymple et al., 2007). The assembly is a compilation of all known information about the sheep genome into a searchable database. The assembly contains a myriad of details, including the location of genes and genetic markers from the linkage, radiation hybrid and physical maps, and the location of SNPs from the Illumina chips and is updated about every 12 months. The currently available version (Oarv3.1) was released in October, 2012. This version will be "frozen" for the next two years to allow an extensive annotation of the assembly by ENSEMBL (Brian Dalrymple, personal communication).

Researchers around the world will benefit from the whole genome assembly for sheep because it will significantly accelerate searches for genetic regions and genes influencing phenotypes in sheep. The assembly will also provide a backbone for the assembly and interpretation of low pass sequences of individual animals expected within the next few years. Without a reference assembly, identification of differences among animals in genome organization, including rearrangements and duplications, will be difficult to interpret.

Application of genomic resources

The availability of a high density SNP chip for sheep has revolutionized researchers' ability to locate genetic regions influencing economically traits. The number of animals needed for identifying a mutation responsible for a single-gene trait has been reduced from more than 100 (with linkage analyses) to 5–13 affected animals along with a similar number of controls. Hundreds of animals are still needed for identifying regions containing quantitative trait loci (QTL) but the time to test markers across a population is reduced from years to just a few months. The cost per genetic marker is also greatly reduced with the SNP chip. However, given the number of animals that are needed for QTL discovery, the cost of a QTL search using the SNP chip may still be prohibitive, particularly for lowly heritable traits, because thousands of animals are likely needed in the study (Notter, 2012). A genotype imputation approach (Hayes et al., 2011) can reduce the cost because a small select group of animals is genotyped with the more expensive high density chip (50,000 SNPs), while the bulk of the animals are genotyped with a cheaper lower density chip (>5000 SNPs). The accuracy of imputation is dependent on the amount of genetic diversity in a breed, with lower genetic diversity resulting in more accurate imputation. After the identification of significant SNPs, the DNA of key individuals could be sequenced and compared to key ancestors within the breed to reveal interesting mutations associated with the analyzed trait.

The volume of data generated within the high density chip (~700,000 genotypes per animal) can be an issue, requiring investments in hardware, software and appropriate analysis expertise. However, the density of markers that are genotyped on a high density SNP chip means that the genetic regions identified in a genetic experiment are of high resolution. This higher resolution helps limit the size of the region in which one needs to search for the causative gene or mutation, leading to a quicker turn-around in the identification and characterization of genomic regions. However, the discovery of genetic markers suitable for genetic selection of quantitative traits is still illusive. At this point in time, there are only a few sheep populations worldwide with appropriate numbers (in the thousands) and appropriate trait measurements. In addition, the genetic markers identified in these populations must be tested in other populations to determine the utility of the markers in across genetic lines.

Due to significant advances in sequencing technology and dramatic cost reductions, it is now feasible to collect whole genome sequence on multiple individuals within a species. This resequencing approach holds considerable promise for livestock animals. One application is the ability to predict genetic values of individual animals within a breeding program. The availability of a whole genome sequences for superior animals is expected to substantially improve accuracies and genetic gain beyond what is currently possible using SNP chips (Meuwissen and Goddard, 2010). In addition, whole genome resequencing of case and control animals segregating for known disease states will enhance the identification of causal mutations. This genetic information will make the sheep increasingly valuable as a biomedical model for human diseases. The sheep genome reference sequence can be used as a template for the alignment and assembly of whole genome resequences.

A database containing information of the sheep transcriptome is another genomic resource that would greatly accelerate research in this species. Cataloging expressed transcripts across tissues of sheep of various developmental ages and environmental conditions will allow better characterization of similarities and differences between sheep and human biological systems. This type of information will be particularly useful when using the sheep as a biomedical model.

Summary

Sheep contribute significantly to food and fiber production across the world through locally and globally distributed meat, milk, and wool markets. In addition, sheep are used in biomedical research as a model organism for several debilitating diseases. A better understanding of the genetic makeup of sheep will lead to improvements in the efficiency of food and fiber production and contribute to a better understanding of health and disease issues in humans. While several genetic regions associated with economically and biologically important traits in sheep have been identified, the number of known causative mutations is relatively small. However, the development and application of a high-density ovine SNP array and the availability of the whole genome assembly for sheep will undoubtedly lead to more discoveries in the near future. It should be noted that the application of genetic markers for selection of quantitative traits in sheep is still in the distant future.

References

Anasari HA, Bosma AA, Broad TE, Bunch TD, Long SE, Maher DW, et al. (1999) Standard G-, Q-, and R-banded ideograms of the domestic sheep (*Ovis aries*) and homology with cattle (*Bos taurus*): report of the committee for the standardization of the sheep karyotype. *Cytogenet Cell Genet* 87, 134–142.

Bernstine RL (1970) A chronic renal model for the fetus. *Lab Anim Care* 20, 949–956.

Broad TE, Hayes H, and Long SE (1997) Cytogenetics: Physical chromosome maps. In: Piper L, Ruvinsky A (eds) *The Genetics of Sheep*. CAB International, Oxford, pp. 241–295.

Burkin DJ, Yang F, Broad T, Wienberg J, Hill DF, and Ferguson-Smith MA (1997) Use of the Indian muntjac idiogram to align conserved chromosomal segments in sheep and human genomes by chromosomal painting. *Genomics* 46, 143–147.

Cake MA, Read RA, Corfield G, Daniel A, Burkhardt D, Smith MM, and Little CB (2012) Comparison of gait and pathology outcomes of three meniscal procedures for induction of knee osteoarthritis in sheep. *Osteoarthritic Cartilage* doi: 10.1016/j.jaco.2012.10.001.

Carter MW, Matrone G, and Metzler G (1965) Estimation of the life span of red blood cells in the growing animal in different nutritional states. *J Gen Physiol* 49, 57–67.

Crawford AM, Montgomery GW, Pierson CA, Brown T, Dodds KG, Sunden SL, et al. (1994) Sheep linkage mapping: nineteen linkage groups derived from the analysis of paternal half-sib families. *Genetics* 87, 271–277.

Crawford AM, Dodds KG, Ede AE, Pierson CA, Montgomery GW, Garmonsway G, et al. (1995) An autosomal genetic linkage map of the sheep genome. *Genetics* 140, 703–724.

Dalrymple BP, Kirkness EF, Nefedov M, McWilliam S, Ratnakumar A, Barris W, et al. (2007) Constructing the virtual sheep genome. *Genome Res* 8, R152.

Di Meo GP, Perucatti A, Floriot S, Hayes H, Schibler L, Rullo R, et al. (2007) An advanced sheep (*Ovis aries*, 2n = 54) cytogenetic map and assignment of 88 new autosomal loci by fluorescence in situ hybridization and R-banding. *Anim Genet* 38, 233–240.

Galli C and Lazzari G (2008) The manipulation of gametes and embryos in farm anaimals. *Reprod Domest Anim* 43, Suppl 2, 1–7.

Gill CA, Davis SK, Taylor JF, Cockett NE, and Bottema CDK (1999) Construction and characterization of an ovine bacterial artificial chromosome library. *Mamm Genome* 10, 1108–1111.

Goldammer T, Brunner RM, Rebl A, Wu CH, Nomura K, Hadfield T, et al. (2009a) A high-resolution radiation hybrid map of sheep chromosome X and comparison with human and cattle. *Cytogenet Genome Res* 125, 40–45.

Goldammer T, Brunner RM, Rebl A, Wu CH, Nomura K, Hadfield T, et al. (2009b) Cytogenetic anchoring of radiation hybrid and virtual maps of sheep chromosome X and comparison of X chromosomes in sheep, cattle and human. *Chromosome Res* 17, 497–506.

Goldammer, T, Di Meo GP, Luhken G, Drogemuller C, Wu CH, Kijas J, et al. (2009c) Molecular cytogenetics and gene mapping in sheep (*Ovis aries*, 2n = 54). *Cytogenet Cell Genomics* 126, 63–76.

Greep RO (1970) Animal models in biomedical research. *J Anim Sci* 31, 1235–1246.

Hayes BJ, Bowman PJ, Daetwyler HD, Kijas JW, and va der Werf JHJ (2011) Accuracy of genotype imputation in sheep breeds. *Anim Genet* 43, 72–80.

International Sheep Genomics Consortium, Archibald AL, Cockett NE, Dalrymple BP, Faraut T, Kijas JW, Maddox JF, et al. (2010) The sheep genome reference sequence: a work in progress. *Anim Genet* 41, 449–453.

Jiang Y (2011) Sequencing and assembly of the sheep genome reference sequence. *Proc. Plant and Animal Genomes XIX Conference*, San Diego, CA. W138.

Jiang Y, Xie M, Dalrymple BP, Kijas J, Talbot R, Archibald A, et al. (2012) The domestic sheep reference genome assembly. *33rd International Society of Animal Genetics*. Cairns, Australia. P1019.

Kijas JW, Townley D, Dalrymple BP, Heaton MP, Maddox JP, McGrath A, et al., International Sheep Genomics Consortium (2009) A genome wide survey of SNP variation reveals the genetic structure of sheep breeds. *PLoS One* 4, e4668.

Kijas JW, Lenstra JA, Hayes B, Boitard S, Porto Neto LR, San Cristobal M, et al., International Sheep Genomics Consortium Members (2012a) Genome-wide analysis of the world's sheep breeds reveals high levels of historic mixture and strong recent selection. *PLoS Biol* 10, e1001258.

Kijas JW, Miller JE, Hadfield T, McCulloch R, Garcia-Gamez E, Porto NEto LR, and Cockett NE (2012b) Tracking the emergence of a new breed using 49,034 SNP in sheep. *PLoS One* 7, e41508.

Kon E, Filardo G, Tschon M, Fini M, Giavaresi G, Marchesini Reggiani L, et al. (2012) Tissue engineering for total meniscal substitution: animal study in sheep model – results at 12 months. *Tissue Eng Part A* 18(15–16), 1573–1582.

Maddox JF and Cockett NE (2007) An update on sheep and goat linkage maps and other genomics resources. *Small Rumin Res* 70, 2–40.

Maddox JF, Davies KP, Crawford AM, Hulme DJ, Vaiman D, Cribiu EP, et al. (2001) An enhanced linkage map of the sheep genome comprising more than 1000 loci. *Genome Res* 11(7):1275–89.

Marshall K, Maddox JF, Lee SH, Zhang Y, Kahn L, Graswer JU, et al. (2011) Genetic mapping of quantitative trait loci for resistance to *Haemonchus contortus* in sheep. *Anim Genet* 40, 262–272.

Matika O, Pong-Wong R, Woolliams JA, and Bishop SC (2011) Confirmation of two quantitative trait loci regions for nematode resistance in commercial British terminal sire breeds. *Animal* 5, 1149–1156.

Meuwissen TH and Goddard ME (2010) Accurate prediction of genetic values for complex traits by whole-genome resequencing. *Genetics* 185, 623–631.

Miller JM, Poissant J, Kijas Jw, and Coltman DW, International Sheep Genomics Consortium (2011) A genome-wide set of SNPs detects population substructure and long range linkage disequilibrium in wild sheep. *Mol Ecol Resour* 11, 314–322.

Montgomery GW, Lord EA, Penrty JM, Dodds KG, Brroad TE, et al. (1993) The Booroola fecundity (FecB) gene maps to sheep chromosome 6. *Genomics* 22, 148–153.

Nefedov M, Zhu B, Thorsen J, Shi CL, Cao Q, Osoegawa K, and de Jong P (2003) New chicken, turkey, salmon, bovine, porcine, and sheep genomic BAC libraries to complement world wide effort to map farm animals genomes. *Plant and Animal Genome XI* (San Diego, CA) P87.

Notter, DR (2012) Application of genomic information for improvement of quantitative traits. *Sheep and Goat Res J* 27(Special Research Symposium):7–12.

Porada CD, Sanada C, Long CR, Wood JA, Desai J, Frederick N, et al. (2010) Clinical and molecular characterization of a re-established line of hseep exhibiting hemophilia A. *J Thromb Haemost* 8, 276–285.

Saidi-Mehtar N, Hors-Cayla MC, Van Cong M, and Benne F (1979) Sheep gene mapping by somatic cell hybridization. *Cytogen Cell Genet* 25, 200–210.

Scherf BD (2000) World watch list for domestic animal diversity. FAO/UNEP, Domestic Animal Diversity Information System (www.fao.org/dad-is/index.asp).

Silva MV, Sonstegard TS, Hanotte O, Mugambi JM, Garcia JF, Nagda S, et al. (2012) Identification of quantitative trait loci affecting resistance to gastrointestinal parasites in a double backcross population of Red Maasai and Dorper sheep. *Anim Genet* 43, 63–71.

Springer A, Kratochwill K, Bergmeister H, Csaicsich D, Huber J, Bilban M, et al. (2012) A combined transcriptome and bioinformatics approach to unilateral ureteral obstructive uropathy in the fetal sheep model. *J Urol* 187, 751–756.

Vajta G and Gjerris M (2006) Science and technology of farm animal cloning: state of the art. *Anim Reprod Sci* 92, 211–230.

Vaiman D, Billault A, Tabet-Aoul K, Schibler L, Vilette D, Oustry-Vaiman A, et al. (1999) Construction and characterization of a sheep BAC library of three genome equivalents. *Mamm Genome* 10, 585–587.

Van der Velden J and Snibson KJ (2011) Airway disease: the use of large animal models for drug discovery. *Pulm Pharmacol Ther* 24, 525–532.

Wilmut I, Schnieke AE, McWhir J, Kind AJ, and Campbell KH (1997) Viable offspring derived from fetal and adult mammalian cells. *Nature* 385, 810–813.

Wilmut I, Sullivan G, and Taylor J (2009) A decade of progress since the birth of Dolly. *Reprod Fertil Dev* 21, 95–100.

Wu CH, Nomura K, Goldammer T, Hadfield TL, Womack JE, and Cockett NE (2007) An ovine whole-genome radiation hybrid panel used to construct an RH map of ovine chromosome 9. *Anim Genet* 38, 534–536.

Wu CH, Nomura K, Goldammer T, Hadfield TL, Dalrymple BP, McWilliam S, et al. (2008) A high-resolution comparative radiation hybrid map of ovine chromosomal regions that are homologous to human chromosome 6 (HSA6). *Anim Genet* 39, 459–467.

Wu, CH, Nomura K, Goldammer T, Hadfield T, Dalrymple BP, McWilliam S, et al. (2009) A radiation hybrid comparative map of ovine chromosome 1 aligned to the virtual sheep genome. *Anim Genet* 40, 435–455.

Zosky GR and Sly PD (2007) Animal models of asthma. *Clin Exp Allergy* 37, 973–988.

Review questions

1. What are the advantages of using the sheep as a biomedical model?

2. Why is the sheep genome reference sequence useful?

3. How is the SNP chip information used to determine the sites of domestication in the sheep?

14 Goat Genetics and Genomic Progress

Mulumebet Worku
Department of Animal Sciences, North Carolina Agricultural and Technical State University, NC, USA

"These sturdy animals may have been the first 'walking larders.'"

MacHugh and Bradley, PNAS *2001 vol. 98 no. 10 5382–5384.*

Introduction

The domestic goat (*Capra hircus*) is an important farm and companion animal species. They are descendants of the bezoar (*Capra aegagrus*) goat. About 800 million goats are distributed globally. Known as "the poor man's cow" they produce milk and dairy products, meat, fiber, and skins. Goats are also used for carrying small loads, land management, and kept as pets.

Goat phenotypic variation and genetics has claimed man's attention since early times (Figure 14.1).

Goat colour genetics is discussed in the Bible, Genesis 30: 25:43. The chimera of ancient Greece mythology a fire-breathing monster with a lion's head had a goat's body and a serpent's tail. Folklore places goats with the discovery of coffee berries by Khaldi and the story of the Billy Goats Gruff teaches the young self-preservation skills.

Genetics and goat domestication

Breeds of goats are primarily classified by use for production of meat, milk, or hair (mohair or cashmere). Globally many goats are considered dual- or multi-purpose animals. The genetic contribution or genotype of a goat results in observable characteristics such as coat color giving the phenotype. **Qualitative traits** fall into discrete categories examples are coat color, presence or absence of horns. These traits are controlled by genes with little impact form the environment. **Quantitative trai**ts include economically important traits such as disease resistance and production. These quantitative traits have **continuous distributions**, are usually affected by many genes (polygenic) and are also affected by **environmental factors**.

Traditional genetic evaluations rely on the collection of on farm performance data such as milk records, classification based on production, and pedigree information. Using phenotypic and genotypic approaches, geneticists reconstruct the genetic landscape. Thus, they study patterns of migration, introgression, how breeds were formed, patterns of diversity, and relationships of breeds in diverse geographical areas.

These studies have benefited from the sequencing of goat mitochondrial DNA. The goat mitochondrial genome is 16,640 bp. It houses genes responsible for 12S and 16S rRNAs, 22 tRNAs and 13 protein-coding regions. Six different haplogroups A, B, C, D, F, and G have been identified. A majority of goats (>90%) are in the A haplogroup. **Mitochondrial DNA** is maternally inherited and no recombination occurs. Mutations serve as useful markers for studies of migration patterns. Analyses of variation in autosomal and Y-chromosomes are also useful for studying domestication and subsequent migrations. Male lineage studies use the Y chromosomes' two haplogroups, Y1 and Y2. On autosomes microsatellites, amplified fragment length polymorphisms, single nucleotide polymorphism (SNP) are being identified.

Genetic evidence indicates that domestication of the goat began about 10,000 years ago in the region of present day Iran. Goat genetic resources were introduced to North America by Spanish explorers (1500s) and English settlers (1600s). Pure Spanish goats survived in the southeastern and southwestern United States. Cross breeding resulted in Spanish goats bred for production of mohair and cashmere hair fibers after the introduction of the Angora breed in Texas (starting in the 1850s). Dairy goat breeds such as the Toggenburg, Saanen, French Alpine, and Nubian were introduced from Europe (1900s).

The meat goat industry is growing in the US. A popular meat goat breed the Boer was imported from South Africa in 1993. Boer and Boer influence goats crossed with Spanish, Tennessee Fainting (see Box 14.1), and other breeds are used to improve goat meat production in the US. Other imported breeds include the Kiko from New Zealand, a breed developed by crossing feral goats with dairy goats, and the Nigerian Dwarf goat from West Africa. Cross breeding programs of the latter with dairy goats is giving miniature goat breeds, such as the LaMancha, which is known for the unique Gopher and Elf ears. In addition to breed registries, efforts are underway to conserve goat germplasm in DNA and tissue banks. To conserve and insure genetic diversity germplasm banks for the Spanish Angora and Tennessee fainting goats are being developed in the US.

Molecular and Quantitative Animal Genetics, First Edition. Edited by Hasan Khatib.

Figure 14.1 Phenotypic manifestations of genetic variation in goats.

Table 14.1 Taxonomic hierarchy of the goat.

Kingdom	Animalia – Animal
Phylum	Chordata – chordates
Subphylum	Vertebrata – vertebrates
Class	Mammalia Linnaeus, 1758 – mammals
Subclass	Theria Parker and Haswell, 1897
Infraclass	Eutheria Gill, 1872
Order	Artiodactyla Owen, 1841 – cloven-even-toed ungulates
Family	Bovidae Gray, 1821 – antelopes, cattle, goats, sheep, bovids
Subfamily	*Caprinae Gray*, 1821
Genus	*Capra Linnaeus*, 1758 – goats

(Adapted from Integrated Taxonomic Information System standard report. URL: www.itis.gov, Accessed August 27, 2012.)

Taxonomy

Goats belong to the family Bovidea and subfamily Caprinaea (Table 14.1). Sequence variation in the 5′ untranslated segment of the sex determining region Y (SRY) region on the Y chromosome has shown that the goat is closely related to the domestic sheep in the family Bovidae. As a member of this family the goat has horns and hoofs. In addition, the goat is an herbivore and has a four-chambered stomach. The scientific name for the domestic goat is ***capra hircus*** ("she-goat goat") from the Latin *Capra* meaning "she- goat" and *hircus* meaning "goat."

Goat chromosome number and structure

Goat linear chromosomal DNA is enclosed in membrane-bound nuclei. Goat metaphase chromosomes are grouped on the basis of their individual length and location of centromere into 30 different pairs (29 autosomes and 1 sex chromosome). The **diploid number is 60**. The sex chromosomes are heteromorphic. Male goats have 58 autosomes, a single X chromosome and one Y chromosome. Female goats have two X chromosomes. The X chromosome is **acrocentric**, has short arms, and is the largest in size. The Y chromosome is **metacentric** (with the centromere in the middle) and is the smallest in size. The autosomes are acrocentric (chromosomes with the centromere very near one end) and vary in length.

Similarity in chromosomal banding patterns and **synteny** has been observed between the goat, man, cow, and sheep. The **Geep** is a product of a mating between sheep (54 chromosomes) and goats (60 chromosomes) indicating genetic compatibility. Sheep-goat natural or artificial hybrids have 57 chromosomes. In the lab, a chimera is produced by fusing the embryos of sheep and goats together. These animals have distinct groups of cell populations. Natural hybrids have in-between features; mosaic hybrids have patches of pure phenotypes of one parent or another.

Changes in chromosomal number or structure may result from deletions, duplications inversions and translocations or abnormal copy number. Such polymorphisms may impact gene structure, expression, and function. Spontaneous ploidial variation and the occurrence of Robertsonian fusion have been reported. Some variations have no phenotypic impact. For example in the Sannen goat breed the long arms of two acrocentric chromosomes (6 and 15) fuse to form one metacentric chromosome due to a Robertosnian translocation with no impact on fertility or health. Coat color variations are associated with copy number variation in the Agouti.

Mapping of the location of goat genes allows for the identification of the linear order of linked genes that are passed along together on a chromosome. The order and distance between economically important quantitative trait loci (QTL) genes on chromosomes is being mapped in goats as in other livestock species. Strategies exploit the short evolutionary distance between goats, cattle, and sheep and similarity in chromosome banding. The availability of sequenced genomes and compatible sequences in species such as man and cattle further accelerated these efforts. Large scale microarray cross hybridization has been used for the identification of variations in the arrangement and number of goat chromosomes. Highly **conserved sequences** and synteny have been observed between chromosomes of man, cattle, and sheep and goats. Differences in gene order give insight into evolution and rearrangements in the different species. Mapping and linkage analysis will aid in selection and migration analysis. In the goat almost 500 loci have been cytogenetically mapped. A male genetic map that comprises 307 markers is available. Linkage groups have been identified and microsatellite has been cytogenetically mapped.

Box 14.1

The Tennessee "fainting goat" is a multipurpose US goat breed, introduced from Canada in the 1880s. Also called mytonic or stiff legged goats, these goats have polymorphisms in the **CLCN1 (chloride Chanel 1) gene**. This gene tells muscle cells to relax and is important in flexing and relaxing of skeletal muscle. Gene polymorphism causes stiffness and muscle tensing resulting in falling over or "fainting."

Patterns of inheritance

Traditional genetic evaluations rely on the collection of on farm performance data such as milk recording, classification and the animal's pedigree information data such as breed registration, body weight, milk yield, type classification, and disease are recorded. Variations in DNA sequence are associated with defects/disorders, disease predispositions, production traits and coat color. Genotyping will help relate the DNA profile to the traditional genetic evaluations. Ongoing and new efforts to characterize the goat genome will help define inheritance patterns, genetic variation, and the complex interaction between goat genes.

Mendelian traits and exceptions

In the goat most phenotypes are inherited autosomal-recessive or dominant. Variations in the goat genome are associated with adaptation of goats in diverse environments. The variations are in autosomes and appear autosomal recessive or dominant and may be sex influenced. Sex limited traits include phenotypes such as milk production. Mendelian traits such as having a beard can be sex influenced. Beards on goats are determined by a gene that is dominant in males but recessive in females. Genes such as the gene for polledness have **epistatic** effects on sex determining genes (see Box 14.2).

Traits such as red blood cell antigens and coat color are encoded by **multiple genes** that may be influenced by epistatic action resulting in diverse phenotypes. The agouti gene controls the patterns of deposition of phaeomelanin and eumelanin. There are at least 14 different alleles that produce all of the other color patterns. Coat color is controlled by alleles on three loci (A, B, and S). Genes on the extension locus can modify the effect of these alleles. The most dominant allele produces a goat with only pheomelanin (black or dark brown) color and the recessive allele eumelanin. Spotting is determined by different genes.

Quantitative trail loci (QTL)

Economically important traits such as parasite resistance in goats are encoded by many genes and influenced by environmental factors. In goats environmental factors such as temperature and photoperiod affect estrus, age, and factors such as stress affect disease susceptibility and production. Evidence exists for breed differences in trypanotolerance (West African Dwarf goat) and resistance to gastro-intestinal parasites (Small East African and Creole goat). Marker assisted selection has been used against diseases like scrapie, Caprine Arthritis Encephalitis Virus (CAEV), and Johne's disease and for casein genes. Variations in casein genes impact milk protein content and yield. Genome assisted selection is helping in the selection of bucks for breeding based one alpha-S1 casein genes in dairy goats. In goats, identification of QTL for economically important traits such as fleece weight and fiber diameter is needed.

A majority of traits are caused by single nucleotide substitutions or deletions (Table 14.2).

Progress in goat genomics

The ability to amplify DNA using the **polymerase chain reaction** (PCR) has facilitated multiple approaches to identify gene variants. Sequencing candidate genes and amplification using PCR for fragment length polymorphism analyses helps identify variations in DNA sequence that may cause phenotypic change (Table 14.3). Similar approaches include PCR-single strand conformation analysis and allele-specific PCR. High throughput methods include real-time PCR and whole genome sequencing with ovine and caprine 50K SNP chips. This will be useful for discovering further causative variants.

Understanding of goat genes is increasing as evidenced by the number of articles and sequences being deposited with the **National Center for Biotechnology Information** (NCBI) and other databases. New goat specific databases are also becoming available for genetic evaluation one example can be found at www.goatgenetics.ca database. International collaborations are resulting in genetic tools such as the new 50K SNP panel for genotyping. Genotyping will increase the accuracy of goat genetic evaluation to maintain superior and best adapted genetics. Deep sequencing has helped in the identification of novel aspects of the goat genome such as micro RNAs. Such studies will help goat specific and cross-cutting issues such as mastitis, hair biology, milk production and the impacts of environment (diet, climate) on life. Reproductive technologies will help multiply these benefits.

Box 14.2

Polled Intersex Syndrome (PIS)

The Polled Intersex Syndrome (PIS) gene is a cause of intersexuality in goats. Affected does show abnormal horn development and intersexuality, resulting in sterility. Goats have a dominant Mendelian gene coding for the "polled" (hornless) character. The PIS locus on goat chromosome (CHI) 1q43 is determined by a deletion of 11.7 kb. Mapping indicates genetic linkage between this locus and four microsatellite markers. This gene is an autosomal sex-determining gene. The PIS gene regulates the transcription of two genes involved in gonadal development: PIS-regulated transcript 1 and forkhead transcription factor gene. This regulatory function is lost in animals homozygous for the PIS gene. Thus, affected XX embryos are intersex.

Table 14.2 Impact of select gene variants on goat phenotypes.

Gene name	Phenotype
Chloride channel 1 (CLCN1)	Myotonia
N-acetylglucosamine-6 sulphatase	Mucopoly-saccharidosis IIID
Lysosomal beta A mannosidase	Beta-mannosidosis
Polled/intersex syndrome locus	Polled/intersex syndrome (PIS)
Thyroglobulin	Congenital goiter and hypothyroidism
Alpha s1 casein	Different levels of alpha s1 casein (CSN1S1*01, N,E,F,G)
Alpha s2 casein	Absence of alpha s2 casein (CSN1S2*0)
PRNP **PRioN** Protein.	Susceptibility to scrapie in different breeds
TLR Toll like Receptor	Disease resistance
DQA2 MHC	Disease resistance

Table 14.3 Examples of genomic progress from the overview of the domestic goat in the National Center for Biotechnology.

1	*Cashmere goat hair*	Identification and characterization of modulators of hair growth and hair biology, in cashmere goats. Conserved and Novel MicroRNAs are being identified by Deep Sequencing
2	*Dairy goats and mastitis*	Gene expression profiling in caprine milk somatic and white blood cells upon intra-mammary challenge with Staphylococcus aureus
3	*Mammary gland metabolism*	Casein gene polymorphism and metabolic profiling using bovine arrays
4	*Diet*	Impact of food-deprivation on mammary transcriptome of lactating goats
5	*Climate*	Transcriptiome analysis of 10 tissues transcriptome to compare differential gene expression in goats in different climatic conditions from different geographical locations

Source: www.ncbi.nlm.nih.gov/genome/10731

Biotechnologies and goat genetics

Reproductive biotechnologies are used to maximize the use of male and female genetic resources through artificial insemination, using sexed semen, *in vitro/in vivo* fertilization, interspecies *in vitro* fertilization, multiple ovulation, and embryo transfer. Goats are cloned through approaches such as somatic cell nuclear transfer, and transgenic animal production is aiding in the production of pharmaceuticals and novel technology.

In goats, AI technology is well developed and accepted for use. Female-based technologies such as embryo transfer and the induction of multiple ovulations to increase the number of progeny maximize female genetics. In Australia, **embryo transfer** of angora goats has been practiced for over 75 years. According to the International Embryo Transfer Society, the number of embryo transfers carried out in goats is much lower than in sheep (1102 goats in 2008 compared to 5226 sheep). Embryos can also be produced *in vitro* for transfer.

Cloning

In goats as in other mammals maturing oocytes are meiotically arrested at the diplotene stage of prophase then resume meiosis and are arrested again at metaphase-II after ovulation. Cell cycle analysis, nuclear reprogramming, and manipulation have become very important with advances in animal cloning such as Somatic Cell Nuclear Transfer. These asexual approaches are used to copy superior goat genetics, conserve elite individuals while providing gender predictability. Commercial goat gene and tissue banking offers genetic insurance to maintain cells and tissue of superior animals injured or neutered prior to exhibiting superior genetics. CapriGen (http://caprigencloning.com/) is the first privately owned company in the United States to produce cloned goats with service offered to the public. Services include gene banking, to the delivery of a healthy cloned goat.

Transgenic goats and those with unique or desirable genetics are being cloned. Cloning is an important tool in preservation of endangered goats such as the Pashmini goat. This special breed of Cashmere goat lives on the Tibetan plateau and produces Cashmere wool generating about $80 million a year.

Transgenic animals

Transgenic goats are used either to carry an exogenous gene or to have a "knock out" gene from their own genome. These genes result in the secretion of recombinant pharmaceutical proteins in their milk for pharming.

Goats are considered cost-effective and useful for the production of biomedical and other proteins. In 2009, the first protein produced in the milk of transgenic goats ATryn a recombinant human anticoagulant used to prevent blood clots received approval for use in humans. Other examples of the uses of transgenic goats for transgene production include goats synthesizing alpha-fetoprotein, monoclonal antibodies, granulocyte colony stimulating factor, butyryl-cholinesterase, malaria antigens, and spider silks. With many more in the pipeline molecular approaches need better definition of the goat genome to evaluate transgene introgression definitively.

Such efforts will aid in the conservation of goat genomic diversity for sustained animal production and welfare.

Summary

Progress in goat genetics and genomic analysis is essential to addressing global needs for food security, conservation of biodiversity and to increase our understanding of animal domestication, human civilizations and the evolution of life. Long considered the poor man's cow, the goat is distributed globally and adapted to diverse environments. Modern biotechnology approaches have resulted in transgenic and cloned goats. Molecular technologies, genomics tools and databases are aiding in gene identification and genomic analysis to harness genetic variation. The sequencing of the goat genome and systems approaches to genetic evaluation will improve animal production and welfare.

Further reading

Adalsteinsson S, Sponenberg DP, Alexieva S, and Russel AJ. 1994. Inheritance of goat coat colors. *J Hered.* 85(4):267–272.

Alexander B, Mastromonaco G, and King WA. 2010. Recent advances in reproductive biotechnologies in sheep and goat. *J Veterinar Sci Technol* 1:101.

Baker RL, Mugambi JM, Audho JO, Carles AB, and Thorpe W. 2004. Genotype by environment interactions for productivity and resistance to gastrointestinal nematode parasites in Red Maasai and Dorper sheep. *Animal Science* 79:343–353.

Beck CL, Fahlke C, and George A. Jr. 1996. Physiology molecular basis for decreased muscle chloride conductance in the myotonic goat. *Proc. Natl. Acad. Sci. USA* 93:11248–11252.

Boulanger L, Passet B, Pailhoux E, and Vilotte JL. 2012. Transgenesis applied to goat: current applications and ongoing research. *Transgenic Res.* 21(6):1183–1190.

Christman CJ, Sponenberg DP, and Bixby DE. *A Rare Breeds Album of American Livestock*. pp. 39–40. Wendell Berry (ed.). American Livestock Breeds Conservancy.

Di Meo GP, Perucatti A, Floriot S, Incarnato D, Rullo R, et al. 2005. Chromosome evolution and improved cytogenetic maps of the Y chromosome in cattle, zebu, river buffalo, sheep and goat. *Chromosome Research* 13(4):349–355.

Embryo Transfer Newsletter, *IETS*, September 2009.

FAOSTAT. 2009. Preliminary data for production of Live Animals in 2009 in the World. FAOSTAT website. Available: http://faostat.fao.org/ Accessed July 4, 2012.

Fehilly CB, Willadsen SM, Tucker EM. 1984. Interspecific chimaerism between sheep and goat. *Nature* 307:634–636.

Fontanesi L, Martelli Pl, Beretti F, Riggio V, Dall'Olio S, Colombo M, Casadio R, Russo V, and Portolano B. 2010. An initial comparative map of copy number variations in the goat (Capra hircus) genome. *BMC Genomics* 11:639 doi:10.1186/1471-2164-11-639.

Hassanin A. 2010. Comparisons between mitochondrial genomes of domestic goat (*Capra hircus*) reveal the presence of NUMTS and multiple sequencing errors. *Mitochondrial DNA* 21(3–4):68–76.

MacHugh DE and Daniel GB. 2001. Livestock genetic origins: Goats buck the trend. *PNAS* 98:5382–5384.

Maddox JF. 2004. A presentation of the differences between the sheep and goat genetic maps. *Genet. Sel. Evol.* 37 (Suppl. 1) (2005) S1–S10 S1 c_ INRA, EDP Sciences, 2004 DOI: 10.1051/gse:2004030

Manfredi, E. 2003. The tale of goat αs1-casein. *Proceedings of the 3rd International Workshop on Major Genes and QTL in Sheep and Goats.* CD-ROM Communication No. 2–31. Toulouse, France.

Mondal NK and Chakrabarti S. 2007. A simpler, cheaper and quicker method to study somatic chromosomes from goat, *Capra hircus* (L.) *Cytologia.* 72(4):419–425.

Naderi S, Rezaei HR, Taberlet P, Zundel S, Rafat SA, et al. 2007. Large-scale mitochondrial DNA analysis of the domestic goat reveals six haplogroups with high diversity. *PLoS One* 2: e1012.

Pannetier M, Elzaiat D, Thépot E, and Pailhoux S. 2012. Telling the story of XX sex reversal in the goat: Highlighting the *Sex-Crossroad in Domestic Mammals. Ex Dev.* 6:33–45.

Parma P, Feligini M, Greeppi G, and Enne G. 2004. The complete nucleotide sequence of goat (*Capra hircus*) mitochondrial genome. *DNA Seq.* 15(5–6):378.

Peletto S, Bertolini S, Maniaci MG, Colussi S, Modesto P, Biolatti C, et al. 2012. Association of an indel polymorphism in the 3'UTR of the caprine SPRN gene with scrapie positivity in the central nervous system. *J Gen Virol.* 93(7):1620–1623.

Raja A, Vignesh AR, Mary BA, Tirumurugaan KG, Raj GD, Kataria R, et al. 2011. Sequence analysis of toll-like receptor genes 1–10 of goat (*Capra hircus*). *Vet Immunol Immunopathol.* 140(3–4):252–258.

Shrestha JNB and Fahmy MH. 2007. Breeding goats for meat production. 3. Selection and breeding strategies. *Small Ruminant Research* 67(2–3):113–125.

Solomon A and Kassahun A. n.d. Genetic Improvement of Sheep and Goats. In *Sheep and Goat Production Handbook for Ethiopia.* Yami A and Merkel RC (eds) www.esgpip.org/Handbook.php.

Van der Werf JHJ. 2007. Marker assisted selection in sheep and goats. In: *Marker-Assisted Selection: Current Status and Future Perspectives in Crops, Livestock, Forestry and Fish.* Guimarães E, Ruane J, Scherf B, Sonnino A and Dargie E (eds), J. FAO, Rome, Italy, pp. 230–247.

Zhou H, Hickford JG, and Gong H. 2008. Allelic variation of the caprine TLR4 gene identified by PCR-SSCP. *Mol Cell Probes* 22(1):65–66.

Review questions

1. In goats, the polled gene is a dominant gene. Provide the genotype and explain the phenotype (horns) of a homozygous positive, homozygous negative, and heterozygous goat. Use male and female offspring to explain the sex influenced expression.

2. Using examples, describe how beards are inherited in goats. Does this affect genotypic ratios when looking at a dihybrid crosses with another independently assorting autosomal allele?

3. Transgenic goats are being used for the production of biomedical and industrial products such as spider's silk. Why is it important to understand goat genetic make up to assure the safety of the food supply?

4. Global gene sequencing and gene expression analysis holds promise for progress in our understanding of goat genetics. Describe applications and the approaches being used in such studies.

5. Understanding goat genetics shows promise in contributing knowledge to human disease and genetics. Give three examples that illustrate this.

6. Give examples of chromosomal variations associated with quantitative traits in goats. Using one of these genes explain how the variation impacts the goat.

7. What transcription factors are impacted by the PIS deletion? Which goat becomes intersex (female or male goats)?

8. Give an example of a goat gene that is affected by multiple alleles and is also epistatic?

9. Giving examples discuss the need for DNA banking and germplasm conservation.

10. Identity at least three unique features of the goat genome or genetic knowledge.

Section 3

Molecular Genetics of Production and Economically Important Traits

15 Bioinformatics in Animal Genetics

José A. Carrillo and Jiuzhou Song
Department of Animal and Avian Sciences, University of Maryland, MD, USA

Introduction

Emergence of new biotechnologies is an accelerating worldwide phenomenon. Several high-throughput approaches have caused revolutionary changes in biology (e.g., Next Generation Sequencing: NGS). Animal breeding and evolution are not excluded from this fact. Dealing with living organisms requires knowledge of genetic principles that allows comprehension of how the genetic material is transmitted through generations and the possible outcomes associated with this process.

Innovative genome-wide technologies and related techniques generate an enormous amount of data that depends considerably on precise statistical analysis to be efficient. The whole process, from data storage to making inferences, requires copious computational resources, presenting a major challenge to informatics workers in their search for new algorithms and methods. For these, the field demands well-trained professionals in computer science and statistics.

Adequate management of high-throughput data demands high-level statistics knowledge and programming skills; such characteristics are hard to find in biologists and non-statistics specialists. For many people, bioinformatics is a diffuse subject that includes biology, evolution, biological modeling, biophysics, math and statistics; however, there are some individuals that define it strictly as a computational science (Hogeweg, 2011). Each of the previously mentioned topics are, *per se*, broad study areas but only a well-balanced combination of these topics provides a framework to properly analyze some events that occur in nature. Currently bioinformatics offers opportunities in different areas for people that possess the right capabilities. Considering that research in general is increasingly diverse, challenging and multidisciplinary, this chapter will try to summarize some practical ideas and methods that are employed in genomic studies.

Important: Biology is still what is driving the science. Whether a statistician, a computer scientist, or molecular geneticist is using bioinformatics to learn from data, animal genetics researchers need to understand the biology to ask questions with statistics, bioinformatics, or molecular genetics.

Bioinformatics and animal genetics

New high-throughput data consist of complicated and noisy information that requires adaptation of traditional statistics or development of new methods fitting for its conditions. DNA sequencing and alignment, gene expression quantification techniques and more complex modeling studies such protein folding patterns generate data sets that can extend to terabytes, making their analysis and interpretation impossible without modern computational and machine learning capabilities. The implementation of such analytical methods requires mathematical and statistical knowledge to apply them correctly to the study of biological systems. Molecular biology skills are advantageous, although the extent to which they are required will depend on the hypothesis and objective of the study.

Animal genetics includes the study of heredity in animals. Throughout history people have wondered why progeny look like their ancestors, what maintains their individual characteristics and how some diseases are transmitted through generations. Mendel was the first recognized person who demonstrated that genes are transmitted systematically across generations. Experiments with around 29,000 pea plants were performed from 1856 to 1863, giving birth to the Mendelian laws of inheritance. Presently, it is known that heredity is represented in the form of DNA molecules with sequences of four nucleotides (Adenine, Thymine, Guanine, Cytosine) that are transmitted to the next generation. The complete sequence information of an organism is the **genome**, which is the blueprint for all cellular structures and activities. **Genomics** is the study of the genome. It investigates the molecular information in order to understand natural variation and diseases.

Genome size is highly variable, ranging from 0.551 mega base pairs (Mb) in the algae *Guillardia theta* to 139,000 (Mb) in the marbled fish *Protopterus aethiopicus*; but even analysis of the smallest genome is impossible to perform without the use of bioinformatics. Machine learning, which provides computers the ability to learn from the data without being specifically programmed, is a suitable approach for exploration and integration of extensive

Molecular and Quantitative Animal Genetics, First Edition. Edited by Hasan Khatib.

and diverse data. Another challenge that the field of bioinformatics faces is the computational limitation. Although the vast majority of people think that informatics is extremely advanced, many problems in different scientific areas cannot be solved due to computational limitations (e.g., calculation of all possible interactions in the genome). High performance computing (HPC) technology using super computers that perform massive numbers of calculations extremely fast is helping the genomic field to address this problem.

Bioinformatics is an area where the number of scientific publications is increasing exponentially. It assists in identification of novel or target genes and understanding their function, regulation, and interactions under different circumstances. The pressure to find cures for different diseases such cancer, tuberculosis, HIV, and many others fuels this field of research. In the animal field, an increasing demand for food, the necessity of improvement in production efficiency and the urgency to reduce the negative environmental impact give reasons for scientists to exploit bioinformatics techniques. Genomic selection in dairy cattle is an example of how people directly involved in agricultural production could take advantage of bioinformatics to translate algorithms and statistical procedures in real production systems improvements.

Knowing of genetic concepts is necessary to choose the right analysis method, understand its limitations, make inferences about the results and recognize their biological implications, while dealing with high-throughput data.

The importance of bioinformatics in genomics research

Most common bioinformatics processes performed during genomic data analysis are described next. These may be used independently or combined, depending on the specifics of the project. Knowing each of them in details requires extensive interdisciplinary knowledge and effort but to work in genomics is essential to have basic ideas from the all implicated areas. Some of these areas look very distant. Normally, people that study biology are less likely to use mathematics for explaining processes or pathways under investigation. In many occasions, combination of different approaches has been useful to describe what could occur or predict outcomes under similar conditions (e.g., protein folding pattern modeling). This chapter serves as an introduction of the bioinformatics processes and provides an extensive list of references for further readings.

Genome sequencing

DNA sequencing is the process of determining the order of the nucleotides in the DNA of living organism. Sequencing is critical for many genomic experiments. The integrity of any information generated downstream of sequencing will depend on the quality of the sequence; therefore, the sequencing process should be as accurate as possible.

There are different methods available to determine the sequence of nucleotides in a genome fragment but the most famous are Maxam and Gilbert and Sanger and Coulson. Currently the most frequently used method is the sequencing by synthesis (SBS) by Illumina. SBS method supports massively parallel sequencing employing a reversible terminator-based technique that permits the identification of single bases as they are incorporated into elongating DNA strands. Pyrosequencing is another method based on the detection of the pyrophosphate release during the addition of nucleotides to the DNA chain. This process is repeated for the four nucleotides until the correct complementary nucleotide is detected by light emitted upon incorporation of the base to the chain.

Ion semiconductor sequencing, or pH detection sequencing, detects hydrogen ions that are released throughout the DNA elongation. The DNA strand to be sequenced is exposed to a single species of deoxyribonucleotide triphosphate (dNTP). Only the dNTP that is complementary to the DNA template will be added to the complementary growing chain. The release of hydrogen ions is captured by an ion-sensitive field-effect transistor (ISFET), which serves as an indicator that the reaction has occurred. This technology differs from others because it employs non-modified nucleotides and omits the use of optics through the entire process.

The chain-terminator method (Sanger method) uses dyed labeled dideoxynucleotide triphosphates (ddNTPs) that, when incorporated to a molecule being sequenced, terminate the elongation of the sequence (Figure 15.1). The ddNTPs are recognized by the sequencing machine, which assigns the respective nucleotide and the quality score to the DNA sequence.

The chain termination method is suitable for obtaining the sequence of relatively short DNA fragments, so that additional methods are employed for longer DNA molecules such as chromosome walking and shotgun sequencing. The latter is a complex method, which applies the chain termination approach to short fragments obtained by splitting larger DNA molecules. After sequencing, the small fragments are re-assembled to obtain the whole sequence. A schematic representation of shotgun sequencing is shown in Figure 15.2. Shotgun sequencing is fast and presently the most commonly used technique for obtaining whole genome sequences.

The Human Genome Project (HGP) was one of the most remarkable scientific achievements of the last decade. The human genome code was completed in 2003 after 13 years of intensive sequencing, in terms of coverage and accuracy. The project was coordinated by the US Department of Energy and the National Institutes of Health, in conjunction with major partners from France, Japan, Germany, and China. During the project, 3 billion base pairs of genomic sequence were determined and the information was stored in databases for continuing analysis. The projected cost for the HGP was predicted to be $3 billion but ended up two years ahead on schedule, costing $2.7 billion. The project involves other data-related issues: storage, manipulation, analysis, data protection, and ethical, legal, and social considerations. The scope of the project went beyond genome sequencing, trying to investigate all the possible outcomes that could emerge with the complete deciphering of the human genome. Since then, genomes from different species including livestock animals have been sequenced. It is worth considering the timing and cost of the different genome projects to understand the sequencing technological evolution and its contribution to extend the deciphered genome list. For example, the bovine genome project cost $53 million and was completed in five years (2003–2008); the macaque genome cost $23 million and was completed in 2 years (2004–2006); and the baboon genome project (2007–2008) costs were only $4 million. These values represent a significant reduction in cost and time compared to the HGP. However, sequence information should be followed by further investigations to

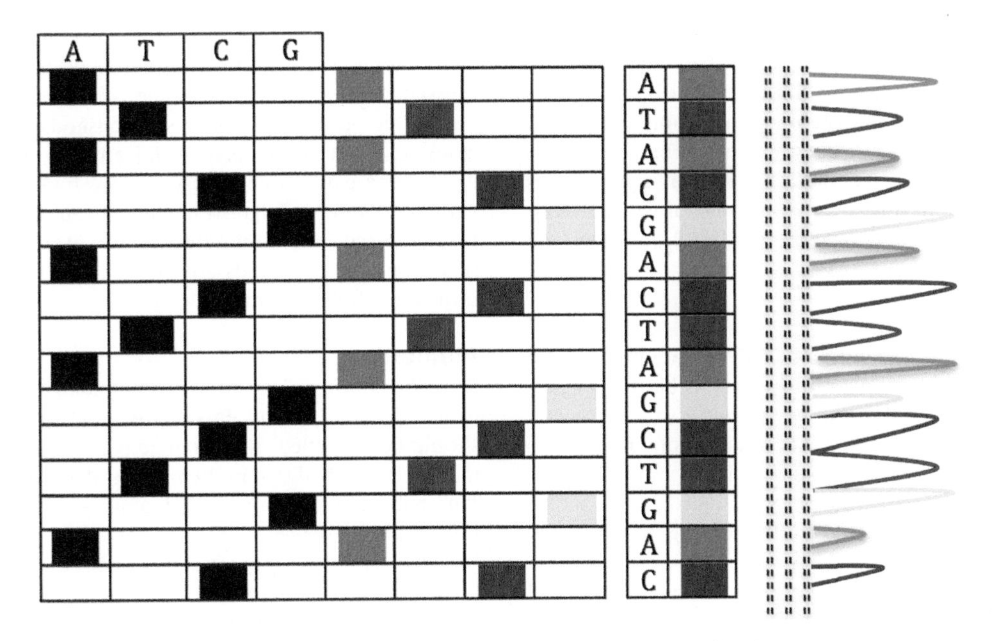

Figure 15.1 DNA Sequencing. Automatic process based on chain termination method using dideoxynucleotides.

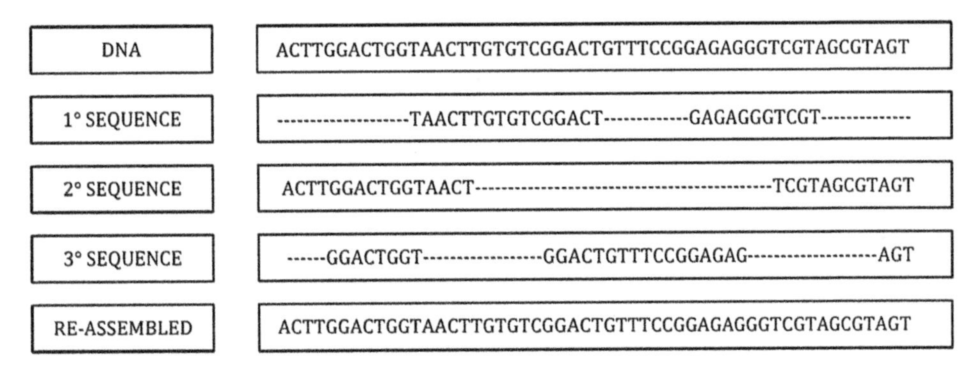

Figure 15.2 Shotgun sequencing. Method usually employed in sequencing long DNA, normally more than 1000 base pairs in length.

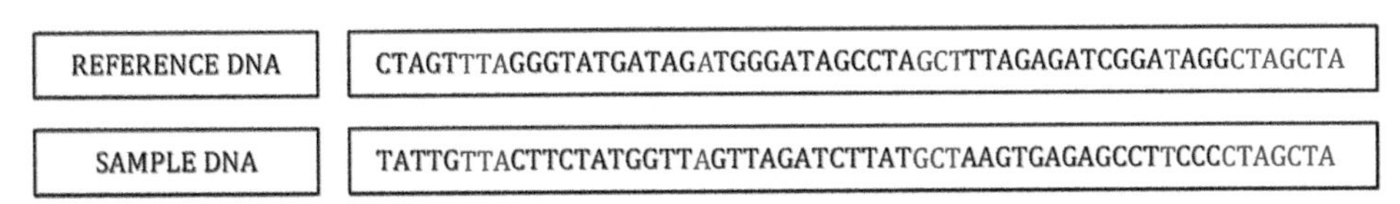

Figure 15.3 Sequence alignment. Comparison of two sequences in order to find common properties between them.

understand more about the nature of genes, their functionality, and disease interactions.

Alignment

Alignment consists of the process where a sequenced segment (read) is compared to a reference, and based on the nucleotide sequence similarity, a physical position in the reference genome is assigned (Figure 15.3).

Pairwise sequence alignment is employed to find similar regions that can explain structural, functional, or evolutionary relationships between samples. Simultaneous multiple alignment is a fairly common technique used in bioinformatics where many sequences of similar length are aligned at the same time. This type of alignment can be divided in two groups: global and local. Global alignment uses the entire length of the reference, while the local algorithm drops parts of the sequence that have no homology with the remainder of the set, representing a more complicated situation because the decision-making step is performed by the algorithm.

There are many software packages that can be used for sequence alignment. The WABA, BLAT, BLAST, and BOWTIE software packages are some common examples. **BLAT** works by keeping in memory the index of a complete genome, and it is normally used to localize a sequence in the genome or to determine the exon structure of an mRNA. **BOWTIE** is an ultrafast aligner with a memory-efficient method used for short reads. It indexes the genome employing a Burrows–Wheeler index keeping the memory footprint small.

BLAST is one of the most-used local alignment tools. It is administered by the National Center for Biotechnology Information (NCBI) and can be accessed through the webpage http://blast.ncbi.nlm.nih.gov/Blast.cgi. The program permits identification of similar regions between sequences, comparing the sample sequence to a database and computing statistical significances of the matches. After entering the website a list of various databases and different basic and specialized BLAST programs are displayed.

The basic BLAST

Nucleotide Blast: Search a nucleotide database employing a nucleotide query. Different algorithms such as blastn, megablast, discontinuous megablast can be selected.

Protein Blast: Compares a protein query to a protein database. The provided algorithms are blastp, psi-blast, phi-blast, and delta-blast.

Blastx: Explores a protein database employing a translated nucleotide query.

Tblastn: In contrast to Blastx, Tblastn searches for a translated nucleotide database using a protein query.

Tblastx: Examines a translated nucleotide database employing a translated nucleotide query.

Other more specific BLAST programs are available on the same page such as **Primer-BLAST**, **SNP flanks**, and **immunoglobulins (IgBLAST)**. In order to select the adequate program a table is provided in the help menu of the initial webpage.

Genome assembly

Genome assembly involves ordering a large number of DNA fragments to reconstruct the original, long DNA sequence from which they were sourced. Although reading a book one letter at a time is possible, reading the whole genome one nucleotide at a time is still unfeasible because of technological limits to read length. DNA sequencing technologies can currently only read DNA fragments between 20 and 1000 nucleotides, leaving biologists with a set of short reads with unknown genomic positions. For this reason, genome assembly continues to be a puzzle with an ambitious mathematical and computational challenges (Pevzner and Shamir, 2011).

The primary strategy is to produce large numbers of reads from many copies of the same genome, which can be represented by a huge overlapping puzzle (Figure 15.4). This complex problem can be simplified at two steps: read generation and fragment assembly, which present biological and algorithmic problems, respectively.

The main challenge in genome assembly is dealing with the large numbers of identical sequences (repeats) that are interspersed throughout the genome. These can be thousands of nucleotides long and repeated thousands of times across the genome, especially in plants and animals with large genomes. Also, some parts of chromosomes are hard to sequence due to variation. Different algorithms have been developed for aligning reads to each other and detecting the overlapping places. Then, sets of overlapping reads are merged and the process continues until the whole genome is deciphered.

Annotation

Sequence information alone has no biological meaning. In order to use the information encoded in nucleotide sequences, genes should be identified. After a gene is recognized, biological information is attached to it in the process of annotation (Figure 15.5). This will permit inference of biological processes based on the nucleotide sequence.

There are some automatic tools that perform genome annotation using computational analysis although manual annotation is also performed but requires an extensive expertise. Ideally, these approaches should be combined to obtain greater enrichment of the data. Programs that compare genes or biological functions across species are also available (e.g., Kerfuffle). They help to identify genes that are conserved or co-localized in different species. Discovery of these common features are helpful to understand the evolution of the species. Some genome annotation databases that are available include Entrez Gene, Ensembl, GENCODE, GeneRIF, RefSeq, Gene Ontology Consortium, and Uniprot. All of them have their respective advantages and limitations and should be considered based on the study requirements.

Gene expression

The gene information contained in the DNA sequence is used to synthetize functional gene products that normally-but not always-are proteins (Schwanhausser et al., 2011) (Figure 15.6).

In non-protein coding genes, the final product is a functional RNA; it could be transfer RNA (tRNA), ribosomal RNA (rRNA), or small nuclear RNA (snRNA) (Morris, 2012). During the gene expression process, the cell can regulate some steps to control several functions and structures. Transcription, translation and post-translational modifications of proteins are examples of such steps. In order to measure gene expression levels, diverse methods

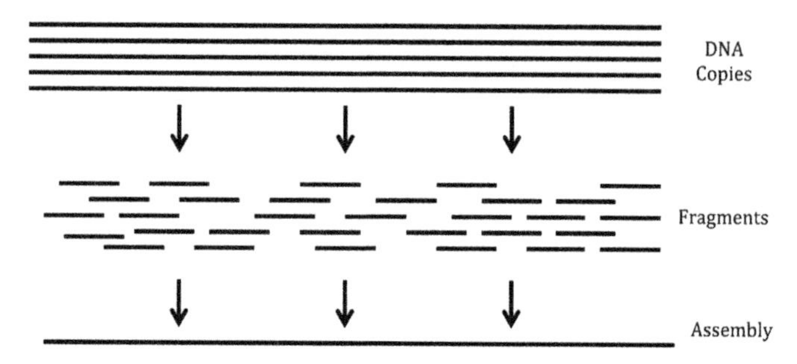

Figure 15.4 Genome assembly. Original DNA sequence is reconstructed from ordering large number of short DNA fragments.

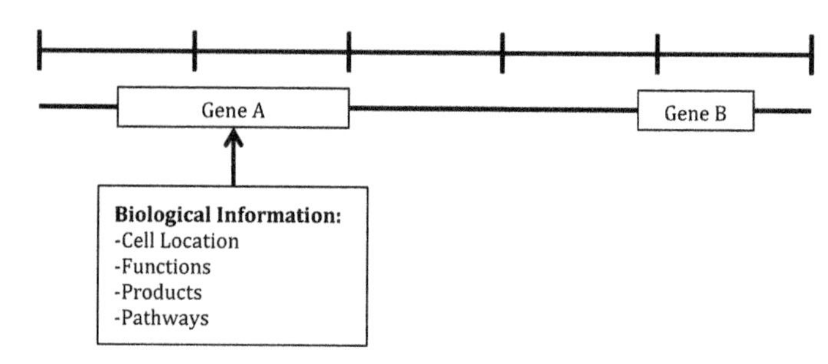

Figure 15.5 Gene annotation. A process where biological or relevant information is linked to a gene or a specific genomic location.

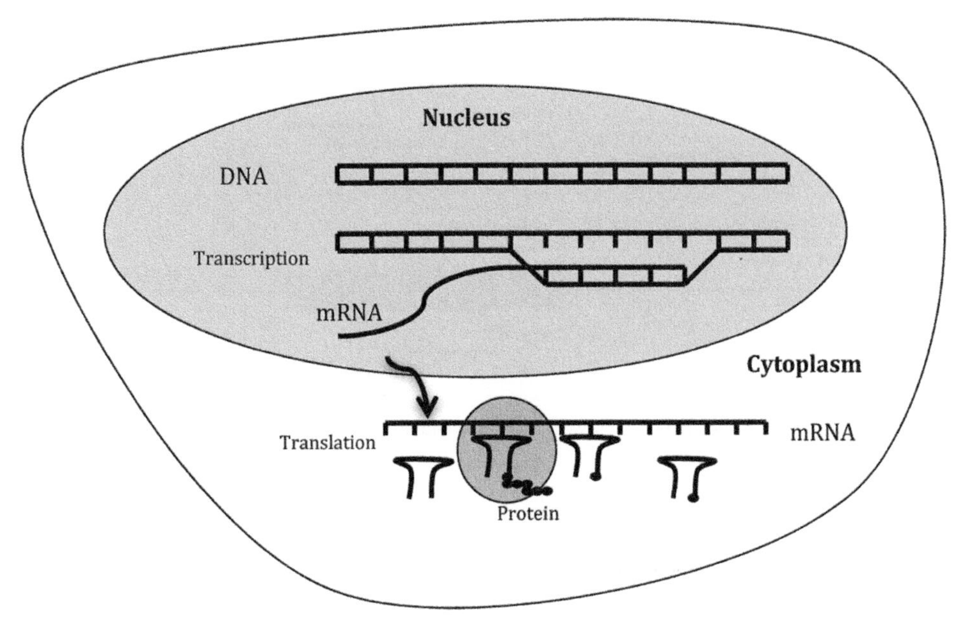

Figure 15.6 Gene expression. Gene products are synthetized based on the information deciphered from the DNA sequence.

have been developed to quantify messenger RNA and some of them will be discussed next.

Microarrays

Microarray analysis is a technology that provides a static measurement (snapshot) of expression levels of a large number of RNA molecules simultaneously, employing hybridization. Expression levels can be understood as the concurrent manifestation of specific biological process at the observed time. Microarrays are regularly used to estimate the genomic expression level difference between two conditions, such as normal and tumorous tissues.

Microarrays require some previous knowledge of the design of the array. Which probes should be included in the chip is the first challenge but not the only one. Quality is another factor that can affect the results if the probes do not meet standard levels for hybridization. To overcome this problem, microarrays manufacturers and governmental institutions have implemented quality assurance programs to guarantee the quality of microarray chips.

The method employed by the machine to read a microarray chip produces signals that are representations of the hybridized molecular abundances. The measures obtained are contained in a set of noisy data. Microarrays have low repeatability; even well-designed experiments that account for infinitesimal details cannot reproduce the same exact results. Small changes in any of several critical steps of the protocol could alter the final result. Changes can be introduced by human error or the machinery involved in the analysis process. Preparation of the mRNA, volume of the sample, washing times, and quantification are all potential sources of error. All of these confound quantification of the true signal, which can be interpreted as the difference between conditions or treatment groups.

Gene expression data present systematic differences when datasets are compared. In order to account for this source of error, a *normalization* procedure is employed to improve sample comparison. Different methods for microarray normalization have been proposed (Van Iterson et al., 2012). Diverse methods could give dissimilar results and lead to different conclusions regarding the circumstances under study.

A major problem in microarray experiments is the sample size. Most statistical analyses work well for a large sample size but the majority of them collapse when sample size is below five. Microarrays produce thousands of phenotypic observations from a small number of samples, sometimes just two or three due to the cost of obtaining them. The statistical analyses for microarrays still present challenges that should be addressed according to the nature of the research.

Cluster analysis, which groups samples depending on their similarities, may be the most utilized method to identify differentially expressed genes. Probabilistic model-based analysis and heuristic methods have been used as alternatives, but all of these methods have advantages and limitations. Choosing which one to use is critical, but there is no neat guide for each situation. There are some loose rules and considerations to consult, but the decision will depend on the researcher's ability and requirements of the project.

RNA-Seq

The genome is composed of DNA, which encodes the information to construct and maintain the cells. For these processes to occur, the DNA information should be transcribed into the correspondent mRNA, known as transcripts. The complete set of transcripts is known as the transcriptome (Yusuf et al., 2012). Knowing and quantifying the transcriptome is fundamental to understand functions, structures, evolution, and physiological stages under specific conditions. An example hypothesis to be tested could be the similarity of the transcriptomes of two genetically identical individuals under divergent imposed or known effects (e.g., inbred animals exposed to different diets). In this case, identifying and quantifying the transcriptome could help to recognize genes and pathways that are related to the correspondent effect.

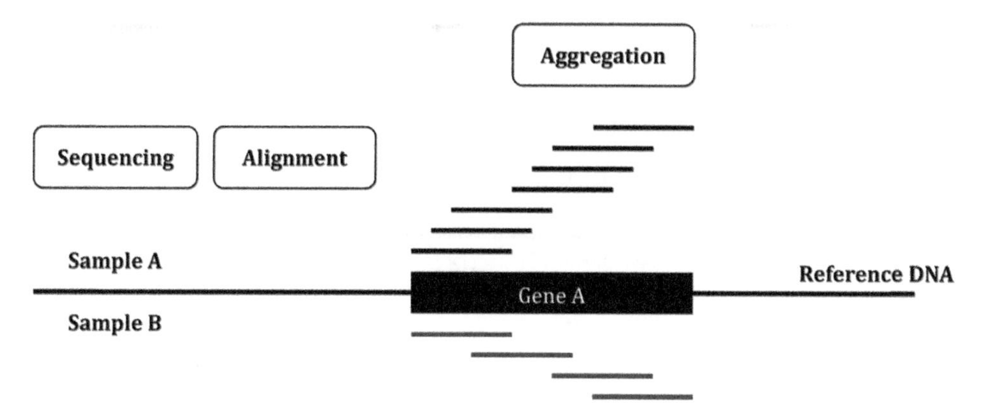

Figure 15.7 RNA-Seq analysis. Frequent steps during a gene expression analysis aimed at quantification of messenger RNAs.

Transcriptome analysis methods

In the pursuit of deciphering the transcriptome more precisely, various methods have been developed, many of them using hybridization approaches that present the problems previously discussed in the microarray section. As an alternative, sequence-based approaches that use the cDNA sequence have been proposed. In the beginning, many methods were expensive, low-throughput and not quantitative; then tag-based procedures like serial analysis of gene expression (SAGE) or cap analysis of gene expression (CAGE) emerged, providing high-throughput and more precise digital expression levels. However, they allowed only a partial analysis of the transcriptome and the identification of the splice-isoforms was impossible, restricting the accurate annotation of the transcripts.

RNA-Seq employs deep-sequencing technologies to obtain sequences from fragments of cDNA that have adaptors in one or both ends (reads). After sequencing, short reads with the same length are obtained. Normally they are around 30–400 base pairs long. The sequencing process is performed in a high-throughput manner, resulting in millions of reads. Later, the reads are aligned to a reference sequence to give a level of expression for each gene. RNA-Seq has other applications like the *de novo* assembly of the genomic transcriptome but these are beyond the scope of this chapter (see Chapter 17).

Use of RNA-Seq does not require prior knowledge of transcript sequences as in microarray technologies. This makes RNA-Seq a suitable method to study complex, not-yet-deciphered transcriptomes. Compared to microarrays, RNA-Seq displays low background signal due to unambiguous mapping of reads to unique genomic regions. Also, the boundaries for expression levels are not limited and enormous fold changes as in the order of 10,000 could be quantified, something that is impossible to accomplish with microarrays due to the lack of sensitivity in both extremes of expression levels. The precision of RNA-Seq is better compared to microarrays and it has been validated using spike-in RNA controls of known concentration measured by quantitative PCR (qPCR). Additional RNA-Seq advantages include a minimal volume of the RNA sample, the greater repeatability of the results compared to microarrays, and the high-throughput manner with lower cost than older technologies (Baginsky et al., 2010). Of course, RNA-Seq presents its own challenges. It is a new technology that has different steps with their own concerns. Bioinformatic analysis of the data introduces various issues, from efficient data storage and transfer methods to adequate statistical analysis of enormous datasets.

RNA-Seq analysis steps

Reads obtained from the sequencing process are checked for quality and then mapped to the reference genome. Many programs may be used for mapping (Bowtie, BLAT, and BFAST are examples); however, the criteria for choosing them and the options setting during the process should be adjusted depending on the researcher's requirements. The fragments are aggregated based on the overlaps and then normalized in order to diminish the noise and make the real signal more detectable. After normalization the fragments are aggregated again and the statistics are calculated (Figure 15.7). For normalizing the data various methods have been proposed in which total count and upper-quartile normalizations are the most commonly used. Regarding models, Poisson and negative binomial distributions are suggested for straightforward read counts. In cases were log counts were employed, normal distribution is proposed because it seems to capture more efficiently the biological variation in large sample sizes.

Gene regulation

Most vertebrates have more than 200 types of cells that differ in specialized functions. Although all the organism's cells share the same genome they can diverge in gene stages: activated or silenced. The determination of the type of cell in which a gene will be transcribed, where the transcription will start and finish, the splicing patterns, when the mRNA will be exported from the nucleus, the timing and frequency of the translation, and the mRNA degradation time are topics included under gene regulation.

An easy way to understand gene regulation is to imagine that genes are turned "on" or "off"; when a product is needed the gene is activated and transcription starts, otherwise it is off. There are other situations where some genes can be more active when their repressors are off (negative regulation). Regulatory mechanisms other than off-on have been reported, where the expression level is gradually modulated from low to high with respect to the conditions. Regulatory systems are complex and usually involve molecules that bind at specific places in the DNA or with

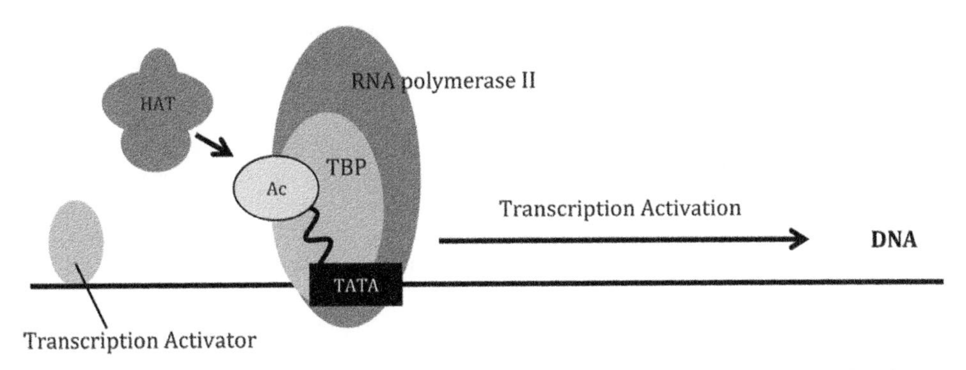

Figure 15.8 Gene regulation. Schematic representation of factors that influence genes activity, making them "active" or "inactive."

other proteins in order to facilitate or suppress an action. When a protein turns on the transcriptional process, it is called a **transcriptional activator protein** (Figure 15.8). Positive and negative regulation systems are not mutually exclusive and can co-exist to act in response to variable cellular conditions. Auto regulated genes are ones whose own products regulate gene transcription, either through positive or negative modulation. Some genes activities can be regulated through the initiation of transcription although others are modulated by the termination of transcription.

All these molecular concepts are necessary to comprehend the bioinformatics methods employed to determine the physical location of the regulators, which molecules are involved in the process, and the network that could describe the parts of the regulatory system and their functionality.

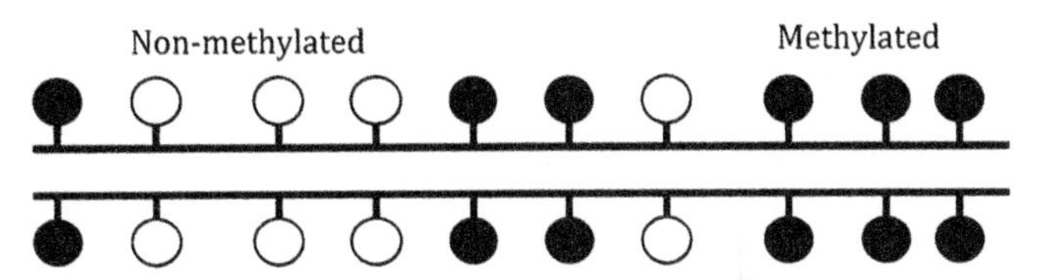

Figure 15.9 Graphical representation of methylation status in a fragment of DNA.

Epigenetics

There are some modifications in the structure of the genome that are independent from the nucleotide sequence but could cause differential gene expression and phenotypes (Martin-Subero, 2011). While the genetic code based on the nucleotide sequence is the same for every cell from an organism, the epigenetic code can be specific for different tissues and cells (Tenzen et al., 2010). The science that studies these events is called Epigenetics (from the Greek: *Epi-* above, over, outer and *Genetics*: the science of genes). Some of these structural transformations can be passed to the progeny as part of an adaptation process that permits the parents' experience to be useful for the next generation (Becker and Weigel, 2012). DNA methylation, acetylation, phosphorylation, and histone modifications are some examples of these changes that can occur at different stages during life and have been related to gene regulation, cellular differentiation, stress events, diseases, aging, and DNA repair (see Chapter 21).

The relationship between these modifications and some phenotypic observations is still a mystery. It has been mentioned that for normal cell differentiation and development DNA methylation is necessary; demethylated and methylated stages are apparent depending on the developmental phase (Figure 15.9). In cancer, genome-wide hypomethylation and local CpG island hypermethylation related to promoters have been observed (Chou et al., 2012). Recognition of these patterns is critical for a better understanding of the physiology and pathogenesis behind the associated diseases in order to develop new therapeutic strategies.

Epigenomics by itself presents many complex issues; however, from a bioinformatics perspective the identification of these alterations is a major challenge demanding better statistical and analytical methods, and a large amount of computational resources. In order to overcome these challenges, different methods have been proposed for analyzing epigenomic data. Several of them are publicly available software packages (bsseq, methylumi, methVisual, MEDIPS); however, some companies offer proprietary analytical tools for their products (Krueger et al., 2012).

Genomic data manipulation

Working with genomic data requires some knowledge of computer languages to prevent introducing errors during data manipulation. The experimental results will be dependent upon the correctness of the data and that is the main reason why any data transformation should be accomplished using computer programs: to protect data integrity. Automatic methods will allow the examination and tracing of any necessary steps performed.

The diversity of processes employed in genomics inspires people to find different approaches and algorithms in order to solve them. Personal preferences in software development have resulted in a wide array of genomic software implemented in many languages, presenting a difficult situation for general users. Various computer languages such as C++, S, R, Perl, Python, among others, have been used to write genomic packages. Some of these programs require specific computing environments and data formats to run properly, leading to the development of specific tools for data conversion and handling, such as BEDtools or SAMtools.

In order to achieve proficient use of all these tools, study of computer languages by geneticists is recommended. Some of these languages, as Perl and Python, are good options due to the amount of existing bioinformatics code and possess the capacity to interact with other languages using pre-written API code as well, facilitating the integration of different ordered processes in one set. This systematic linking of programming steps is often called an analysis pipeline. Genomics studies demand the comparison of results; they should be obtained under the most similar conditions as possible in order to distinguish the real signal from the noise. Pipelines are good strategies for establishing work methodologies, specifically for situations that require the same process exact repetition.

R language

R is a free easily accessible scripting software environment that is used for statistical computing and graphics. It is a GNU project that runs on platforms including Windows, MacOs, and UNIX, and is administered by various CRAN mirrors. R was developed from the S language (S for **statistics**), which originated at Bell Laboratories through the work of John Chambers.

Robert Gentleman and Ross Ihaka at the University of Auckland created R in 1991 but it was not released to the public until 1993. The name R refers to the first letter of both founders' names. Currently, the R Development Core Team maintains R. Usually R employs a command line interface; however some graphical user interfaces (GUI) have been created to increase user-friendliness for non-informaticians. RStudio, StatET, and ESS could be considered as R GUI examples.

R is a very concise and functional language that offers the possibility of combining various commands where each uses the output of the previous one (i.e., pipes). It also allows analysis of complex and very specific processes through modules (packages) that can be tailored depending on user requirements. R offers sophisticated graphic capabilities to visualize intricate data (e.g., ggplot2), a characteristic that is absent in most other statistical packages. There is a community of proactive and enthusiastic users who are willing to help troubleshoot R analysis and to collaborate for improving every aspect of the software. Many discussion threads and useful details are available on the R website: http://cran.r-project.org.

Bioconductor

Bioconductor is an R-based, open source and development software project that provides tools for manipulation, analysis and assessment of high-throughput genomic data. The project started in 2001 and presently is under the supervision of the Bioconductor core team. The team is mainly located at the Fred Hutchinson Cancer Research Center although other national and international institutions participate as active members. Bioconductor can be installed from its site: www.bioconductor.org/install/. Currently most of the genomic data analyses are performed using add-on modules (packages) that can be accessed through www.bioconductor.org. This site also provides example scenarios for which Bioconductor could be employed: microarrays, variants, sequence data, annotation, and high throughput assays are a few. After accessing any of these categories on the Bioconductor website, explanations, scripts and examples for different procedures are provided. The instructions are simple but in order to use them proficiently, knowledge of molecular genomics and R is highly recommended, although the latter is not determinative because R and Bioconductor could be learned concurrently while working through an analysis. Almost all the procedures, descriptions and comments from other users may be obtained on the Internet, making the use and implementation of Bioconductor packages easier than for closed, proprietary programs. Often, adjustment of packages to meet project requirements should be performed to obtain an adequate analysis, and R/Bioconductor has the right flexibility to perform such customizations due to the accessible, modifiable package scripts. Although R has many advantages, it must be used with caution because it does not display any warning when used incorrectly (e.g., if a dataset violates analysis assumptions), a capability that many commercial programs have. This situation demands that users should understand the statistical and programming concepts corresponding to their chosen analysis to avoiding erroneous and misleading results.

Web-based tools

Web-based tools offer an alternative for performing genomic analysis (Feng et al., 2012; Geffers et al., 2012). Some of them simplify the process by providing a GUI, an important feature for biologists that are not familiar with programming; pre-determined settings and parameters usually employed and adopted as proper practices by peers are commonly incorporated as defaults (Jolley et al., 2012; Revanna et al., 2012; Reinhold et al., 2012). These characteristics make web-based tools an easy option for regular users that lack the depth of knowledge demanded by other methods.

DAVID

DAVID stands for Database for Annotation, Visualization, and Integrated Discovery, and is a set of web-based tools developed for genomic analysis. It was developed by Dennis et al. in 2003 and is provided through the National Institute of Allergy and Infectious Diseases (NIAID), National Institute of Health (NIH). The current software version is DAVID Bioinformatics Resources 6.7, which consists of an assortment of functional annotation tools that help to elucidate the biological significance and implications of a set of genotypes.

DAVID is capable of recognizing enriched biological processes using the Gene Ontology (GO) denomination to cluster genes according to their functionality, visualizing genes based on the Kegg and BioCarta pathway maps, converting gene identifiers, and accessing the NIAID Pathogen Annotation Browser, among many other capabilities that could be necessary in a genomic analysis. DAVID can be accessed through the website: http://david.abcc.ncifcrf.gov. On the web page, a detailed description of each analysis and the correct way to use them is provided. Citation requirements and other issues regarding it employment is also well explained.

USCS Genome Bioinformatics

The UCSC Genome Browser was developed and is maintained by the Genome Bioinformatics Group, which is a team formed by members from the Center for Biomolecular Science and Engineering (CBSE) at the University of California, Santa Cruz (UCSC). The website can be accessed at http://genome.ucsc.edu/index.html. The site provides reference sequences and draft assemblies for various organisms' genomes, visualization of the chromosomes with zoom capabilities for close observation of the annotated features, portal access for other genomic projects such as ENCODE and Neandertal, and other helpful web-based tools for genomic analyses (Dreszer et al., 2012; Maher, 2012).

The list of currently available tools is as follows:

BLAT: Used to identify 95% or greater similarity between sequences of 25 nucleotides or more.

Table Browser: Permits retrieval of affiliated data and DNA sequence contained within a track.

Gene Sorter: Can be used to sort related genes based on different criteria like similar gene expression profiles or genomic proximity.

Genome Graphs: Provides different graphical capabilities for visualizing genome-wide data sets.

In-silico PCR: Employed for searching a sequence database with a set of PCR primers, applying an indexing strategy for fast performance.

Lift Genome Annotation: Converts genomic positions and genome annotation files between different assemblies.

VisiGene: Allows the use of a virtual microscope in order to visualize *in situ* images, showing where the gene is employed in the respective organism. Various institutions collaborate for providing and updating the image library.

Other Utilities: Gives a list of links and instructions for other tools and utilities developed by the UCSC Genome Bioinformatics Group, specifically aimed at genomic studies.

Web tools offerings are extensive and deciding which one to use is a personal decision that must be made by the investigator. Of course, some web tools are more suitable than others with respect to theory, although in practice different aspects of the research will determine the choice. It will depend on the study species, cost, availability of antibodies, and many other factors that could be critical in the decision-making process. On the other hand, it is important to develop a consistent methodology when analyzing the same type of data in order to diminish the error due to different analysis strategies and to make results of similar studies comparable. This will allow the incorporation of older methods into a more fine-tuned process with a traceable record of the improvements.

Shared cyber-infrastructure and computational tools

The cost and complexity of problems in genomics requires the maximization of resources. Thus, people around the world are combining economical, intellectual, and infrastructural resources to diminish undesirable effects (e.g., duplicated work, time wasting, increased cost). Methods for standardization of specific procedures provide more robust results from conclusive comparisons between similar studies but concurrently, genomics is very interdisciplinary where individuals have diverse interests. Various cyber-infrastructure consortiums were created to overcome the previous mentioned problems but always considering the common goals of the potential users. Examples are: Galaxy, iPlant/iAnimal. These tools will help individuals to do genomics over developing bioinformatics at the programming level in the future.

Bioinformatics perspectives in animal genetics

Genomics is a relatively new research area where the number and diversity of studies is increasing quickly. This growth presents several challenges regarding the data produced by novel molecular technologies. Different types of studies demand tailored analysis methodologies, fueling the invention of new approaches. Data capture, manipulation, processing, analysis, storage, and utilization are examples of common issues for all high-throughput assays. In other words, to carry out genomics research it is critical to be able to program and apply bioinformatics to some extent. New advances allow faster and inexpensive sequencing, accruing problems that extremely large data sets present. This scenario is a fertile field for ingenious bioinformaticians who can develop new strategies for solving such problems.

High-throughput analyses have been introduced to animal genetics in recent years but eventually they will be used routinely. The study of cancer and other diseases in humans has produced advances in genomics that can be adapted for animal related issues including selection, which can employ genetic markers in order to choose the best animal for a specific condition or to develop disease resistant lines (Eck et al., 2009). New drug development and animal recombinant proteins, which can be used in humans for treating diseases and decreasing animal environmental impact, are examples of opportunities that integrate bioinformatics and animal genetics.

An example of the bioinformatics impact in humans is the discovery of the Retinol-binding protein 4 (RBP4) gene that encodes the RBP4 protein, which is the specific carrier for retinol (vitamin A) in the blood. Clinically becomes important because RBP4 contributes to insulin resistance and play an important role in diabetes and obesity. In animals, Insulin-like growth factor 2 (IGF2) is an imprinted gene that affects the lean muscle content in different species. The direct application for this finding is in bovines, where the bi-allelic expression of this gene is associated with an increase in the rib eye area (REA), a trait that is important for selection in beef cattle.

Bioinformaticians and biologists have the opportunity to apply innovative and creative techniques to these problems, and their efforts will be economically rewarded. Job opportunities in biomedical research are increasing and the boundaries between fields are progressively becoming diffuse, presenting many new options. Experienced professionals can transfer their knowledge to new areas where interdisciplinary knowledge is necessary and junior professionals will have more jobs and learning opportunities that could lead to immediate and long term benefits.

References

Baginsky, S., Hennig, L., Zimmermann, P. and Gruissem, W. Gene expression analysis, proteomics, and network discovery. *Plant Physiology* 152:402–410 (2010).

Becker, C. and Weigel, D. Epigenetic variation: origin and transgenerational inheritance. in *Curr Opin Plant Biol* 15(5): 562–567 (2012 Elsevier Ltd, 2012).

Chou, A.P., Chowdury, R., Li, S. et al. Identification of retinol binding protein 1 promoter hypermethylation in isocitrate dehydrogenase 1 and 2 mutant gliomas. in *J Natl Cancer Inst* 3;104(19):1458–1469 (2012).

Dreszer, T.R., Karolchik, D., Zweig, A.S., et al. The UCSC Genome Browser database: extensions and updates 2011. *Nucleic Acids Research* 40:D918–D923 (2012).

Eck, S.H., Benet-Pages, A., Flisikowski, K., et al. Whole genome sequencing of a single Bos taurus animal for single nucleotide polymorphism discovery. *Genome Biol* 10:6 (2009).

Feng, X., Xu, Y., Chen, Y. and Tang, Y.J. MicrobesFlux: a web platform for drafting metabolic models from the KEGG database. *BMC Syst Biol* 6:94 (2012).

Geffers, L., Herrmann, B. and Eichele, G. Web-based digital gene expression atlases for the mouse. *Mamm Genome* 31:31 (2012).

Hogeweg, P. The roots of bioinformatics in theoretical biology. *PLoS Comput Biol* 7:e1002021 (2011).

Jolley, K.A., Hill, D.M., Bratcher, H.B., et al. Resolution of a meningococcal disease outbreak from whole-genome sequence data with rapid web-based analysis methods. *J Clin Microbiol* 50:3046–3053 (2012).

Krueger, F., Kreck, B., Franke, A. and Andrews, S.R. DNA methylome analysis using short bisulfite sequencing data. *Nat Meth* 9:145–151 (2012).

Maher, B. ENCODE: The human encyclopaedia. *Nature News* 489:46 (2012).

Martin-Subero, J.I. How epigenomics brings phenotype into being. *Pediatr Endocrinol Rev* 9 Suppl 1:506–510 (2011).

Morris, K.V. *Non-Coding RNAs and Epigenetic Regulation of Gene Expression: Drivers of Natural Selection.* Caister Academic Press (2012).

Pevzner, P. and Shamir, R. *Bioinformatics for Biologists.* xxix, 362 p. Cambridge University Press, Cambridge/New York (2011).

Reinhold, W.C., Sunshire, M., Liu, H., et al. CellMiner: A web-based suite of genomic and pharmacologic tools to explore transcript and drug patterns in the NCI-60 cell line set. *Cancer Res* 72:3499–3511 (2012).

Revanna, K.V., Munro, D., Gao, A., et al. A web-based multi-genome synteny viewer for customized data. *BMC Bioinformatics* 13:190 (2012).

Schwanhausser, B., Busse, D., Li, N., et al. Global quantification of mammalian gene expression control. *Nature* 473:337–342 (2011).

Tenzen, T., Zembowicz, F. and Cowan, C.A. Genome modification in human embryonic stem cells. *J Cell Physiol* 222:278–881 (2010).

Van Iterson, M., Duijkers, F.A.M., Meijerink, J.P.P., Admiraal, P., van Ommen, G.J.B., Boer, J.M., van Noesel, M.M., and de Menezes, R.X. et al. A novel and fast normalization method for high-density arrays. in *Statistical Applications in Genetics and Molecular Biology* 11(4):1–31 (2012).

Yusuf, D., Butland, S.L., Swanson, M.I., et al. The transcription factor encyclopedia. *Genome Biology* 13:R24 (2012).

Review questions

1. Describe the importance of bioinformatics in genomic studies.

2. Cite and describe briefly two of the most-used genomic analysis techniques.

3. What are the challenges in microarray technology?

4. What is the purpose of performing an RNA-Seq analysis?

5. How can bioinformatics be used to find gene regulatory mechanisms?

6. Name a programming language. Why is it helpful in bioinformatics?

7. What is “Bioconductor”?

8. Name a web-based tool and explain its usefulness in data analysis.

9. Give three important uses of bioinformatics tools in animal genomics.

16 Genome-wide Association Studies in Pedigreed Populations

Dirk-Jan de Koning

Department of Animal Breeding and Genetics, Swedish University of Agricultural Sciences, Uppsala, Sweden

Introduction

In livestock genomics, many studies have focused on mapping QTL using family based design as testified by the thousands of QTL in the Animal QTL database: www.animalgenome.org/cgi-bin/QTLdb/index. With the availability of genome-wide SNP chips for most livestock species, the focus has moved towards genome-wide association studies (GWAS). QTL analyses require the collection, or targeted breeding, of families that are studied for the traits of interest while GWAS can use the existing population. QTL mapping using linkage analyses using the within family linkage disequilibrium (LD) between markers and QTL, which is extensive. This means that a QTL study with linkage can be done with a few hundred markers. A GWAS uses the LD at the population level and requires tens, or even hundreds of thousands of markers. A linkage based study only exploits a few (often only one) generations of recombination to locate the QTL while a GWAS uses all historical recombinations since the QTN mutation occurred. This means that GWAS studies give a much greater precision than family based QTL studies.

The principles of a genome-wide association study (GWAS) are very straightforward: (1) Identify a target population that is segregating for the trait of interest. (2) Genotype the population sample with sufficient markers to cover the genome. (3) For each marker, test for an association between the marker genotype and the trait of interest. In this section the general principles are presented while in the later parts of this chapter we delve deeper into the scenarios for livestock.

Experimental designs

Roughly speaking we can distinguish three different designs for GWAS: (1) Case control Studies, (2) Cohort studies, and (3) Family-based designs.

Case control

In a Case control study, cases are identified from the general population that show a particular phenotype (often a disease but you could also consider extremes of the phenotypic distribution). These are subsequently matched with controls that do not show this phenotype. These controls must be drawn from the same population and match the cases in terms of age, gender, and so on. The statistical analysis tests for significant differences in allele or genotype frequencies among cases and controls. The most well-known case control effort is that of the Welcome Trust Case Control Consortium (WTCCC) (Burton et al., 2007)who used 14,000 cases and 3000 controls for a range of human diseases. The large number of cases and controls in these studies on human diseases may seem in stark contrast to the figures that are presented for disease mapping in dogs where the recommended sample sizes are in the order of tens to maybe a few hundred (Karlsson and Lindblad-Toh, 2008). Beside the clear differences in population structure between human and dogs, it is important to note that the WTCCC describes the detection of more subtle effects on human disease, while Karlsson and Lindblad-Toh describe the detection of Mendelian disorders (single mutations that determine the phenotype) or loci that at least double the liability to disease (Burton et al., 2007; Karlsson and Lindblad-Toh, 2008). Statistical power of (planned) experiments can be evaluated using the Genetics Power Calculator (Purcell et al., 2003) at: http://pngu.mgh.harvard.edu/~purcell/gpc/.

Cohort study

In a so-called cohort study a population of individuals, often from a distinct geographical location, is observed throughout life with extensive phenotype recording. This approach is not well suited to target a specific disease or phenotype but may be very useful to identify risk factors for complex disease such as smoking, alcohol intake, and body mass index (BMI) in humans. Several resource populations in livestock could be considered as cohort studies.

Family based designs

A third design for GWAS is the family-based design where association between markers and traits is evaluated within families. This can offer statistical robustness because this kind of design avoids problems with stratification at the population level. However, a within-family analysis for association will have reduced power to detect significant effects compared to analyses across families. In livestock, many family based designs are readily available for GWAS, either because they have been created as resource populations for QTL mapping or because of the family structures that are common to livestock breeding.

Molecular and Quantitative Animal Genetics, First Edition. Edited by Hasan Khatib.

Genotyping and linkage disequilibrium

For a GWAS the target population is genotyped for a large number of markers to cover the entire genome. The number of markers should be large enough that any potential QTL in the genome is in sufficient linkage disequilibrium (LD) with one of the SNPs that is genotyped. Linkage disequilibrium is a measure that describes how well the genotype at a known locus predicts the genotype of a neighboring, but unobserved, locus. The calculation of LD and how it relates to (effective) population size easily merits its own chapter. For the purpose of this chapter we provide a concise description of the most common measures of LD, |D'|, and r^2 (see Box 16.1). The current markers of choice are single nucleotide polymorphisms (SNPs) which are very abundant in the genome. So-called SNP chips that evaluate tens, or even hundreds of thousands of SNPs in a single assay are routinely available for most livestock species (chicken, cow, pig, sheep, goat). The SNPs on these SNP chips have usually been selected from (partial) re-sequencing of several breeds within a species and are often chosen to have minor allele frequencies (MAF) > 0.05. Whiles this broadens the number of lines in which the SNP is likely to segregate, it also means that putative QTL that segregate at a low frequency are often not in very strong LD with any of the SNPs on the chip.

Quality control

A crucial step in any GWAS analysis is the quality control (QC) after the genotyping. Using stringent procedures at this stage can prevent disappointment or chasing false positives at a later stage. The quality control needs to be done, both at the level of the individual samples as well as that of the individual SNPs (Chanock et al., 2007; Manolio et al., 2007). At the individual level you need to check that a sufficient proportion of SNPs (e.g., >90%) was successfully called for that sample. You also want to include blind replicates to test whether they are genotyped consistently. Samples with an excessive number of heterozygous calls may be indicative of a contaminated sample that includes DNA from multiple individuals. You may want to check whether the genotype results for sex-linked markers agree with the presumed sex of the individual. Furthermore you may want to check whether the genotype results show unexpected family relationships or, if possible, parentage errors. At the individual SNP level you also want screen the call rate: a sufficient proportion of individuals must have successful genotype calls for a given SNP (e.g. >95 %). Also the use of duplicate samples can indicate whether a SNP produces consistent genotype calls. A specific test for each SNP is whether the alleles segregate according to Hardy–Weinberg equilibrium (HWE). Under HWE it is expected that a locus with allele frequencies p and q segregates with the expected frequencies p^2, 2pq, and q^2 for the three genotype classes, respectively. Deviations from this expectation can be easily tested with a Chi-squared test. In livestock populations, deviations from HWE are often expected because of the family structures, in which case this QC aspect is ignored or implemented with a very stringent threshold for exclusion. If parent information is available, Mendelian errors can be used to screen out suspect SNPs. Often SNPs with a low minor allele frequency (MAF, typically MAF < 0.01) are removed from the analysis. This may not be because the SNP genotyping are deemed unreliable, but simply to prevent small genotype classes in the statistical analyses that could easily give rise to false positive associations. Following the analysis, it is very important to visually inspect the genotyping results for those SNPs that are deemed significant. It is even argued that the genotyping for these SNPs should actually be repeated on an unrelated platform (Chanock et al., 2007; Manolio et al., 2007). Figure 16.1 provides three real life examples on how genotypes are called by the genotyping software (from www.ncbi.nlm.nih.gov/projects/gap/cgi-bin/study.cgi?study_id=phs000001.v3.p1).

Box 16.1

A simple example of linkage disequilibrium

In this example we consider eight individuals with the following haplotypes.

a	g	a	g	A	G	A	G
A	G	A	G	A	g	a	g
a	g	A	g	A	G	A	G
A	G	A	G	a	g	a	g

(a)

These can be arranged in a simple table of haplotype frequencies.

	A	a	
G	8	0	50%
g	2	6	50%
	62.5%	37.5%	

(b)

Here, we introduce two measures of LD: D', and r^2. D' ranges from 0 (no LD) to 1 (complete LD), is not sensitive to marginal alleles frequencies and directly related to the recombination fraction between two loci. R^2 also ranges from 0 to 1, but reflects the correlation between two loci. R^2 is very sensitive to marginal allele frequencies and relates directly to the statistical power of the study. If a marker M and causal gene G are in LD, then a study with N cases and controls which measures M (but not G) will have the same power to detect an association as a study with r^2*N cases and controls that directly measured G. Therefore r^2*N is referred to as the "effective sample size."

For the previous example, let x_{11} be the frequency of haplotype GA, x_{22} is the frequency of haplotype ga, x_{21} is the frequency of haplotype gA and x_{12} the frequency of haplotype Ga. The disequilibrium $D = x_{11}*x_{22} - x_{12}*x_{21}$; $D' = D/Dmax$ and $r^2 = D^2/[p_1 \cdot p_2 \cdot q_1 \cdot q_2]$, where p_i and q_i are the frequencies of allele i for the respective markers. Dmax is $\min[p_1*q_2, p_2*q_1]$ if D is positive or $\max[-p_1*q_1, -p_2*q_2]$ if D is negative. In this example $D' = ((8*6) - 0)/(8*6) = 1$ while $r^2 = (8 \times 6 - 0)^2 / (10 \times 6 \times 8 \times 8) = 0.6$. This illustrates clearly that different measures of LD have different interpretations. For an online example of estimation of LD see: www.slideshare.net/awais77/measures-of-linkage-disequilibrium-10238151.

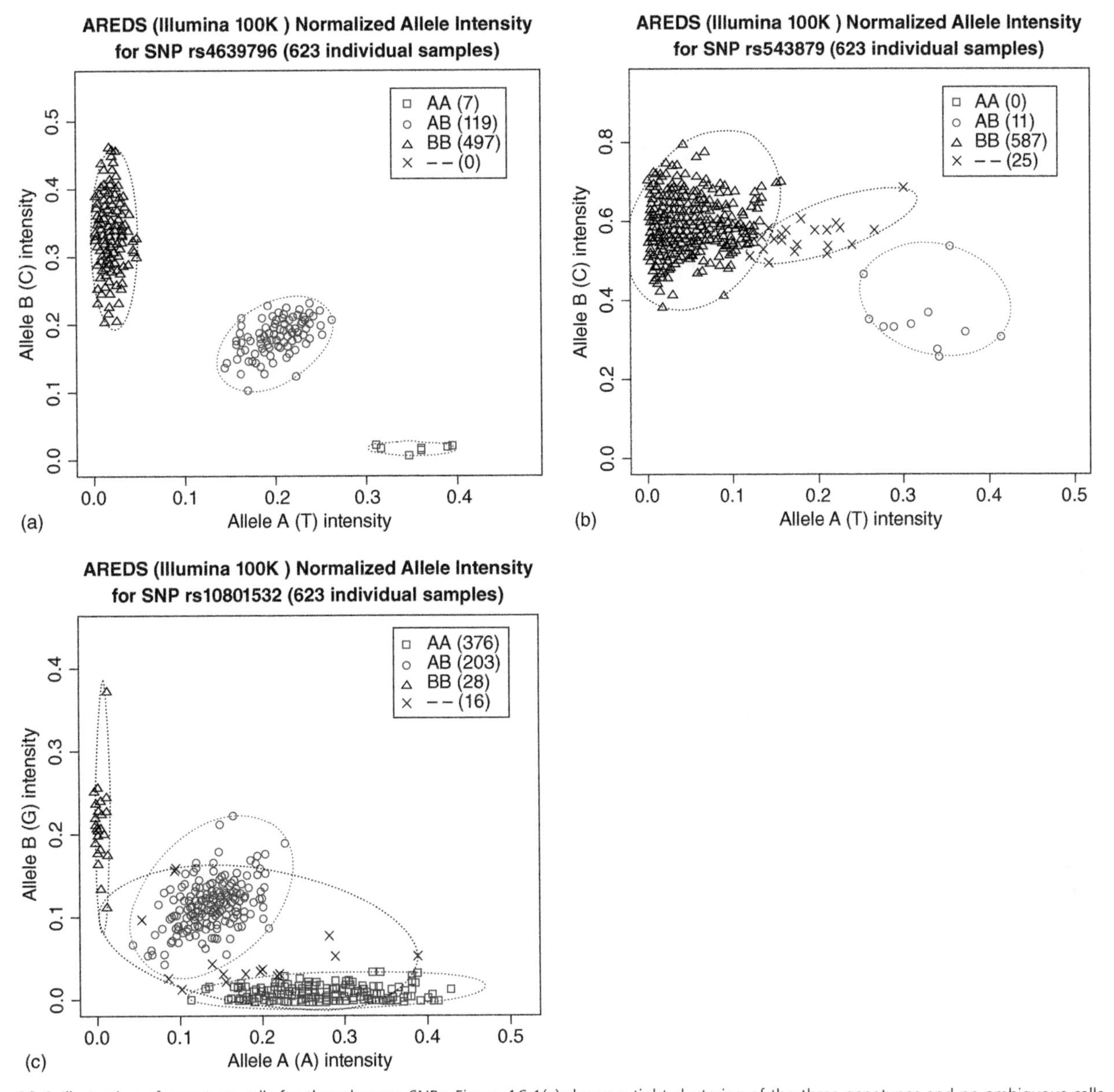

Figure 16.1 Illustration of genotype calls for three human SNPs. Figure 16.1(a) shows a tight clustering of the three genotypes and no ambiguous calls. Figure 16.1(b) assigns only two of the three possible genotypes and misses out an entire cluster of what could be the heterozygous individuals. Figure 16.1(c) shows less tight clustering with some partial overlap between the clusters resulting in some missing genotypes. These figures illustrate why it is important to visually inspect the genotypes for the "winning SNPs."

The statistical analysis

The statistical analysis of each SNP for association with the trait of interest depends on the genetic model as well as the experimental design. For a case-control study a straightforward Chi-squared test can be implemented comparing either the allele or genotype frequencies among the cases and controls to the expected frequencies under the H_0. This analysis takes the form of a contingency table. For an early example see Samani et al. (2007). A contingency table for a given SNP can assume an allelic (additive) model or a genotypic (additive-dominant) model. The odds ratio indicates the increased risk for carrying one or two copies of the risk allele.

When studying a quantitative trait, the analysis is a linear model in which the trait score is the y-variable and the genotype is one of the exploratory variables. In a purely additive, or so-called allelic, model the genotype can be modeled a linear covariate related to the number of copies of a given allele (e.g., the C allele), taking the values 0, 1, and 2 for genotypes for SNP genotypes AA, AC, and CC, respectively. In an additive-dominance model, or genotypic model, the three genotypes are modeled as three classes of a fixed effect. A case-control study can also be analyzed in a generalized linear model using, for instance, logistic regression. A major advantage of using a generalized linear model over the contingency table is that the linear facilitates the inclusion of other fixed effects, random effects, or co-variables that may affect the trait under study. An example of a GWAS for a quantitative trait is the study of uric acid levels in two population cohorts in Italy (Li et al., 2007). The authors performed a linear regression of inverse normalized levels of uric acid against the number of alleles (additive model). Sex, age, and age^2 were fitted as covariates Li et al., 2007). The effect of the most significant SNP is given in Figure 16.2. Note that the genotypic effect seems to be consistent between the two cohorts (Figure 16.2).

Significance testing

Significance thresholds have been debated very widely in the context of GWAS studies. The most conservative approach is to do a Bonferroni correction for the total number of SNPs in the experiment. This would overestimate the effective number of tests as many SNPs are correlated through LD. Permutation approaches have also been suggested but these can be very time-consuming for large studies with many SNPs. For human genetic studies a consensus threshold of $P < 0.5*10^{-7}$ has been adopted as a yardstick for significance. Given the lower effective population size in livestock compared to humans as well as differences in the genome-size and recombination fraction, the effective number of tests in a livestock study can be expected to be lower compared to a human GWAS. This would suggest that a slightly more liberal *P* value could be adopted for significance testing. It must be stressed that the emphasis should not be on the actual *P* value but rather on replication of the SNP effect in an (as much as possible) independent sample of the population.

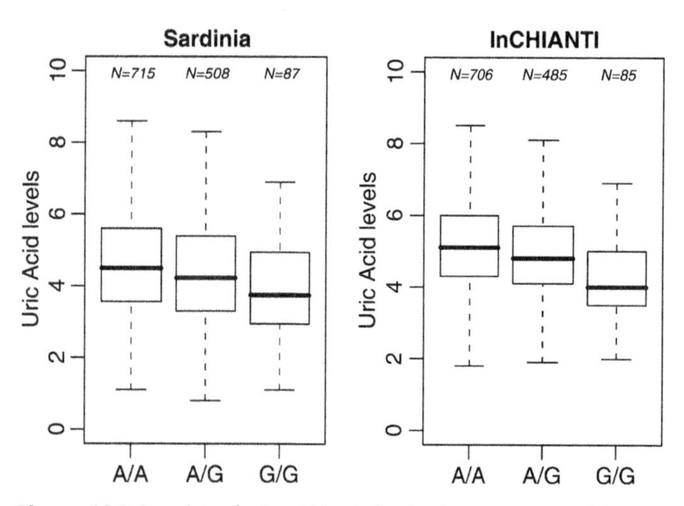

Figure 16.2 Box plots of uric acid levels for the three genotypes of the most significant SNP in a GWAS of two Italian cohorts (Li et al., 2007). Reproduced from Li et al. (2007) under the creative commons license (http://creativecommons.org/licenses/by/2.5/).

Inspection of GWAS results

The results of a GWAS can be summarized in a so-called "Manhattan plot." In such a graph the *P* values for each SNP are plotted against their genome location. The *P* values are transformed as $-10\text{Log}(P)$ so that higher values means stronger significance. Figure 16.3 shows the Manhattan plot for the GWAS on uric acid levels (Li et al., 2007).

Another way to plot the results is a so-called QQ plot. In general QQ plots are comparisons between two distributions where one is the expected distribution and the other the realized distribution. In GWAS studies the QQ plot often shows the realized and sorted *P* values for all SNPs plotted against their expected values under H_0. If the slope of the corresponding line is >1 this suggests there is some underlying structure in the data that cause inflated test statistics. Potential causes for substructure in the data are different breeds/races or family structures. The excellent review by Mark McCarthy provides examples of typical QQ plots (McCarthy et al., 2008) (Figure 16.4).

Dealing with population structure

Several approaches have been put forward to account for inflated test statistics due to population structure: Originally proposed by Devlin and Roeder in 1999, genomic control deals directly with inflated test statistics by adjusting them downwards: genomic control estimates the inflation factor λ from the QQ plot and adjusts the *P* values accordingly (Devlin and Roeder, 1999). This approach is easy to implement but it uses the same adjustment for all *P* values and only deals with a linear correction. Another option is to explicitly identify the substructure within the population and include the subgroups in the association model (Yu et al., 2006). Using the actual genotyping data clusters or subpopulations can be identified using population genetics software like STRUCTURE (http://pritch.bsd.uchicago.edu/software/structure2_1.html). For each individual, the subgroup to which it clusters can be included in the analyses. A similar, but much faster, approach was suggested by Price et al. in 2006. In the so-called EIGENSTRAT approach (part of EIGENSOFT at: http://genepath.med.harvard.edu/~reich/Software.htm) the genotypic relationship between all individuals is captured in a genome-wide identity-by-state (IBS) matrix. Subsequently vector loadings of the principal component (PCP) of this matrix are included in the model to account for substructure within the population. The number of PCP to be included in the model can be decided on the basis of a scree plot where the effect of each consecutive PCP on the eigenvalue of the IBS matrix is plotted. There are several approaches that use some form of the IBS matrix to detect clusters within the data while others use phylogenetic approaches to do the same things. Box 16.2 illustrates the

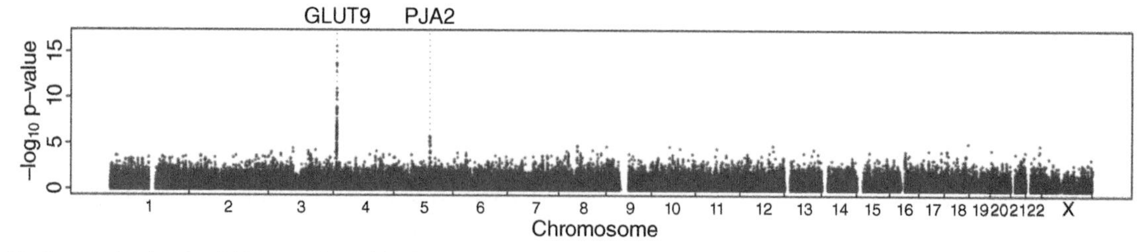

Figure 16.3 Manhattan plot for the GWAS on uric acid levels in two Italian Cohorts (Li et al., 2007). Reproduced from Li et al. (2007) under the creative commons license (http://creativecommons.org/licenses/by/2.5/).

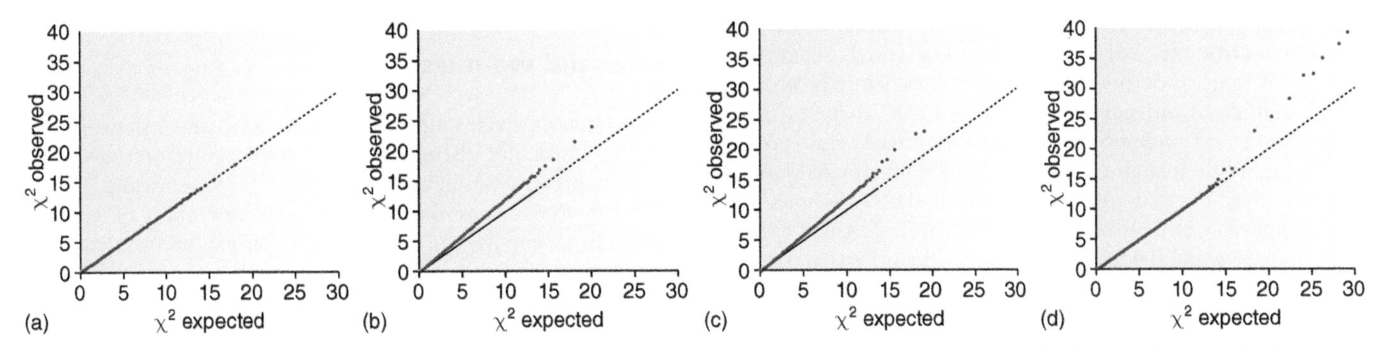

Figure 16.4 Illustrative examples of QQ plots from McCarthy et al. (8). Plot (a) shows a QQ plot where there is no overall inflation of P values. Plot (b) shows an overall inflation of P values. Plot (c) suggests overall inflation as well as some real significant results. Plot (d) shows no evidence for inflation but strong evidence for real effects. Reprinted by permission from Macmillan Publishers Ltd: *Nature Reviews Genetics*. May; 9(5):356–369, copyright (2008) www.nature.com/nrg/.

effect of stratification of the test statistic and how PCP can mitigate this (Box 16.2, Chen et al., 2009) while genomic control does not.

Methods and tools for GWAS in pedigreed populations

In livestock populations there is no getting away from the fact that there are extensive pedigree links amongst most breeding animals. Rather than try and make experiment designs that minimize the genetic links between individuals, we might as well as embrace the fact that we have potentially large family structures and use this to our advantage. When planning for case control type studies or sampling phenotypic extremes in, for instance, pigs or poultry, it is strongly recommended to pick a full sib as a control. This provides a very good level of control from family stratification within the data. In GWAS for quantitative traits it is important to fit the family or the pedigree as part of the statistical model. A general linear mixed model for a GWAS in livestock is:

$$y = W\alpha + X\beta + Zu + e \qquad (16.1)$$

Where $\boldsymbol{y}$ is the vector of phenotypes of interest, where $\boldsymbol{\alpha}$ is the vector with marker effects, $\boldsymbol{\beta}$ is the vector of relevant fixed effects and covariates, $\boldsymbol{u}$ is the vector of random animal effects and $\boldsymbol{e}$ is the random residual. $\boldsymbol{W}$, $\boldsymbol{X}$, and $\boldsymbol{Z}$ are incidence matrices. At this stage it must be noted that the definition of phenotype is important. It suffices to state here that estimated breeding values (EBV) are **not** good phenotypes for GWAS. You must use de-regressed EBV (Garrick et al., 2009) or at least adjust the EBV for the value of the mid-parent. Several tools are available for GWAS analysis using some form of mixed linear models. The main advantage of using these software packages is that they often include tools that deal with the QC and are automated to run analyses across all SNPs. Next I will outline the main tools that are applicable to livestock data.

PLINK

Originally coined as **p**opulation-wide **link**age analysis, PLINK provides and extensive suite of analysis (http://pngu.mgh.harvard.edu/~purcell/plink/) (Purcell et al., 2007). PLINK is mostly used for case control studies but can also handle quantitative phenotypes. PLINK includes the possibility to include PCP to correct for substructure in the data and also features a family based association test. PLINK is not best suited to deal with all the complexities of livestock populations but is still a very good starting point for GWAS analyses.

TASSEL

TASSEL is developed by the group of Ed Buckler working on maize genetics (www.maizegenetics.net/tasselx). The methods were developed to test for associations across a large panel of inbred maize lines. Because of the complex historical relationships between different maize lines the method contains two random effects: Q for structure and K for family relationships, both estimated on the basis of marker data (Yu et al., 2006). This combination makes the design suitable for a wide range of population structures. More recently, Zhang et al. added some improvements to make the analyses faster and more suitable for GWAS (Zhang et al., 2010).

EMMA/EMMAX

EMMA(X) stands for Efficient Mixed-Model Association (eXpedited) (http://genetics.cs.ucla.edu/emmax/). It was initially developed for the analysis of associations across a panel of inbred mouse lines, not unlike the analyses of maize for which TASSEL was designed. EMMAX models genetic relationships in the data on the basis of a marker-derived relationship matrix or a phylogenetic model (Kang et al., 2008). The computational advantages come from fast maximum likelihood routines and more recently, the introduction of a pre-correction step (Kang et al., 2010). Given similar computation performance, TASSEL can handle more complexity compared to EMMAX.

GenABEL

GenABEL (www.genabel.org/) provides a large suite of routines for GWAS analyses and is almost exclusively developed by Yurii Aulchenko. Of the many possibilities within GenABEL, the initial development of the GRAMMAR analysis was an important development for livestock GWAS (Aulchenko et al., 2007). In order to avoid running a mixed model for every individual SNP, Aulchenko et al proposed to pre-correct the phenotype for polygenic effects prior to GWAS analyses (Aulchenko et al., 2007). It should be noted that TASSEL and EMMAX implemented similar approaches in 2010 (Kang et al., 2010; Zhang et al., 2010). Within GenABEL, the polygenic effects can be estimated using a marker-based relationship matrix. An advantage of marker based relationship matrices is that we do not rely on accurate pedigree recordings. The initial disadvantages of the GRAMMAR analyses, over-conservative and underestimating the SNP effects, was recently overcome by the introduction of a correction factor (Grammar-lambda) (Svishcheva et al., 2012). For all the GWAS software that is applicable to livestock studies, GenABEL is easily the most comprehensive and flexible. This comes at the price that it is also quite involved and the researcher needs to find their way among a large number of routines. Users need to be experienced users of R or choose to become so during the process.

Things to remember about analysis

(1) Most domestic and wild animal samples have both "breed" and pedigree effects. It may be useful to incorporate both effects in your model. (2) A marker derived IBD matrix can be explored from several angles and you can learn a lot about your data. (3) Clustering and PCA can further help the understanding of your data and improvement of your model. (4) Use mixed models to account for family relationships.

Probably the most important take-home message: Doing the actual GWAS is the best way to do QC, also the most frustrating!

What did we miss?

This chapter provides the basics about GWAS and a starting point on the journey of discovery. Many aspects were not covered in order to preserve simplicity throughout. A very exciting development in livestock is the ongoing implementation of genomic selection (Meuwissen, et al., 2001), spearheaded by dairy cattle but more recently spreading to other species. Genomic Selection and GWAS have their overlaps and in particular, the so-called "variable selection" methods for Genomic Selection can be applied to GWAS as well. A very recent review on Genomic Selection is provided by de Los Campos et al. (2013).

Here, I only discussed the analysis of single SNPs, ignoring the area of haplotype mapping. An emerging area is the move from single SNPs or small haplotypes towards "region heritabilties" where the effects of a large number of SNPs or an entire chromosome are estimated as a random effect. With the advances of sequencing techniques there is a growing interest in the area of imputation to increase the marker density of chip genotyped individuals and by doing so increasing the power and precision. Many of these require some form of haplotyping for which a wide array of tools has been developed. All of these topics would have made nice sections in this chapter but may have to wait until the next version.

Box 16.2

Effects of stratification on type I error and use of principal components to mitigate the problem

This is a brief summary of the article "Genetic Structure of the Han Chinese Population Revealed by Genome-wide SNP Variation" by Chen et al. (2009).

Chen et al. (2009) studied the genetic structure of the Chinese population using >350000 SNPs on more than 6000 Han Chinese from 23 different provinces. Among other findings, they show a clear differentiation between Han Chinese from the north and south of China (Chen et al., 2009). They use this information to simulate a case control study with different amounts of admixture. 500 randomly selected individuals from the north were assigned as the case while another 500 were selected as controls with different proportions of individuals from the south mixed in to create admixture. The admixture levels varied from 0% (only northern individuals as controls to 100% (only southern individuals as controls) with intermediate levels of 20%, 40% and 60%. The results are illustrated here (Figure 1), which is a reproduction of Figure 4 of the original paper (Chen et al., 2009).

The figure clearly illustrates that Genomic Control can correct small amounts of inflated type I error due to stratifications (e.g., 20% in the current example with lambda of 1,17) but that greater levels of admixture require approaches like principal component correction (Chen et al., 2009).

Box material reprinted from *The American Journal of Human Genetics*, Vol. 85, Chen et al., Genetic structure of the Han Chinese population revealed by genome-wide SNP variation, pp. 775–785 Copyright (2009), with permission from The American Society of Human Genetics.

Figure 1 QQ plots of the p values from the simulated association analyses with or without correction for population stratification using a simulated Case-Control study on the basis of Han Chinese individuals (Chen et al., 2009). The columns correspond to the Q-Q plots of the uncorrected, GC-corrected, and PCA-corrected *p* values. The rows correspond to 20%, 40%, 80%, and 100% stratification of the simulated case and control samples.
(A–C) 20% stratification: 500N cases, 400N and 100S controls.
(D–F) 40% stratification: 500N cases, 300N and 200S controls.
(G–I) 80% stratification: 500N cases, 100N and 400S controls.
(J–L) 100% stratification: 500N cases and 500S controls.
Originally published as Figure 4 in Chen et al. (2009). Reprinted from *The American Journal of Human Genetics*, Vol. 85, Chen et al., Genetic Structure of the Han Chinese Population Revealed by Genome-wide SNP Variation, pp. 775–785 Copyright (2009), with permission from The American Society of Human Genetics.

Box 16.2 *(Continued)*

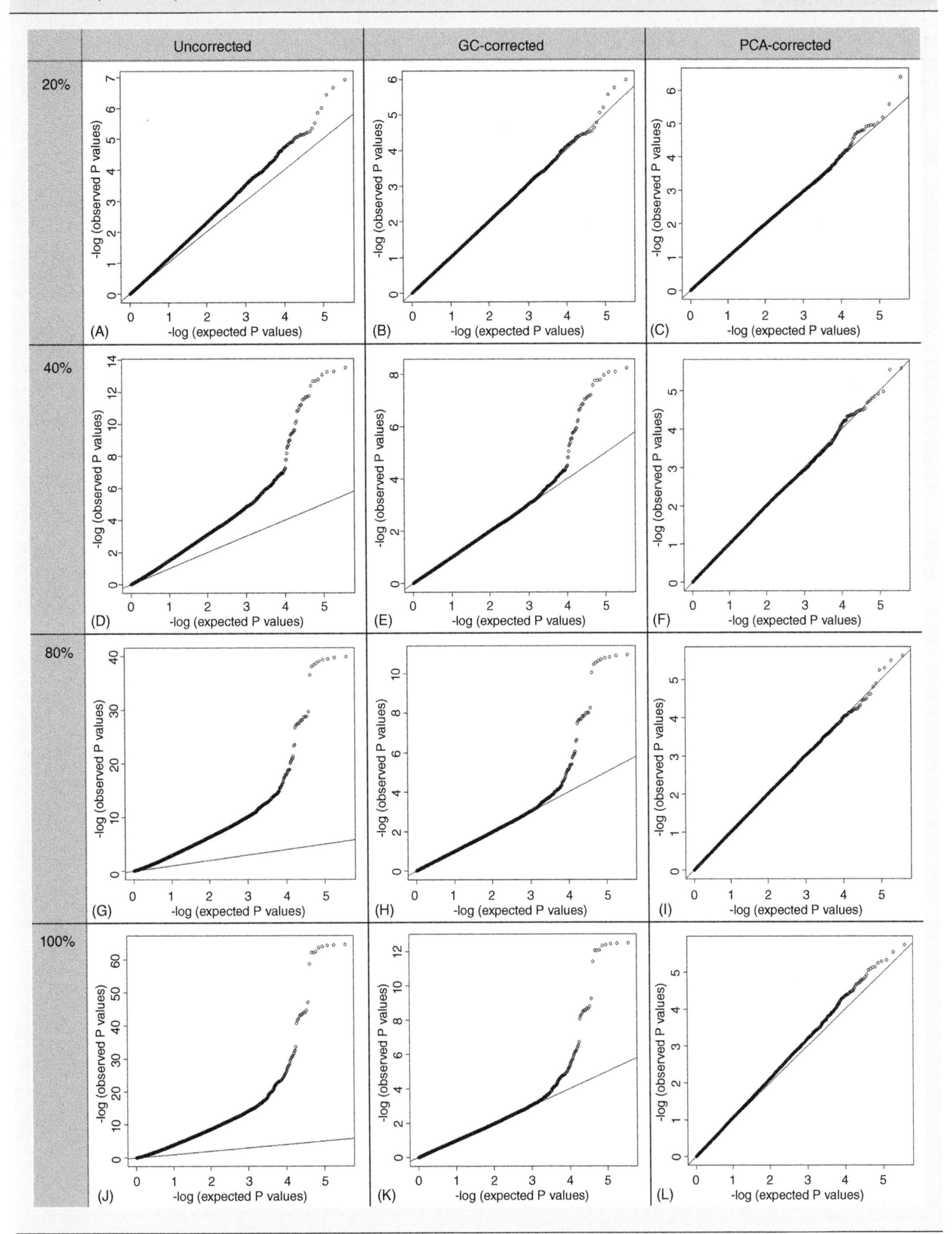

Acknowledgments

Much of the material in this chapter was based on lectures that in their turn were shared, borrowed or simply taken from colleagues. I need to name input from Chris Haley, Suzanne Rowe, Henk Bovenhuis, and Albert Tenesa. (Any likeness to any of your slides or texts is purely coincidental and can only be an indication that great minds think alike!) Last but not least I need to thank the captive (or captivated in some cases) audiences that consumed the contents of this chapter in lecture form. Your questions and blank stares have helped me greatly to reduce the ambition of what to include in a single unit of teaching.

References

Aulchenko YS, De Koning DJ, Haley C. GRAMMAR: a fast and simple method for genome-wide pedigree-based quantitative trait loci association analysis. *Genetics* 2007; 107.

Burton PR, Clayton DG, Cardon LR, Craddock N, Deloukas P, Duncanson A, et al. Genome-wide association study of 14,000 cases of seven common diseases and 3,000 shared controls. *Nature* 2007 Jun;447(7145):661–678.

Chanock SJ, Manolio T, Boehnke M, Boerwinkle E, Hunter DJ, Thomas G, et al. Replicating genotype–phenotype associations. *Nature* 2007 Jun;447(7145):655–660.

Chen J, Zheng H, Bei J-X, Sun L, Jia W, Li T, et al. Genetic Structure of the Han Chinese Population Revealed by Genome-wide SNP Variation. *The American Journal of Human Genetics* 2009 Dec;85(6):775–785.

de Los Campos G, Hickey JM, Pong-Wong R, Daetwyler HD, Calus MPL. Whole genome regression and prediction methods applied to plant and animal breeding. *Genetics* 2013 Feb;193(2):327–345.

Devlin B, Roeder K. Genomic control for association studies. *Biometrics* 1999 Dec;55(4): 997–1004.

Garrick DJ, Taylor JF, Fernando RL. Deregressing estimated breeding values and weighting information for genomic regression analyses. *Genetics Selection Evolution* 2009;41(1):55.

Kang HM, Sul JH, Service SK, Zaitlen NA, Kong S-Y, Freimer NB, et al. Variance component model to account for sample structure in genome-wide association studies. *Nat. Genet.* 2010 Apr;42(4):348–354.

Kang HM, Zaitlen NA, Wade CM, Kirby A, Heckerman D, Daly MJ, Eskin E. Efficient control of population structure in model organism association mapping. *Genetics* 2008 Mar;178(3): 1709–1723.

Karlsson EK, Lindblad-Toh K. Leader of the pack: gene mapping in dogs and other model organisms. *Nature Reviews Genetics* 2008 Sep;9 (9):713–725.

Li S, Sanna S, Maschio A, Busonero F, Usala G, Mulas A, et al. The GLUT9 gene is associated with serum uric acid levels in Sardinia and Chianti cohorts. *PLoS Genetics* 2007;3(11):e194.

Manolio TA, Rodriguez LL, Brooks L, Abecasis G, Ballinger D, Daly M, et al. New models of collaboration in genome-wide association studies: the Genetic Association Information Network. *Nat. Genet.* 2007 Sep;39(9):1045–1051.

McCarthy MI, Abecasis GR, Cardon LR, Goldstein DB, Little J, Ioannidis JPA, Hirschhorn JN. Genome-wide association studies for complex traits: consensus, uncertainty and challenges. *Nature Reviews Genetics* 2008 May;9 (5):356–369.

Meuwissen T, Hayes B, Goddard M. Prediction of total genetic value using genome-wide dense marker maps. *Genetics* 2001 Apr;157(4): 1819–1829.

Price AL, Patterson NJ, Plenge RM, Weinblatt ME, Shadick NA, Reich D. Principal components analysis corrects for stratification in genome-wide association studies. *Nat. Genet.* 2006 Jul;38(8):904–909.

Purcell S, Cherny SS, Sham PC. Genetic Power Calculator: design of linkage and association genetic mapping studies of complex traits. *Bioinformatics* 2003 Jan;19(1):149–150.

Purcell S, Neale B, Todd-Brown K, Thomas L, Ferreira MAR, Bender D, et al. PLINK: a tool set for whole-genome association and population-based linkage analyses. *Am. J. Hum. Genet.* 2007 Sep;81(3):559–575.

Samani NJ, Erdmann J, Hall AS, Hengstenberg C, Mangino M, Mayer B, et al. Genomewide association analysis of coronary artery disease. *New England Journal of Medicine* 2007 Aug;357 (5):443–453.

Svishcheva GR, Axenovich TI, Belonogova NM, van Duijn CM, Aulchenko YS. Rapid variance components–based method for whole-genome association analysis. *Nat. Genet.* 2012 Sep;44 (10):1166–1170.

Yu J, Pressoir G, Briggs WH, Vroh Bi I, Yamasaki M, Doebley JF, et al. A unified mixed-model method for association mapping that accounts for multiple levels of relatedness. *Nat. Genet.* 2006 Feb;38(2):203–208.

Zhang Z, Ersoz E, Lai C-Q, Todhunter RJ, Tiwari HK, Gore MA, et al. Mixed linear model approach adapted for genome-wide association studies. *Nat. Genet.* 2010 Apr;42(4): 355–360.

17 Molecular Genetics Techniques and High Throughput Technologies

Wen Huang
Department of Genetics, North Carolina State University, NC, USA

Central dogma of molecular biology

DNA, RNA, and proteins are three major classes of biological molecules that regulate cellular functions and carry genetic information passed from parents to offspring. The central dogma of modern molecular genetics states that, with several exceptions, the biological information encoded in DNA generally flows from DNA to RNA, then to proteins (Figure 17.1). This process provides a means for cells to translate sequence information in DNA into biochemically functional units. In particular, the nucleic acids DNA and RNA are the central and most commonly studied biological molecules in animal genetics because:

- DNA encodes what "versions" of genes (alleles) an organism inherited from parents. It determines the relatedness and uniqueness of an organism with respect to its peers.
- RNA, as a necessary intermediate from DNA to functional proteins, is a window to the activity of genes. It tells us when and where a gene becomes active and how active it is.
- DNA and RNA are relatively simple molecules and are easy to analyze.

In this chapter, we are primarily concerned with the analytical techniques used to identify and manipulate DNA and RNA in animal molecular genetics. Current advances in molecular genetics make the present the best time to study animal molecular genetics. Since the beginning of this century, we have determined the sequences of many animal genomes that are relevant to our daily life including cattle, sheep, pigs, and chickens, among others (see a comprehensive list of sequenced genomes at the National Center for Biotechnology Information (NCBI) website: www.ncbi.nlm.nih.gov/nucgss/?term=sequenced%20genomes). We have the necessary technologies to quickly and accurately analyze almost all aspects of biological molecules. We also have powerful computers and software tools to help us interpret molecular data.

Review of properties of nucleic acids

DNA (**D**eoxyribo**N**ucleic **A**cid) molecules within cells are highly organized and extremely long polymers whose building blocks are small monomers called deoxyribonucleotides (Figure 17.2). For example, the X chromosome of cattle is a chain of approximately 149 million deoxyribonucleotides. There are three major components of a deoxyribonucleotide, a phosphate group, a pentose sugar deoxyribose, and a nitrogenous base. The phosphate and sugar form the backbone of a deoxyribonucleotide and are the same for all nucleotides. Different nucleotides only differ by the nitrogenous bases they contain. There are four major bases found in DNA, adenine (abbr. A), cytosine (C), guanine (G), and thymine (T). In fact, deoxyribonucleotides are often loosely referred to by the bases they carry. Deoxyribonucleotides are joined end to end through phosphodiester bonds, forming a strand of DNA (Figure 17.2). In cells, DNA exists as a pair of strands that are complementary to each other. When DNA strands are paired, the phosphate and pentose backbones face outwards, while bases are inside and facing each other. Because of their chemical conformation, A always pairs with T, and C always pairs with G through non-covalent hydrogen bonds. These base-pairing rules ensure that strands of DNA faithfully complement each other and that identical information is copied when DNA is replicated. The two ends of a DNA strand are not uniform, thus DNA strands have directionality. The phosphate end of a DNA strand is called the 5′ end because the phosphate group is attached to the fifth carbon of the sugar ring, whereas the hydroxyl end of a DNA strand is called the 3′ end (Figure 17.2).

RNA (**R**ibo**N**ucleic **A**cid) shares a very similar chemical composition with DNA except that there is an additional hydroxyl group on the pentose ring (ribose instead of deoxyribose because of the extra oxygen atom) of the monomer ribonucleotide and that uracil base instead of thymine is found in RNA. Nonetheless, uracil has a similar structure to thymine so it forms hydrogen bonds with adenine. Unlike DNA, RNA typically exists as a single stranded molecule, though extensive intra-molecule base-pairing often occurs, thereby causing "folds" and "hairpins" of the RNA molecule. Because of the phosphate group they carry, both DNA and RNA are acids and are negatively charged when they are in a neutral solution such as water.

Cells are compartmentalized, and biological molecules also have their unique cellular distribution related to their functions. The majority of DNA in a cell is located within the nucleus, where it is replicated and transcribed to make RNA. On the other hand, RNA molecules are made and processed in the nucleus and transported to the cytoplasm, where they are translated into proteins.

Molecular and Quantitative Animal Genetics, First Edition. Edited by Hasan Khatib.

As proteins perform many important biological functions, they are distributed throughout the cell according to their roles.

Analytical techniques of biological molecules such as nucleic acids usually take advantage of various aspects of their properties and fall into one of the three broad categories: purification, quantification, and identification.

Purification of nucleic acids from cells

The extraction and purification of nucleic acids from cells involve two major steps. First, cells are broken apart to release molecules they contain, including DNA, RNA, and proteins. Subsequently, nucleic acids are separated from other cellular components and purified (**Figure 17.3**).

Cell membranes are primarily made of phospholipids, which can be disrupted by either mechanical force or detergents. Almost all nucleic acid extraction procedures begin with vigorously shaking cells resuspended in a lysis solution that contains a detergent such as SDS (sodium dodecyl sulfate), a buffer such as Tris base (to keep the pH constant), and salt to provide necessary ion strength. When in contact with detergent, cell membranes, as well as nuclear membranes are dissolved, releasing nucleic acids, proteins, and lipids as a mixture into the lysis solution.

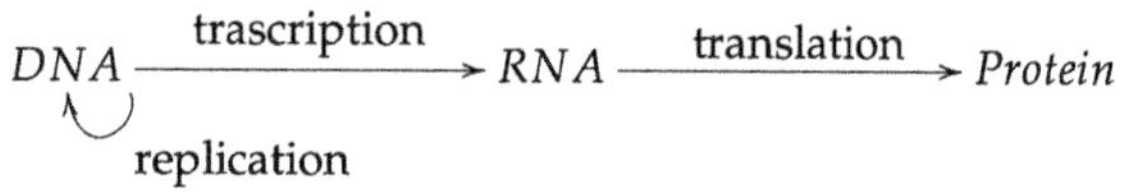

DNA —trascription→ *RNA* —translation→ *Protein*
replication

Figure 17.1 Central dogma of molecular biology. The central dogma of molecular biology is a synthesis made by Francis Crick based on previous discoveries of the steps taken to make protein from DNA. It elegantly describes the direction of information flow in biology. DNA is replicated so information is copied to itself, as depicted by the arrow pointing to itself. RNA is made from DNA through a process called transcription. Eventually, functional proteins are synthesized by translating information in the RNA. There are exceptions to this route of information flow. For example, many viruses including HIV use a type of enzyme called reverse transcriptase that can make DNA molecules using their RNA as a template. Animal genomes also contain elements known as retrotransposons that are only capable of moving themselves in the genome by using the reverse transcription of an intermediate RNA product. However, many retrotransposons lose their mobility due to mutations and are often found as dormant remnants within the genome.

There are several ways to separate nucleic acids from other components of the cellular mixture. The conventional way is through differential solubility extraction. It takes advantage of the fact that the solubility of nucleic acids, proteins, and lipids is different in water and in the organic solvent, phenol/chloroform (the mixture of phenol and chloroform). For example, to extract DNA from the cellular mixture, neutral or slightly alkaline (pH 7.5–8) phenol/chloroform is added. After mixing and centrifugation, the mixture separates into two phases. On the top is the

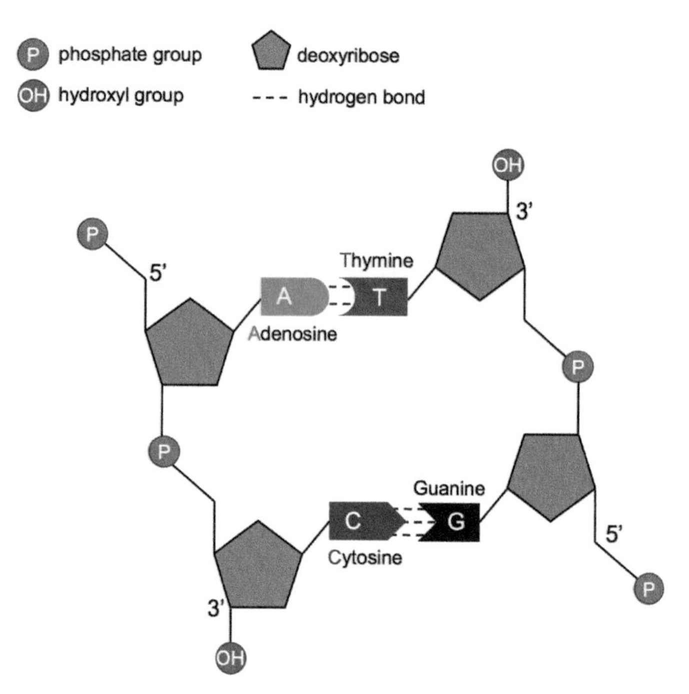

Figure 17.2 Structure of DNA. A DNA strand is made of a stretch of deoxyribonucleotides. Each deoxyribonucleotide is composed of a deoxyribose in the center, a phosphate group attached to the 5th carbon of the deoxyribose ring, a hydroxyl group to the thirrd carbon of the sugar ring, and a nitrogenous base to the first carbon. DNA strands pair up to form a double stranded structure when condition permits hydrogen bond formation. Two hydrogen bonds are formed between the pairing of an adenosine and a thymine while three hydrogen bonds are formed between a cytosine and a guanine. Because there are more hydrogen bonds between a C-G pair, the pairing is chemically stronger than an A-T pair.

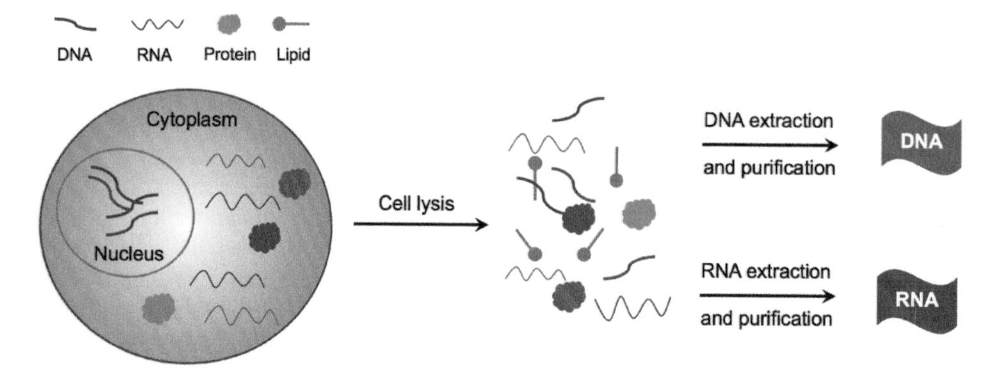

Figure 17.3 Extraction and purification of nucleic acids. Cells are lysed to release a mixture of DNA, RNA, proteins, and lipids. The mixture can then be separated and purified into different classes of biomolecules. Note that lipids are structural components of membranes in cells and that monomer lipids are only shown after membranes are broken.

lighter aqueous phase while the denser organic (phenol/chroloform) phase settles on the bottom. Nucleic acids are polar molecules, so they are much more soluble in the hydrophilic aqueous phase than in the hydrophobic organic phase. On the other hand, lipids and many proteins are hydrophobic so they are more soluble in phenol/chroloform. Other proteins that contain both hydrophilic and hydrophobic groups will precipitate at the interface, or middle boundary, of the two phases. The differential solubilities and chemical properties allow DNA and RNA to stay in the aqueous phase while proteins and lipids are extracted in the organic phase or precipitated at the interface. After removing and storing the aqueous layer, DNA can then be precipitated by adding an equal volume of cold isopropyl alcohol and salt to the aqueous solution. The addition of salt is to neutralize charges on DNA molecules such that the molecules do not repel each other. This allows them to aggregate and precipitate out of solution. The efficiency of RNA precipitation is lower thus DNA is enriched and relatively purified. For most applications, residual RNA does not interfere. When the purity of DNA is essential, it is not uncommon to remove RNA by RNase digestion.

RNA extraction is very similar to that of DNA except that a low pH phenol is used. The acidity and ion strength of phenol is critical in determining the solubility of DNA in phenol. As pH decreases, the solubility of DNA in phenol increases. Hence when a low pH phenol is used, the aqueous phase will contain much more RNA than DNA.

Alternatively, nucleic acids can be selectively absorbed to certain silica surfaces at specific pH ranges. Therefore, when the cellular mixture is passed through a solid silica filter, nucleic acids bind to the filter whereas proteins, lipids, and other cellular components flow through. The bound nucleic acids on the silica filter can then be eluted using a solvent (such as water) at a different pH, in which nucleic acids are soluble.

Determining the quantity and purity of nucleic acids

Following nucleic acid purification, the quantity and purity of nucleic acids must be assessed before they are further analyzed. In theory, one can weigh dry nucleic acids to determine the quantity; however, this is rarely practiced because they often come in extremely small amounts. For example, there is only approximately 6 μg of DNA in 1 million bovine cells (6×10^{-12} g amounting to 6 pg per cell). In a laboratory setting, the quantity of nucleic acids is almost always indirectly determined by measuring the concentration of a nucleic acid solution. Mass can then be calculated by multiplying concentration and volume.

UV spectrophotometry

Molecules absorb light based on chemical properties. Molecules with different chemical compositions absorb light at different wavelengths. The pattern of light absorption over a range of wavelengths is called an absorbance spectrum and can be measured by a spectrophotometer (Figure 17.4). DNA and RNA both absorb light efficiently at a wavelength around 260 nm but absorb less light at 230 and 280 nm (the range of ultraviolet [UV] light). The amount of nucleic acids present in the solution is proportional to the absorbance by a constant factor. A solution of DNA or RNA that absorbs one unit of UV at 260 nm (A_{260}) contains approximately 50 ng/μL of DNA or 40 ng/μL of RNA respectively. This allows one to quantitatively determine nucleic acid concentration using UV spectrophotometry (Figure 17.4).

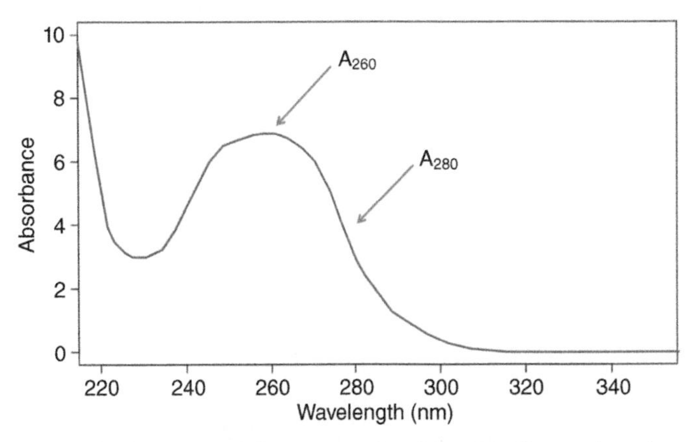

Figure 17.4 UV spectrophotometry of nucleic acids. The UV absorption spectrum of pure nucleic acid solution has a peak at 260 nm and low absorption at 230 and 280 nm. The concentrations of DNA and RNA can be calculated using the formula: [DNA] = $A_{260} \times$ 50 ng/μL, and [RNA] = $A_{260} \times$ 40 ng/μL.

In addition to nucleic acids, contaminants such as proteins and organic solvents carried over from the purification procedure also absorb UV light. Contaminants interfere with quantitation (artificially inflating A_{260}) and downstream analyses that require purity of nucleic acids. UV spectrometry therefore provides a simple way to also check the purity of nucleic acid preparations. For example, proteins absorb a significant amount of UV light at a wavelength of 280 nm but not 260 nm. If a DNA solution contains proteins, its A_{260}/A_{280} ratio will be smaller. It has been empirically determined that a pure DNA solution should have an A_{260}/A_{280} ratio around 1.8 and a pure RNA solution around 2.0 because RNA absorbs more light. Therefore, the deviation of the A_{260}/A_{280} ratio of a nucleic acid preparation from its expected value is indicative of contamination.

Fluorometry

Alternatively, there are fluorescent dyes that specifically bind DNA and RNA. For example, PicoGreen is a proprietary double stranded DNA (dsDNA) specific fluorescent dye that does not bind RNA or proteins. Like UV absorbance, the amount of fluorescence is proportional to the amount of DNA in the solution. One can measure fluorescence of a DNA solution using a fluorometer after addition of PicoGreen and determine its concentration accordingly.

Gel electrophoresis

Finally, when nucleic acid samples are abundant, gel electrophoresis can be used to determine the concentration and quality (Figure 17.5). Nucleic acid molecules are charged and move in a predictable way when electric field is present. In an agarose gel, which is a solid matrix of highly connected agarose molecules, smaller molecules move faster between the pores of the agarose matrix than larger molecules. This difference in velocity of movement separates molecules according to their sizes. For detection and visualization, nucleic acids can be stained with fluorescent dyes such as ethidium bromide, which specifically detects nucleic acids (Figure 17.5). The amount of fluorescence increases as the

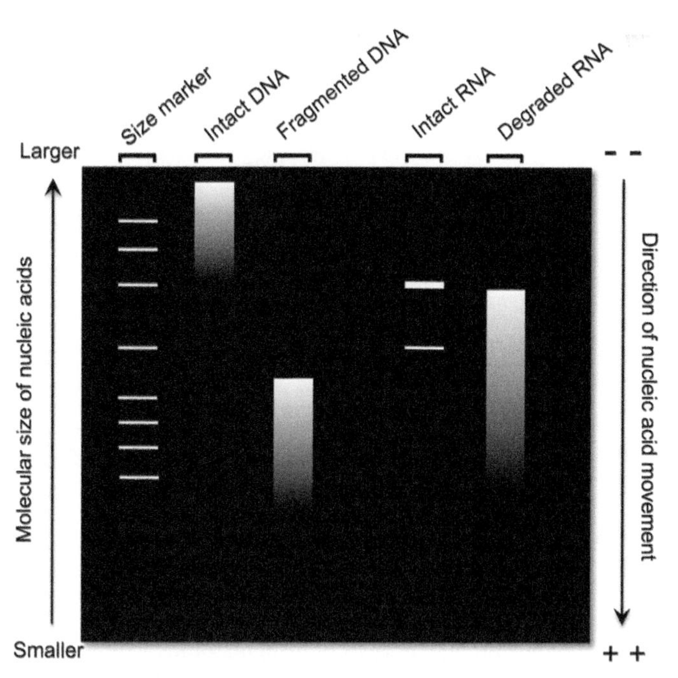

Figure 17.5 Gel electrophoresis of nucleic acids. In a gel electrophoresis, nucleic solutions are added to one end of an agarose gel. An electric field is applied such that negatively charged nucleic acid molecules move from the negatively charged cathode to the negatively charged anode. A size marker is usually included by the side of the nucleic acid samples as a reference for interring sizes of nucleic acids. The gel is typically stained with ethidium bromide, which is a fluorescent dye that binds to double stranded nucleic acids. In a laboratory, the stained gel is visualized in a dark room under a UV light so that only nucleic acids are visible. The bright bands and smears on the dark background represent nucleic acids in the gel. A tight band is many nucleic acids molecules of the same size packed at the same position of the gel. A smear along the path of nucleic acid movement is simply many bands that cannot be easily distinguished. For example, sizes of molecules in genomic DNA or degraded RNA vary extensively so they are manifested as smears. Although RNA are single stranded molecules, there are extensive internal base-pairing, which makes them detectable by ethidium bromide.

amount of nucleic acids increases. One can then compare the amount of fluorescence of a nucleic acid sample to some reference sample whose concentration is known to estimate concentration.

Gel electrophoresis is also commonly used to quality check nucleic acids. Qualities of nucleic acid preparations are crucial for their analyses. Sizes of nucleic acids are very informative of their quality. For example, genomic DNA consists of very long molecules and moves very slowly in the gel. If fragmentation occurs, DNA molecule sizes decrease thus they move faster in the gel (Figure 17.5). Because the majority (>95%) of cellular RNA molecules is cytoplasmic ribosomal RNA, ribosomal RNAs dominate the representation of RNA in a gel as two bands (two subunits of ribosomal RNA). When RNA molecules are degraded during the preparation, their sizes decrease and appear as a small, continuous smear on the gel.

Polymerase chain reaction (PCR)

Polymerase chain reaction or PCR is one of the most revolutionary inventions in the history of molecular biology. PCR is often an intermediate step to enrich for a critical mass of homogeneous DNA molecules carrying a specific sequence (Figure 17.6). The principle of PCR is as follows. First, double stranded DNA templates are denatured under high temperature. This process breaks hydrogen bonds between the base pairs and turns DNA into single stranded molecules. Second, the temperature of the reaction is reduced to permit DNA base-pairing again, a process called annealing. However, by adding many magnitudes of excess of short DNA sequences, called primers, template DNA is much more likely to base pair with primers than its initial partner. There are two primers added, each will bind to one of the DNA strands in opposite directions. Third, as soon as primers form stable base pairs with single stranded template, DNA polymerase begins adding free deoxynucleotides to the end of the primer, extending the newly synthesized DNA strand. Only the nucleotide that can pair with the next base (i.e., A with T, or C with G) on the complementary strand can be added sequentially. These three basic steps, denaturation, annealing, and extension constitute one cycle of a PCR reaction and double the amount of DNA sequences encompassed by the primers. The "chain reaction" amplification is achieved by repeating this cycle many times. Therefore in the next cycle, DNA products from the previous cycle can serve as new templates, further doubling the amount of specific DNA. If N cycles of PCR are performed, there will be 2^N times as many molecules as in the starting DNA materials when substrates needed to make new DNA remain abundant. In other words, the growth of the amount of amplified DNA is exponential. The specificity of PCR is provided by the specific sequences of deoxynucleotides in primers. By designing sequences of primers, PCR can be used to amplify only the regions flanked by the two primers during the annealing step.

Conventional PCR reactions complete all necessary cycles before the yield of DNA is evaluated by gel electrophoresis. In recent years, a new technology called real-time PCR is also commonly used to monitor PCR reaction and obtain quantitative measurements of the amount of DNA templates. Real-time PCR is essentially a combination of PCR amplification and nucleotide quantitation. It measures the quantity of DNA by using double stranded DNA specific fluorescent dyes such as SYBR Green. SYBR Green is chemically similar to PicoGreen (as introduced previously for DNA quantification) except that it is resistant to high temperatures, making it more suitable for PCR. A real-time PCR instrument uses an optical monitor to record the amount of fluorescence continuously, providing a real time survey of DNA quantity in the reaction (Figure 17.7).

The growth of DNA quantity in a PCR reaction typically has three phases: (1) an exponential phase in which DNA quantity doubles in each cycle; (2) a non-exponential phase when DNA appears to grow linearly due to reduced efficiency of PCR when primers and free nucleotides are consumed; (3) and a plateau phase when no new DNA is made because primers and free nucleotides are depleted (Figure 17.7).

At the beginning of a PCR reaction, the amount of DNA present is very small, therefore the SYBR Green fluorescence cannot be easily detected. At a certain number of cycles, the fluorescence exceeds a critical threshold when they can be reliably detected. This number is called the threshold cycle (C_t) (Figure 17.7). The C_t is predictive of the amount of DNA present in the sample initially. The smaller the C_t, the more DNA the starting sample contains. In this way, real time PCR can be used to determine relative amounts of DNA in samples by comparing their C_ts.

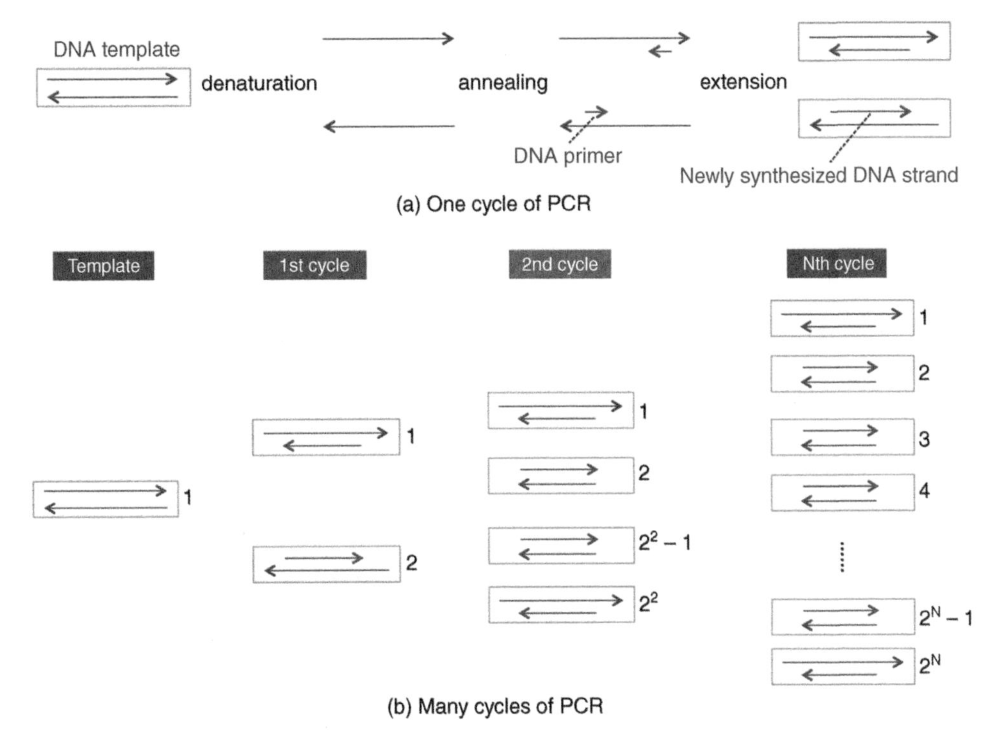

Figure 17.6 Polymerase chain reaction. (a) One cycle of a PCR reaction. Double stranded DNA is denatured to become single stranded and allowed to bind with primers. DNA polymerase extends the DNA strands by adding free nucleotides to the ends of primers. **(b)** Amplification occurs in cycles where newly synthesized DNA from the previous cycle also serves as templates for the next cycle. Therefore the amount of DNA flanked by the two primers grows exponentially.

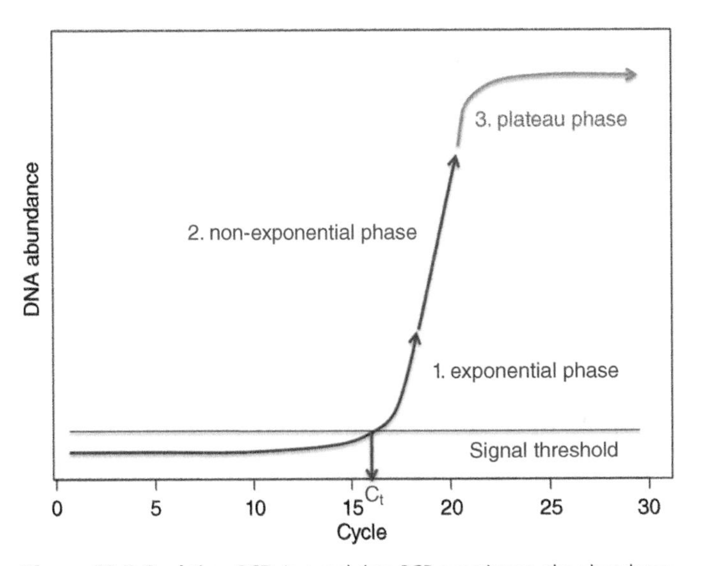

Figure 17.7 Real-time PCR. In a real time PCR experiment, the abundance of DNA is monitored through fluorescence detection. The threshold cycle (C_t) is determined by setting a threshold line right above the baseline fluorescent signal and finding the number of the cycle that intersected with the threshold.

Real time PCR has become increasingly popular for quantitation of specific DNA molecules and finds itself in many molecular genetics applications. For example: by coupling real-time PCR with reverse transcription, a process that turns RNA into DNA, it can be used to rapidly quantify the expression of specific genes.

Determining the identity of DNA

The identity of a DNA molecule is the specific arrangement of its nucleotide bases, or simply its sequence. Determination of the identity of DNA is one of the most important analyses in genetics. The DNA sequence tells us what a gene looks like, how versions of genes differ between individuals and between species, and what product is to be expected from the gene, among many other types of information.

Chain termination method

The chain termination sequencing (also called Sanger sequencing) of DNA was developed by Frederick Sanger in the 1970s. It takes advantage of the sequential manner of template-based DNA polymerization, that is, a nucleotide can only be added until the one preceding it has been added to the chain. Sanger sequencing is similar to a PCR reaction but is unique in several ways. It requires a DNA template, the sequence of which is to be determined; just one primer (as opposed to two in PCR) coming from the strand of the template to be sequenced, free nucleotides for extension from the end of the primer, and DNA polymerase (Figure 17.8). Unlike in a PCR reaction, in which all free nucleotides are deoxynucleotides, Sanger sequencing uses a small fraction of dideoxynucleotides that are fluorescently labeled. Each of the four fluorescent dideoxynucleotides (A, T, C, G) emits a unique color when fluorescing. While deoxynucleotides possess a 3′ hydroxyl group (Figure 17.2), permitting addition of nucleotides to the 3′ end of an elongating DNA chain, dideoxynucleotides lack this hydroxyl group. Because only a small fraction of free nucleotides are dideoxynucleotides, DNA chains are still

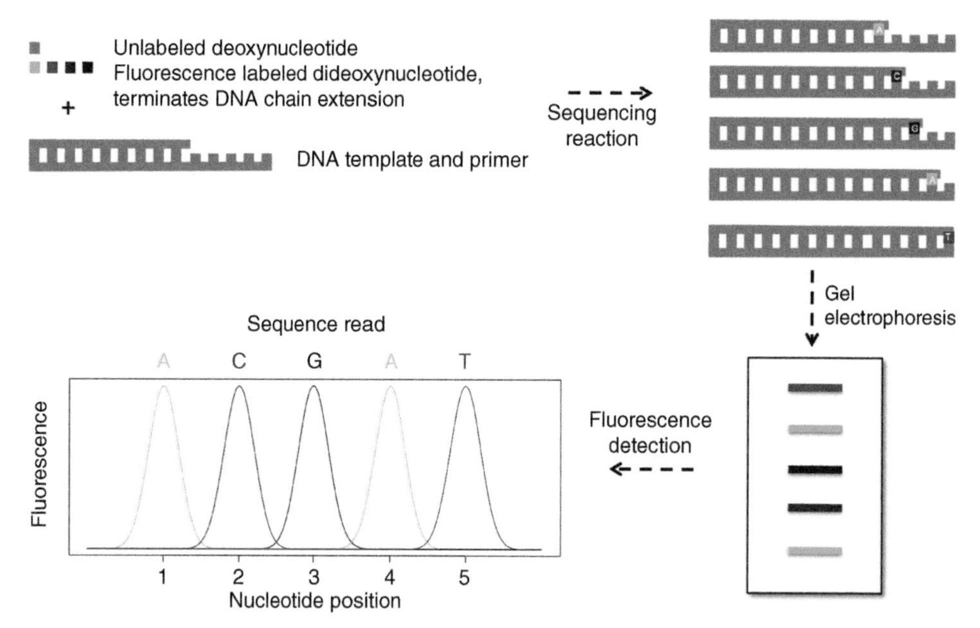

Figure 17.8 Chain terminator sequencing of DNA. Chain terminator (Sanger) sequencing takes place in a mixture of (1) DNA template to be sequenced; (2) DNA primer from where the sequence begins; (3) free deoxynucleotides and fluorescently labeled dideoxynucleotides; (4) DNA polymerase; and (5) necessary pH buffering agents and ion strength. Shown is a DNA fragment with the sequence ACGAT following the primer. The sequencing reaction results in five unique DNA fragments, each representing one of the bases in the sequence and is labeled at the end by fluorescent dideoxynucleotides. The fluorescently labeled fragments are resolved by gel electrophoresis. The fluorescent signals can then be read and processed into a sequence read.

allowed to extend until a dideoxynucleotide is incorporated to the 3'end, which terminates the DNA chain extension (Figure 17.8). This generates a mixture of DNA chains of variable lengths, each stopping at a fluorescently labeled dideoxynucleotide. This mixture can then be fractionated by gel electrophoresis, which separates DNA fragments according to their sizes. For example, a DNA chain that stops at its 100th nucleotide position will migrate farther than the chain that stops at its 101st position. In the meantime, the florescent colors of the bands in the gel correspond to the identity of the nucleotide bases. Therefore we can determine both the relative positions and the identities of nucleotide bases.

Restriction Fragment Length Polymorphism (RFLP) for variant genotyping

In most animal genetics applications, what matters is not the full sequence of DNA but rather differences among individuals in a population. Among individuals within the same species, genome sequences (or the complete DNA sequence of the animal) only differ by less than 1% on average. This means the majority of DNA sequences are shared among individuals, thus the full sequences of an animal's genome contains a lot of redundant information. There is a simple and inexpensive method to accurately assign DNA bases to individuals at positions where they are known to differ. Such positions are said to be polymorphic in a population and the most abundant type is single nucleotide polymorphisms or SNPs where individuals differ by single nucleotides (as opposed to consecutive changes of two or more nucleotides). Restriction enzymes are bacterial endonucleases that cut DNA internally at specific sequence motifs called restriction sites. When a polymorphism such as SNP overlaps with a restriction site, it changes the restriction site such that it can no longer be recognized by the restriction enzyme. Therefore, when DNA molecules flanking the SNP from diverse individuals are digested, the DNA base polymorphism is turned into a restriction fragment length polymorphism (Figure 17.9).

Concept of parallelization and high throughput assays

A typical large mammalian genome, such as the cattle genome, has approximately 3 billion bases, and contains more than 20,000 genes. Although automatic equipment and simplified and perfected assays have greatly improved the efficiency of analyzing biological molecules, traditional methods of analyzing one molecule at a time do not scale up practically. For example, a Sanger sequencing assay can read up to 1000 DNA bases in 2 hours. It takes 3,000,000 non-overlapping sequencing assays or 6,000,000 hours to read one genome only once.

There are at least two ways one can accelerate such processes. One is to develop assays that run extremely fast so that it takes a shorter time to analyze individual molecules. The other is to run millions of assays simultaneously thus within a unit time, more molecules can be analyzed. The latter is called parallelization and is an important concept in modern high throughput assays. It turns out that it is a lot easier to parallelize assays than to shorten them with our current technologies. In fact, speed is often sacrificed to achieve better parallelization. Parallelization of molecular biology assays shares great analogy with parallel computing in which computational jobs are distributed to a large number of computing units to execute.

However, the size or scale of an assay limits parallelization. Even if one could parallelize millions of assays, an immediate problem is how to fit such a large number of assays into a reasonable workspace. Moreover, the amount of reagents needed to run

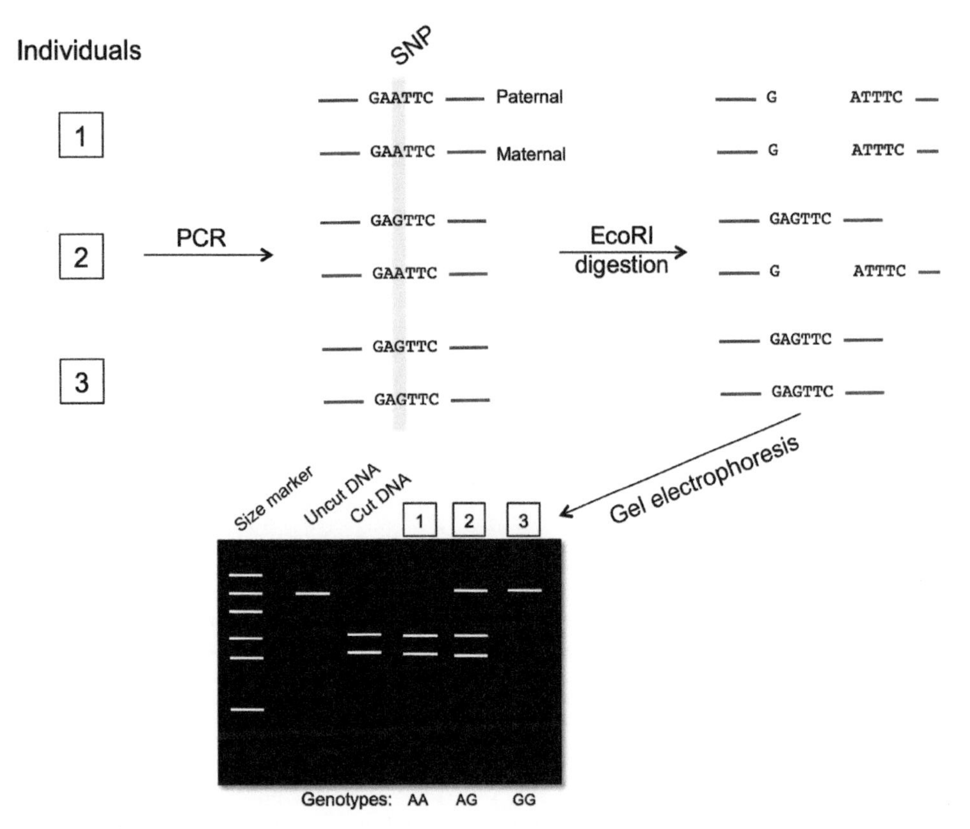

Figure 17.9 RFLP genotyping of DNA. Using PCR, DNA flanking a polymorphism, in this case a SNP, is amplified for three individuals. Each individual has two DNA fragments amplified, each coming from either the paternal or maternal chromosome. The DNA have identical sequence among the individuals except at the SNP site, where the first individuals have a genotype AA, that is, A for both the paternal and maternal chromosomes, individual 2 AG and individual 3 GG. The DNA molecules are digested by EcoRI, which recognizes a six-base motif GAATTC. The digestion products are resolved by electrophoresis. When the SNP genotype is AA, both the paternal and maternal alleles are digested, giving rise to two digested fragments. When the SNP genotype is AG, only one of the alleles is digested therefore both cut and uncut DNA are present. When the SNP genotype is GG, neither of the alleles is digested so the DNA stays intact.

so many assays would be enormous and expensive. An obvious solution is to reduce the size of the assays and use less reagents. This again shares analogy with development in the computing industry where more and more computing power is manufactured into smaller and smaller units.

The remainder of this chapter will introduce how parallelization and scale reduction, particularly within microarrays and next generation sequencing technologies, have been applied to transform biological research.

Microarray technology

Let us take a look at how one could quantitatively measure the messenger RNA expression of one gene in a homogeneous population of cells before microarray technology was introduced. We have briefly mentioned previously that real-time PCR on complementary DNA can be used to measure RNA expression and that is very commonly done. Nonetheless, we will introduce another hybridization-based method that shares more similarity with microarrays. First, RNA is extracted from cells and purified. Such an RNA sample is in fact a mixture of all mRNAs of all genes and other structural RNAs such as ribosomal RNAs. To separate RNAs, we can fractionate them by gel electrophoresis (Figure 17.10). The size separated RNAs are then transferred and immobilized onto a membrane. This step attaches RNA molecules to the membrane but exposes their bases. Once RNAs are immobilized, we can spread synthetic DNA called probes that are labeled with radioactive elements onto the membrane. The synthetic DNA probes carry a sequence of bases that are complementary to the RNA of interest. When the DNA probes hybridize with complimentary RNA molecules, they stay on the membrane. The more specific RNA molecules for a given gene on a membrane, the more the DNA probes will hybridize thus the higher the radioactivity. This method is called Northern blot and still considered the gold standard of detecting and quantifying specific RNA sequences.

Now imagine that we want to measure expression of 20,000 genes. A Northern blot assay typically takes overnight to complete. If we were to do 20,000 blots sequentially, we might want to do 10 simultaneously every day, it will take more than five years to analyze all of the genes. What if we can do all 20,000 blots simultaneously? A hybridization based DNA microarray does exactly that in principle.

A DNA microarray is a thumb sized square glass chip covered by an array of tens of thousands to millions of spots. Each spot contains short, but sufficiently specific, single stranded DNA oligonucleotides called probes. Different spots bear probes with different sequences. In essence, each spot is equivalent to a Northern blot assay of extremely small scale, such that many small assays can be parallelized on a microarray. If we hybridize a chemically labeled nucleic acid sample with a microarray, each

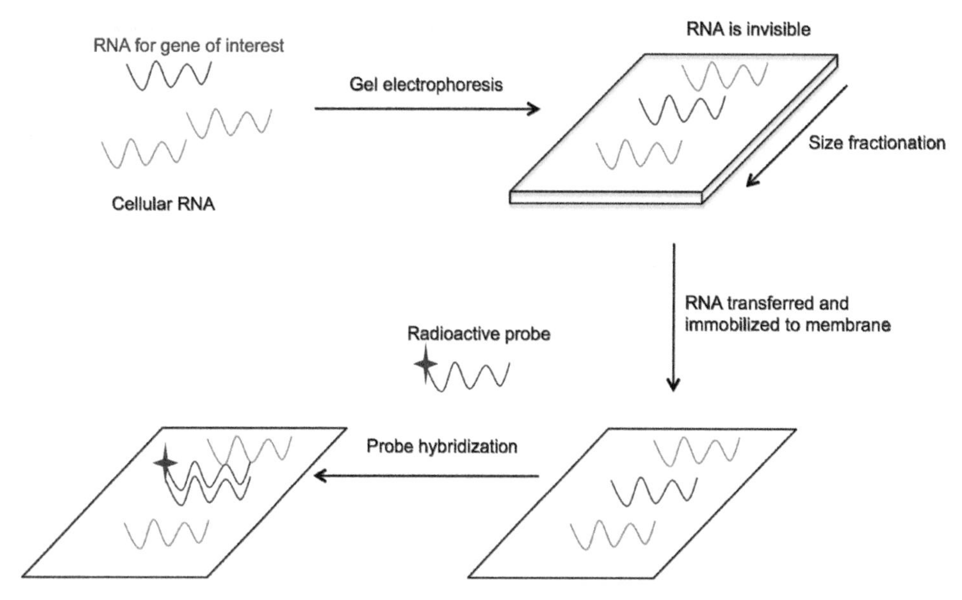

Figure 17.10 Northern blot of RNA. RNA is extracted and size fractionated by gel electrophoresis followed by transfer to a membrane. RNA is immobilized on the membrane and subject to hybridization with radio-labeled DNA probes. The radioactivity can then be visualized by exposing a film or by a phosphor imager.

spot is going to bind to sequences that are complementary to the probes it contains. By measuring how much nucleic acid is bound on each spot, we can then infer how much a certain sequence is present in the sample.

Let us consider a gene expression experiment as an example to understand how microarrays work. Our goal is perform 20,000 gene expression measurements on a microarray.

Array fabrication

Thanks to genome sequencing projects, we have determined the sequences of all genes we know of in the genome, allowing us to design probes to target all of them. DNA probes are chemically synthesized *in situ*, or attached onto the microarray using a special kind of "printer." The probes are arranged as a two dimensional array on the microarray, and we have to remember where we put the probes targeting each gene (Figure 17.11). This step of array fabrication requires expensive equipment and is typically done by companies specialized in microarray manufacturing, such as Affymetrix, Agilent, Illumina, and Nimblegen, among others. Such industrialized manufacturing also improves uniformity of arrays and is particularly good for comparing results done with the same or similar arrays.

Labeling RNA

A key difference between a traditional Northern blot and the tiny Northern blot on a microarray is that the probes are immobilized instead of the RNA sample. This is important because the identity of a sequence corresponds to its physical location on the microarray, which is pre-determined in the array fabrication step. Consequently, instead of labeling probes as in a traditional Northern blot, we label the RNA sample. The RNA molecules we extract from cells are chemically unmodified so we need to attach some measurable signal to them to make them visible. This is performed by two steps of complementary copying. First, complementary DNA (cDNA) is synthesized using the RNA sample as a template. This is similar to taking a mirror image of the RNA sample, albeit the image is coded in DNA so thymines replace uracils in the sequence of the complement. Because the DNA is synthesized using the RNA molecule as a template, the amount of cDNA for a particular gene is proportional to the amount of messenger RNA present in the RNA sample. Finally, the cDNA molecules can then be used as templates to synthesize through *in vitro* transcription of complementary RNAs (cRNAs) that look identical to the cDNA sequence except that the nucleotides are chemically modified to contain fluorescent dyes. The amount of cRNA synthesized will be proportional to the amount of cDNA template present, thus will be ultimately proportional to the amount of RNA present in the original sample. Labeling of the RNA molecules accomplishes two things: first, it converts chemically unmodified RNA into an easily detectable, labeled RNA molecule. Second, it proportionally amplifies the amount of RNA so it is easier to work with and detect.

Hybridization of cRNA

The chemically modified cRNA can now be analyzed on the microarray. We suspend them in solution and spread them on the microarray. The chemically modified cRNA molecules are allowed to move freely on the array surface until they hit a complementary probe at one of the spots on microarray. cRNAs are often randomly fragmented to ease movement on the array and to compensate the fact that only part of the original RNA molecule has been used to synthesize probes. We then wash off extra cRNAs that do not hybridize to any of the probes as well as any molecules that hybridized very loosely to probes with low sequence similarity. If a gene is highly expressed, more cRNA molecules from that gene will be present and hybridize to the probes. This is exactly the same principle used in Northern blots,

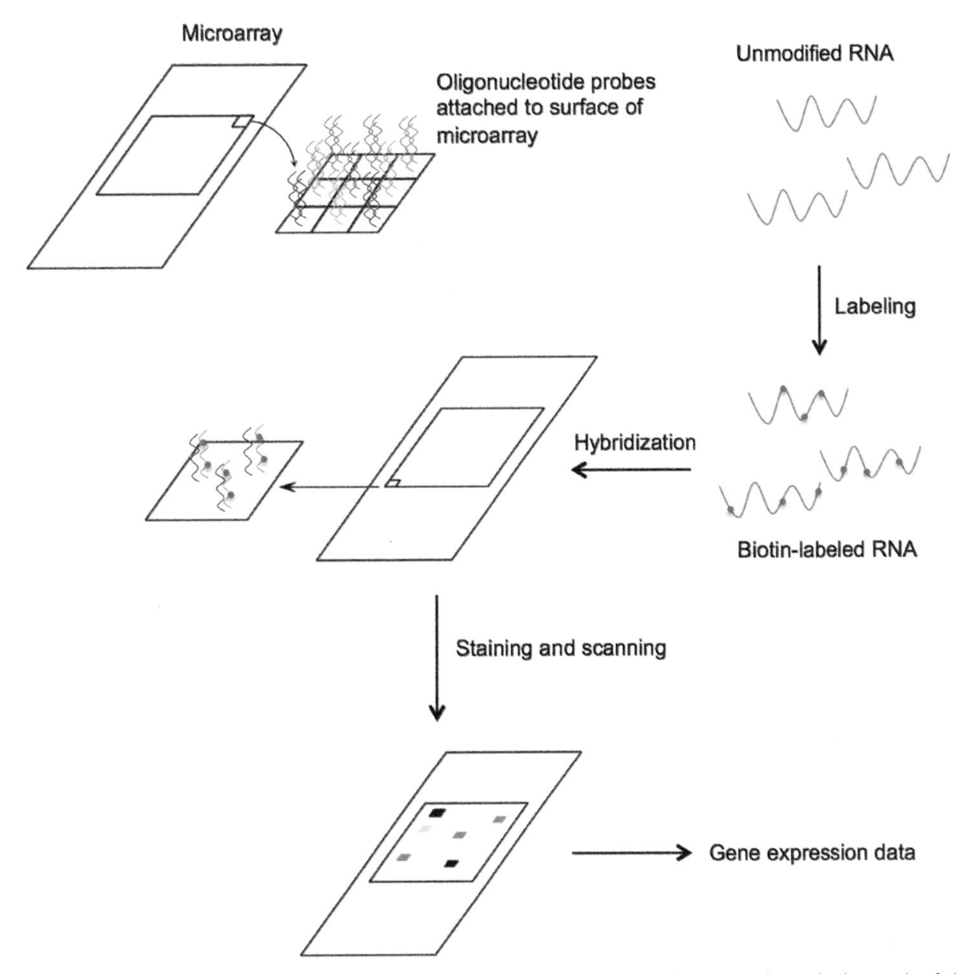

Figure 17.11 Gene expression profiling by microarrays. Specific oligonucletoide probes are synthesized and attached to each of the thousands of spots on the surface of microarrays. RNA samples are labeled and hybridized with the microarray surface. The signal intensity of each spot is then determined and used to estimate the amount of RNA hybridization (thus gene expression). The simultaneous hybridization and detection of many spots on the microarray allow quantification of expression of many genes.

where the level of gene expression is proportional to the amount of hybridization that would occur.

Quantitation of hybridization

To determine the genome-wide gene expression profile, we just need to determine the level of hybridization at each spot as there is a one-to-one relationship between spots on the microarray and genes in the genome. If we have modified the cRNA with fluorescent dye, we can simply measure the intensity of fluorescence at each spot on the microarray. As an alternative to direct dye-labeling, if the cRNA has been modified with a chemical group such as biotin, the microarray can be stained with fluorescent dyes conjugated to streptavidin, a molecule that recognizes and binds to biotin. Measurement of fluorescence can be achieved by taking a high resolution digital photo when the fluorescent groups are excited to emit lights. The digital photo can then be analyzed by specialized computer software to extract fluorescence intensity at each spot.

Other applications of microarrays

The applications of microarrays are not limited to measuring RNA expression. When probe sequences are properly designed, microarrays can be used to genotype DNA with very high throughput. This takes advantage of the fact that signal on each spot of a microarray is determined by the strength of hybridization. Therefore, when a probe is designed to match one allele of DNA, it will not hybridize well with another allele, and vice versa. This signal difference allows one to determine what alleles a DNA sample carry at a particular site in the genome. And many sites can be genotyped simultaneously by parallelizing such hybridization reactions on the microarrays. Additionally, microarrays can also be used to measure the amount of specific DNA bound by certain proteins in chromatin immunoprecipitation assays (termed ChIP-chip).

Next generation sequencing technology

Introduction to next generation sequencing

A representative human genome was sequenced at the beginning of the twenty-first century. That unprecedented effort cost billions of dollars, more than ten years of time, and the cooperation of thousands of scientists from around the world. About a decade

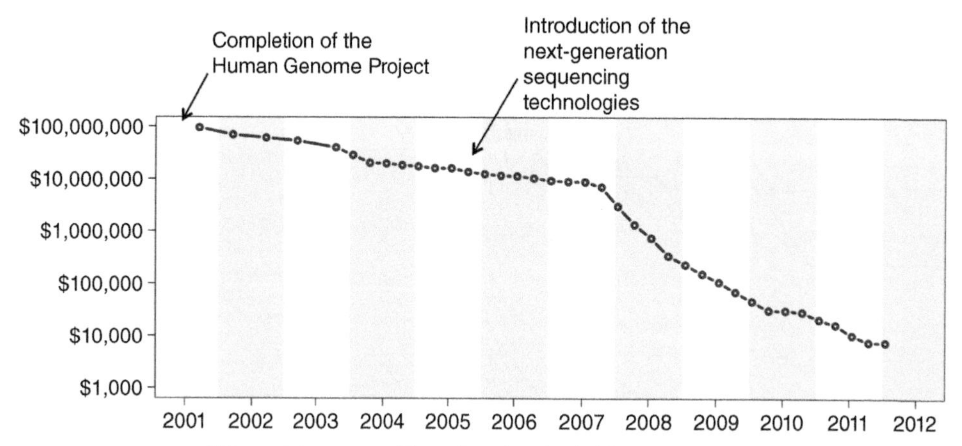

Figure 17.12 Cost to sequence a genome. This graph is generated from data downloaded from www.genome.gov/sequencingcosts/. The cost is plotted at log scale over time. The time marks for the critical events are approximate. The initial human genome sequences were announced in 2000 and published in 2001. At the time, the cost to sequence a human genome was 100 million dollars. Note that the total cost for the Human Genome Project, which started many years before 2001, was much higher than the price in 2001. The first next-generation sequencer (454) was introduced in 2004 and data from that technology was published in 2005. However, next generation sequencing did not become popular until after 2007, when the cost for sequencing a genome began to plunge.

later, we can now sequence a large genome within weeks for less than US$10,000 and the cost is still rapidly declining (Figure 17.12). This has been made possible by the next-generation sequencing (the first generation being the Sanger sequencing) technologies.

There are currently three leading technologies SOLiD, HiSeq/MiSeq, and 454, offered by ABI, Illumina, and Roche, respectively (see a comprehensive poster explaining the chemistry here: www.nature.com/nrg/posters/sequencing/Sequencing_technologies.pdf). While these technologies have their unique sequencing chemistry, they all share the same feature of being massively parallel. These platforms all manage to perform up to hundreds of millions of microscopic scale sequencing reactions on surfaces that are square inches in size. For example, the Illumina HiSeq system uses a process called bridge PCR to generate hundreds of millions of clusters on a surface called a flow cell. Each cluster would then constitute a small sequencing assay, enabling simultaneous sequencing of hundreds of millions of DNA molecules (Figure 17.13). A key step in any of the high throughput sequencing technologies is to make sure that only a single unique DNA molecule is sequenced in one sequencing unit (e.g., a cluster on the flow cell). This is usually achieved by aggressively diluting DNA samples followed by a separation procedure designed to isolate unique DNA molecules from each other. The SOLiD and Roche 454 systems separate DNA molecules by emulsion PCR, in which single DNA molecules are captured by microdroplets of lipids containing primers and sequencing reagents. Illumina sequencing platforms use bridge PCR to generate clusters of molecules. Each cluster contains copies of the original molecule and clusters are spread from each other on the surface of the flowcell.

Unique challenges of next generation sequencing

Next generation sequencing technologies are still being rapidly developed and optimized. The rapid rise in popularity of these technologies in recent years is unprecedented. However, there exist many unsolved issues that are unique to next generation sequencing technologies.

Higher error rate

As compared to traditional Sanger sequencing, the error rate for next generation sequencing is higher. There are generally more steps in a next generation sequencing assay than Sanger sequencing, leaving more room for errors to occur. Additionally, signals from the microscopic scale sequencing reactions in next generation sequencing technologies are more difficult to detect and decode.

Shorter reads

Next generation sequencing platforms often produce short sequence reads. A typical Sanger sequencing assay can produce 800–1000 bases of sequence. With the exception of the 454 platform, which produces sequences around 500 bases long, Illumina and SOLiD platforms typically produce sequences shorter than 200 bases. Short sequences are harder to use for downstream analyses, such as alignments, because it is difficult to determine the genomic origins of shorter sequences.

Data storage and analysis burden

The amount of data next generation sequencers can generate is enormous, amounting to Gigabytes or even Terabytes disk space. This rate of data generation has already presented a significant burden on data storage, especially in large sequencing centers. This has stimulated very active technology development aiming to find better approaches of storing sequence data. Such large data also require extremely powerful computer hardware and software, in addition to skilled bioinformatics analysis personnel.

Applications of next generation sequencing technologies

When coupled with various sample preparation techniques, next generation sequencing technologies find their applications in many areas. A basic and common principle behind the applications is to convert biological signal into DNA sequences, which can then be subsequently decoded rapidly and economically.

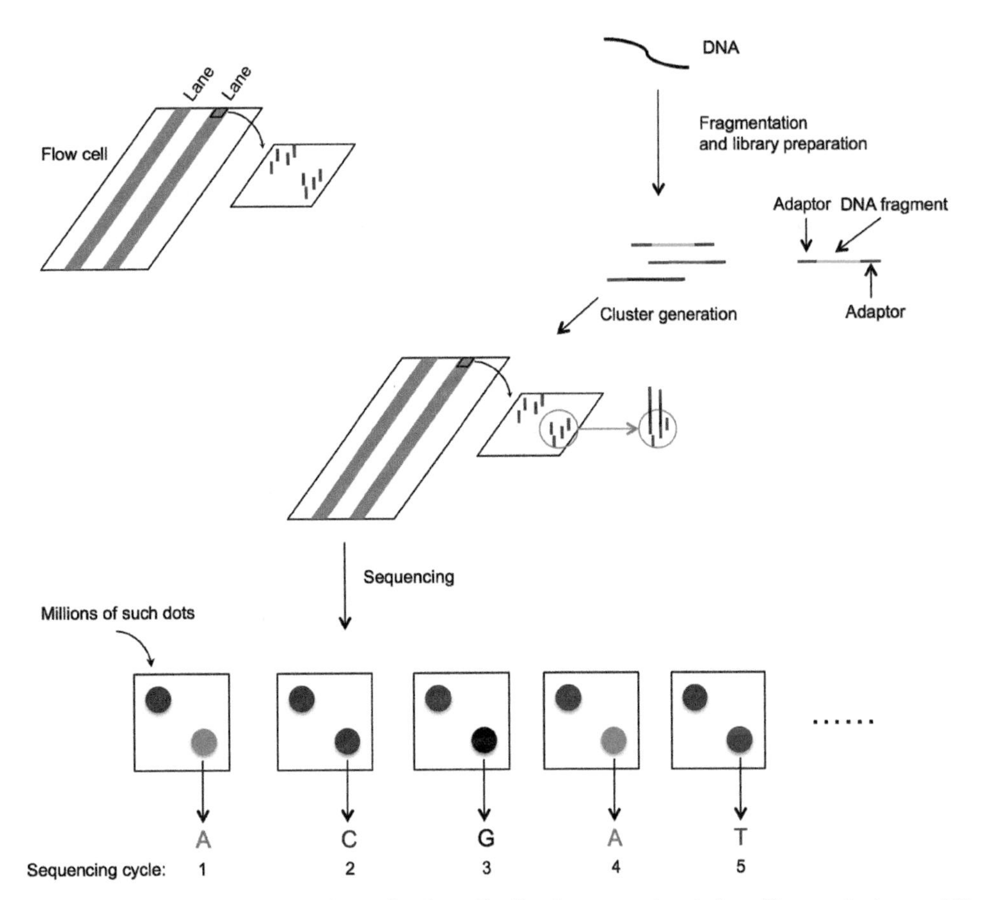

Figure 17.13 Illustration of a next-generation sequencing technology. The Illumina sequencing platform (Genome Analyzer or HiSeq system) is illustrated. Sequencing is operated on a glass slide called flow cell. Each flow cell contains several distinct units called lanes to accommodate multiple samples. There are billions of oligonucleotides attached to the surface of the flow cell. These oligonucleotides are complementary to adaptor sequences that are added to the ends of fragmented DNA samples. DNA fragments are amplified by PCR using primers complementary to adapters and extensively diluted to make a sequencing library, a large collection of fragmented unique DNA molecules. The sequencing library can then be loaded onto the flow cell to be sequenced. Sequencing begins with a process called cluster generation, in which clusters of identical sequences are generated using single molecules as template. This results in scattered spots on the flow cell, where each spot contains many molecules with identical sequence and spots are distant from each other. Illumina sequencing is achieved by sequentially adding fluorescence labeled nucleotides. The nucleotides are chemically modified such that only one nucleotide can be added to the end of a DNA chain at a time. Once the fluorescence labeled nucleotides are added, the flow cell is digitally scanned. The color of the fluorescence determines what nucleotide has just been added. Subsequently, the terminator at the end of the newly added nucleotide is removed to allow extension of additional nucleotide. This process is repeated by a certain number of cycles, thus the sequential arrangements of bases on the template can determined.

Genome sequencing

The most straightforward application of next generation sequencing is for determination of genome sequences. However, most genomes contain chromosomes that are millions of nucleotides in size while sequencers can only read up to several hundreds of them. Therefore, scientists have developed ways to break genomes into small pieces ("shotgun sequencing") and stitch their sequences together with the assistance of computers. There are two distinct approaches in genome sequencing. The first is called *de novo* assembly, in which DNA fragments are arranged by finding overlaps among each other (Figure 17.14). *De novo* assembly using short reads produced by next generation sequencing is particularly challenging as overlap is limited by the length of sequences. The second is called genome resequencing, where a highly similar reference genome of high quality is available and DNA fragments are aligned to the reference genome to determine their origin (Figure 17.14). Genome resequencing is technically easier than *de novo* assembly because the scaffold of the genome is already determined.

Transcriptome sequencing

Commonly referred to as RNA-Seq, transcriptome sequencing has emerged as an alternative to microarrays for measuring genome-wide RNA expression. After RNA is proportionally reverse transcribed into cDNA, they can be sequenced and aligned to the genome. The number of sequences coming from a gene in the genome represents the amount of RNA present in the initial RNA sample, allowing one to quantify the transcriptional activity. Unlike microarrays, RNA-Seq does not require one to know the sequences of genes *a priori*; therefore, it offers great potential to discover novel transcription in a genome. Alternatively, when reference sequences are not available, RNA-Seq sequences can be assembled *de novo* to construct the transcriptome.

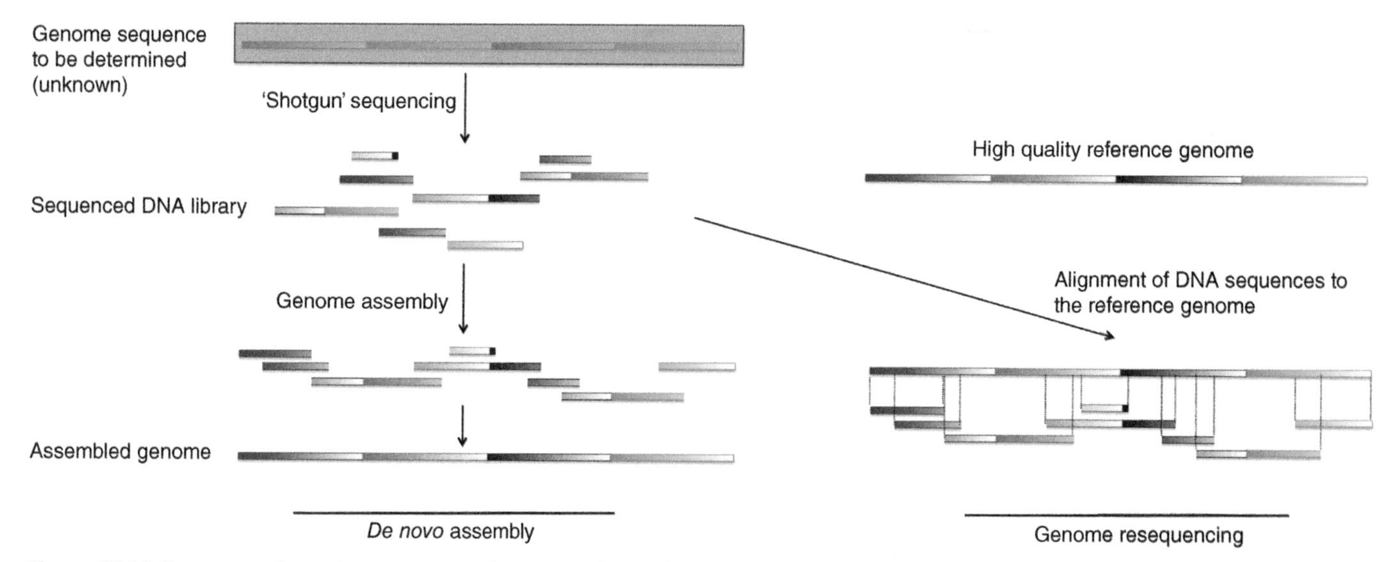

Figure 17.14 Two approaches using next-generation sequencing to determine genome sequences. Genome sequence of an animal can be determined by two approaches. Both approaches begin with fragmenting DNA randomly to generate a library of sequences ("shotgun" sequences). For *de novo* assembly, sequences are merged together in a specific order based on their overlaps. These merged sequences produce an assembly of the unknown genome sequence. Alternatively, when a high quality reference genome is available (such as cattle), shotgun sequences can be aligned to the reference. This takes advantage of the fact that genome sequences of additional individuals are highly similar to the reference genome. The aligned sequences can then be used to identify variations in the genomes of individuals.

Protein-nucleic acid interaction

Through immunoprecipitation of proteins, nucleic acids bound by proteins can be purified. For example, transcription factors interact with DNA sequences to activate transcription. By using antibodies that target specific transcription factors, one can enrich for DNA that co-precipitate with the transcription factors. The amount of a specific sequence of DNA bound by the transcription factors is indicative of the binding affinity and activity. These DNA can then be sequenced to determine their identities and aligned to the genome to determine their origins. This technique is called chromatin immunoprecipitation – sequencing (ChIP-Seq) and is often used to identify regions of the genomes that are being controlled by certain transcription factors.

Summary

Nearly 60 years after Francis Crick and James Watson solved the double helix structure of DNA, molecular genetics research in animal species has advanced significantly. We have at hand almost all the necessary tools to answer important biological questions. In this chapter, principles of several routinely used molecular genetics techniques that exploit the chemical properties of nucleic acids have been introduced. As biology moves from investigating one gene at a time into the era of high throughput investigations, both array and sequencing technologies are also introduced.

Further reading

Only fundamental principles of the molecular genetics techniques are introduced in this chapter. All of the techniques are covered in great detail in more specialized laboratory manuals, textbooks, and review articles, some of which are provided below for further references.

Crick, F. (1970). Central dogma of molecular biology. *Nature*, 227(5258):561–563.

Mardis, E. R. (2011). A decade's perspective on DNA sequencing technology. *Nature*, 470(7333):198–203.

Metzker, M. L. (2010). Sequencing technologies – the next generation. *Nat Rev Genet*, 11(1): 31–46.

Morozova, O., and Marra, M. A. (2008). Applications of next-generation sequencing technologies in functional genomics. *Genomics*, 92(5):255–264.

Rio, D. C., Ares Jr, M., Hannon, G. J., and Nilsen, T. W. (2010). *RNA: A Laboratory Manual*. Cold Spring Harbor Laboratory Press, Cold Spring Harbor, NY.

Sambrook, J. and Russell, D. (2001). *Molecular Cloning: A Laboratory Manual*, 3rd edn. Cold Spring Harbor Laboratory Press, Cold Spring Harbor, NY.

Review questions

It is important to understand chemical properties of nucleic acids and relate them to the analytical techniques. Several key questions are provided to help you understand the materials in this chapter.

1. What properties of DNA ensure the fidelity of replication?

2. What electric charges do nucleic acids carry? Why? In what direction do they migrate when an electric field is applied?

3. List two methods that can be used to determine DNA concentration in a solution and briefly explain.

4. Why is the growth of DNA molecules in a PCR reaction exponential?

5. How is the parallelization in microarray and next generation sequencing achieved respectively?

18 Single Genes in Animal Breeding

Brian W. Kirkpatrick
Department of Animal Sciences, University of Wisconsin–Madison, WI, USA

Introduction

"Single genes" in the title of this chapter is a reference to individual genes whose allelic variants have a sufficiently large effect as to cause discrete categories of phenotypes. Such genes may also be referred to as "major" genes. Phenotypic categories may be two or three in number depending on mode of gene action. Single genes with **dominant** and **recessive** allelic variants will have two phenotypic categories; for example, normal or affected phenotypes for a gene with a recessive allele causing disease in the **homozygote**. Single genes with **additive** or incompletely dominant effects will have three phenotypic categories corresponding to the three possible genotypes (homozygous for allele 1, heterozygous, homozygous for allele 2). Recessive alleles have effects that are masked by dominant alleles and thus require a homozygous genotype for the associated phenotype to be observed. Additive alleles, in contrast, incrementally increase or decrease the phenotype. The remainder of this chapter describes how single genes are mapped and identified (or tagged by linked genetic markers) and provides examples of beneficial or detrimental single genes that have been or may be applied in animal breeding.

Mapping and identifying single genes

Strategies for mapping and identifying single genes will depend greatly on mode of gene action. Mode of gene action must first be assessed by analyzing the available pedigree information for the animals exhibiting the desired (or undesired) phenotype. For example, a phenotype whose appearance skips generations in a pedigree and occurs in individuals whose pedigrees contain inbreeding loops (i.e., the affected individual is related to a common ancestor through both parents) is likely due to the effects of a **recessive** allele. In contrast, the appearance of multiple affected individuals from a common sire with unrelated dams suggests an allele with **dominant**, **incompletely dominant**, or **additive** effect. The putative mode of gene action, frequency of the allele of interest, and availability of lines or breeds fixed for alternative alleles will determine the animals and family structure most useful for effectively mapping the causative gene.

First, let's consider the case of a major gene whose phenotype is associated with an allele that is relatively common, that is, breeds or lines exist in which the alternative alleles occur with high frequency. In this case, the gene can be effectively mapped by phenotyping and genotyping members of a family where one or both parents (as illustrated in Figure 18.1) are heterozygous for the gene of interest.

Heterozygous F_1 generation parents may be created by the crossing of two lines or breeds that are presumed to have alternative homozygous genotypes at the **locus** (i.e., gene) of interest. Mating two F_1s produces an F_2 generation that will display all possible genotypes and phenotypes at the gene of interest. The gene can then be mapped by genotyping the F_2 and the preceding parental and grandparental animals with multiple genetic markers and statistically testing association between marker genotypes and phenotypes. Genotypes of genetic markers on the chromosome containing the gene of interest will show a stronger association with the phenotypic categories of the F_2 offspring, and the strongest association among those is indicative of the likely location of the gene. Alternatively, matings can be made in which one parent is heterozygous and the other homozygous at the gene of interest (a backcross in the case of contrasting breeds or lines or a half-sib family within a population). Results of the mapping analysis of this type for a major gene look like a contour plot where the peak of the mountain corresponds to the most likely location of the gene (Figure 18.2).

A good example of this type of approach and the subsequent analysis of a **positional candidate gene** (gene suspected to cause the phenotype both because of its function and its location relative to the mapping results) is provided by the mapping of the myostatin gene in cattle that is responsible for the double muscling phenotype (Figure 18.3).

Certain breeds of cattle, such as the Belgian Blue and Piedmontese, have a high frequency of an allele at the myostatin locus that causes double muscling, or muscular hypertrophy. Researchers used a backcross design to map the location of the gene that causes the muscular hypertrophy phenotype (Charlier et al., 1995). Their approach exploited the difference in allele frequency between the Belgian Blue (fixed for the muscular hypertrophy allele) and the Friesian (fixed for the wild type allele) breeds. F_1

Molecular and Quantitative Animal Genetics, First Edition. Edited by Hasan Khatib.

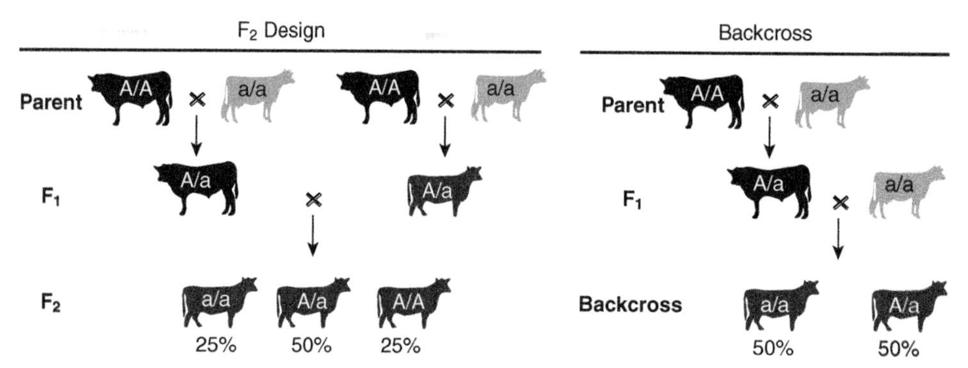

Figure 18.1 Planned mating designs used to map single genes. Different parent breeds are denoted by black and light gray shading and by homozygous genotypes in the top row. Expected frequency of the various single gene genotypes in the F2 and backcross progeny are shown at the bottom.

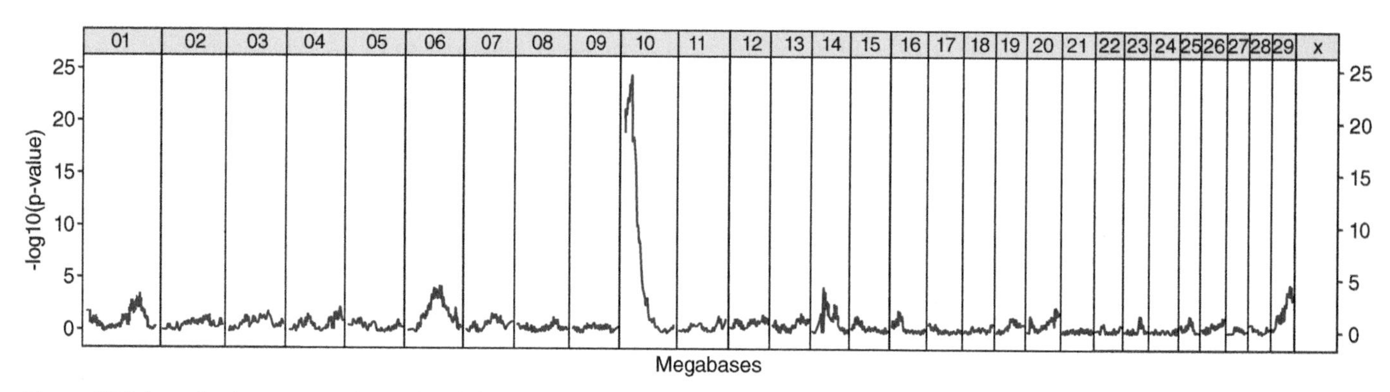

Figure 18.2 Example of a gene mapping analysis (linkage analysis) to identify the chromosomal location of a major gene. The *y*-axis in the figure corresponds to p-value from a statistical test of marker-trait association and the *x*-axis corresponds to marker location. Each vertical panel corresponds to a different chromosome, such that the whole figure represents the entire genome (excepting the X chromosome). A large bovine half-sib family (136 daughters of the same sire) were evaluated for the trait (ovulation rate) and genotyped with a SNP (single nucleotide polymorphism) panel that included 2701 autosomal SNP markers. In simple terms, the analysis groups daughters by alternative SNP alleles inherited from their sire at a given SNP and asks the question, do the two groups of daughters differ in ovulation rate? For SNPs closest to the gene of interest, groups of daughters inheriting alternative SNP alleles from the sire will differ most greatly in ovulation rate. The highest peak (on Chromosome 10) corresponds to the location of the gene being mapped. Source: B.W. Kirkpatrick.

Belgian Blue x Friesian females were mated to Belgian Blue sires and the resulting backcross offspring categorized as normal or double muscled. Genotyping with **microsatellite markers** throughout the genome and testing marker-trait association revealed the gene's location on bovine chromosome 2. The next step of moving from a mapped chromosomal region to identifying the actual gene responsible for the major gene phenotype can be fraught with difficulty; however, in this specific case previous research with mice made the task simpler. Researchers studying growth factors in mice had previously used the technique of gene **knockouts** to study how removing a gene from the mouse genome affected mouse phenotype. Knockout mice for one of these growth factors, myostatin, had an interesting phenotype (McPherron et al., 1997): extreme muscular development (Figure 18.4).

By comparing the bovine gene map with the mouse gene map, the scientists who mapped double muscling in cattle were able to deduce that the region of bovine chromosome 2 containing the double muscling gene corresponded to the location of the myostatin gene in mice. Given the knowledge of myostatin's relationship with muscle development from the mouse knockout studies, and its corresponding location with the double muscling locus in cattle, it became a very strong positional candidate gene for double muscling in cattle. Subsequent examination of the myostatin gene in cattle indeed revealed multiple mutations across breeds with double muscling phenotypes that either altered the gene's expression or resulted in non-functional gene products, either of which would result in a situation equivalent to the mouse myostatin knockout and account for the muscular hypertrophy phenotype.

Sometimes the trait of interest and the allele responsible are relatively rare or the phenotype of interest has a lethal or detrimental effect that precludes the fixation of the causative allele in the population. In these cases, and assuming the phenotype is due to a recessive allele for which affected individuals are homozygous, the trait can be mapped using only a modest number of affected individuals through **homozygosity mapping**. Homozygosity mapping exploits the fact that most individuals with an affected phenotype for a trait due to a rare, recessive allele will appear in pedigrees in which inbreeding has occurred, such that there is a common ancestor that is a carrier of the recessive allele and from whom the affected individual has

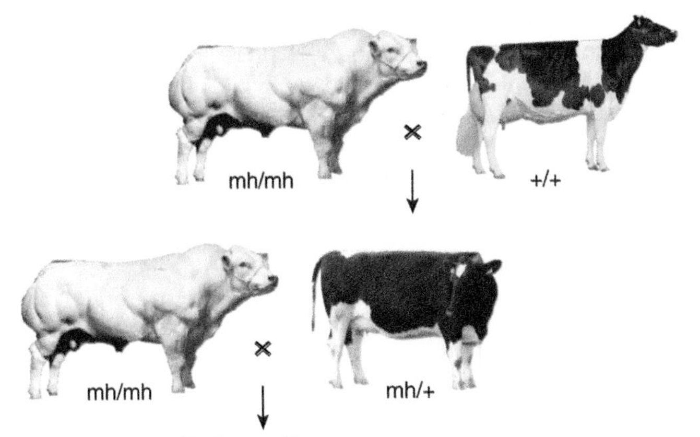

Figure 18.3 Backcross design used in mapping the gene responsible for double muscling in cattle (Charlier et al., 1995). The F_1 parent in the second generation of the pedigree passes either the muscular hypertrophy allele (mh) or normal, wild-type allele (+) on to offspring who are distinguishable by degree of muscularity (double muscled vs. normal). Genotyping genetic markers throughout the genome and subsequent analysis of association between marker genotype and muscling phenotype led to the identification of a region on bovine chromosome 2 containing the gene for muscular hypertrophy.

inherited the allele through both the paternal and maternal sides of the pedigree (Figure 18.5).

The affected individual must be homozygous for the recessive allele at the causative gene and will also be homozygous for a region surrounding the gene. Genotyping multiple affected individuals will lead to identification of partially overlapping regions of homozygosity when comparing across individuals; the gene's location can be quickly narrowed down to that region in which *all* affected individuals are homozygous. Less than a dozen affected individuals with common or related carrier ancestors are sufficient for homozygosity mapping provided the affected individuals are not highly inbred. Being highly inbred increases the length and number of regions of homozygosity, reducing the effectiveness of the mapping effort. Being less inbred implies that the common carrier ancestor is separated by more generations from the affected individual; with each generation recombination events accumulate breaking down the extent of homozygosity. The recent discovery of a mutation causing a genetic disorder in Angus cattle in the USA, arthrogryposis multiplex, provides an example of homozygosity mapping to map a gene with a recessive lethal allele (Figure 18.6).

Ideally, linkage mapping, as in the double muscling example, or homozygosity mapping would result in identification of a narrow genomic region harboring the gene of interest. Consideration of

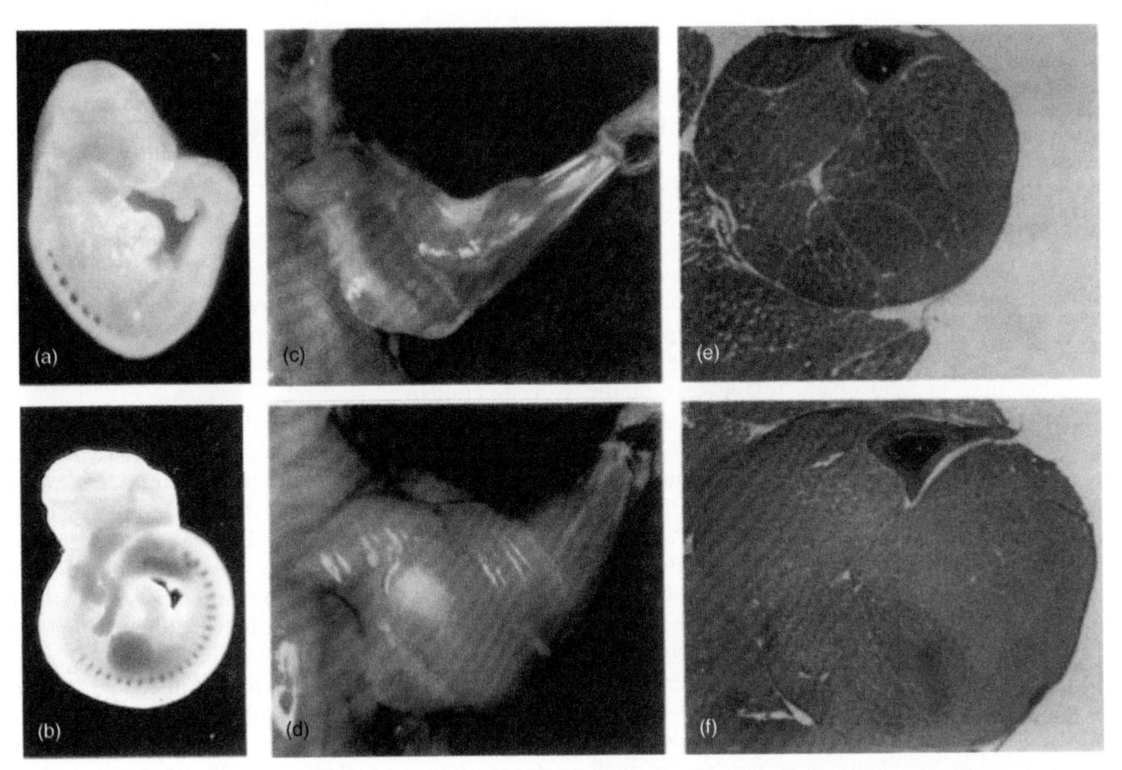

Figure 18.4 Myostatin expression and gene knockout in a mouse model. Myostatin expression (a, b; blue stain) during fetal development on day 9.5 and day 10.5 is specific to the cell types that will become muscle in the adult mouse. Mice missing the myostatin gene (d,f) exhibit dramatic increase in muscularity compared to normal mice (c,e) with a functional myostatin gene. All images from McPherron et al. (1997). Reprinted by permission from Macmillan Publishers Ltd: *Nature* Alexandra C. McPherron, Ann M. Lawler, Se-Jin Lee. Regulation of skeletal muscle mass in mice by a new TGF-beta superfamily member. *Nature*, 387, 83–90. Copyright 1997.

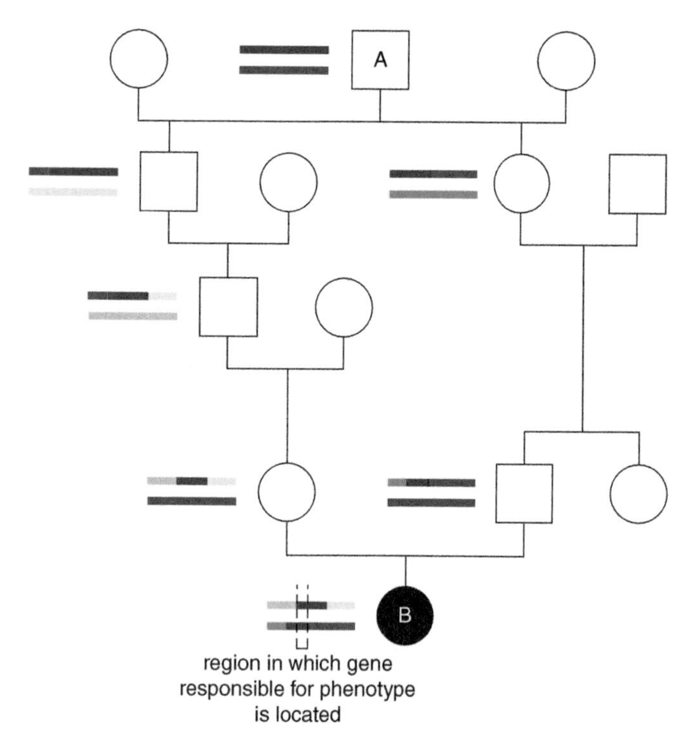

Figure 18.5 Pedigree with inbreeding loop and affected individual (B) useful in homozygosity mapping. The ancestor (A) to which B traces through both paternal and maternal sides of the pedigree is a carrier (blue chromosome) of a detrimental recessive allele. The gene causing the detrimental phenotype associated with the recessive allele in individual B must be located in the narrow region, which is blue in both chromosomes.

the genes in this region would lead to few or possibly a single candidate gene. Subsequent sequence analysis of affected and unaffected or carrier and non-carrier animals will identify the causative DNA alteration. Sometimes, as in the double muscling example, this path is straightforward and that is in fact the result. However, complications can occur which make this process more problematic. It is possible that the narrowed genomic region may reveal no obvious candidate gene or multiple candidates. In the case of the former, it may be that even though our knowledge of the genome sequence may be complete in the species under study, our knowledge of coding sequences, regulatory elements or non-coding genes may be inadequate. In the case of the latter, sequence analysis of the multiple candidate genes in affected and unaffected or carrier and non-carrier animals may reveal multiple DNA alterations (**polymorphisms**) putatively related to the phenotype but with no one clearly identifiable as *the* causative change.

If the causative gene and polymorphism are identified, the causative polymorphism would be used directly as the basis for genetic testing. However, if the causative gene and polymorphism are uncertain, it is still possible to use a linked polymorphism as the **genetic marker**. This is possible so long as the marker is within the same region as the causative locus (up to approximately one million basepairs of DNA) and in high **linkage disequilibrium** with it (close proximity does not ensure high linkage disequilibrium). Linkage disequilibrium can be simply described as a state of non-random association between alleles at two loci. If the linkage disequilibrium is very high (e.g., allele "A" at the genetic marker is always co-inherited with mutant allele at the major gene and allele "B" is always co-inherited with the wild-type allele), then the genetic marker can serve as a useful proxy in selection.

What types of DNA sequence alterations create single gene effects?

Basically, any sequence variation that results in a change in gene expression or functionally significant change in the protein or RNA encoded by a gene could result in a profound phenotypic effect (see Figure 18.7 for a schematic diagram of a gene with some features noted).

Selected examples of these are listed in Table 18.1.

Point mutations, a substitution of one base for another, if in the coding sequence may alter the encoded amino acid, and if so are termed **non-synonymous mutations**. Such non-synonymous mutations may create a less functional or non-functional gene product. Most deleterious are point mutations that cause a codon that previously encoded an amino acid to now encode a stop codon. Such changes result in production of a truncated protein that is most likely non-functional. Not all point mutations are deleterious, however. Because of the redundancy in codon sequences, an amino acid may be encoded by multiple codons (most frequently the third base of the codon may vary); consequently, a point mutation may result in no change in the encoded amino acid (a **synonymous mutation**). Non-synonymous mutations are categorized as **missense** mutations if they result in replacement of one amino acid with another and **nonsense** mutations is they result in a stop codon in place of an amino acid.

Point mutations in non-coding regions can also have profound effects on phenotype by altering gene expression. A point mutation in the boundary area between exon and intron may change RNA splicing leading to a non-functional protein product (splice site mutations could be either in coding or non-coding regions as indicated by Figure 18.7). Gene expression is controlled in part by sequences preceding the gene that promote transcription of an RNA complement of the gene, with transcription promotion being affected by enhancers or repressors that increase or decrease transcription. These DNA sequences are specific and function by binding transcription factors or repressors; alterations to these sequences can cause significant under- or over-expression of a gene's mRNA.

Yet other mutations may affect translation of the gene by altering sequences that are bound by microRNAs. MicroRNAs are short RNA sequences transcribed from microRNA genes whose function is to bind target mRNA sequences leading to their destruction (cleavage) by the cell's RNA interference (RNAi) mechanism. Mutations that alter an existing microRNA binding site or result in the creation of a site can then affect gene expression by altering translation (either by destruction of mRNA or by binding with and tying up the mRNA).

Insertions or deletions of bases can also create alterations with profound effect on phenotype. Insertion or deletion of bases within the coding sequence, not in multiples of three, are **frameshift mutations** that alter the subsequent downstream codon reading for the remainder of the gene. Frameshift mutations dramatically alter the amino acid sequence of the protein and often truncate it as well because of a frameshift-induced stop codon. In addition to small insertions or deletions there can also

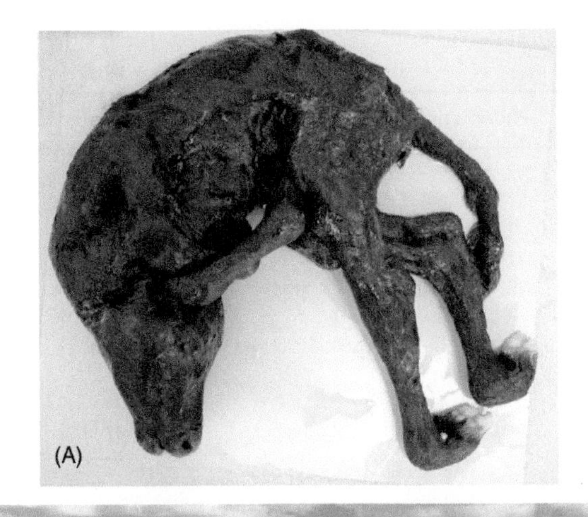

Band 602 of Ideal 928 72 #7530738AMF
Band 234 of Ideal 3163 #8505294
Rito 149 of Ideal 443 7892403
Tehama Bando 155 #9891499AMF
Tehama 72 Rito 330 8404282
Tehama Blackcap G373 8933961
Tehama Blackcap E475 8398440
Band 234 of Ideal 2118 8505264
Rito 9J9 of B156 7T26 #9682589AMC
Rito 149 of J845 7T26 9238034
9J9 G A R 856 10895323
P S Power Play #8974207GDF
Blackbird G A R 833 10461201
Blackbird G A R 137 9162818
Pathfinder

Production						Maternal					
CED Acc	BW Acc	WW Acc	YW Acc	YH Acc	SC Acc	CEM Acc	Milk Acc	MkH MkD	MW Acc	MH Acc	$EN
+10 .92	+1.7 .97	+30 .96	+72 .95	+0 .95	-1.02 .95	+10 .91	+19 .95	1957 5180	+33 .91	+.5 .91	+9.11

Carcass					
CW Acc	Marb Acc	RE Acc	Fat Acc	Carc Grp/Pg	Usnd Grp/Pg
+23 .84	+.39 .85	+.41 .84	-.010 .83	72 256	2188 5341

$Values					
$W	$F	$G	$QG	$YG	$B
+19.69	+19.20	+28.57	+20.92	+7.65	+55.23

(C)

Figure 18.6 By 2008 it became apparent to Angus breeders in the USA that a genetic defect was becoming more prevalent in the breed. The defect manifested as stillborn calves with gross skeletal deformities (**A**: photo courtesy of Laurence Denholm, NSW Dept. of Primary Industries) leading to the common name of Curly Calf syndrome, more properly arthrogryposis multiplex. The disease manifested itself because of the heavy use of a sire, GAR Precision 1680 (B), popular as a trait leader for $Beef value (highlighted in green in C). By using a recently developed genetic marker chip incorporating >50,000 single nucleotide polymorphism (SNP) markers and homozygosity mapping, Dr. Jonathon Beever at the University of Illinois was able to map the genetic defect and identify its basis (a deletion spanning two genes) in less than four months. The defect did not originate in GAR Precision 1680, but could be traced back to his maternal grandsire (carrier ancestors highlighted in yellow). Rapid development of a genetic marker test permitted identification of non-carrier offspring of this elite sire and related animals for subsequent use in breeding.

Schematic diagram of a stereotypical gene

exon intron

—//— CCAAT — TATAA — [ ] — [MAG|GT — YAG|G] —

enhancer sequence | CAT box | TATA box | Donor splice site | Acceptor splice site

Figure 18.7 Schematic diagram of a stereotypical eukaryotic gene showing some of the typical features. Enhancer sequences can be located at variable distances relative to the gene and are variable in sequence motif. CAT and TATA boxes are conserved sequences important for promotion of gene transcriptions (promoter region). Exons denote the regions of the gene that carry the code that will be translated into a protein and are separated by non-coding, intron sequences. The entirety of the gene is initially transcribed into an RNA sequence, which is then processed (spliced) to remove the intronic regions; consensus donor and acceptor splice sites are indicated with M corresponding to either an A or C base and Y corresponding to either a C or T at that position.

be large scale duplications or deletions of chromosomal segments that result in duplication or deletion of genes with consequences on the amount of gene product produced.

Examples of single genes in animal breeding

Single genes relevant to animal breeding include both cases of beneficial and detrimental alleles. Detrimental alleles, also referred to as genetic defects, are typically recessive alleles causing loss of gene function. Often the effects of these alleles appear after the intensive use of a popular breeding animal (typically an artificial insemination sire) such that over time there is use of related animals on both paternal and maternal sides of a pedigree. Such recessive lethal or detrimental alleles have historically been problematic as it was not possible to distinguish between homozygous wild-type and carrier animals without expensive and time-consuming breeding trials between a prospective carrier

Table 18.1 Selected examples of DNA alterations associated with single gene traits.

Gene	Polymorphism	Phenotype	Reference
Myostatin (MSTN)	Point mutation altering amino acid (missense) 11 bp deletion creating premature stop codon Point mutations in coding sequence creating premature stop codons (nonsense)	Muscular hypertrophy in cattle	Grobet et al. (1998)
	Point mutation creating micro RNA binding site	Muscular hypertrophy in sheep	Clop et al. (2006)
von Willebrand factor (VWF)	Point mutation altering splice site creating frame shift and premature stop codon	Von Willebrand's disease (clotting disorder) in dogs	Rieger et al. (1998)
Bone morphogenetic protein 15 (BMP15)	Point mutation in coding sequence creating premature stop codon (nonsense) Point mutation altering amino acid (missense)	Increased ovulation rate in sheep	Galloway et al. (2000)
Cytochrome P450, family 2, subfamily D, polypeptide 6 (CYP2D6)	Gene duplication	Ultra-rapid drug metabolism in humans	Ingelman-Sundberg (2001)
	Gene deletion	Slow drug metabolism in humans	
Solute carrier family 4, anion exchanger, member 3 (SLC4A3)	Single base insertion creating frame shift and premature stop codon	Progressive retinal atrophy in dogs	Downs et al. (2011)

sire and known carrier dams. However, recent development of complete genome sequences and high density genetic marker panels for livestock and companion animal species now makes mapping and identification of these alleles in a short time frame feasible through homozygosity mapping as described earlier. A comprehensive listing of single gene defects identified in companion animal and livestock species will not be provided here as the list is quite long and growing. However, comprehensive information can be found at the Online Mendelian Inheritance in Animals (OMIA) website (www.ncbi.nlm.nih.gov/omia). This is a searchable database that provides information on known genetic defects in a multitude of species. The reader is encouraged to explore this searchable database for a current and comprehensive view. To illustrate a single gene defect we will look at a blood clotting disorder in dogs called von Willebrand disease.

The von Willebrand factor (vWF, to denote the protein; VWF to denote the gene) is just one of several proteins that play crucial roles in the blood clotting mechanism that prevents uncontrolled bleeding; deficiencies in any one of several factors can be sufficient to cause a clotting disorder. vWF serves the specific roles of binding clotting Factor XIII and aiding in adhesion of platelets to the subendothelium of blood vessels. Von Willebrand disease in dogs occurs in three types corresponding to different mutations in VWF. Type-1 is the least severe form of the disease and is due to a recessive allele that causes lower than normal concentrations of vWF, but not its absence. Hence, clotting is less effective but not absent in homozygotes. Heterozygotes have a slightly lower concentration of vWF but exhibit normal clotting. This is an example of how mode of gene action is dependent on phenotype being considered. On a molecular level (blood concentration of vWF) the mode of gene action is additive, while on the whole animal level it would be defined as recessive/dominant (homozygotes for the mutated allele have poor clotting, while heterozygotes or homozygotes for the wild-type allele have normal clotting). The polymorphism responsible for type-1 von Willebrand disease in dogs is a splice site mutation (Holmes et al., 1996) (see Figure 18.7), which leads to incorrect splicing in some, but not all, transcripts. As a result, approximately 5% of the time the correct protein is produced from the mutant allele, while the remainder of the time a non-functional protein is produced.

Type-2 von Willebrand disease in dogs is a more severe form that is due to a missense mutation in exon 28 of the gene (Kramer et al., 2004). The result of this mutation is alteration of the encoded amino acid sequence in a region of the protein (A2 domain) that is critical for binding of vWF molecules with each other and for proteolytic digestion of vWF. Rather than being functional individually, vWF molecules are functional when bound together as a group of multiple molecules (multimers). While type-2 vWD in dogs is associated with a decrease in vWF multimers, it is not clear if this results from a failure to form multimers or an increase in proteolytic digestion. The trait is inherited as an autosomal recessive in dogs (Gavazza et al., 2012), though both dominant and recessive forms of type-2 vWD are observed in humans.

Type-3 vWD is the most severe form and is due to a single base deletion in the VWF coding sequence that causes a frame shift leading to a premature stop codon (Venta et al., 2000). As a result the protein produced is severely truncated, including none of the normal mature protein (i.e., only part of the signal peptide and none of the functional protein is produced). Individuals that are homozygous for the mutant allele are devoid of vWF and suffer severe bleeding. Heterozygous individuals have sufficient vWF to exhibit normal blood clotting, even though the level of vWF is less than in homozygous wild-type genotypes. The discovery of the mutant allele for type-3 vWD in Scottish Terrier dogs is an example of an *a priori* candidate gene analysis. Unlike the positional candidate gene analysis that followed an initial mapping effort in the earlier example of the myostatin gene for double muscling in cattle, in this case geneticists working with dogs built on the knowledge of the VWF gene sequence and role in clotting disorders in humans. This information was used to facilitate sequencing of the canine VWF, specifically VWF **complementary DNA (cDNA)** made from the **messenger RNA (mRNA)** of Scottish Terrier dogs known to be mutant homozygotes, heterozygotes (carriers) and wild-type homozygotes. Analysis of this sequence data revealed the truncated nature of the mRNA from the mutant allele and the single base deletion.

Examples of single genes with beneficial effects are more limited, and a few examples will be provided here. Among these are single genes affecting reproduction, muscling, mature body size, and feathering in birds. Coat color would also fall under this heading but is the topic of a separate chapter in this text (Chapter 19).

Several single genes affecting ovulation rate and litter size in sheep have been identified and used. These include the bone morphogenetic protein receptor 1B locus (BMPR1B or "Booroola" gene), the bone morphogenetic protein 15 locus (BMP15 or "Inverdale" gene) and growth and differentiation factor 9 (GDF9). The Booroola gene was identified phenotypically in a commercial Australian Merino flock in which the owners had recognized certain animals with higher litter size and had selected a subset of the flock for litter size. Subsequent study characterized the phenotypic effect of the gene as an increase of roughly 1.5 ova and 1 lamb per copy of the allele, going from homozygous wild-type to heterozygote to homozygous for the Booroola allele (Piper et al., 1985). The search for the gene was engaged by groups in New Zealand, France and Scotland with all identifying BMPR1B as the likely causative gene in 2001 (Mulsant et al., 2001, Souza et al., 2001, Wilson et al., 2001). The genetic basis for the phenotype is a missense mutation in BMPR1B.

The Inverdale gene was recognized several years later in Romney sheep in New Zealand but was mapped more readily and the putative causative gene identified more quickly in part because the associated high litter size phenotype displayed inheritance indicative of an X-linked gene. This immediately narrowed the field of search from the whole genome to the X chromosome. The scientists in New Zealand who mapped the Inverdale gene were able to provide very strong evidence that BMP15 is responsible, because a second flock of sheep (Hannah) also displaying an X-linked pattern of inheritance for a high litter size phenotype was identified and sequencing of BMP15 in both Inverdale and Hannah sheep revealed independent mutations that would render the respective alleles non-functional (functionally relevant missense mutation for the Inverdale allele and nonsense mutation for the Hannah allele) (Galloway et al., 2000). In contrast to BMPR1B, homozygosity for the high ovulation rate BMP15 allele leads to infertility (non-functional ovaries) due to arrested follicular development.

The third major gene, GDF9, was known to have a sequence and pattern of expression similar to BMP15 (**paralogs**, meaning both were derived from a common ancestral gene during the course of evolution through a gene duplication event) and was examined as a candidate gene in several sheep breeds known to have high litter size (Hanrahan et al., 2004); as in the type-3 von Willebrand disease example given earlier, this was an *a priori* examination of a candidate gene, rather than a positional candidate gene analysis following mapping. Sequence analysis of GDF9 in some of these highly prolific sheep breeds did reveal mutations that could account for the observed high litter sizes. Interestingly, as with BMP15, homozygosity of the mutant GDF9 allele was commonly associated with infertility due to non-functional ovaries (again due to arrested follicular development at an early stage).

These three alleles for high ovulation rate and high litter size have been used in commercial sheep production, particularly where more intensive management permits the necessary changes in ewe nutrition and lambing management. Such high litter size alleles and genotypes are not compatible with extensive sheep management systems, that is, range sheep production, where lamb losses and challenges to ewe re-breeding make a single offspring more desirable.

Myostatin and double muscling in cattle has already been described previously, but the use of muscular hypertrophy phenotypes is exploited in other species as well. In sheep, a mutation in the myostatin gene that is the basis for muscular hypertrophy in Texel sheep has been identified as a single base change in the 3′ untranslated region of the gene that creates a micro-RNA binding site (Clop et al., 2006). Because of the micro-RNA binding site created in the Texel myostatin allele, myostatin mRNA corresponding to the allele is rapidly degraded leading to reduced levels of myostatin. As seen before, reduced expresson of myostatin leads to muscular hypertrophy (double muscling).

Myostatin mutations have also been observed in dogs and horses where non-functional alleles are associated with increased muscularity and faster racing times. A two-base pair deletion in the canine myostatin gene in the Whippet breed creates a premature stop codon and non-functional gene product. Dogs carrying one copy of this allele are both more muscular and faster in competitive races (Mosher et al., 2007). Dogs carrying two copies of the allele are "grossly overmuscled" in the words of the author and presumably at a disadvantage with regard to racing. Research with Thoroughbred horses has similarly identified polymorphisms in the myostatin gene that have a profoundly significant association with racing distance (ie. at what distance does a horse best compete, short versus mid- to long-distance races). The association between these polymorphisms, expression of the myostatin gene, and optimal race distance have not been described (Hill et al., 2010). Marker-assisted selection and use of specific matings to produce the desired genotype would be advantageous to breeders in this situation.

Just as the Broad Breasted White turkey, which is too extreme in muscularity and body conformation to mate naturally, requires the use of artificial insemination for its commercial application, exploitation of muscular hypertrophy phenotypes in other species in some cases require modification to management systems. Cattle which are carriers of alleles for the double muscling trait offer beneficial increased muscling and improved carcass leanness, but production of homozygous, double muscled calves from homozygous, double muscled mothers present serious challenges regarding dystocia, and as a consequence delivery of calves via Cesarean section is typical (Kolkman, 2010). Producers who choose to produce double muscled cattle for the premium received for their exceptionally lean carcasses in some markets must be prepared to manage accordingly for the anticipated complications.

But myostatin is not the only gene for which alleles causing muscular hypetrophy have been identified. At least two additional genes of this type, name callipyge (Cockett et al., 1996) and Carwell (Nicol et al., 1998) have been identified. The callipyge gene is imprinted, meaning to affect the phenotype the allele has to have been inherited from a specific parent (e.g., from sire vs dam). Callipyge is the subject of another chapter in this textbook and will not be discussed here other than to say that it increases muscularity in the loin and hindquarters by about 30%, but also increases toughness of the meat. The Carwell or REM gene (for **r**ib-**e**ye **m**uscling) has been mapped to ovine chromosome 18 in close proximity to the callipyge gene. The causative gene and polymorphism has not yet been identified and it is not clear if Carwell/REM is an allele of callipyge or a different gene. Carwell's effect on muscular hypertrophy is limited to the *longissimus dorsi* muscle, but it is a more useful genetic variant compared to callipyge from the standpoint that it has no detrimental effect on meat tenderness.

Several single genes have been used or considered for use in poultry production. Among those are two Z-linked genes, one for feather type which permits easier sexing of birds, and another for body size that could potentially improve efficiency of production and heat tolerance. The first of these is a mutation that causes differences in the rate of feather development. The gene as identified through its phenotypic effects, has historically been referred to as the K locus. It has recently been characterized as a tandem duplication on the Z chromosome in which the prolactin receptor (PRLR) and sperm flagellar protein 2 (SPEF2) genes are duplicated (Elferink et al., 2008). Given the location of this locus on the sex chromosome (Z) common to males and females, matings can be made such that the resulting male offspring will exhibit fast feathering and the females slow feathering (Figure 18.8). The feathering phenotype is visible at one day of age and provides a simpler means of determining chick gender than vent sexing.

Figure 18.8 Use of a sex-linked gene for feathering to aid in determining sex of chicks. A male line fixed for the slow feathering allele (k) is bred to a female line hemizygous for the fast feathering allele (K). Among resulting offspring, those exhibiting fast feathering are males and those exhibiting slow feathering are females.

The second Z-linked gene, which has received more limited use, is one affecting growth and mature size of birds. Of interest in this case is a recessive allele that causes dwarfism. The objective would be to produce females **hemizygous** for the dwarfism allele (meaning one copy due to its location on a sex chromosome as in a gene on the X chromosome in male mammals or the Z chromosome in female birds) that have smaller size, conferring advantages in maintenance cost, denser stocking rate and heat tolerance (Islam, 2005, Merat, 1984). The allele has received limited use, however, because its effects are not completely recessive, causing a slight reduction in growth rate in offspring of crosses that would exploit the allele (Figure 18.9). The allele causing the dwarf phenotype results from a 1773 basepair deletion at the 3′end of the growth hormone receptor (GHR) gene which both deletes the normal stop codon and results in replacement of the last 27 amino acids of the normal gene product plus the addition of 26 amino acids (Agarwal et al., 1994). The resulting altered protein has undetectable growth hormone-binding activity.

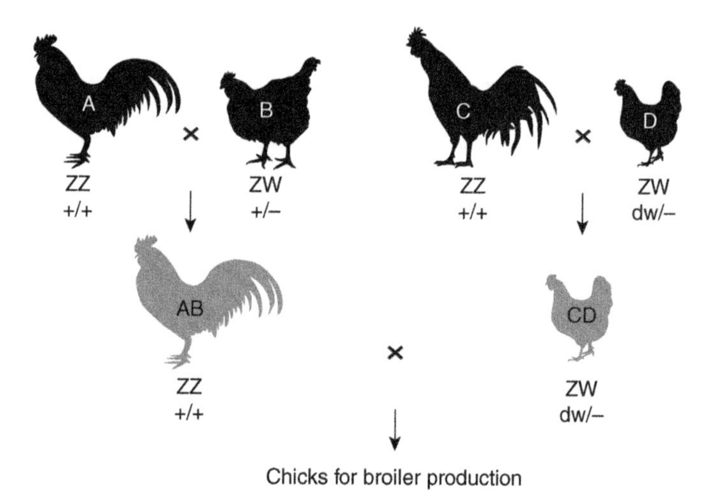

Figure 18.9 Use of a sex-linked gene for dwarfism to enhance efficiency of broiler production in chickens. Z and W denote sex chromosomes; dw and + denote alleles at a gene for dwarfism on the Z chromosome. Two lines (A, B) are crossed to produce a paternal line that is crossed with a maternal line derived from two other lines (C, D), one of which is fixed for a sex-linked allele for dwarfism (dw). In the second generation AB hybrids are crossed with CD hybrids to produce broiler chicks with nearly normal growth rate. The CD broiler hens are smaller in size due to the dwarf allele and thus have lower maintenance cost, greater heat tolerance and higher stocking rate in housing.

Another genetically based feathering phenotype with relevance to poultry breeding is the naked neck phenotype. Birds with this allele have an absence of feathers on their neck that confers enhanced heat tolerance. The putative causative polymorphism has recently been identified as a chromosomal rearrangement, specifically the insertion of a segment from chicken chromosome 1 into chromosome 3 (Mou et al., 2011). The effect of this insertion is to alter the regulation of the growth differentiation factor 7 gene (GDF7), leading to selective loss of feathering on the neck due to higher production of retinoic acid (vitamin A) in skin tissue of the neck and its interaction with GDF7. Identification of the genetic polymorphism underlying this trait will facilitate selective breeding to transfer this trait to different chicken lines.

Single genes have and will continue to play a role in animal breeding in the future. Recessive genetic defects will continually be revealed over time as specific breeding animals attain popularity and become heavily used within a population. Fortunately, the development of genome sequence information and dense panels of genetic markers now make the task of mapping and developing markers for these tractable. Also noteworthy is that identification of major genes is sometimes a matter of definition of phenotype. As the definition of a trait is narrowed, it becomes more likely to identify genes that account for greater effects (e.g., ovulation rate vs litter size, circulating concentration of insulin-like growth factor-1 vs yearling weight, etc., level of gene expression as a phenotype).

Summary

The term "single genes" in this chapter refers to genes whose allelic variants cause profound differences in animal phenotype. In some cases the effect is potentially beneficial and may be exploited through specifically planned matings and/or altered management systems that accommodate the profound difference in phenotype. In other cases the altered allele has a decidedly detrimental effect, as is the case for recessive alleles associated with lethality or diminished health. In either case, recent genome sequencing and genetic marker development efforts have resulted in genomic tools that greatly speed the process of mapping and in some cases identifying the genes and polymorphisms responsible for the dramatic phenotypic effects. This information in turn facilitates the selection for or against these alleles. It is noteworthy that identification of major genes is sometimes a matter of definition of phenotype. As the definition of a trait is narrowed, it becomes more likely to identify genes that account for greater effects (e.g., ovulation rate vs litter size, circulating concentration of insulin-like growth factor-1 vs yearling weight, etc.).

References

Agarwal, S. K., Cogburn, L. A. and Burnside, J. 1994. Dysfunctional growth hormone receptor in a strain of sex-linked dwarf chicken: evidence for a mutation in the intracellular domain. *J Endocrinol*, 142, 427–434.

Charlier, C., Coppieters, W., Farnir, F., Grobet, L., Leroy, P. L., Michaux, C., et al.. 1995. The MH gene causing double-muscling in cattle maps to bovine Chromosome 2. *Mamm Genome*, 6, 788–792.

Clop, A., Marcq, F., Takeda, H., Pirottin, D., Tordoir, X., Bibe, B., et al. 2006. A mutation creating a potential illegitimate microRNA target site in the myostatin gene affects muscularity in sheep. *Nat Genet*, 38, 813–818.

Cockett, N. E., Jackson, S. P., Shay, T. L., Farnir, F., Berghmans, S., Snowder, G. D., et al. 1996. Polar overdominance at the ovine callipyge locus. *Science*, 273, 236–238.

Downs, L. M., Wallin-Håkansson, B., Boursnell, M., Marklund, S., Hedhammar, Å., Truvé, K., et al. 2011. A frameshift mutation in golden retriever dogs with progressive retinal atrophy endorses SLC4A3 as a candidate gene for human retinal degenerations. *PLoS One*, 6, e21452.

Elferink, M. G., Vallee, A. A., Jungerius, A. P., Crooijmans, R. P. and Groenen, M. A. 2008. Partial duplication of the PRLR and SPEF2 genes at the late feathering locus in chicken. *BMC Genomics*, 9, 391.

Galloway, S. M., McNatty, K. P., Cambridge, L. M., Laitinen, M. P., Juengel, J. L., Jokiranta, T. S., et al. 2000. Mutations in an oocyte-derived growth factor gene (BMP15) cause increased ovulation rate and infertility in a dosage-sensitive manner. *Nat Genet*, 25, 279–283.

Gavazza, A., Presciuttini, S., Keuper, H. and Lubas, G. 2012. Estimated prevalence of canine Type 2 Von Willebrand disease in the Deutsch-Drahthaar (German Wirehaired Pointer) in Europe. *Res Vet Sci*, 93, 1462–146.

Grobet, L., Poncelet, D., Royo, L. J., Brouwers, B., Pirottin, D., Michaux, C., et al. 1998. Molecular definition of an allelic series of mutations disrupting the myostatin function and causing double-muscling in cattle. *Mamm Genome*, 9, 210–213.

Hanrahan, J. P., Gregan, S. M., Mulsant, P., Mullen, M., Davis, G. H., Powell, R. and Galloway, S. M. 2004. Mutations in the genes for oocyte-derived growth factors GDF9 and BMP15 are associated with both increased ovulation rate and sterility in Cambridge and Belclare sheep (Ovis aries). *Biol Reprod*, 70, 900–909.

Hill, E. W., McGivney, B. A., Gu, J., Whiston, R. and MacHugh, D. E. 2010. A genome-wide SNP-association study confirms a sequence variant (g.66493737C>T) in the equine myostatin (MSTN) gene as the most powerful predictor of optimum racing distance for Thoroughbred racehorses. *BMC Genomics*, 11, 552.

Holmes, N. G., Shaw, S. C., Dickens, H. F., Coombes, L. M., Ryder, E. J., Littlewood, J. D. and Binns, M. M. 1996. Von Wille-brand's disease in UK dobermanns: possible correlation of a polymorphic DNA marker with disease status. *J Small Anim Pract*, 37, 307–308.

Ingelman-Sundberg, M. 2001. Genetic susceptibility to adverse effects of drugs and environmental toxicants. The role of the CYP family of enzymes. *Mutat Res*, 82, 11–19.

Islam, M. A. 2005. Sex-linked dwarf gene for broiler production in hot-humid climates. *Asian-Australian Journal of Animal Science*, 18, 1662–1668.

Kolkman, I. 2010. *Calving problems and calving ability in the phenotypically double muscled Belgian Blue breed*. Ph.D. Dissertation, University of Gent.

Kramer, J. W., Venta, P. J., Klein, S. R., Cao, Y., Schall, W. D. and Yuzbasiyan-Gurkan, V. 2004. A von Willebrand's factor genomic nucleotide variant and polymerase chain reaction diagnostic test associated with inheritable type-2 von Willebrand's disease in a line of German shorthaired pointer dogs. *Vet Pathol*, 41, 221–218.

McPherron, A. C., Lawler, A. M. and Lee, S. J. 1997. Regulation of skeletal muscle mass in mice by a new TGF-beta superfamily member. *Nature*, 387, 83–90.

Merat, P. 1984. The sex-linked dwarf gene in the broiler chicken industry. *World's Poultry Science Journal*, 40, 10–18.

Mosher, D. S., Quignon, P., Bustamante, C. D., Sutter, N. B., Mellersh, C. S., Parker, H. G. and Ostrander, E. A. 2007. A mutation in the myostatin gene increases muscle mass and enhances racing performance in heterozygote dogs. *PLoS Genet*, 3, e79.

Mou, C., Pitel, F., Gourichon, D., Vignoles, F., Tzika, A., Tato, P., et al. 2011. Cryptic patterning of avian skin confers a developmental facility for loss of neck feathering. *PLoS Biol*, 9, e1001028.

Mulsant, P., Lecerf, F., Fabre, S., Schibler, L., Monget, P., Lanneluc, I., et al. 2001. Mutation in bone morphogenetic protein receptor-IB is associated with increased ovulation rate in Booroola Merino ewes. *Proc Natl Acad Sci U S A*, 98, 5104–5109.

Nicol, G. B., Burkin, H. R., Broad, T. E., Jopson, N. B., Greer, G. J., Bain, W. E., et al. 1998. Genetic linkage of microsatellite markers to the Carwell locus for rib-eye muscling in sheep. *Proc. 6th World Cong. Genet. Appl. Livest. Prod.*, 26, 529–532.

Piper, L. R., Bindon, B.M., and Davis, G.H. 1985. The single gene inheritance of the high litter size of the Booroola Merino. In: *Genetics of Reproduction in Sheep*, Land, R.B., Robinson, D.W. (eds). Butterworths, London. 115–125.

Rieger, M., Schwarz, H. P., Turecek, P. L., Dorner, F., van Mourik, J. A., and Mannhalter, C. 1998. Identification of mutations in the canine von Willebrand factor gene associated with type III von Willebrand disease. *Thromb Haemost*, 80, 332–337.

Souza, C. J., MacDougall, C., Campbell, B. K., McNeilly, A. S. and Baird, D. T. 2001. The Booroola (FecB) phenotype is associated with a mutation in the bone morphogenetic receptor type 1 B (BMPR1B) gene. *J Endocrinol,* 169, R1–6.

Venta, P. J., Li, J., Yuzbasiyan-Gurkan, V., Brewer, G. J., and Schall, W. D. 2000. Mutation causing von Willebrand's disease in Scottish Terriers. *J Vet Intern Med,* 14, 10–19.

Wilson, T., Wu, X. Y., Juengel, J. L., Ross, I. K., Lumsden, J. M., Lord, E. A., et al. 2001. Highly prolific Booroola sheep have a mutation in the intracellular kinase domain of bone morphogenetic protein IB receptor (ALK-6) that is expressed in both oocytes and granulosa cells. *Biol Reprod,* 64, 1225–1235.

Review questions

1. To which is using a half-sib family for linkage mapping more equivalent: using an F_2 family or a backcross? Why?

2. Consider Figure 18.2 and the information in the legend again. The SNP panel included approximately 200 markers on the X chromosome. Why are no linkage analysis results shown for the X-chromosome?

3. Consider the pedigree that follows in the figure: for what proportion of the genome will individual B be homozygous for identical chromosomal segments inherited from common ancestor A (equivalently, if common ancestor A is heterozygous with genotype m1*/m2* at a specific locus, what is the probability that individual B has either genotype m1*/m1* or m2*/m2*)? Next, assume six individuals with similar pedigrees were used in homozygosity mapping: for what proportion of the genome would all six be expected to exhibit homozygosity for the **same** chromosomal segments inherited from their mutually common ancestor, A?

4. Why would an insertion or deletion of three base pairs in a coding region be expected to have less severe consequences than one single base pair insertion or deletion?

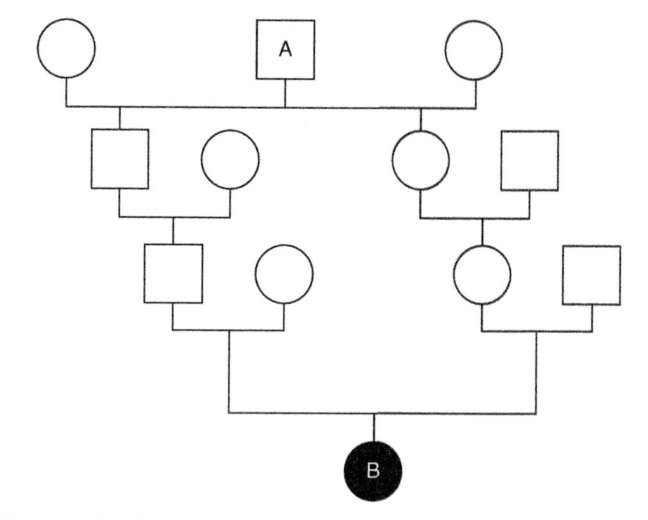

Figure 18.10 For question **2** and **3**.

5. Search the Online Mendelian Inheritance in Animals (OMIA) database (http://omia.angis.org.au/home/) for the trait "mannosidosis." How many types of mannosidosis are listed in the database, and for how many species are there entries for mannosidosis? What is mannosidosis and which genes have been identified as causing mannosidosis?

19 Molecular Genetics of Coat Color: It is more than Just Skin Deep

Samantha Brooks
Department of Animal Science, Cornell University, NY, USA

Introduction

Changes in coat color are one of the first hallmarks of domestication (Cieslak et al., 2011). To date, genetic research (much of which is conducted in mice and rats) has discovered variants at more than 300 loci influencing mammalian coat color (Cieslak et al., 2011). Despite this extensive knowledge base the pathways controlling pigmentation are still not fully understood. In domesticated animals pigmentation traits can impact value through both aesthetic and physiological mechanisms. For example, a light colored coat may be attractive, but can also reflect more radiant heat and is a beneficial adaptation in hot, sunny climates (Fadare et al., 2012, Hansen, 1990). In other cases, color adaptation is strictly a result of a human phenomenon. Producers may have a personal preference for a color, or tradition may influence the favored coat of an animal in a given function. Many registries establish permitted coat colors as part of their breed standard and may ban individuals based solely on nonconformance for this trait, despite many other valuable qualities in the animal. Thus, a good working knowledge of the genetics of coat color can translate to improved marketability of most species of livestock, and potential profits for the producer.

In considering the etiology of color traits it is important to understand the pathways through which any given mutation can exert a measurable effect. Melanocytes, the pigment-producing cells, must first successfully migrate across the developing embryo and colonize the skin. Secondly, the melanocytes must receive and translate the necessary cellular signals directing them to begin pigment production. Finally, the melanocyte needs a full complement of enzymatic pathways and a functional melanosome, the organelle in which pigment is made.

With the completion of the first human and animal genome sequences we discovered that just ~20,000 genes were sufficient to produce all of the 100,000s of proteins in a mammalian species. This is accomplished through an efficient system of alternative regulatory and splicing variants that tailor a particular gene to multiple functions. Thus one gene sequence can be "recycled" multiple times. While this system is remarkably efficient in terms of sequence length, it requires complex regulatory networks, most of which we do not yet fully understand. In an effect known as "**pleiotropy**," a single mutation can impact seemingly unrelated functions of a gene, resulting in diverse phenotypic traits. Thus benign changes in pigmentation genes can occasionally have severe impacts on animal health. Understanding of the multifunctional nature of pigmentation genes as well as careful study of pleiotropic phenotypes associated with desired alleles are necessary in order to design effective and safe effective selection schemes for coat color.

Pathways of melanocyte migration and differentiation from the neural crest

Melanoblasts, the precursor to the pigment producing melanocyte, originate at the neural crest and migrate first dorsolateraly then ventrally during embryonic development. In order to successfully make this journey the melanoblasts must follow a carefully orchestrated map of chemotaxic signals all the way to their destination in the skin. In Figure 19.1 melanoblasts expressing *KIT* are labeled by transgenic expression of a *DCT–lacZ* construct and can be seen descending from the neural crest. Therefore it is not surprising that disruption of these signaling pathways results in patches of skin lacking pigment in the adult animal. The extent of pigmentation depends on the gene altered, severity of mutation and to some degree, chance.

The *KIT* gene encodes a transmembrane receptor important for detection of just one of these signals by the melanoblast. *KIT* gene polymorphisms are among the most common causes of white spotting in mammals. Common phenotypes at this locus include roaning, spotting, and dominant white. In the pig, multiple full-length gene duplication events and an exon-skipping polymorphism (I^P and I) result in the **patch** and **dominant white** phenotypes valued in specific breeds, and in the case of white, for leather production (Marklund et al., 1998, Moller et al., 1996). Yet, structural variation at this locus is not unique to these porcine alleles. A ~45MB inversion of a region on ECA3 results in the attractive Tobiano spotting pattern in the horse, probably by disruption of a distant regulatory region of *KIT* (Brooks et al., 2007). Recent work described two serial translocations of a genomic region containing or impacting the *KIT* gene, both resulting from circular intermediates and causing the common spotting phenotype **color-sided** in cattle (Figure 19.2) (Durkin et al., 2012). SNP polymorphisms are also common. In the horse

Molecular and Quantitative Animal Genetics, First Edition. Edited by Hasan Khatib.

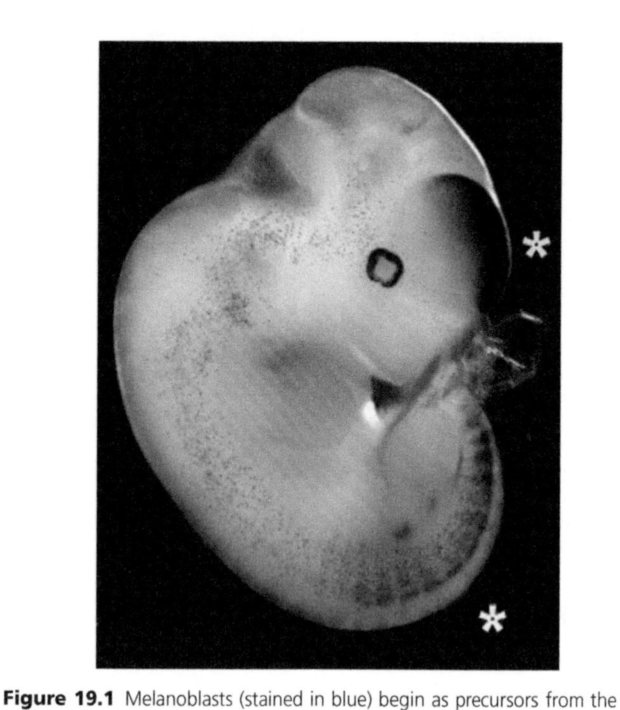

Figure 19.1 Melanoblasts (stained in blue) begin as precursors from the neural crest and migrate across the body of the embryo.

Reproduced with permission from MacKenzie, M. A., Jordan, S. A., Budd, P. S. and Jackson, I. J. Activation of the receptor tyrosine kinase *Kit* is required for the proliferation of melanoblasts in the mouse embryo. *Developmental Biology*, 192, 99–107. Copyright © 1997, Elsevier.

for example, SNPs resulting in 17 unique alleles for dominant white have been observed to date (Haase et al., 2007, 2009, 2011; Holl et al., 2010). The *KIT* locus in the horse is a good example of **genetic heterogeneity**, or multiple alleles causing the same phenotype. **Genetic heterogeneity** can complicate genetic studies and lead to confusion in diagnostic genotyping.

However, the phenotypes resulting from *KIT* gene variants are not always positively regarded. Complete depigmentation of the skin and hair has its own set of challenges as it can leave animals without protection from the sun. Most *KIT* alleles are dominant, reflecting the importance of this receptor to the pigmentation system, although the penetrance of coat color phenotypes may vary and make selection based on color occasionally difficult. In mutant mice pleiotropic phenotypes are observed in gametogenesis, hematopoiesis, tumorigenesis, and immunity, although the function of this gene is not entirely understood in livestock species (Eppig et al., 2012). Indeed, in many species *KIT* gene mutations are homozygous sterile or lethal. In the horse, no living homozygote has been identified for any of the **Dominant White** alleles, suggesting that it may in fact be embryonic lethal in the homozygous state (Haase et al., 2010). Although pigs are heavily selected for health and efficient production individuals with the *I/I* genotype still exhibit mild anemia compared to their wild-type counterparts (Johansson et al., 2005). An allele for white-spotting in the domesticated cat is mapped at or near the *KIT* locus (Cooper et al., 2006). Alteration of the ligand for the *KIT* receptor, *KITLG* (aka **MGF, mast cell growth factor**) produces phenotypes often similar to those of the *KIT* locus, as one might predict. The *roan* and *white* coat colors of Shorthorn and Belgian Blue cattle are due to a missense mutation at this locus (Seitz et al., 1999).

Melanocyte migration is also directed through signaling of the **endothelin receptor type B**. A dinucleotide change in the coding sequence of *EDNRB* leads to the attractive ***frame overo*** spotting pattern in the heterozygous horse (Metallinos et al., 1998; Santschi et al., 1998; Yang et al., 1998). *Frame overo* pattern is a good example of **incomplete penetrance**. Rather than conforming to a strictly dominant mode of inheritance, the degree of white spotting resulting from the *frame overo* allele is variable in heterozygous carriers. Indeed carrier individuals are occasionally reported as lacking the spotting pattern. Due to **incomplete penetrance** at this locus is important to rely on genotyping tests, rather than phenotype alone, to identify carriers. Importantly, melanocytes are not the only cell type to use the *EDNRB* signal during differentiation from the neural crest. Horses homozygous for this change are completely white, but affected with severe megacolon as a result of the loss of enteric ganglia. Cells important for normal innervation of the gut and maintenance of peristalisis also fail to migrate without *EDNRB* signaling. Thus horse owners desiring the *frame overo* coat color must carefully weigh the consequences of positive selection for this locus. Fortunately, commercial testing for this allele is widely available and can be very effectively used to avoid carrier to carrier matings.

Many of the genes important for melanocyte migration are regulated through the action of the *MITF* transcription factor. Thus *MITF* phenotypes are important spotting traits in a number of domesticated species. For example, in the Japanese quail introduction of a premature stop codon by a deletion in the coding sequence of *MITF* produces both **silver** and **white** phenotypes by an additive mode of inheritance (Minvielle et al., 2010). The authors of this work also produced evidence that this locus is also the cause of the rare **blue** color in the chicken, although additional work will be needed to conclusively prove this hypothesis. Yet *MITF* has important roles in several processes other than pigmentation. In the white Japanese quail, homozygous for a *MITF* mutation, subclinical osteopetrosis can be observed by x-ray (Kawaguchi et al., 2001). Likewise, a single base change leading to conversion of a arginine amino acid to an isoleucine in the protein product of this locus results in white color, deafness, and ocular abnormalities in the Fleckvieh breed of cattle (Philipp et al., 2011). The *MITF* locus is also implicated in control of the degree of white spotting of the Holstein breed (Liu et al., 2009). In the domestic dog *MITF* controls an attractive white spotting phenotype in Boxers, Beagles, and Newfoundlands (Leegwater et al., 2007; Rothschild et al., 2006).

Melanocyte signaling and regulation

Once the melanoblasts have successfully arrived at their final destination in the skin they require diverse intercellular signals in order to function. In most mammals, various shades of color result from modulation of just two types of pigment, red **phaeomelanin** and black **eumelanin**. When and where pigment will be produced is primarily controlled by signaling though the melanocortin 1 receptor (*MC1R*). Upon stimulation by melanocyte stimulating hormone from the hypothalamus, *MC1R* sets off a cascade in intracellular pathways resulting in up-regulation of pigment producing enzymes. *MC1R* is particularly important for directing eumelanin production, thus loss-of-function variants create red coat colors and gain of function results in dark or black

Figure 19.2 Color-sided in cattle is a unique spotting pattern resulting from a structural variation that encompasses the KIT gene. Color-sided can be found in the Belgain Blue (a) and Brown Swiss breeds (b and c) and is semi-dominant, as can be observed in these heterozygous (c) and homozygous (b) individuals.

phenotypes. Indeed *MC1R* polymorphisms are among some of the most frequently identified color variants, affecting diverse species including the guinea pig, alpaca, jaguar, and even the extinct woolly mammoth (Rompler et al., 2006).

Among livestock and other domesticated species a number of notable *MC1R* alleles exist. Dominant black, an important economic trait for fiber production, is a result of an amino acid change in the *MC1R* of sheep (Vage et al., 1999). Various porcine breeds exhibit a number of *MC1R* alleles for both dominant black and recessive red coloration, including a unique two base pair insertion that results primarily in a red background, but also black spots on the body though somatic reversion to the wild type (Kijas et al., 1998; 2001). Likewise, multiple alleles exist in the goat, as well as the domestic cat and dog, again resulting in variable black and/or red phenotypes (Fontanesi et al., 2009a; Peterschmitt et al., 2009; Schmutz and Melekhovets, 2012). **Dominant black** and **red** colors in cattle are the result of *MC1R* polymorphisms (Joerg et al., 1996, Klungland and Vage, 1999). A unique brindle (alternating black and red patches or stripes) phenotype is the result of deletion of two and change of a third amino acid in the rabbit (Fontanesi et al., 2010). Yet *MC1R* signaling is not simply a control of pigmentation. In mice and man *MC1R* variants (specifically those causing recessive red phenotypes) are associated with increased sensitivity to pain (Delaney et al., 2010; Mogil et al., 2005). Anecdotally, **chestnut** (recessive *MC1R* loss of function mutation) horses and **red** Cocker Spaniels are reported to be excitable and sensitive, suggesting that there may be some subtle pleiotropic behavioral effects of this locus, perhaps related to *MC1R* mediated pain reception. Certainly, given concerns in welfare and management of livestock species further research in to the perception of pain in *MC1R* mutants may be warranted.

MC1R activation by its ligand, melanocyte stimulating hormone, can be attenuated or blocked by a competitive inhibitor named agouti signaling protein (*ASIP*). Binding of the *ASIP* protein to the *MC1R* receptor results in a decrease in eumelanin production within the melanocyte. *ASIP* signaling is important for establishing patterning across the body, and along the hair shaft, in many wild-type coat colors. For example, in the horse *ASIP* directs production of primarily eumelanin in the extremities and phaeomelanin across the body. This results in the familiar **bay** color (Figure 19.3). In mice (Furumura et al., 1996), goats (Fontanesi et al., 2009b), and dogs (Dreger and Schmutz, 2011) temporal changes in *ASIP* expression results in banding of red and black pigment along the hair shaft creating a mottled coat that is advantageous as camouflage.

A variety of *ASIP* polymorphisms are important for domesticated species. A large 190 kb duplication encompassing *ASIP*

Figure 19.3 Disruption of the MC1R receptor leads to loss of eumelanin production, as can be seen in these chestnut (on the left, homozygous recessive e/e) and bay (E/-)mares.
(photo credit S. Brooks)

results in the *white* trait in sheep, an important economic characteristic for wool production. In this case, the structural change places an *ASIP* sequence behind a promoter intended for the *ITCH* gene and results in constitutive expression of *ASIP*. Recessive black is due to an 11 bp deletion and a non-synonymous change within *ASIP* in horses and dogs, respectively (Kerns et al., 2004; Rieder et al., 2001). Normande cattle possess a unique insertion of a LINE repeat element just upstream of this locus (Girardot et al., 2006). The authors documented extensive overexpression of *ASIP* transcripts which likely cause the characteristic brindle coat color of this breed. Yet overexpression of *ASIP* has consequences beyond coat color in other model species. The **viable yellow** allele in the mouse also produces overexpression of *ASIP* due to an insertion 5′ of the gene and has a mottled yellow color. However, the viable yellow mouse also develops obesity and type II diabetes in the adult. Comparatively, this increase in *ASIP* expression may have similar effects in the Normande cattle, yet for a dairy animal this tendency towards increased adiposity may be a selective advantage.

Melanin production and transport

Healthy melanocytes manufacture pigment within specialized organelles called **melanosomes**. These organelles contain the enzymes necessary for pigment production and carefully regulate the required internal pH (Ancans et al., 2001). Within the melanosome tyrosine is converted in to the melanin pigments by several specialized enzymes (Schiaffino, 2010). Once mature and full of pigment granules the melanosome is transported across the cytoskeleton to the ends of specialized dendritic extensions of the melanocyte. Keratinocytes seem to acquire pigment by engulfing vesicles from the ends of the pseudopodia (Van Den Bossche et al., 2006). These vesicles are transported to the center of the cell where they can perform their primary function, photoprotection of the DNA within the nucleus (Boissy, 2003). Keratinocytes may retain these vesicles maintaining pigmentation of the dermis and epidermis, or can package them as part of the hair shaft, giving color to the hair coat (Slominski et al., 2005).

Structural integrity of the melanosome is in part maintained by a specialized glycoprotein product of the *PMEL* (or *PMEL17*) locus (Fowler et al., 2006). At least two separate alleles have been observed at this locus in cattle. Simmental and Herford breeds possess recessive mutations in the *PMEL* coding sequence leading to a dilute coat color, as well as a hypotrichosis, a congenital deficiency of the hair coat (Jolly et al., 2008). Highland cattle possess another allele, a three base-pair deletion resulting in semi-dominant color dilution but no hair abnormalities (Schmutz and Dreger, 2013). Three separate alleles are also present in the chicken, resulting in the ***dominant white***, **smoky**, and **dun** plumage colors (Kerje et al., 2004). *Dominant white* is the allele responsible for the characteristic white color of the Leghorn chicken, a popular breed for egg production. Remarkably pleiotropic effects of the *PMEL* locus in the chicken are often behavioral phenotypes. *Dominant white* individuals are both less prone to feather picking and less explorative behavior (Natt et al., 2007, Karlsson et al., 2010). ***Silver***, a popular color in the Rocky Mountain Horse and Shetland Pony breeds, is characterized by a chocolate brown body color and flaxen mane and tail (Brunberg et al., 2006). This dilution of the coat only effects eumelanin production, and is therefore not visible on genetic backgrounds incapable of producing eumelanin. Multiple Congenital Ocular Anomalies, a severe semi-dominant eye condition, is associated with the *Silver* trait in the horse although it is not yet clear if the two phenotypes are caused by the same polymorphism or just in tight LD. In contrast, the *Merle* trait in the dog is clearly the cause of a color phenotype as well as ocular and auditory abnormalities (Clark et al., 2006). *Merle* (Figure 19.4) is the result of a retrotransposon insertion in an intron of *PMEL*. Spontaneous loss of this insertion creates clonal spots of wild-type color on the *Merle* background.

Creation of pigment from the amino acid tyrosine is a multistep process beginning with the enzyme tyrosinase (*TYR*). Tyrosinase catalyzes the first two steps of melanin synthesis, therefore complete loss of this enzyme results in oculocutaneous albinism (Spritz, 1994). Spontaneous albinos occur in virtually all vertebrates and although many genes can cause this trait, mutations altering *TYR* are frequently the culprit. *TYR* alleles for albino are known in the cat, chicken, cattle, sheep, and rabbit (including the common laboratory strains of rabbit) (Aigner et al., 2000; Imes et al., 2006, Rowett and Fleet, 1993; Tobita-Teramoto et al., 2000). Two SNPs resulting in amino acid changes in the tyrosinase enzyme produce the temperature-sensitive coat dilution of the Siamese (c^s) and Burmese (c^b) cat breeds (Figure 19.5) (Lyons et al., 2005b). Albino and Siamese (c^s) cats, like many animals with *TYR* mutations, possess a variety of vison abnormalities including strabismus, loss of retinal pigmentation, and disorganization within the optical cortex (Kaas, 2005). Recessive white, an important economic plumage trait in poultry, is due to insertion of a large retroviral element within an intron of the *TYR* gene in the chicken (Chang et al., 2006). Modulation of tyrosinase activity can also be achieved through alteration of accessory enzymes, tyrosinase related protein (*TYRP1*) for example. Although *TYRP1* mutations can also occasionally lead to albinism, in livestock species this locus in more frequently found in individuals with true brown color. Brown alleles exist in the ***chocolate*** and ***cinnamon*** domestic cat, Weimaraner breed of dog, *dun*

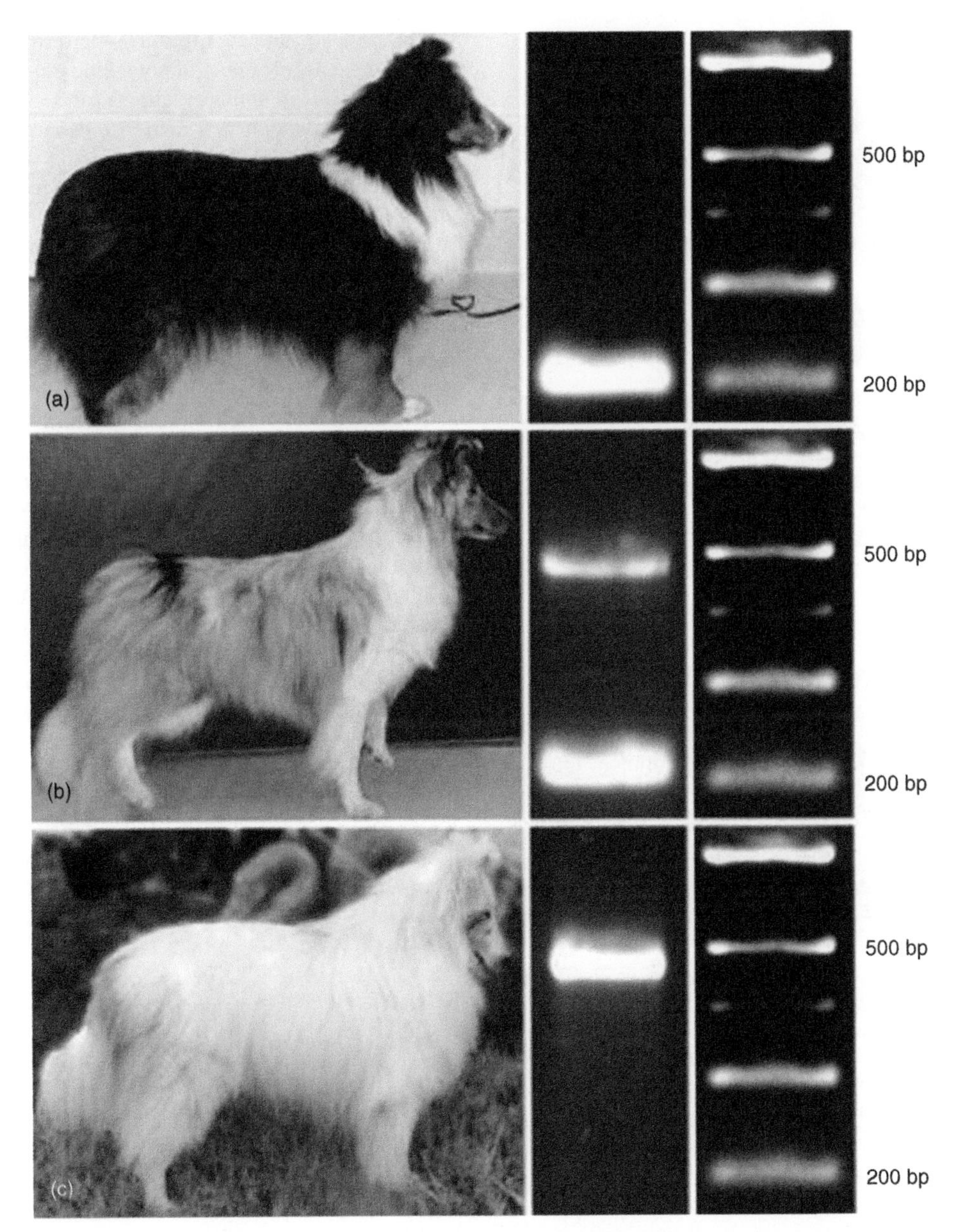

Figure 19.4 Insertion of a retrotransposon element results in the merle color in dogs, as well as ocular and auditory health issues. The insertion is readily apparent in the product size produced by PCR amplification of this region (wild-type: a, heterozygous: b, homozygous: c).

Reproduced with permission from Clark, L. A., Wahl, J. M., Rees, C. A. and Murphy, K. E. 2006. Retrotransposon insertion in SILV is responsible for merle patterning of the domestic dog. *Proceedings of the National Academy of Sciences of the United States of America*, 103, 1376–1381. Copyright (2006) National Academy of Sciences, U.S.A.

Dexter cattle and the Soay wild sheep (Berryere et al., 2003, Gratten et al., 2007; Lyons et al., 2005a, Schmutz et al., 2002,).

Transport of mature melanosomes to the periphery of the dendritic extensions along actin filaments is accomplished by a myosin motor (*MYO5A*) and a complex of adaptor proteins including Rab27a and melanophilin (*MLPH*) (Fukuda et al., 2002). All three of these proteins possess important pleiotropic functions leading to health impacts of genetic changes in their sequence. As a result, mutations in genes of the melanosome transport complex are rare in livestock. A rare recessive lethal allele in the horse is the result of a single base-pair mutation in *MYO5A* (Brooks et al., 2010). Resulting in color dilution and severe neurologic defects, **Lavender Foal Syndrome** illustrates the consequences of pleiotropic phenotypes at this locus. Figure 19.6 shows the *Lavender* dilution in the chicken, a recessive trait due to a point mutation in the *MLPH* gene that is not associated with any neurologic deficits (Vaez et al., 2008). The attractive ***blue*** and ***cream*** colors in the cat are due to a premature stop codon in *MLPH* (Ishida et al., 2006). Similar dilutions in the dog are also a result of this locus, and are sometimes associated with alopecia and skin inflammation (Drogemuller et al., 2007). A particularly attractive dilution, *Silver* is a key economic trait in the fur producing mink and is likely due to a yet undescribed change in *MLPH* (Anistoroaei and Christensen, 2007).

Figure 19.5 In the cat, two TYR alleles result in temperature-sensitive dilution of the coat color, cs (homozygote on left) and the milder cb (homozygote on the right). Compound heterozygotes (cs/cb) exhibit a moderate degree of coat dilution. Reproduced with permission from Lyons, L. A., Imes, D. L., Rah, H. C. and Grahn, R. A. 2005b. Tyrosinase mutations associated with Siamese and Burmese patterns in the domestic cat (*Felis catus*). *Anim Genet*, 36, 119–126. Copyright © 2005, John Wiley & Sons, Inc.

Figure 19.6 Lavender (B), a recessive dilution of the underlying plumage color (A) is the result of a single base change in the MLPH gene sequence in the chicken. Reproduced with permission from Vaez, M., Follett, S. A., Bed'hom, B., Gourichon, D., Tixier-Boichard, M. and Burke, T. 2008. A single point-mutation within the melanophilin gene causes the lavender plumage colour dilution phenotype in the chicken. *BMC Genetics*, 9, 7.

Conclusions

Color traits are common in nearly all domesticated mammals and birds. Aside from their aesthetic appeal, they are often used as identification for individual animals and are a defining characteristic of some breeds. Certainly, it is human nature to select unusual and appealing individuals for breeding stock. Beyond these conscious efforts, appearance of color variants could also be a result of the release of negative selective pressure imposed by predation on visually conspicuous individuals. Indeed appearance of color variants may in some cases be a direct consequence of selection for tame behavior during domestication (Keeler et al., 1970). Both color intensity and white spotting share some pathways important for neurodevelopmental processes, thus genetic changes for tame behavior may have pleiotropic effects on color. Yet additional work will be required in order to determine if the association between domestication and coat color is causative, and not just coincidental.

For the producer or herdsman a basic understanding of the inheritance schemes for color traits relevant to their breed or function of interest are key. The *frame overo* spotting pattern of the horse is a prime example. The pattern itself is dominant, but with some epistatic effect of polygenic modifiers. In this case, use of the available genetic test allows for definitive identification of carriers, and avoidance of carrier to carrier matings that bear the risk of producing a non-viable foal. Balancing selection at this locus, acting positively on the desirable pattern in the heterozygote, but negatively on the lethal homozygote, will drive the frequency of both homozygous states down. In the end it may be difficult for producers relying on closed populations to find non-carrier breeding stock. Furthermore, this situation also creates an ethical debate. In some registries, and even some countries, breeding of animals carrying "genetic defects" is forbidden. Yet due to pleiotropy the same allele can be regarded both as a positive economic trait and a defective or even lethal disorder, depending on the point of view.

Certainly the availability of genetic tests for many coat color alleles has enabled very efficient selection of breeding stock and prediction of the phenotype in future offspring. Beyond the importance of color as an economic trait, polymorphisms leading to altered pigmentation illustrate a cautionary tale for artificial selection. We now understand the "multipurpose" nature of most gene sequences, and the potential negative impacts of genetic changes on animal health. Certainly the use of genotype assisted selection hold great promise for animal industries. Yet selection on QTL loci without understanding of the molecular mechanisms behind these phenotypes may lead to an unintentional increase in the frequency of some associated negative pleiotropic phenotypes. Thus, while color phenotypes have historically been the corner stone of Mendelian genetics in many species, it seems coat colors may yet have many lessons to teach us about the nature of complex loci.

Summary

Coat color traits have origins as old as the domestication process itself. Variations in color are important economic traits, advertising a particular breed or attracting attention in the show ring. Yet many genes have more than one function, and those involved in pigmentation are no exception. As a result, polymorphisms impacting color often have important consequences for animal health. The pigmentation process can be divided in to three general steps, migration and differentiation of melanocytes, intercellular signaling and production in the melanosome. Alterations within these pathways create spotting patterns, color dilutions and changes in hue. Understanding the basics of inheritance for color traits in a species of interest is key to maintaining the profitability of any animal agribusiness, as well as the health of the animals themselves.

Key terms

Pleiotropy
Melanoblast
Genetic Heterogeneity
Incomplete Penetrance
Melanosome
Phaeomelanin
Eumelanin
Balancing Selection

References

Aigner, B., Besenfelder, U., Muller, M. and Brem, G. 2000. Tyrosinase gene variants in different rabbit strains. *Mamm Genome*, 11, 700–702.

Ancans, J., Tobin, D. J., Hoogduijn, M. J., Smit, N. P., Wakamatsu, K. and Thody, A. J. 2001. Melanosomal pH controls rate of melanogenesis, eumelanin/phaeomelanin ratio and melanosome maturation in melanocytes and melanoma cells. *Experimental Cell Research*, 268, 26–35.

Anistoroaei, R. and Christensen, K. 2007. Mapping of the silver gene in mink and its association with the dilution gene in dog. *Cytogenetic and Genome Research*, 116, 316–318.

Berryere, T. G., Schmutz, S. M., Schimpf, R. J., Cowan, C. M. and Potter, J. 2003. TYRP1 is associated with dun coat colour in Dexter cattle or how now brown cow? *Animal Genetics*, 34, 169–175.

Boissy, R. E. 2003. Melanosome transfer to and translocation in the keratinocyte. *Exp Dermatol*, 12 Suppl 2, 5–12.

Brooks, S. A., Gabreski, N., Miller, D., Brisbin, A., Brown, H. E., Streeter, C., et al. 2010. Whole-genome SNP association in the horse: identification of a deletion in myosin Va responsible for Lavender Foal Syndrome. *PLoS Genetics*, 6, e1000909.

Brooks, S. A., Lear, T. L., Adelson, D. L. and Bailey, E. 2007. A chromosome inversion near the KIT gene and the Tobiano spotting pattern in horses. *Cytogenetic and Genome Research*, 119, 225–230.

Brunberg, E., Andersson, L., Cothran, G., Sandberg, K., Mikko, S. and Lindgren, G. 2006. A missense mutation in PMEL17 is associated with the Silver coat color in the horse. *BMC Genetics*, 7, 46.

Chang, C. M., Coville, J. L., Coquerelle, G., Gourichon, D., Oulmouden, A. and Tixier-Boichard, M. 2006. Complete association between a retroviral insertion in the tyrosinase gene and the recessive white mutation in chickens. *BMC Genomics*, 7, 19.

Cieslak, M., Reissmann, M., Hofreiter, M. and Ludwig, A. 2011. Colours of domestication. *Biological Reviews of the Cambridge Philosophical Society*, 86, 885–899.

Clark, L. A., Wahl, J. M., Rees, C. A. and Murphy, K. E. 2006. Retrotransposon insertion in SILV is responsible for merle patterning of the domestic dog. *Proceedings of the National Academy of Sciences of the United States of America*, 103, 1376–1381.

Cooper, M. P., Fretwell, N., Bailey, S. J. and Lyons, L. A. 2006. White spotting in the domestic cat (Felis catus) maps near KIT on feline chromosome B1. *Anim Genet*, 37, 163–165.

Delaney, A., Keighren, M., Fleetwood-Walker, S. M. and Jackson, I. J. 2010. Involvement of the melanocortin-1 receptor in acute pain and pain of inflammatory but not neuropathic origin. *PLoS One*, 5, e12498.

Dreger, D. L. and Schmutz, S. M. 2011. A SINE insertion causes the black-and-tan and saddle tan phenotypes in domestic dogs. *J Hered*, 102 Suppl 1, S11–18.

Drogemuller, C., Philipp, U., Haase, B., Gunzel-Apel, A. R. and Leeb, T. 2007. A noncoding melanophilin gene (MLPH) SNP at the splice donor of exon 1 represents a candidate causal mutation for coat color dilution in dogs. *The Journal of Heredity*, 98, 468–473.

Durkin, K., Coppieters, W., Drogemuller, C., Ahariz, N., Cambisano, N., Druet, T., et al. 2012. Serial translocation by means of circular intermediates underlies colour sidedness in cattle. *Nature*, 482, 81–84.

Eppig, J. T., Blake, J. A., Bult, C. J., Kadin, J. A. and Richardson, J. E. 2012. The Mouse Genome Database (MGD): comprehensive resource for genetics and genomics of the laboratory mouse. *Nucleic Acids Research*, 40, D881–886.

Fadare, A. O., Peters, S. O., Yakubu, A., Sonibare, A. O., Adeleke, M. A., Ozoje, M. O. and Imumorin, I. G. 2012. Physiological and haematological indices suggest superior heat tolerance of white-coloured West African Dwarf sheep in the hot humid tropics. *Trop Anim Health Prod*, 45, 157–65.

Fontanesi, L., Beretti, F., Riggio, V., Dall'olio, S., Gonzalez, E. G., Finocchiaro, R., et al. 2009a. Missense and nonsense mutations in melanocortin 1 receptor (MC1R) gene of different goat breeds: association with red and black coat colour phenotypes but with unexpected evidences. *BMC Genetics*, 10, 47.

Fontanesi, L., Beretti, F., Riggio, V., Gomez Gonzalez, E., Dall'olio, S., Davoli, R., et al. 2009b. Copy number variation and missense mutations of the agouti signaling protein (ASIP) gene in goat breeds with different coat colors. *Cytogenet Genome Res*, 126, 333–347.

Fontanesi, L., Scotti, E., Colombo, M., Beretti, F., Forestier, L., Dall'olio, S., et al. 2010. A composite six bp in-frame deletion in the melanocortin 1 receptor (MC1R) gene is associated with the Japanese brindling coat colour in rabbits (Oryctolagus cuniculus). *BMC Genetics*, 11, 59.

Fowler, D. M., Koulov, A. V., Alory-Jost, C., Marks, M. S., Balch, W. E. and Kelly, J. W. 2006. Functional amyloid formation within mammalian tissue. *PLoS Biology*, 4, e6.

Fukuda, M., Kuroda, T. S. and Mikoshiba, K. 2002. Slac2-a/melanophilin, the missing link between Rab27 and myosin Va: implications of a tripartite protein complex for melanosome transport. *The Journal of Biological Chemistry*, 277, 12432–12436.

Furumura, M., Sakai, C., Abdel-Malek, Z., Barsh, G. S. and Hearing, V. J. 1996. The interaction of agouti signal protein and melanocyte stimulating hormone to regulate melanin formation in mammals. *Pigment Cell Res*, 9, 191–203.

Girardot, M., Guibert, S., Laforet, M. P., Gallard, Y., Larroque, H. and Oulmouden, A. 2006. The insertion of a full-length Bos taurus LINE element is responsible for a transcriptional deregulation of the Normande Agouti gene. *Pigment cell research/sponsored by the European Society for Pigment Cell Research and the International Pigment Cell Society*, 19, 346–355.

Gratten, J., Beraldi, D., Lowder, B. V., Mcrae, A. F., Visscher, P. M., Pemberton, J. M. and Slate, J. 2007. Compelling evidence that a single nucleotide substitution in TYRP1 is responsible for coat-colour polymorphism in a free-living population of Soay sheep. *Proceedings. Biological Sciences/The Royal Society*, 274, 619–626.

Haase, B., Brooks, S. A., Schlumbaum, A., Azor, P. J., Bailey, E., Alaeddine, F., et al. 2007. Allelic heterogeneity at the equine KIT locus in dominant white (W) horses. *PLoS Genetics*, 3, 2101–2108.

Haase, B., Brooks, S. A., Tozaki, T., Burger, D., Poncet, P. A., Rieder, S., et al. 2009. Seven novel KIT mutations in horses with white coat colour phenotypes. *Animal Genetics*, 40, 623–629.

Haase, B., Obexer-Ruff, G., Dolf, G., Rieder, S., Burger, D., Poncet, P. A., et al. 2010. Haematological parameters are normal in dominant white Franches-Montagnes horses carrying a KIT mutation. *Veterinary Journal*, 184, 315–317.

Haase, B., Rieder, S., Tozaki, T., Hasegawa, T., Penedo, M. C. T., Jude, R. and Leeb, T. 2011. Five novel KIT mutations in horses with white coat colour phenotypes. *Animal Genetics*, 42, 337–339.

Hansen, P. J. 1990. Effects of coat colour on physiological responses to solar radiation in Holsteins. *Vet Rec*, 127, 333–334.

Holl, H., Brooks, S. and Bailey, E. 2010. De novo mutation of KIT discovered as a result of a non-hereditary white coat colour pattern. *Animal Genetics*, 41, 196–198.

Imes, D. L., Geary, L. A., Grahn, R. A. and Lyons, L. A. 2006. Albinism in the domestic cat (Felis catus) is associated with a tyrosinase (TYR) mutation. *Anim Genet*, 37, 175–178.

Ishida, Y., David, V. A., Eizirik, E., Schaffer, A. A., Neelam, B. A., Roelke, M. E., et al. 2006. A homozygous single-base deletion in MLPH causes the dilute coat color phenotype in the domestic cat. *Genomics*, 88, 698–705.

Joerg, H., Fries, H. R., Meijerink, E. and Stranzinger, G. F. 1996. Red coat color in Holstein cattle is associated with a deletion in the MSHR gene. *Mamm Genome*, 7, 317–318.

Johansson, A., Pielberg, G., Andersson, L. and Edfors-Lilja, I. 2005. Polymorphism at the porcine Dominant white/KIT locus influence coat colour and peripheral blood cell measures. *Animal Genetics*, 36, 288–296.

Jolly, R. D., Wills, J. L., Kenny, J. E., Cahill, J. I. and Howe, L. 2008. Coat-colour dilution and hypotrichosis in Hereford crossbred calves. *N Z Vet J*, 56, 74–77.

Kaas, J. H. 2005. Serendipity and the Siamese cat: the discovery that genes for coat and eye pigment affect the brain. *ILAR J*, 46, 357–363.

Karlsson, A. C., Kerje, S., Andersson, L. and Jensen, P. 2010. Genotype at the PMEL17 locus affects social and explorative behaviour in chickens. *British Poultry Science*, 51, 170–177.

Kawaguchi, N., Ono, T., Mochii, M. and Noda, M. 2001. Spontaneous mutation in Mitf gene causes osteopetrosis in silver homozygote quail. *Developmental Dynamics: An Official Publication of the American Association of Anatomists*, 220, 133–140.

Keeler, C., Mellinger, T., Fromm, E. and Wade, L. 1970. Melanin, adrenalin and the legacy of fear. *The Journal of Heredity*, 61, 81–88.

Kerje, S., Sharma, P., Gunnarsson, U., Kim, H., Bagchi, S., Fredriksson, R., et al. 2004. The Dominant white, Dun and Smoky color variants in chicken are associated with insertion/deletion polymorphisms in the PMEL17 gene. *Genetics*, 168, 1507–1518.

Kerns, J. A., Newton, J., Berryere, T. G., Rubin, E. M., Cheng, J. F., Schmutz, S. M. and Barsh, G. S. 2004. Characterization of the dog Agouti gene and a nonagoutimutation in German Shepherd Dogs. *Mammalian Genome: Official Journal of the International Mammalian Genome Society*, 15, 798–808.

Kijas, J. M., Moller, M., Plastow, G. and Andersson, L. 2001. A frameshift mutation in MC1R and a high frequency of somatic reversions cause black spotting in pigs. *Genetics*, 158, 779–785.

Kijas, J. M., Wales, R., Tornsten, A., Chardon, P., Moller, M. and Andersson, L. 1998. Melanocortin receptor 1 (MC1R) mutations and coat color in pigs. *Genetics*, 150, 1177–1185.

Klungland, H. and Vage, D. I. 1999. Presence of the dominant extension allele E(D) in red and mosaic cattle. *Pigment Cell Res*, 12, 391–393.

Leegwater, P. A., Van Hagen, M. A. and Van Oost, B. A. 2007. Localization of white spotting locus in Boxer dogs on CFA20 by genome-wide linkage analysis with 1500 SNPs. *The Journal of Heredity*, 98, 549–552.

Liu, L., Harris, B., Keehan, M. and Zhang, Y. 2009. Genome scan for the degree of white spotting in dairy cattle. *Animal Genetics*, 40, 975–977.

Lyons, L. A., Foe, I. T., Rah, H. C. and Grahn, R. A. 2005a. Chocolate coated cats: TYRP1 mutations for brown color in domestic cats. *Mammalian Genome: Official Journal of the International Mammalian Genome Society*, 16, 356–366.

Lyons, L. A., Imes, D. L., Rah, H. C. and Grahn, R. A. 2005b. Tyrosinase mutations associated with Siamese and Burmese patterns in the domestic cat (*Felis catus*). *Anim Genet*, 36, 119–126.

Mackenzie, M. A., Jordan, S. A., Budd, P. S. and Jackson, I. J. 1997. Activation of the receptor tyrosine kinase *Kit* is required for the proliferation of melanoblasts in the mouse embryo. *Developmental Biology*, 192, 99–107.

Marklund, S., Kijas, J., Rodriguez-Martinez, H., Ronnstrand, L., Funa, K., Moller, M., et al. 1998. Molecular basis for the dominant white phenotype in the domestic pig. *Genome Research*, 8, 826–833.

Metallinos, D. L., Bowling, A. T. and Rine, J. 1998. A missense mutation in the endothelin-B receptor gene is associated with Lethal White Foal Syndrome: an equine version of Hirschsprung disease. *Mammalian Genome: Official Journal of the International Mammalian Genome Society*, 9, 426–431.

Minvielle, F., Bed'hom, B., Coville, J. L., Ito, S., Inoue-Murayama, M. and Gourichon, D. 2010. The "silver" Japanese quail and the MITF gene: causal mutation, associated traits and homology with the "blue" chicken plumage. *BMC Genetics*, 11, 15.

Mogil, J. S., Ritchie, J., Smith, S. B., Strasburg, K., Kaplan, L., Wallace, M. R., et al., A. 2005. Melanocortin-1 receptor gene variants affect pain and mu-opioid analgesia in mice and humans. *Journal of Medical Genetics*, 42, 583–587.

Moller, K., Fussing, V., Grimont, P. A., Paster, B. J., Dewhirst, F. E. and Kilian, M. 1996. *Actinobacillus minor* sp. nov., *Actinobacillus porcinus* sp. nov., and *Actinobacillus indolicus* sp. nov., three new V factor-dependent species from the respiratory tract of pigs. *International Journal of Systematic Bacteriology*, 46, 951–956.

Natt, D., Kerje, S., Andersson, L. and Jensen, P. 2007. Plumage color and feather pecking – behavioral differences associated with PMEL17 genotypes in chicken (*Gallus gallus*). *Behavior Genetics*, 37, 399–407.

Peterschmitt, M., Grain, F., Arnaud, B., Deleage, G. and Lambert, V. 2009. Mutation in the melanocortin 1 receptor is associated with amber colour in the Norwegian Forest Cat. *Anim Genet*, 40, 547–552.

Philipp, U., Lupp, B., Momke, S., Stein, V., Tipold, A., Eule, J. C., et al. 2011. A MITF mutation associated with a dominant white phenotype and bilateral deafness in German Fleckvieh cattle. *PLoS One*, 6, e28857.

Rieder, S., Taourit, S., Mariat, D., Langlois, B. and Guerin, G. 2001. Mutations in the agouti (ASIP), the extension (MC1R), and the brown (TYRP1) loci and their association to coat color phenotypes in horses (*Equus caballus*). *Mammalian Genome: Official Journal of the International Mammalian Genome Society*, 12, 450–455.

Rompler, H., Rohland, N., Lalueza-Fox, C., Willerslev, E., Kuznetsova, T., Rabeder, G., et al. 2006. Nuclear gene indicates coat-color polymorphism in mammoths. *Science*, 313, 62.

Rothschild, M. F., Van Cleave, P. S., Glenn, K. L., Carlstrom, L. P. and Ellinwood, N. M. 2006. Association of MITF with white spotting in Beagle crosses and Newfoundland dogs. *Animal Genetics*, 37, 606–607.

Rowett, M. A. and Fleet, M. R. 1993. Albinism in a Suffolk sheep. *J Hered*, 84, 67–69.

Santschi, E. M., Purdy, A. K., Valberg, S. J., Vrotsos, P. D., Kaese, H. and Mickelson, J. R. 1998. Endothelin receptor B polymorphism associated with lethal white foal syndrome in horses. *Mammalian Genome: Official Journal of the International Mammalian Genome Society*, 9, 306–309.

Schiaffino, M. V. 2010. Signaling pathways in melanosome biogenesis and pathology. *Int J Biochem Cell Biol*, 42, 1094–1104.

Schmutz, S. M., Berryere, T. G. and Goldfinch, A. D. 2002. TYRP1 and MC1R genotypes and their effects on coat color in dogs. *Mammalian Genome: Official Journal of the International Mammalian Genome Society*, 13, 380–387.

Schmutz, S. M. and Dreger, D. L. 2013. Interaction of MC1R and PMEL alleles on solid coat colors in Highland cattle. *Animal Genetics*, 44(1), 9–13.

Schmutz, S. M. and Melekhovets, Y. 2012. Coat color DNA testing in dogs: Theory meets practice. *Molecular and Cellular Probes*, 26(6), 238–242.

Seitz, J. J., Schmutz, S. M., Thue, T. D. and Buchanan, F. C. 1999. A missense mutation in the bovine MGF gene is associated with the roan phenotype in Belgian Blue and Shorthorn cattle. *Mammalian Genome: Official Journal of the International Mammalian Genome Society*, 10, 710–712.

Slominski, A., Wortsman, J., Plonka, P. M., Schallreuter, K. U., Paus, R. and Tobin, D. J. 2005. Hair follicle pigmentation. *J Invest Dermatol*, 124, 13–21.

Spritz, R. A. 1994. Molecular genetics of oculocutaneous albinism. *Human Molecular Genetics*, 3 Spec No, 1469–1475.

Tobita-Teramoto, T., Jang, G. Y., Kino, K., Salter, D. W., Brumbaugh, J. and Akiyama, T. 2000. Autosomal albino chicken mutation (ca/ca) deletes hexanucleotide (-deltaGACTGG817) at a copper-binding site of the tyrosinase gene. *Poult Sci*, 79, 46–50.

Vaez, M., Follett, S. A., Bed'hom, B., Gourichon, D., Tixier-Boichard, M. and Burke, T. 2008. A single point-mutation within the melanophilin gene causes the lavender plumage colour dilution phenotype in the chicken. *BMC Genetics*, 9, 7.

Vage, D. I., Klungland, H., Lu, D. and Cone, R. D. 1999. Molecular and pharmacological characterization of dominant black coat color in sheep. *Mammalian Genome: Official Journal of the International Mammalian Genome Society*, 10, 39–43.

Van Den Bossche, K., Naeyaert, J. M. and Lambert, J. 2006. The quest for the mechanism of melanin transfer. *Traffic*, 7, 769–778.

Yang, G. C., Croaker, D., Zhang, A. L., Manglick, P., Cartmill, T. and Cass, D. 1998. A dinucleotide mutation in the endothelin-B receptor gene is associated with lethal white foal syndrome (LWFS); a horse variant of Hirschsprung disease. *Human Molecular Genetics*, 7, 1047–1052.

Review questions

1. Explain the concept of pleiotropy and why it's important from an industry standpoint.

2. Name at least three processes integral to the final appearance of pigment in the coat.

3. What gene products are active within the melanosome as opposed to on the surface of the melanocyte?

4. Several coat colors in cattle are characteristic of particular breeds, can you name three examples?

5. What theories could explain the appearance of coat colors among the first traits following domestication?

6. As a breeder of Australian Sheep Dogs, known for their *merle* color, you would like to improve the marketability of your pups by ensuring that all of them have an attractive coat pattern. You do not, however, want to increase the incidence of eye problems. What breeding strategies could help you achieve your goals?

20 Molecular Genetics-Nutrition Interactions in Ruminant Fatty Acid Metabolism and Meat Quality

Aduli E.O. Malau-Aduli[1,2] and Benjamin W.B. Holman[2]

[1] School of Veterinary and Biomedical Sciences, Faculty of Medicine, Health and Molecular Sciences, James Cook University, Queensland, Australia
[2] Animal Science and Genetics,Tasmanian Institute of Agriculture, School of Land and Food, Faculty of Science, Engineering and Technology, University of Tasmania, Hobart, Australia

Introduction

Since the 1960s, it was already known that nutrition was among the most significant environmental factors regulating the action of genes and the subsequent exhibition of phenotypes in animals (Jacob and Monod, 1961). Ordovas (2008) concluded that "it is of paramount importance that genes are considered in the context of nutrition and that nutrition is considered within the context of genes." This notion is the foundational building block of the concept of genetics-nutrition interactions in molecular genomics. In other words, genetics-nutrition interaction describes the modulatory effect of a dietary component on a specific phenotype by a genetic variant. Alternatively, this notion refers to the dietary modification of the effect of a genetic variant on a phenotypic trait. However, the mechanisms responsible for individual and breed-specific differences in dietary responses are very complex and poorly understood. This lack of understanding necessitates a comprehensive examination of genetics-nutrition interactions in livestock at the molecular level, to unravel the role of genetic factors contributing to differences in response to supplementary nutrients for the metabolic alterations of fatty acids that affect meat quality.

A major dilemma that undergraduate students face in animal science is understanding the complex scientific concepts of genetics by nutrition interaction relationships in pasture-based systems obtainable in the sheep, dairy, and beef cattle industries (Malau-Aduli et al., 2012a). It has been suggested that in order to challenge old assumptions and break new ground in teaching and learning in the animal sciences, a paradigm shift from the traditional **teacher-focus** to a modern **student-centered** learning approach is necessary for establishing synergy (systematic working together in concert) between **research-led** teaching and **inquiry-based** learning to enhance students' understanding of both science content and scientific practices (Malau-Aduli and Lane, 2012).

Australian data conclusively demonstrated that the implementation of an innovative, inquiry-based learning, and research-led teaching approach made a significant contribution to the student learning experience in animal science through the development of students' critical thinking and scholarly values of genetics-nutrition interactions in livestock (Malau-Aduli et al., 2012a). This strategy in addition to the theoretical concepts taught in class, included student exposure to hands-on genetics-nutrition experimental growth trials with sheep (Holman and Malau-Aduli, 2013, Holman et al., 2012, Malau-Aduli and Akuoch, 2012, Malau-Aduli and Holman, 2010, Malau-Aduli et al., 2012b), laboratory experiments on intramuscular fat extraction, fat melting point, sensory evaluation of meat eating qualities, statistical data analysis, livestock industry field visits, scientific journal article critiques, and seminar presentations.

This chapter deals with our experience with inquiry-based, student-focused teaching and research approach to genetics-nutrition interactions in ruminant livestock at the University of Tasmania. An in-depth literature review on the manipulation of muscle and adipose tissue fatty acid composition and molecular marker associations with meat quality is also presented. Examples of our laboratory practicals on intramuscular fat extraction, fat melting point, meat quality and molecular genetics techniques in gene sequencing homology are also included.

Genetics-nutrition interactions in ruminants

Understanding the concept of genetics-nutrition interactions must begin with a grasp of the basics. "Genetics" simply refers to the research field that seeks to understand gene function and organization influencing the inheritance pattern of traits from parents to their progeny. The four nucleotide bases adenine (A), cytosine (C), guanine (G), and thymidine (T) constitute the genetic code or DNA (deoxyribonucleic acid) and are arranged into three base sequences (codons) that encode for amino acids that make proteins. Throughout the DNA, genes are organized in relatively small clusters with large tracks of non-coding DNA intervening between each gene. This non-coding DNA was initially and erroneously thought to be non-functional "junk," but in reality, encodes for microRNAs that influence gene expression (Figure 20.1).

Molecular and Quantitative Animal Genetics, First Edition. Edited by Hasan Khatib.

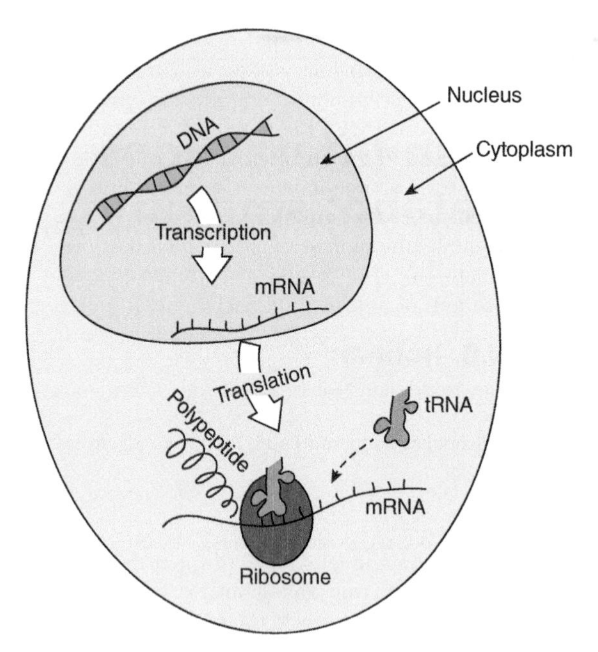

Figure 20.1 DNA transcription and translation process diagram.

Nutrition provides a variety of cellular signals that trigger the transcription of DNA into RNA (ribonucleic acid). This response to a variety of signals is gene-specific and ultimately converges on transcription factors which bind to promoter regions in the 5′ region ("upstream") of a given gene. The interactions between genes are facilitated by the helical structure of the DNA which is often coiled and wrapped around histone proteins such that the promoter regions (which are very close to the start sequence for a gene), interact with enhancers which can be many thousands of base pairs upstream. After a cellular signal triggers a specific or clustered pattern of gene transcription, the primary transcript undergoes a series of modifications including capping, polyadenylation, and splicing out of the introns to make a mature mRNA (messenger RNA) that is exported out of the nucleus. The mRNA transcript is "read" (translated) by a transfer RNA (tRNA) in the presence of ribosomes in the cytoplasm or on the rough endoplasmic reticulum. The process of translation requires ribosomal RNA, charged amino acids, tRNA, initiation, elongation, and termination factors. The original DNA code is read from the complimentary mRNA through the anticodon on the tRNA to insert the correct amino acid corresponding to the specific codon. These complex processes combine to dictate the phenotypic expression of the genes in the animal which can be measurable in terms of growth, wool, milk, or meat production.

Ruminant livestock such as goats, sheep, and cattle are naturally endowed with a complex digestive system in which rumen-resident microbes assist with the breakdown of cellulose, hemicelluloses, lignin, and other plant-derived feeds. The ability of ruminants to respond to feeding and grow, produce wool, milk, or meat of superior quality is primarily determined by the genetic make-up inherited from the parental generation. However, optimal, measurable, and quantifiable performance can only be maximized in a conducive and enabling set of environmental conditions. Regardless of the superiority of an inherited genetic potential, a less than optimal nutrition will be counterproductive to the full expression of the animal's capability.

Figure 20.2 Teaching and research sheep yard: University of Tasmania, Cambridge TAS, Australia.

Educating Australian undergraduate students in molecular genetics-nutrition interactions in ruminants

The role of genotype (inherited genetic make-up) in the determination and prediction of future performance and response of livestock under diverse farming conditions (phenotypic performance in physically observable and measurable characteristics) is poorly understood, and less so by undergraduate students.

An approach we have utilized in teaching genetics-nutrition interactions to Australian students is one that involves the use of genetically diverse sheep breeds derived from Merino ewes (to minimize maternal perturbations and turbulences) maintained under the same grazing and management conditions. The sheep are all weaners (same age) and supplemented with a variety of commonly available feeds such as canola, lupins, and micro-algal drench of *Spirulina* at the University of Tasmania Farm in Cambridge, Hobart, Tasmania. Students get hands-on experience measuring, recording and analyzing liveweight, daily gain, body condition, and wool growth (Figures 20.2–20.5) in a series of field practicals that run for about 8–10 weeks of the semester. Livestock industry visits (Figures 20.6–20.8) further enhance student understanding and engagement with practicing farmers. In the molecular genetics laboratory, the students utilize the sheep blood and wool as well as their own cheek cells for a comparative gene sequence homology analysis (Figures 20.8–20.10).

We utilized a combination of scientific research inquiry and student-based teaching approach in teaching the concept of genetics-nutrition interactions to our undergraduate students. Scientific inquiry includes (Keys and Bryan, 2001):

- Identifying and posing questions,
- Designing and conducting investigations,
- Analyzing data and evidence,
- Using models and explanations, and
- Communicating findings.

Unlike the classical **teacher-focused** approach that emphasizes transmission of research knowledge to a student audience (Healey, 2000; 2003), the **student-focused** approach emphasizes students constructing their own knowledge through active,

Figure 20.3 Teaching undergraduate students body condition scoring in genetically divergent lambs: University of Tasmania, Cambridge TAS, Australia.

Figure 20.4 Animal science students working in groups with experimental sheep flock during a field practical: University of Tasmania, Cambridge TAS, Australia.

Figure 20.5 Sheep nutrition-genetics interactions feeding trial with students: University of Tasmania, Cambridge TAS, Australia.

Figure 20.6 Meat quality student field trip: Longford Meatworks TAS, Australia.

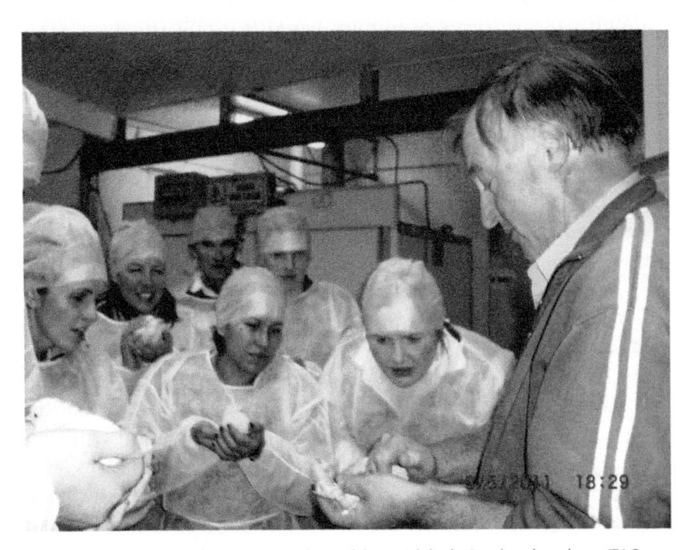

Figure 20.7 Animal science student visit to Nichols Poultry hatchery TAS, Australia.

Figure 20.8 Explaining Robotic Milking with owner John Van Adrichem at Togari TAS, Australia.

inquiry-based participation. Such inquiry-based experiences have been demonstrated to provide valuable opportunities for students to improve their understanding of both science content and scientific practices because inquiry-based learning is essentially a question-driven, open-ended process, and students must have personal experience with scientific inquiry to understand this fundamental aspect of science (Edelson et al., 1999). It has also been suggested by Healy (2005) that research-based learning structured around inquiry is one of the most effective ways for students to benefit from the research that occurs in departments. This is primarily because undergraduate students are likely to gain most benefit from research in terms of depth of learning and understanding when they are involved actively, particularly through various forms of inquiry-based learning (Healy, 2005). This approach has been demonstrated to enhance student learning, including the development of graduate attributes, and potentially leads to increased student enrolments and completions in graduate research programs (Blackmore and Cousin, 2003).

From our experience, the primary objective of inquiry-based learning and research-led teaching approach to enhance students' critical thinking and target their learning needs through active participation in hands-on growth experimental research with sheep was achieved. The curiosity-stimulated, research-led teaching approach actively engaged the undergraduate students in Animal Production Systems Unit to reflectively ask questions, think innovatively about what they needed to do to get answers and motivated them to search and retrieve relevant information from published literature. This approach in combination with the livestock industry visits, seminar presentations, field trips and laboratory experiments was very effective in giving them a rich learning experience as indicated by empirical and quantitative evidence from the Student Evaluation of Teaching and Learning (SETL) scores (Figures 20.11–20.12).

The concept of "learning by doing" is an effective way for students to benefit from staff research (McLachlan, 2006). Further supporting evidence comes from the work of Blackmore and Cousin (2003) demonstrating that students involved in research-based inquiries acquire a more sophisticated level of intellectual development. This is because active learning is more likely to encourage students to adopt a deep approach to learning, than is the teacher-focused transmission model which may encourage a surface approach (Brew and Boud, 1995). This research-led teaching approach and experimental exercise with sheep undoubtedly made a significant contribution to the student learning experience in animal science and genetics through the development of students' critical thinking skills, statistical analytical skills and scholarly values going by the following student comments on the best aspects of teaching and support they received:

> The field practicals were very enjoyable, made me understand the subject very well. Assignment support was very helpful in that it gave me a diverse understanding of how to tackle animal science written report.
>
> Field trips were an excellent component of this Unit. They were an opportunity to meet & hear from producers and very useful in gaining a practical knowledge of animal production industries.
>
> BBQ and field trips to the Uni Farm – best aspects for me.
>
> Felt that the seminars were a great way to learn.
>
> Aduli is very diligent about giving people the best chance to learn and show their knowledge.
>
> Very good at explaining topics, knew material very well and made subject matter interesting.
>
> Encyclopaedic knowledge and an ability to explain clearly and precisely. Good interaction with students.
>
> Plenty of diagrams and photos to develop good mental picture of concepts.
>
> Good mid-semester test, helpful for learning.
>
> Lectures and field experiments were really inspiring, lab helped understand material.
>
> Excursions made it interesting.

Some of the laboratory-based experiments we have engaged our students with and the detailed step-by-step procedures include:

1. Fats and beef quality, by determination of intramuscular fat and melting point in a beef cut (Appendix 20.1).
2. Sensory evaluation, of meat quality in grain-fed versus grass-fed cattle (Appendix 20.2).
3. Total lipid extraction, from a beef cut using fatty acid methylation and analysis (Appendix 20.3).
4. Molecular genetics, using DNA extraction from sheep blood and wool and students' cheek cells, DNA quantification using Nanodrop-8000, polymerase chain reaction assay, gel electrophoresis, PCR assay clean-up, ethanol precipitation clean-up of labeled dye terminator of DNA sequencing products, and analyses (Appendix 20.4).

Review of fatty acids and their manipulation, metabolism, and effect on quality in ruminants

The nutritional quality of meat is largely related to its fatty acid (FA) content and composition. Evidence from numerous studies

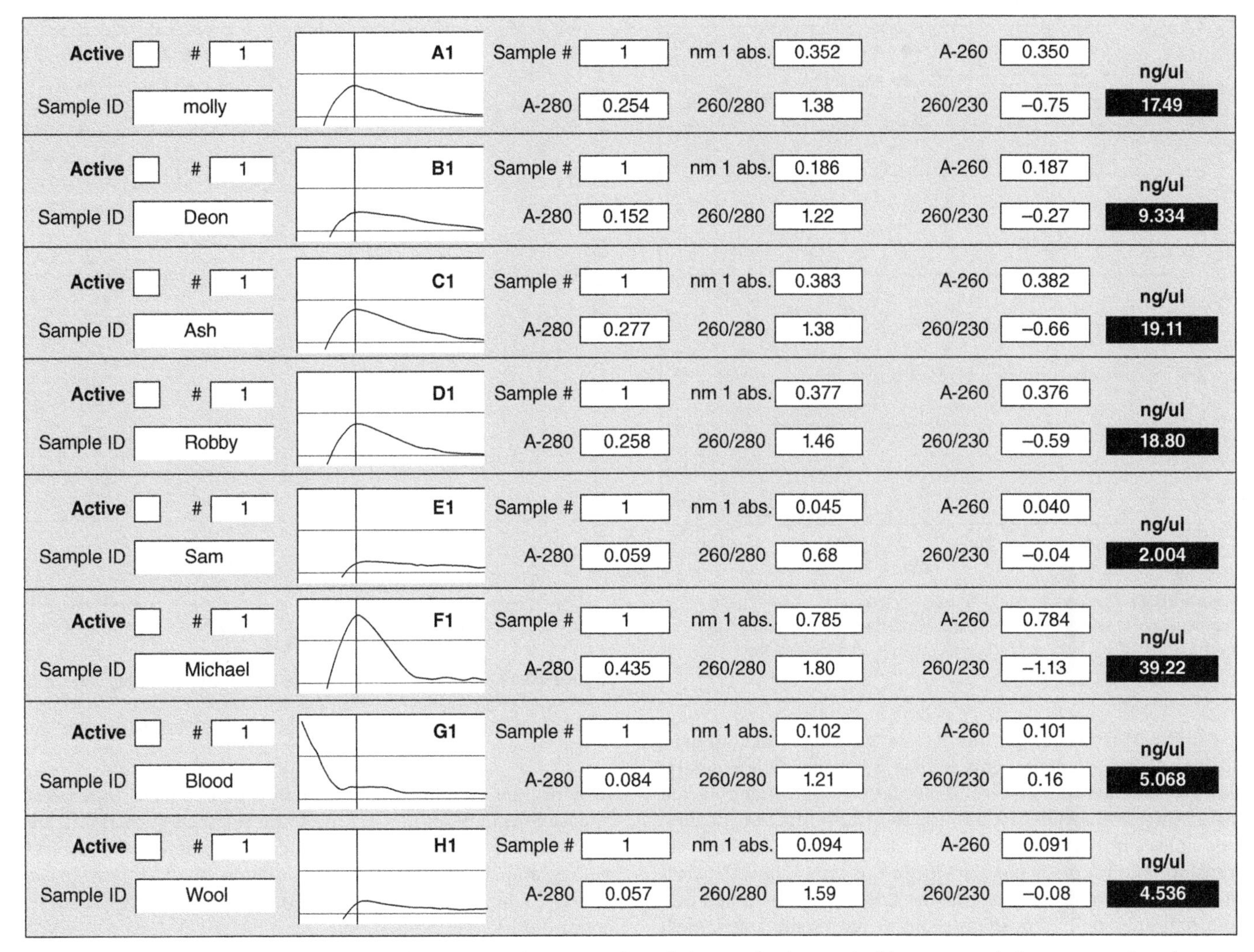

Active #	Well	Sample ID	Sample #	nm 1 abs.	A-260	A-280	260/280	260/230	ng/ul
1	A1	molly	1	0.352	0.350	0.254	1.38	−0.75	17.49
1	B1	Deon	1	0.186	0.187	0.152	1.22	−0.27	9.334
1	C1	Ash	1	0.383	0.382	0.277	1.38	−0.66	19.11
1	D1	Robby	1	0.377	0.376	0.258	1.46	−0.59	18.80
1	E1	Sam	1	0.045	0.040	0.059	0.68	−0.04	2.004
1	F1	Michael	1	0.785	0.784	0.435	1.80	−1.13	39.22
1	G1	Blood	1	0.102	0.101	0.084	1.21	0.16	5.068
1	H1	Wool	1	0.094	0.091	0.057	1.59	−0.08	4.536

Figure 20.9 DNA quantification using Nanodrop by UTAS animal science students during a molecular genetics laboratory practical on sequence homology between sheep blood, wool, and human cheek cells.

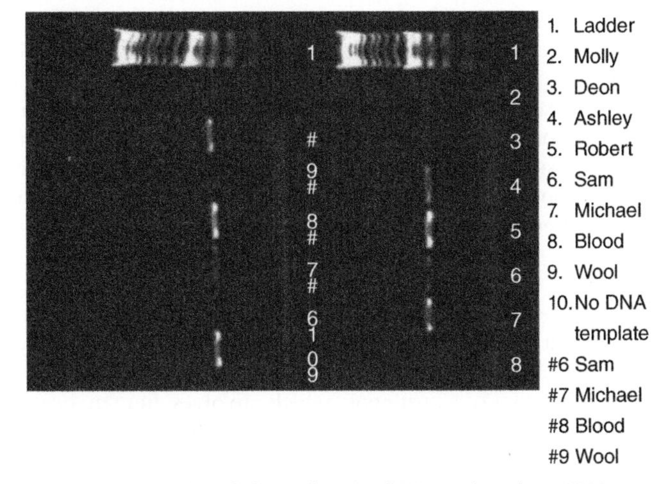

Figure 20.10 Sample gel electrophoresis of PCR products from UTAS animal science students during a molecular genetics laboratory practical on sequence homology between sheep blood, wool, and human cheek cells.

(Baeza et al., 2012; Conte et al., 2010; Orrù et al., 2011; Schennink et al., 2008; Taniguchi et al., 2004) suggest that nutritional and genetic factors as well as their interactions are key regulators of fatty acid metabolism and lipogenesis in cattle. In terms of FA desaturation, experimental evidence shows that stearoyl-CoA desaturase (SCD) is one of the most promising candidate genes to explain FA variability and regulation. Understanding and manipulating these regulators can provide a viable means to tailor fatty acid composition to achieve desirable meat qualities. An overview of published literature on research of nutritional and genetic effects on ruminant fatty acid profiles and related meat quality is presented.

Fatty acids

Lipids include fats, waxes, sterols, and many other derivatives. The primary biological function of lipids is energy storage, but they are also vital components of cellular membranes and signaling molecules (Campbell and Farrell, 2012). Lipids and fats have become synonymous with negative health effects and have been widely discriminated against by consumers for decades. However, this generalization is in many instances, inaccurate.

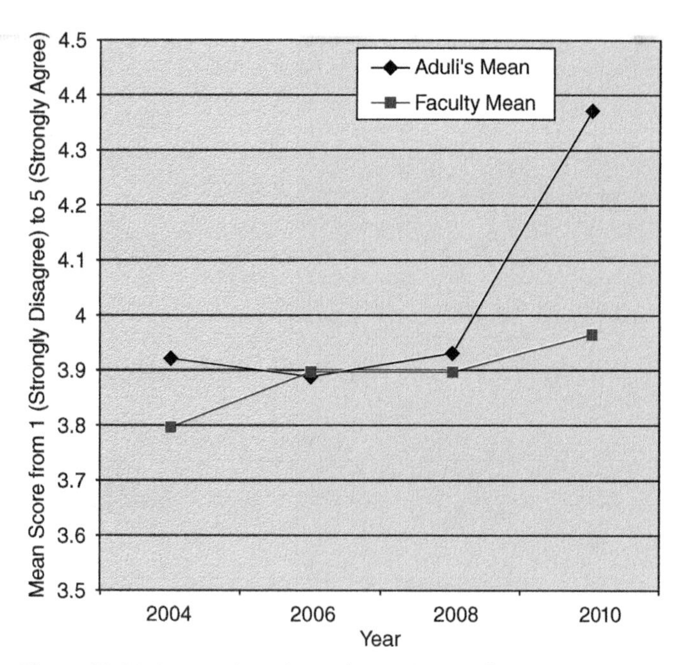

Figure 20.11 Comparative unit "Student Evaluation of Teaching and Learning" (SETL) scores between Animal Production Systems and Faculty means (2004–2010) involving 125 students with an 83% response rate.

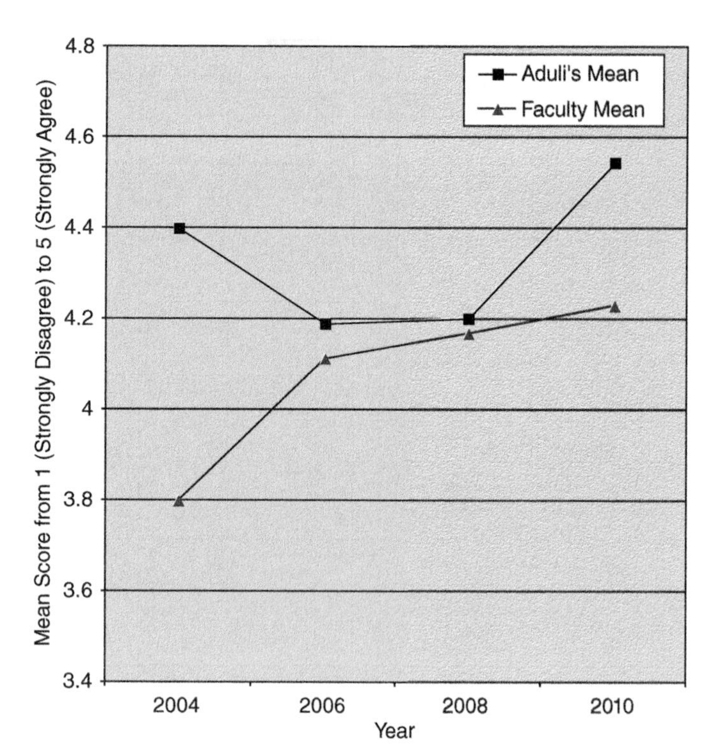

Figure 20.12 Comparative unit "Student Evaluation of Teaching and Learning" (SETL) evaluation between Animal Production Systems and Faculty means (2004–2010) involving 125 students with an 83% response rate.

Fatty acids (FAs) are the subunits or "building blocks" of lipids and are found as components of triacylglycerols, wherein three FAs are attached to a glycerol molecule (Campbell and Farrell, 2012; Gurr and James, 1972). It is the transformation of these free lipids into triacylglycerols that allows energy storage. The transformation is made possible due to the amphiphilic structure of FAs arising from their hydrophilic carboxyl group attachment to a hydrophobic hydrocarbon chain or tail (Webb and O'Neill, 2008). It is the length and bonds between carbon atoms of these hydrocarbon chains that are used to differentiate between long- or short-chain and saturated or unsaturated FAs. Long-chain FAs have hydrocarbon chains ranging from 13 to 21 carbon atoms in length. Saturated FAs (SFAs) have hydrocarbon chains with no double bonds unlike unsaturated FAs (UFAs) that can either be monounsaturated (MUFA) or polyunsaturated (PUFA) if only one double bond or several double bonds exist, respectively (Gurr and James, 1972). It is these characteristics which directly influence lipid functionality and traits.

FAs can be divided into two major groups, namely essential and non-essential FAs. Essential FAs cannot be synthesized *de novo* by mammals and must be provided in the diet. These include: (1) α-linolenic acid, an omega-3 FA; and (2) linoleic acid, an omega-6 FA (Webb and O'Neill, 2008). Non-essential FAs do not require dietary sourcing as they can be synthesized principally by the liver.

Fatty acid deposit sites in ruminants

FAs are located in ruminant muscle and adipose tissues and organs. Muscle FA deposits have major implications on meat quality because of intrinsic relationships with sensory and nutritional eating quality. Adipose FA deposits are the preferable sites for energy storage and they are dominated by triacylglycerols (Malau-Aduli et al., 1997a; Scollan et al., 2001a). Health conscious consumers trim the adipose tissue from meat cuts prior to consumption. Nonetheless, adipose tissues have widespread applications in other indirect ruminant products such as lard and food additives. Organs with significant FA deposits include the kidney and liver. While the consumption of these FA deposits is limited in many Western societies, they are eaten in other communities.

Differences between FA profiles have been found between anatomical deposit sites in ruminants (Malau-Aduli et al., 1997b). For instance, muscle FA profiles are largely long- chained (C20–22) (Aurousseau et al., 2007), but the adipose tissue has comparatively much greater amounts of short-chain FAs (Malau-Aduli et al., 1998a; Wood et al., 2008;). Bas and Morand-Fehr (2000) demonstrated that on-the-whole, adipose and muscle FA deposit profiles are similar with only minor differences (Wood et al., 2008). These differences are thought to stem from the preferential incorporation of essential FAs into muscle deposits than adipose tissue (Wood et al., 2008), for instance linoleic acid is more predominant in ruminant muscle compared to adipose tissue. Differences between organ FA profiles remain relatively unexplored in key ruminant species.

Fatty acid effect on quality

Meat quality can be divided into two fractions; (1) Sensory, which describe consumer's tangible sense stimulation upon consumption; and (2) Nutritional, which involves human health benefits with consumption. Both these are affected by tissue FA composition and content.

Sensory quality

FAs are acknowledged as key determinants of ruminant meat quality (Bas and Morand-Fehr, 2000; Scollan et al., 2001a), especially in meat appearance, aroma, flavor, firmness and shelf-life than for other key sensory quality traits (Wood et al., 2004; 2008). Both tenderness and juiciness are fundamental to meat sensory quality and are affected by total lipid content, especially intramuscular fat or marbling, rather than lipid profile (Hopkins et al., 2007).

The acceptability of ruminant meat depends on its appearance, as a consumer's "first taste is with the eye." Zhou et al. (1993) found white subcutaneous fat to be associated with high SFA content. As UFA levels increase, subcutaneous fat becomes yellow in color. The genetics of fat color and associations with fatty acids and beta-carotene in cattle have been published (Kruk et al. 1997a,b, Kruk et al., 1998, 2004; Siebert et al., 2000, 2006). Fat firmness is higher when SFA content is higher than UFA due to differences in melting temperatures. Malau-Aduli et al. (2000a) and Perry et al. (1998) showed that SFAs melted at higher temperatures than UFAs. SFAs have higher and better resistance to oxidation compared to UFAs and hence protect against peroxides and free radicals damaging proteins, enzymes and other biological molecules (Webb and O'Neill, 2008). Oxidation of FAs can undermine meat shelf-life by promoting the release of "rancid" odors and flavors.

Flavor is one of the underpinning traits of meat consumer satisfaction. It is defined as the combined taste and smell sensations stimulated by meat consumption. FA profile has been shown to affect both these sensations contributing to flavor. With aroma for instance, Lorenz et al. (2002) suggested that increased linoleic acid content resulting from grain-feeding to German Simmental bulls caused a "soapy" aroma once cooked. Likewise Elmore et al. (1999) found that higher PUFA levels in beef steaks resulted in increased lipid oxidation, thus altering aromatic profile. Steaks from cattle fed fish oil have rancid and "fishy" flavors. Meat taste has also been linked with FA profile. Nute et al. (2007) found feeding lambs different oil supplements influenced lamb flavor intensity, with α-linolenic acid content positively correlated with flavor intensity. Similarly, methyloctanoic acid and 4-methylnonoic acid branched chain FAs have been associated with sheep's distinctive taste (Watkins et al. 2012). In goats, Wong et al. (1975) showed that hircinoic acid was a contributing factor to goat meat's "goatniness."

Nutritional quality

Ruminant meat nutritional quality revolves around increased human health benefits and simultaneous reduction of antagonists with consumption. SFAs have been widely acknowledged as key causal factors of cardiovascular disease in developed nations. Producers are actively seeking for ways to reduce SFA content in meat by resorting to leaner meat production. Therefore, interest in omega-3 (n-3) PUFAs has been increasing because of the strong association between n-3 PUFAs and improved human health through reduced coronary heart disease and the promotion of mental health and infant development. Increased n-3 intakes can be achieved through ruminant meat consumption, although it is a relatively lesser n-3 source compared with fish (Scollan et al., 2001a). Interestingly, n-3 content is widely reported as a ratio with Omega-6 (n-6) PUFA. This stems from n-6 countering action to n-3 utilization through its competition for rate-limiting enzymes. A healthy human diet should contain both n-6 and n-3, ideally at a 4:1 ratio. Sourcing n-6 to maintain this ratio is not as challenging as n-3, therefore this ratio provides an accurate means to monitor and compare FA profiles between ruminant species.

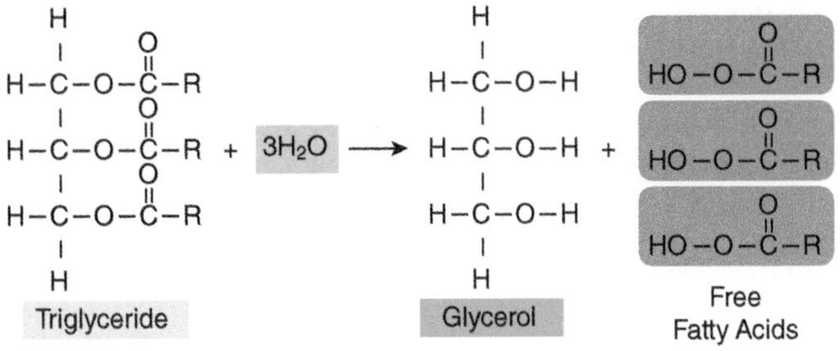

Triglyceride + $3H_2O$ → Glycerol + Free Fatty Acids

Figure 20.13 Hydrolysis of triacylglycerides in ruminants.

Linoleic acid has significant biological roles in reducing serum cholesterol and subsequent promotion of cardiovascular health. Ruminant meat is a rich source of linoleic acid, albeit as conjugated linoleic acid (CLA). This is sourced from dietary FAs and is synthesized *de novo* in the adipose tissue. CLA obtained from the ingestion of ruminant meat has been linked with reduced risks of cancer and diabetes and a positive effect on human immune function (Quinn et al., 2000).

Ruminant fatty acid metabolism.

In ruminants, ingested FAs vastly differ from those absorbed (Bas and Morand-Fehr, 2000; Doreau et al., 2011) due to microbial biohydrogenation. However, a higher trend of absorption to ingestion occurs in ruminants unless they are consuming a lipid-rich diet (Doreau and Ferlay, 1994; Lock et al. 2006). Lipid metabolism in the rumen is sequential in nature. In the initial phase, the majority of consumed lipids are lipolyzed or "broken down" into free FAs through rumen microbial action, particularly by *Anaerovibrio lipolytica*, *Butyrivibrio fibrosolvens* and several other microbial strains of lipase and esterase producing bacteria (Jenkins, 1993; Lourenco et al., 2010). Along with free FAs, other products of lipolysis include glycerol and galactose, which are both fermented into propionic and butyric acid end-products. After lipolysis, FAs undergo biohydrogenation and isomerization (Figure 20.13).

Biohydrogenation and isomerization have been the topics of numerous review papers (Jenkins, 1993). Both biohydrogenation and isomerization are the processes in which free FAs are hydrogenated to SFAs through ruminal microbial and protozoan activity, the latter's role through predation of microbes rather than by direct involvement (Doreau et al., 2011). The first step of biohydrogenation depends on the isomerization reactions of isomerase. Isomerase is only functional when a FA has a free carboxyl group, or a PUFA has a *cis*-12 diene double bond configuration (Jenkins, 1993). It acts to shift FA double bonds in free UFAs so that the UFAs become trans-11 isomers. These isomers can then be hydrogenated or saturated through microbial reductase activity. The theoretical end-product of biohydrogenation and isomerization is stearic acid and other SFAs. Numerous MUFA isomers also pass to the duodenum for absorption (Jenkins, 1993). For instance, CLA isomers, such as health promoting c9t11CLA and t10c12CLA (O'Quinn et al., 2000), are formed through incomplete hydrogenation of linoleic acid, itself a primary substrate for biohydrogenation (Lourenco et al., 2010).

Fatty acid protection from biohydrogenation within the rumen

Ruminal microbial enzymes are pH sensitive, in that their activity declines at low pH and improves at high pH levels. Van Nevel and Demeyer (1996) reported that during lipolysis, lipase the necessary precursor for biohydrogenation, is inhibited by low pH. Hence, management strategies that promote high pH could be used to improve meat quality as typified by research with grain supplementation to ruminants. Grain supplementation has been shown to increase rumen pH (Dixon and Stockdale, 1999) and in severe cases, induce acidosis. Aurousseau et al. (2007) found that in Ile-de-France pasture-fed lambs which were stall finished on concentrates, both CLA isomers *cis*-9 and *trans*-11 decreased. Similarly, Daley et al. (2010) in a comparison of grain- and pasture-fed cattle concluded that both CLA and n-3 meat content are comparatively reduced in grain-fed cattle. However, Descalzo et al. (2005) found that in Argentine beef, the total FA content of grain-fed cattle was greater than in grass-fed cattle. This increase was attributed to increased dietary lipid content.

French et al. (2000) reported an increase in total FA, SFA, n-6:n-3 ratio and a decrease in CLA and PUFA:SFA ratio with increasing concentrate proportions within continental crossbred steer diets in agreement with previous reports by Malau-Aduli et al. (1998a,–d). Similarly, Ryan et al. (2007) found that Boer goats fed concentrates had higher total FA content, SFA, MUFA, and n-6 FAs within *Longissimus dorsi* muscle cuts compared to those solely grass-fed.

Differences in dietary FA profiles have profound effects on FA digestion and meat quality. For example, Demirel et al. (2004) found that the nutritional quality of lamb meat could be improved by increasing PUFA intake using several feed types. Scollan et al. (2001b), Mandell et al. (1997), and Shingfield et al. (2003) found similar results in cattle using fish meal supplementation, although the latter was in Finnish Ayrshire dairy cows. Fish meal is rich in n-3 PUFA and its supplementation was shown to improve beef n-3 contents of eicosapentaenoic acid (EPA) and decosahexaenoic acid (DHA), while arachidonic acid (ADA) decreased. Both EPA and DHA have inhibitory effects on the final phases of biohydrogenation (Ashes et al., 1992; Raes et al., 2004) as they affect group B ruminal microbes which hydrogenate trans-18:1 FA to stearic acid (Shingfield et al., 2003). Ruminant FA digestibility is thought to increase with increasing hydrocarbon chain length (Schmidely et al., 2008). Consequently, it is the FA profile and content of this long chain FA in ruminant feeds that affects rumen passage. Scollan et al. (2001b) reported that in steers supplemented with linseed, fish oil, megalac, and fish oil/linseed mix, the level of biohydrogenation varied depending on the level of long chain PUFAs in the supplement.

Lipogenesis

The *de novo* synthesis of FAs is collectively termed as lipogenesis and relationships between genetic variation, lipid metabolism and nutritional genomics have been published by Corella and Ordovas (2005), Ordovas and Corella (2004a,b, 2005) and Ordovas (2008). In ruminants, lipogenesis occurs predominantly within the adipose tissue as opposed to the liver in monogastrics (Ingle et al., 1972a,b). FA synthesis occurs through an extramitochondrial system in which acetyl-CoA is transformed into palmitate in adipose cellular cytosol (Laliotis et al., 2010). Hence, acetyl-CoA is the primary substrate and limiting product for ruminant lipogenesis. Acetyl-CoA is sourced from acetate, a basic sugar, which is oxidized within the mitochondria to an intermediate (pyruvate) before reaching its final form (Laliotis et al., 2010). The use of acetate is characteristic of ruminants as glucose is primarily used by monogastrics (Hood et al., 1972). This process of lipogenesis is highly complex and controlled by an intricate network of factors, including genes controlling the expression of promotants and suppressive lipolytic enzymes.

Sire breed and sex effects on ruminant fatty acid profile

Breed comparisons and differences in fatty acid profile have been the topic of numerous studies investigating meat quality in different ruminant species. Choi et al. (2000) found that *Longissimus dorsi* muscle cuts of Welsh Black steers had more n-3 PUFAs and less n-6 PUFAs compared to Holstein Friesians. Likewise, Malau-Aduli et al. (1998a) found greater EPA and lower ADA and αLA content within the triceps *brachii* muscle of Jersey compared to Limousin cows all maintained as a single herd. Numerous research papers reporting substantial breed differences and gender variations in fatty acid composition in cattle have been published (Deland et al. 1998; Malau-Aduli et al., 2000a,–d, 1995, 1996, 1998b; Pitchford et al., 2001, 2002; Siebert et al. 1998, 1999).

Breed differences are also evident in ovine species. For instance, Welsh Mountain lambs were shown to have higher palmitic and oleic FA and lesser PUFA content in the *semimembranosus* tissue than Suffolk and Soay lambs (Fisher et al., 2000). Furthermore, Mezöszentgyörgyi et al. (2001) found that Suffolk subcutaneous adipose tissues had lower proportions of C12:0 and C14:0 SFAs compared to Pannon and Booroola Merinos and the proportions of saturated FAs were higher in female lambs than males of the same age.

Molecular marker association with ruminant fatty acid composition and meat quality

Malau-Aduli et al. (2011) investigated the genetic association between polyunsaturated fatty acids (PUFA), delta-6 desaturase (FADS2) and fatty acid binding protein (FABP4) gene clusters in Australian crossbred sheep. Thirty-one single nucleotide polymorphisms (SNP) were genotyped in *Longissimus dorsi* muscle samples from 362 crossbred prime lambs sired by five genetically divergent rams. They reported that FAPB4 SNP was significantly associated with 18:4n-3 (stearidonic acid), while FADS2 SNP was significantly associated with intramuscular levels of eicosapentaenoic (C20:5n-3) and docosahexaenoic (C22:6n-3) acids. Their results suggested that these SNP variants may have a direct effect on functional lipid synthesis pathway associated with higher delta-six desaturase activity. For the first time, the study provided evidence for an association between genetic variants of FADS2, FABP4, and omega-3 PUFA in sheep muscle. They concluded that these SNPs could potentially be novel markers of choice for prime lamb producers to effectively select for enhanced muscle omega-3 long-chain fatty acid content in their breeding flock.

Stearoyl-CoA desaturase encoded by the ***stearoyl-CoA desaturase*** (*delta-9-desaturase*) (*SCD*) gene and fatty acid synthase encoded by the ***fatty acid synthase*** (*FASN*) gene are two enzymes that play important roles in determining the fatty acid profile of ruminant tissues (Kim and Ntambi, 1999). Stearoyl-CoA desaturase plays a rate-limiting role in the synthesis of unsaturated fatty acids by insertion of a *cis*-double bond in the Δ^9 position of fatty acids, with palmitate (16:0) and stearate (18:0) proposed as the preferred substrates that are converted to palmitoleate (9*c*-16:1) and oleate (9*c*-18:1), respectively. The "CC" genotype of the *SCD* SNP was initially reported to be associated with a higher amount of MUFA that included 9*c*-14:1, 9*c*-16:1 and 9*c*-18:1, and lower melting point in the intramuscular fat of Japanese Black Cattle (Taniguchi et al., 2004). The association of the "CC" genotype with a higher content of 9*c*-14:1 was also found in brisket adipose tissue of 223 Canadian Angus and Charolais-based commercial cross-bred beef steers (Li et al., 2012). However, no significant associations were detected for palmitate (16:0), stearate (18:0), palmitoleate (9*c*-16:1) or oleate (9*c*-18:1). Ohsaki et al. (2009) found that the CC genotype was significantly associated with a lower concentration of 18:0 and a higher concentration of 18:1 (presumably 9*c*-18:1) in both peri-renal and intramuscular adipose tissues of Japanese Black Cattle, but the associations with 16:0 or 16:1 (presumably 9*c*-16:1) were not significant.

In a population of Fleckvieh bulls, Barton et al. (2010) detected that the CC genotype was significantly associated with a lower concentration of 18:0 in both intramuscular and subcutaneous fats and a higher concentration of 9*c*-18:1 in the muscle fat, but the associations with 16:0 or 9*c*-16:1 were not significant. A previous study found that the CC genotype of the SNP was associated with a lower concentration of 9*c*-16:1 in both the intramuscular and subcutaneous fat of yearling bulls of northern Spanish Asturiana de los Valles and Asturiana de la Montana breeds (Li et al. 2010). The inconsistent associations of the SNP for palmitate (16:0), stearate (18:0), palmitoleate (9*c*-16:1) or oleate (9*c*-18:1) suggest that other SNPs of *SCD* or other genes may also play a role in the desaturation of palmitate (16:0) and stearate (18:0) in beef tissues. On the other hand, the "CC" genotype of the SNP was significantly associated with a higher concentration of 14:1 (presumably 9*c*-14:1) in both the peri-renal and intramuscular adipose tissues of Japanese Black Cattle (Ohsaki et al., 2009), in both the muscle and subcutaneous fat of Fleckvieh bulls (Barton et al., 2010), in the intermuscular, intramuscular and subcutaneous adipose tissues of yearling bulls of northern Spanish Asturiana de los Valles and Asturiana de la Montana breeds (Li et al., 2010), and in the brisket adipose tissue of Canadian commercial steers. The remarkable consistency of the association of the *SCD* SNP with 9*c*-14:1 reported to date indicates a predominating effect of the *SCD* on the desaturation of 14:0 in beef cattle, and the results also suggest that the *SCD* SNP may be causative or is closely linked to the causative mutations for the desaturation of 14:0.

Taniguchi et al. (2004) identified a single nucleotide polymorphism [C/T] in the fifth exon of bovine *SCD* gene (c:878C>T) that causes an amino acid change from alanine to valine and found that the SNP was associated with the percentage of monounsaturated fatty acids (MUFA) and melting point of intramuscular fat in Japanese Black steers. Zhang et al. (2008) reported three SNPs in the coding regions of the bovine *FASN* that encode the thioesterase (TE) domain and found that the g.17924A>G SNP that causes an amino acid change from threonine to alanine was associated with some of the individual and groups of fatty acids in the *Longissimus dorsi* muscle of American Angus cattle (Zhang et al., 2008). Li et al. (2012) examined the associations of the two SNPs of *SCD* and *FASN* with the fatty acids of brisket adipose tissue in a population of 223 Canadian Angus and Charolais-based commercial cross-bred beef steers. It was found that the "CC" genotype of the *SCD* SNP was significantly associated with lower concentrations of saturated fatty acids (SFA) including 10:0, 14:0, and 20:0, higher concentrations of monounsaturated fatty acids including 9*c*-14:1, 12*c*-16:1, and 13*c*-18:1, higher concentrations of polyunsaturated fatty acids (PUFA) including 9*c*,15*c*-18:2, 10*c*,12*c*-18:2, 11*c*,13*t*-18:2, and 12*c*,14*t*-18:2, but lower concentrations of other PUFA of 9*c*,13*t*/8*t*,12*c*, and 20:2n-6 ($P < 0.05$). The "AA" genotype of the *FASN* SNP was significantly associated with higher concentrations of SFAs of 10:0, 12:0, 13:0, 14:0, and 15:0, lower concentrations of unsaturated fatty acids of 9*c*-18:1 and 20:3n-6, and higher concentrations of unsaturated fatty acids of 9*c*-14:1 and 12*c*-16:1 ($P < 0.05$). Significant epistatic effects between the *SCD* and *FASN* SNP genotypes were also found for several fatty acids including 10:0, 23:0, 6*t*/7*t*/8*t*-18:1, 12*t*-18:1, 13*t*/14*t*-18:1, 16*t*-18:1, total *trans*18:1, and 9*c*,13*t*/8*t*,12*c*-18:2 ($P < 0.05$). These results further suggest that *SCD* and *FASN* are strong candidate genes influencing fatty acid composition in beef cattle.

Concluding remarks

- In teaching the concepts of genetics-nutrition interactions with ruminants to undergraduate students in animal science, the use of hands-on "learning by doing" approach enhances a deeper understanding of taught principles.
- The use of inquiry-based, student-focused teaching approach facilitates a grasping of science content and scientific principles as students are able to then construct their own knowledge through curiosity and deep thinking.
- Basic molecular genetics skills are acquired when students are taken through a comparative sequence homology practicals in the laboratory based on field trials. Extracting DNA from their own cheek cells and comparing it with the DNA sequence of sheep they had monitored in the field makes it an exciting experience.
- Incorporation of field visits to industries with genetically diverse livestock gives the students an opportunity to link concepts taught in class with face-to-face interactions with practicing farmers. Such experiences "stick" to their memories better than lectures alone.
- One of the best ways to follow-up laboratory practicals on intramuscular fat extraction in beef is to organize a BBQ using beef cuts from grass-fed versus grain-fed cattle with the students serving as sensory evaluators of meat eating qualities of marbling, flavor, tenderness, juiciness, and aroma.
- Several weeks of growth trials in the field gives the students an opportunity to be involved in experimental designs, data collection and analysis using computer software such as SAS or SPSS. This develops confidence and skills in data management, editing and statistical analyses. These are key skills they will later need as graduate students.
- In-building a seminar presentation component based on their findings in the field and laboratory enhances student communication skills and confidence to respond to questions from the audience.
- Working in groups fosters team spirit and relationships between students in which each member contributes to the group.
- Enthusiasm from the lecturer for genetics and nutrition and respect for students being mentored is a necessary tool for sustained interest in the subject. From our experience, it can be positively "infectious."
- Fatty acid profile has key roles in both sensory and nutritional quality aspects regarding ruminant products.
- Ruminant metabolism and lipogenesis of fatty acids is unique due to the interaction of ruminal microbes.

• Nutrition and molecular genetic markers can be used in managing ruminant fatty acid profiles, and subsequent quality traits.

Appendix 20.1: Fats and beef quality laboratory practicals

Experiment 1.1: Determination of intramuscular fat percentage in a beef cut

1. Weigh 5 g of the meat sample.
2. Homogenize the meat sample with a Ronson homogenizer.
3. Weigh out exactly 2 g of the homogenized meat sample.
4. Measure 5.5 ml of chloroform: methanol (2: 1) solvent into a 45 ml plastic tube.
5. Add the meat sample to the solvent and shake vigorously for about 5 min.
6. Use a filter paper to separate the filtrate and re-homogenize the residue again with 5.5 ml of solvent.
7. Add 5.5 ml of 10%KCl to the combined filtrates and shake. Allow to sediment.
8. Weigh a clean, dry, ceramic crucible.
9. Carefully remove the upper layer using pipettes while retaining the lower layer
10. Transfer the lower lipid layer into the weighed ceramic crucible.
11. Evaporate in a laminar fume hood over a heating block.
12. Place the crucible in a dessicator for 15–20 min to dry properly.
13. Weigh the crucible with the dried intramuscular fat sediment.
14. Calculate the amount of intramuscular fat (IMF) as: Final wt – Initial wt
15. Convert the intramuscular fat to percentage: IMF wt(g)/2 (g) × 100.

Experiment 1.2: Determination of the melting point of subcutaneous fat in a beef cut

1. Cut about 100g of subcutaneous or intramuscular fat from the beef sample into a petri dish.
2. Put the petri dish into an oven at 100°C for about 10–15 min to melt the fat.
3. Using air suction with your finger, suck the melted fat into thin capillary tubes.
4. Refrigerate the tubes for about 10–15 min to allow the fat to solidify.
5. Remove the samples from the fridge and mark the fat level with a pen.
6. Attach each capillary tube to a thermometer using a string.
7. Place this in a beaker containing water and heat over a heating block.
8. Keep observing closely until the fat melts and "slips."
9. Record the temperature at which this slip occurs. This is your melting point.

Appendix 20.2: Sensory evaluation of meat quality in grain-fed versus grass-fed beef

You are given barbequed meat cuts from beef cattle that were fed grain or grass to see, taste and evaluate for eating quality attributes. All meat cuts were Halal-certified.

Grain-fed beef cut

1. Meat color
( ) Very attractive ( ) Attractive ( ) Not attractive ( ) Off-color
2. Flavor and aroma
( ) Very good ( ) Good ( ) Fair ( ) Bad ( ) Off-flavor
3. Juiciness
( ) Very juicy ( ) Juicy ( ) Medium ( ) Dry
4. Tenderness
( ) Very tender ( ) Tender ( ) Slightly tough ( ) Tough ( ) Very tough
5. Marbling
( ) Highly marbled ( ) Moderately marbled ( ) Lean ( ) Very lean

Grass-fed beef cut

1. Meat color
( ) Very attractive ( ) Attractive ( ) Not attractive ( ) Off-color
2. Flavor and aroma
( ) Very good ( ) Good ( ) Fair ( ) Bad ( ) Off-flavor
3. Juiciness
( ) Very juicy ( ) Juicy ( ) Medium ( ) Dry
4. Tenderness
( ) Very tender ( ) Tender ( ) Slightly tough ( ) Tough ( ) Very tough
5. Marbling
( ) Highly marbled ( ) Moderately marbled ( ) Lean ( ) Very lean

Appendix 20.3: Total lipid extraction from a beef cut for fatty acid analysis

Experiment 3.1: Total lipid extraction from a beef cut for fatty acid analysis

1. Cut approximately 0.1 g of the meat sample.
2. Snap-freeze in liquid nitrogen in a stainless steel mortar and grind to powder
3. Transfer the ground sample using a spatula into a 20 ml test tube.
4. Add 3 ml of methanol containing butylated hydroxyl toluene (BHT) to the ground sample and vortex for 1 min.
5. Add 6 ml of chloroform and 1.5 ml of water.
6. Place on a shaker for 15 min.
7. Centrifuge for 5 min at 3000 rpm (i.e., 800 × g).
8. Carefully transfer the lower layer (lipid extract) into a clean test tube.
9. Re-extract the top aqueous layer with 5 ml of chloroform and repeat steps 6 and 7.
10. Dry the combined lipid extracts at 40°C in a heating block under a steady stream of Nitrogen gas in a fume hood.
11. Re-constitute the extracted lipid in 100μl of hexane in a screw-capped tube.

Experiment 3.2: Fatty acid methylation

1. Add 1.5 ml of methanol: chloroform: hydrochloric acid (10: 1: 1, v/v/v) to the extracted lipids.
2. With the tube firmly capped, heat the extracted lipids in an oven at 80°C for 2 h.

3. Add 3 ml of water and extract three times with hexane: chloroform (4:1, v/v) to obtain fatty acid methyl esters (FAME)
4. Dry the FAME at 40°C in a heating block under a steady stream of nitrogen gas in a fume hood.
5. Reconstitute the dried FAME in 100 µl of Iso-octane (Trimethyl Pentane) in vials and refrigerate until ready for gas chromatographic analysis for fatty acids.

Fatty acid analysis using GC-MS

An internal injection standard (19:0 FAME) is added to the FAME before analysis by gas chromatography (GC) using an Agilent Technologies 7890A GC (Palo Alto, California, USA) equipped with a Supelco Equity-1 fused silica capillary column (15 m × 0.1 mm). Helium is used as the carrier gas. Samples are injected by a split/splitless injector and an Agilent Technologies 7683B Series auto-sampler and operated in splitless mode, at an oven temperature of 120°C. After 1 minute, the oven temperature is raised to 270°C at 10°C per minute and finally to 300°C at 5°C per min and held for 5 min.

Peaks of individual fatty acids are then quantified by Agilent Technologies GC ChemStation software (Palo Alto, CA, USA). Individual fatty acid identification is confirmed by mass spectral data by comparing retention time data with those the standards. GC–mass spectrometric analyses are usually performed on a Finnigan Thermoquest GCQ GC–mass spectrometer fitted with an on-column injector using Thermoquest Excalibur software (Austin, TX, USA). The GC we have is fitted with a capillary column of similar polarity to that described above. GC peak areas can be used as percentages or converted to mg/100 g using the 19:0 FAME internal injection standard prior to statistical analysis. This way, the fatty acids can be categorized into total saturated, monounsaturated and polyunsaturated fatty acids.

Appendix 20.4: Molecular genetics laboratory practical

Experiment 4.1: DNA extraction from cheek, blood and wool tissues

Aim

The objective of this laboratory practical is to enable students to acquire some molecular techniques in extracting genomic DNA from animal tissues. DNA is essentially the animal's genetic "blueprint" that can be used for various purposes including parentage verification, genotyping, QTL mapping or sequencing among many other downstream applications. Amplification of the DNA by PCR and resolution of fragments by gel electrophoresis using oxidase cytochrome gene primers would be followed through in the next couple of weeks. Today, we will use the MOBIO Ultraclean Blood, Tissue, and Cell DNA extraction kits for the practical using sheep wool and blood, and human cheek tissues.

Safety precautionary measures

Laboratory coats and gloves must be worn at all times. DNA can easily get contaminated, therefore, you need to be clean and use only sterilized environments. Be very careful with liquid nitrogen. Don't let any drop touch your skin. Demonstrators will help you. Use wet ice in an esky at all times to put your tubes and samples to maintain the DNA's integrity. Hand towels are provided to keep the work area clean at all times. Used pipette tips are to be collected in a waste container for disposal. You will be working in three groups each led by a demonstrator.

Step-by-Step protocol – Cheek and wool tissue

1. Label your 2 ml tubes with the sample ID (sheep number) and your name (cheek cells).
2. For the wool sample, take a small quantity and add to liquid nitrogen for a minute. Then quickly grind it into powder using pestle and mortar. Transfer the ground wool powder into eppendorf tubes.
3. Shake to mix **Solution TD1**. *Solution TD1 is a high concentration salt solution required for tissue cell lysis and binding the genomic DNA.*
4. For the wool sample, add 900 µl of Solution TD1 to the eppendorf tubes containing the ground samples. Add 20 µl of Proteinase K and incubate at 65°C for 30 min.
5. Transfer the whole contents into the 2 ml Bead tubes provided.
6. For your cheek cells, take a swab and give it a good swirl in a 2 ml Bead Tube containing 700 µl of Solution TD1.
7. Secure both the wool and cheek-containing tubes in a bead beater and shake at maximum speed for 10 s. *This is to ensure complete homogenization and cell lysis through mechanical shaking.*
8. Remove the tubes from the shaker and spin in a centrifuge at 10,000 rpm for 1 min at room temperature. *This ensures that the cellular debris is sent to the bottom of the tube while DNA remains in the supernatant.*
9. While avoiding the beads, transfer the entire supernatant to a Spin Filter (provided) into a clean and labelled 2 ml microcentrifuge tube and centrifuge at 10,000 rpm for 30 s at room temperature. *This is to ensure that the DNA is selectively bound to the silica membrane of the Spin Filter while contaminants pass through.*
10. Discard the flow through. *This is the non-DNA organic and inorganic waste.*
11. Add 400 µl of **Solution TD2** and centrifuge at 10,000 rpm for 30 s at room temperature. *Solution TD2 is an ethanol based wash solution that is needed to clean the DNA bound to the silica membrane of the Spin Filter. It removes residual salts, cellular debris and proteins while allowing the DNA to stay bound to the membrane.*
12. Discard the flow through. *This is the non-DNA material washed away by Solution TD2.*
13. Centrifuge again at 10,000 rpm for 1 minat room temperature to remove residual Solution TD2. *This second spin removes all traces of the wash solution because the ethanol can interfere with many downstream applications such as PCR, restriction enzyme digests and gel electrophoresis.*
14. Carefully place Spin Filter in a new clean 2 ml Collection Tube (provided). *Avoid splashing any Solution TD2 onto the spin filter.*
15. Add 50 µl of **Solution TD3** to the center of the white filter membrane. *Solution TD3 is an elution buffer and placing it in the center of the small white membrane ensures that it is uniformly and completely wet for a more efficient and complete release of DNA from the silica Spin Filter membrane. When Solution TD3 passes through the silica membrane, DNA that was bound in the presence of high salt is now selectively released since Solution TD3 lacks salt.*
16. Centrifuge at 10,000 rpm for 30 s at room temperature.
17. Discard the Spin Filter. DNA in the 2-ml Collection Tube is now ready for any downstream application. Store in the −20°C to −80°C freezer until needed for PCR.

Step-by-Step protocol – Blood

1. Label your 2 ml tubes with the sample ID (sheep number)
2. Add 200 μl of whole blood to a 2 ml Collection Tubes provided and add 10 μl of Proteinase K. *Proteinase K breaks down the cell wall and helps the process of cell lysis.*
3. Add 200 μl of **Solution B1** and mix by vortexing for 15 s. *Solution B1 is a lysis agent containing salt and detergent.*
4. Incubate the sample at 65°C for 10 min. Centrifuge briefly to collect the lysate. *The heat helps to denature the proteins and completes the lysis.*
5. Add 200 μl of **Solution B2** and vortex for 15 s. Centrifuge briefly to collect the sample from the lid. *Solution B2 contains 100% ethanol and provides optimal conditions for DNA binding.*
6. Transfer the lysate to the Spin Filter and centrifuge for 1 min at 13,000 rpm. *DNA binds to the silica membrane in the Spin Filter while the liquid flow through contains unwanted cell materials such as denatured proteins and RNA*
7. Transfer Spin Filter to new 2 ml Collection Tube (provided).
8. Add 500 μl of **Solution B3** to the Spin Filter. Centrifuge for 30 s at 13,000 rpm. *Solution B3 is a salt-based wash solution that cleans the DNA bound to the Spin Filter.*
9. Remove the Spin Filter and discard the flow through. Place Spin Filter back into the same 2 ml Collection Tube.
10. Add 500 μl of **Solution B4** to the Spin Filter. Centrifuge for 30 s at 13,000 rpm. *Solution B4 is an ethanol-based wash solution that cleans the DNA bound to the Spin Filter and removes all residual salt contaminants.*
11. Remove Spin Filter and discard the flow through. Place Spin Filter back into the same 2 ml Collection Tube.
12. Centrifuge again for 30 s at 13,000 rpm to dry the Spin Filter membrane. *This second spin dries off the ethanol from the Spin Filter for maximal DNA release during elution.*
13. Carefully remove the Spin Filter and transfer to a new 2 ml Collection Tube (provided).
14. Add 100–200 μl of **Solution B5**.
15. Centrifuge at 13,000 rpm for 1 min.
16. Remove the Spin Filter unit and close tube lid. Genomic DNA in the 2 ml Collection Tube is now ready for any downstream application. Store the DNA in the −20 to −80°C freezer until needed for PCR.

Experiment 4.2: DNA quantification and polymerase chain reaction (PCR) assay

Aim

The objective of this laboratory practical is to quantify the DNA extracted last week and conduct an integrity and quality check using the Nanodrop-8000. The second objective is to amplify (make multiple copies of) the DNA extracted last week through polymerase chain reaction (PCR) technique. PCR involves a repeated cycle of DNA denaturation, annealing and extension (heating the DNA to a high temperature to untangle its double stranded conformation into single strands, cooling and using oligonucleotide primers to bind to the ends of the strands and copy the sequence, and using Taq polymerase to restore the double strands). Today, we will use the PCR mix to set up the assay and run the polymerase chain reaction.

Safety precautionary measures

- Laboratory coats and hand gloves must be worn at all times.
- Use wet ice in an esky to put your samples to maintain the DNA's integrity at all times.
- Hand towels would be provided to keep the work-bench clean at all times.
- Used pipette tips should be collected in a waste beaker for eventual disposal.
- You will work in three groups of six students each led by a demonstrator. We will wait for each other so that all the samples can go into the thermocycler (PCR machine) together.

Step-by-Step protocol – DNA quantification using the Nanodrop-8000

1. Look for the "ND-8000" logo on the computer desktop and click.
2. Select the "Nucleic Acid" option. The menu comes up with an 8-sample default option.
3. To initialize the instrument, pipette 5 μl water onto the pedestal when indicated by the software.
4. After initialization, wipe both the upper and lower pedestals with a clean Kimwipe and pipette 2 μL of your blank solution and click "continue."
5. Select manual sample ID entry and enter the number of samples. Type the sample ID's, click "Next Well" and then "Finished."
6. Wipe both upper and lower pedestals and pipette 2 μL of your sample into each lower pedestal and lower the arm gently.
7. Select "Measure" on the top left corner.
8. Print your results.
9. Wipe samples away and pipette 5 μl water onto each pedestal. Lower the pedestal arm and let it sit for 2–3 min, then wipe away.

Step-by-Step protocol – PCR assay

The following procedure needs to be carried out with the utmost care for a number of reasons:

- to prevent contamination of the reactions with foreign DNA.
- to accurately pipette very small volumes.
- to limit waste of very expensive reagents.

All reagents and reactions should be kept on ice as much as possible. Use a new pipette tip for every step and always wear gloves. Remember to label your tubes. UV-treat your PCR and Mastermix tubes in the UV cross-linker (this cross-links any foreign DNA that may be in the tubes so it cannot be amplified). For each genomic DNA sample (cheek cells, wool, blood) prepare the following PCR reaction. You also need to prepare a "No Template Control" (NTC) as a negative control. To this reaction add water instead of template. This control reaction tests for contamination of the PCR reactions.

Reagent components

- MyTaqHS Red Mix – containing buffer, Taq polymerase, $MgCl_2$, dNTP's, red dye – (25 μL)
  - The Taq is bound unto an antibody, and is therefore inactive until the antibody is denatured. This limits non-specific amplification.
- Forward Primer – 10μMLCO1490 (4 μL)
- Reverse Primer – 10μMHCO2198 (4 μL)
- Template DNA (× μL)
  - This is extracted gDNA from blood, wool or cheek.
- DNase free Water (19-× μL)
- Total volume per tube should equal 50 μL

1. You need to add between 50 and 300 ng gDNA to your reaction. Calculate the number of μl you need to add to the PCR reaction from your Nanodrop quantification.
2. Pipette these reagents to make a 50 μL final volume in a 200 μL PCR tube. For many samples, a Master-mix tube can be made (*a master mix limits the handling of many small volumes, so is generally more accurate and less prone to contamination*). You will need to work with your demonstrator to calculate the master-mix reagent volumes.
3. Each group should have the following PCR reactions: 1 cheek reaction for each person, plus 1 blood and 1 wool.
4. Input thermocycler (PCR machine) settings:
  a. Initial Denaturation: 95°C for 1:00 min
  b. Denaturation: 95°C for 20 s
  c. Annealing: 55°C for 20 s
  d. Extension: 72°C for 30 s
  e. Go to step b an additional 29× (30 cycles in total)
  f. Final extension: 72°C for 5 min
  g. End/holding temperature 10°C
  h. Set the final volume for 50 μL and do not change lid temperature (which is 105°C).
 Primer sequences
  a. LCO 1490: GGT CAA CAA ATC ATA AAG ATA TTG G
  b. HCO 2198: TAA ACT TCA GGG TGA CCA AAA AAT CA

Experiment 4.3: Gel electrophoresis of PCR products

Aim
To utilize the amplified DNA from Polymerase Chain Reaction (PCR) products to resolve the DNA fragments through gel electrophoresis. It essentially involves the introduction of an electric current and the subsequent migration of DNA fragments on the basis of their molecular weights. The lighter fragments migrate further away from the wells while the heavier fragments are nearer to the wells. After staining the gel with a dye, the gel can be read and photographed under ultraviolet (UV) light rays with the banding patterns attributable to the different primers/markers being visible (see Figure 20.14).

Step-by-Step protocol – Gel electrophoresis
1. Pour 1X TAE Buffer over prepared 1% agarose gel until just covered.
  a. 50× Stock TAE Buffer: 242 g Tris Base, 57.1 ml Glacial Acetic Acid, 18.6 g EDTA dissolved in 900 ml dH2O; adjust pH to 8.0; make up to 1 l. Dilute 20 ml in 1 l dH2O to get 1× working solution
2. Carefully transfer 5 ul of each PCR reaction into each into a well in the gel.
3. Load DNA ladder into adjacent lane (or both adjacent lanes for more accurate product size estimation).
4. Connect to power pack (connect red to red, black to black!).
5. Run at 80 V for 40 min.
6. Visualize on Gel Doc System and print the image.

Experiment 4.4: Purification of PCR product

Aim
To remove the excess primers, dNTPs, enzyme, and buffer from the pure PCR product. This pure product can then be sequenced.

Step-by-Step protocol – PCR product purification
1. Shake to mix the SpinBind before use. Add 5 volumes of **SpinBind** to the PCR reaction. Example: add 500 μl to a 100 μl PCR reaction. *What's happening: SpinBind is a buffered salt solution. By mixing it with your PCR reaction product, you are creating a pH-buffered high salt condition. pH is critical after the addition of*

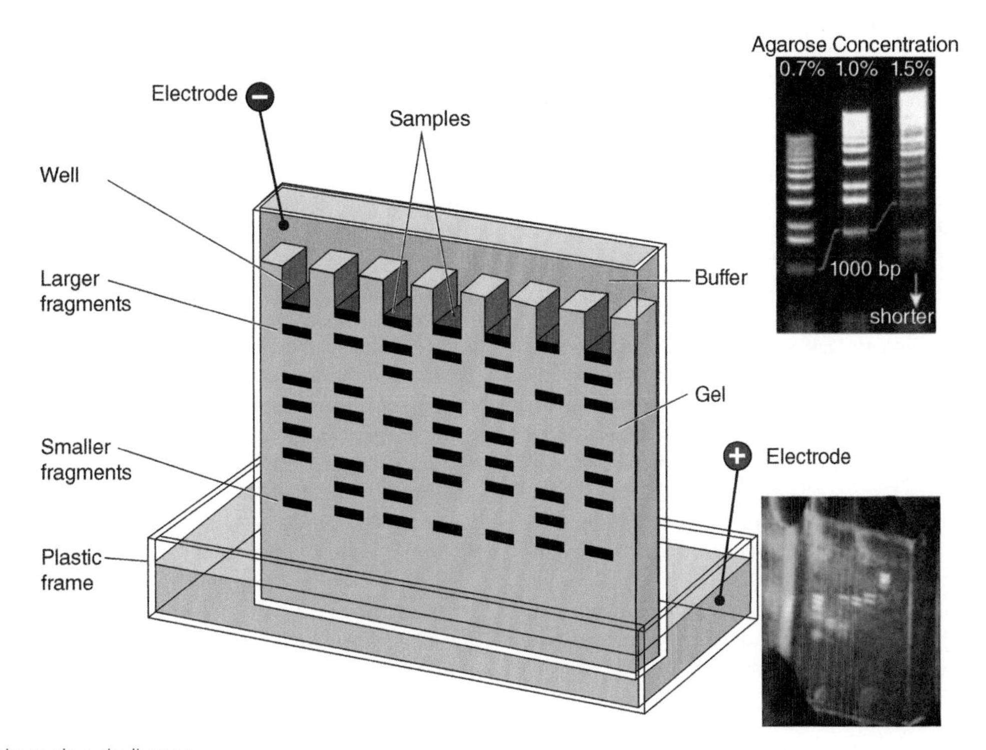

Figure 20.14 Gel electrophoresis diagram.

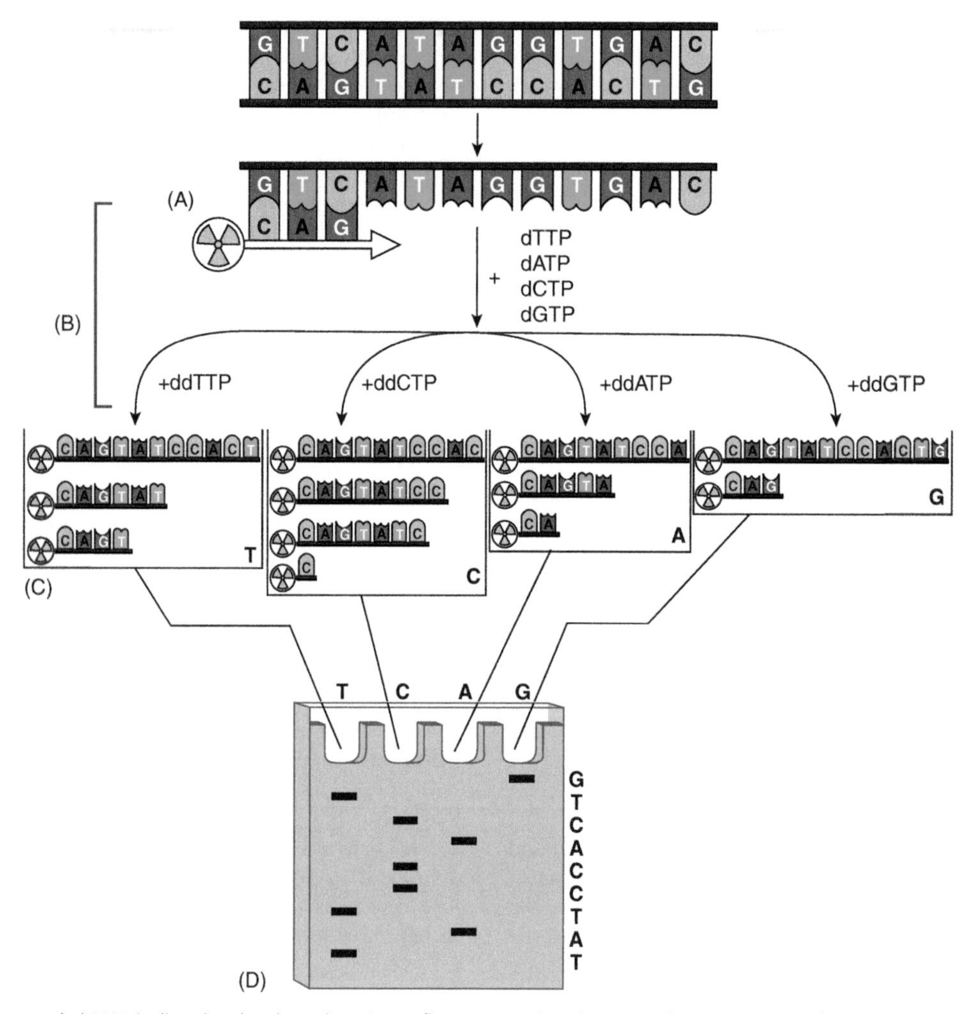

Figure 20.15 Double-stranded DNA is dissociated and a radioactive or fluorescent primer is annealed to the DNA (A); four separate reactions are performed to synthesize new DNA, each reaction contains all four deoxynucleotides and a small portion of one of the dideoxynucleotide bases (B); DNA is synthesized, terminating each time a ddNTP is incorporated (C); DNA from all four reactions is separated on a gel in side-by-side lanes to produce a sequence ladder (D); the sequence is read from the bottom up, and is the compliment (opposite) of the base identified in the gel.

SpinBind. pH above 8 results in low DNA recovery. pH below 5 results in primers co-purifying with your sample DNA. Optimal pH range is 6–7.5 for high recovery and total primer removal.

2. Mix well by pipetting. If an oil overlay was used, you will now have two layers. The top layer is oil. *What's happening: Mixing well creates a homogeneous salt concentration throughout the sample tube.*

3. Transfer PCR/**SpinBind** mixture to a Spin Filter unit, while avoiding the transfer of oil. *What's happening: Although the oil does not affect DNA binding, it can be very messy if you do not remove it at this step. If you used a wax overlay, try to avoid carrying over any during this step.*

4. Centrifuge 10-30 seconds at a minimum 10,000 × g (approximately 13,000 rpm) in a table-top micro-centrifuge. *What's happening: DNA in the size range from 60 bp to 10 kb will bind to the white silica spin filter membrane at the bottom of the silica spin filter unit. The liquid that passes through the membrane will contain unwanted components of the PCR reaction such as: PCR primers, dNTPs, enzyme, and buffer constituents. The desired PCR product DNA will bind to silica under high salt conditions.*

5. Remove the Spin Filter basket and discard the liquid flow-through from the tube by decanting. *What's happening: This is flow through waste. This contains Guanidine HCl, so be sure to check with your lab safety office for proper disposal and handling.*

6. Replace the Spin Filter basket in the same tube.

7. Add 300 μl of **SpinCleanTM Buffer** to the **Spin Filter**. *What's happening: SpinCleanTM is a wash solution. It will remove any traces of unwanted contaminants while allowing the desired PCR product DNA to stay bound to the silica spin filter membrane. SpinCleanTM is an ethanol based wash solution. It contains less than 80% ethanol. This solution is flammable so use caution near flames.*

8. Centrifuge 10–30 s at a minimum 10,000 × g. *What's happening: As the SpinCleanTM passes through the spin filter membrane, it cleans the PCR product DNA.*

9. Remove Spin Filter basket and discard liquid flow through by decanting then replace basket back into the same tube.

10. Centrifuge 30–60 s at minimum 10,000 × g. *What's happening: This step will remove any last traces of SpinCleanTM. The ethanol in the SpinCleanTM can have a negative effect on the DNA purity.*

11. Transfer **Spin Filter** to a clean **2 ml Collection Tube**

12. Add 50 μl of **Elution Buffer** (10 mM Tris) solution provided or sterile water directly onto the center of the white **Spin Filter** membrane. The choice of using Tris or water at this point will not affect yield. DNA is more stable for storage in Tris. *What's happening: Placing the Elution Buffer in the center of the small white silica membrane will make sure the entire membrane is wetted. This will result in more efficient release of the desired DNA.*

13. Centrifuge 30–60 s at a minimum 10,000 × g. *What's happening: As the Elution Buffer passes through the silica membrane, DNA is released, and it flows through the membrane, and into the collection tube. The DNA is released because it can only bind to the silica spin filter membrane in the presence of salt. Elution Buffer is 10 mM Tris pH. 8 and does not contain salt.*

14. Remove Spin Filter basket from the **2 ml Collection Tube**. Seal tube and store DNA at −20°C. *What's happening: Purified DNA is now in the collection tube. The DNA is UltraClean and free of all reaction components such as primers or linkers, enzyme, salts and dNTPs. The DNA is now ready to use for any application. The DNA is in a 50 μl volume. To concentrate it, see the Hints and Troubleshooting Guide.*

15. Quantify your pure PCR product solution on the Nanodrop as described last week

Experiment 4.5: Set up of sequence reactions

Double-stranded DNA is dissociated and a radioactive or fluorescent primer is annealed to the DNA; four separate reactions are performed to synthesize new DNA, each reaction contains all four deoxynucleotides and a small portion of one of the dideoxynucleotide bases (B); DNA is synthesized, terminating each time a ddNTP is incorporated (C); DNA from all four reactions is separated on a gel in side-by-side lanes to produce a sequence ladder (D); the sequence is read from the bottom up, and is the compliment (opposite) of the base identified in the gel (see Figures 20.15 and 20.16).

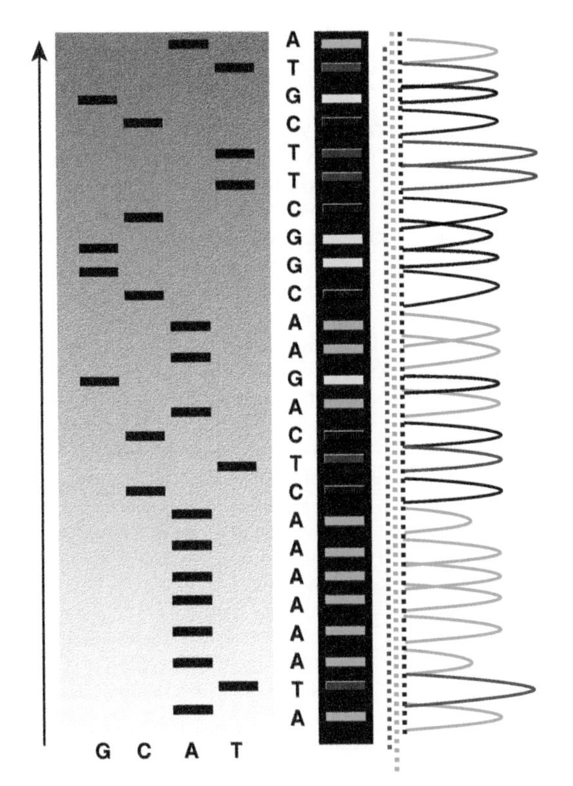

Figure 20.16 Sequence ladder by radioactive sequencing compared to fluorescent peaks.

Reagent components:

- CEQDTCS Quick Start Master Mix (4 μL)
- MQ dH2O (× μL)
- 1.6 mM Sequencing Primer (2 μ)
- 30 ng Purified PCR product (× μ l)
- Total should equal 10 μL

Thermocycler program for sequencing:

- Denaturation (20 s @ 90°C)
- Annealing (20 s @ 50°C)
- Extension (4 min @ 60°C)
- Repeat (× 30)
- Hold (@10°C indefinitely)

Experiment 4.6: Ethanol precipitation clean-up of labelled dye terminator DNA sequence products

Aim

Post-reaction clean-up of labelled DNA sequencing products to remove both the unincorporated dye terminators and residual salts is a critical step for all capillary electrophoresis-based automated sequencing technologies. This is because:

- Un-incorporated dye terminators tend to form "dye blobs" (usually within the first 80 bases) compromising the accuracy of base calling, and
- Excess salts and other ion-carrying molecules in the sequencing mix act to lower signal intensity by competing with the dye-labelled DNA sequencing products for migration into the capillaries during the electro-kinetic injection sample loading process.

Ethanol precipitation is an efficient and cost-effective way of removing unincorporated dye terminators and salts from sequencing reactions. This method provides savings in cost per sample, operation time and produces higher signal intensities due to better DNA recovery.

Step-by-Step protocol – Ethanol precipitation

1. Prepare a labelled, sterile 1.5 mL microfuge tube for each sample

2. This step to be done by lab demonstrators: Prepare fresh Stop Solution/Glycogen mixture as follows (per sequencing reaction): 2 μl of 3 M Sodium Acetate (pH 5.2), 2 μl of 100 mM Na_2-EDTA (pH 8.0) and 0.5 μl of 20 mg/ml of glycogen (supplied with kit).

3. To each of the labelled tubes, add 4.5 μl of the Stop Solution/Glycogen mixture.

4. Transfer the sequencing reaction to the appropriately labelled 1.5 ml microfuge tube and mix thoroughly.

5. Add 60 μl of cold 95% (v/v) ethanol/dH_20 from −20°C freezer and mix thoroughly.

6. Immediately centrifuge at 14,000 rpm at 4°C for 15 min.

7. The pellet should be visible. Carefully remove the supernatant with a micropipette. *Note: For multiple samples, always add the cold ethanol/dH_20 immediately before centrifugation.*

8. Rinse the pellet twice with 200 μl 70% (v/v) ethanol/dH_20 from −20°C freezer. For each rinse, centrifuge immediately at 14,000 rpm at 4°C for a minimum of 2 min. After centrifugation, carefully remove all of the supernatant with a micropipette.

9. Vacuum-dry for 10 min (or until dry).

10. Re-suspend the sample in 30 μL of the Sample Loading Solution.

11. Transfer the sample into a well in the sequencing plate. Cover with 1 drop of mineral oil.

References

Ashes, J.R., Siebert, B.D., Gulati, S.K., Cuthbertson, A.Z., Scott, T.W., 1992. Incorporation of n-3 fatty-acids of fish oil into tissue and serum-lipids of ruminants. *Lipids* 27, 629–631.

Aurousseau, B., Bauchart, D., Galot, A., Prache, S., Micol, D., Priolo, A., 2007. Indoor fattening of lambs raised on pasture: 2. Influence of stall finishing duration on triglyceride and phospholipid fatty acids in the *longissimus thoracis* muscle. *Meat Science* 76, 417–427.

Baeza, M.C., Corva, P.M., Soria, L.A., Pavan, E., Rincon, G., Medrano, J.F. 2012. Genetic variants in a lipid regulatory pathway as potential tools for improving the nutritional quality of grass-fed beef. *Animal Genetics* (Early View June 13, 2012) DOI: 10.1111/j.1365–2052.2012.02386.x

Barton, L., Kott, T., Bures, D., Rehak, D., Zahradkova, R., Kottova, B. 2010. The polymorphisms of stearoyl-CoA desaturase (SCD) and sterol regulatory element binding protein-1 (SREBP-1) genes and their association with the fatty acid profile of muscle and subcutaneous fat in Fleckvieh bulls. *Meat Science* 85, 15–20.

Bas, P., Morand-Fehr, P., 2000. Effect of nutritional factors on fatty acid composition of lamb fat deposits. *Livestock Production Science* 64, 61–79.

Blackmore, P., Cousin, G. 2003. Linking teaching and research through research-based learning. *Educational Developments* 4, 24–27.

Brew, A., Boud, D. 1995. Teaching and research: Establishing the vital link with learning. *Higher Education* 29, 261–173.

Campbell, M.K., Farrell, S.O., 2012. *Biochemistry*, 7th edn. Brooks/Cole Cengage Learning, Belmont, CA.

Choi, N., Enser, M., Wood, J., Scollan, N., 2000. Effect of breed on the deposition in beef muscle and adipose tissue of dietary n-3 polyunsaturated fatty acids. *Animal Science* 71, 509–520.

Conte, G., Mele, M., Chessa, S., Castiglioni, B., Serra, A., Pagnacco, G., Secchiari, P. 2010. Diacylglycerol acyltransferase 1, stearoyl-CoA desaturase 1, and sterol regulatory element binding protein 1 gene polymorphisms and milk fatty acid composition in Italian Brown cattle. *Journal of Dairy Science* 93, 753–63.

Corella, D, Ordovas, J.M. 2005. Single nucleotide polymorphisms that influence lipid metabolism: interaction with dietary factors. *Annual Review of Nutrition* 25, 341–390.

Daley, C.A., Abbott, A., Doyle, P.S., Nader, G.A., Larson, S., 2010. A review of fatty acid profiles and antioxidant content in grass-fed and grain-fed beef. *Nutrition Journal* 9, (10 March 2010) – (2010 March 2010).

Deland, M.P.B., Malau-Aduli, A.E.O., Siebert, B.D., Bottema, C.D.K., Pitchford, W.S. 1998. Sex and breed differences in the fatty acid composition of muscle phospholipids in crossbred cattle. *Proceedings 6th World Congress on Genetics Applied to Livestock Production*, Armidale, Australia, 25, 185–188.

Demirel, G., Wachira, A.M., Sinclair, L.A., Wilkinson, R.G., Wood, J.D., Enser, M., 2004. Effects of dietary n-3 polyunsaturated fatty acids, breed and dietary vitamin E on the fatty acids of lamb muscle, liver and adipose tissue. *British Journal of Nutrition* 91, 551–565.

Descalzo, A., Insani, E., Biolatto, A., Sancho, A., Garcia, P., Pensel, N., Josifovich, J., 2005. Influence of pasture or grain-based diets supplemented with vitamin E on antioxidant/oxidative balance of Argentine beef. *Meat Science* 70, 35–44.

Dixon, R.M., Stockdale, C.R., 1999. Associative effects between forages and grains: consequences for feed utilisation. *Australian Journal of Agricultural Research* 50, 757–774.

Doreau, M., Bauchart, D., Chilliard, Y., 2011. Enhancing fatty acid composition of milk and meat through animal feeding. *Animal Production Science* 51, 19–29.

Doreau, M., Ferlay, A., 1994. Digestion and utilisation of fatty acids by ruminants. *Animal Feed Science and Technology* 45, 379–396.

Edelson, D.C., Gordin, D.N., Pea, R.D. 1999. Addressing the challenges of inquiry-based learning through technology and curriculum design. *Journal of Learning Science* 8, 391–450.

Elmore, J.S., Mottram, D.S., Enser, M., Wood, J.D., 1999. Effect of the polyunsaturated fatty acid composition of beef muscle on the profile of aroma volatiles. *Journal of Agricultural and Food Chemistry* 47, 1619–1625.

Fisher, A., Enser, M., Richardson, R., Wood, J., Nute, G., Kurt, E., Sinclair, L., Wilkinson, R., 2000. Fatty acid composition and eating quality of lamb types derived from four diverse breed× production systems. *Meat Science* 55, 141–147.

French, P., Stanton, C., Lawless, F., O'Riordan, E., Monahan, F., Caffrey, P., Moloney, A., 2000. Fatty acid composition, including conjugated linoleic acid, of intramuscular fat from steers offered grazed grass, grass silage, or concentrate-based diets. *Journal of Animal Science* 78, 2849–2855.

Gurr, M.I., James, A.T., 1972. *Lipid Biochemistry: An Introduction*. Cornell University Press, Ithaca, NY.

Healey, M. 2000. Developing the scholarship of teaching in higher education: A discipline based approach. *Higher Education Research and Development* 19, 169–189.

Healey, M. 2003. The scholarship of teaching: issues around an evolving concept. *Journal of Excellence in College Teaching* 14, 5–26.

Healy, M. 2005. Linking research and teaching to benefit student learning. *Journal of Geography in Higher Education* 29, 183–201.

Holman, B.W.B., Kashani, A., Malau-Aduli, A.E.O. 2012. Growth and body conformation responses of genetically divergent Australian sheep to Spirulina (*Arthrospira platensis*) supplementation. *American Journal of Experimental Agriculture* 2(2), 160–173.

Holman, B.W.B., Malau-Aduli, 2012. A review of sheep wool quality traits. *Annual Reviews and Research in Biology* 2(1), 1–14.

Holman, B.W.B., Malau-Aduli, A.E.O. 2013. Spirulina as a livestock supplement and animal feed. *Journal of Animal Physiology and Animal Nutrition* DOI: 10.1111/j.1439–0396.2012.01328.x.

Hood, R., Thompson, E., Allen, C., 1972. The role of acetate, propionate, and glucose as substrates for lipogenesis in bovine tissues. *International Journal of Biochemistry* 3, 598–606.

Hopkins, D.L., Stanley, D.F., Toohey, E.S., Gardner, G.E., Pethick, D.W., van de Ven, R., 2007. Sire and growth path effects on sheep meat production 2. Meat and eating quality. *Australian Journal of Experimental Agriculture* 47, 1219–1228.

Ingle, D., Bauman, D., Garrigus, U., 1972a. Lipogenesis in the ruminant: *in vivo* site of fatty acid synthesis in sheep. *The Journal of Nutrition* 102, 617–623.

Ingle, D.L., Bauman, D., Garrigus, U., 1972b. Lipogenesis in the ruminant: *in vitro* study of tissue sites, carbon source and reducing equivalent generation for fatty acid synthesis. *The Journal of Nutrition* 102, 609–616.

Jacob, F., Monod, J. 1961. Genetic regulatory mechanisms in the synthesis of proteins. *Journal of Molecular Biology* 3, 318–356.

Jenkins, T., 1993. Lipid metabolism in the rumen. *Journal of Dairy Science* 76, 3851–3863.

Keys, C.W., Bryan, L.A. 2001. Co-constructing inquiry-based science with teachers: Essential research for lasting reform. *Journal of Research in Science Teaching* 38, 631–645.

Kim, Y.C., Ntambi, J.M. 1999. Regulation of stearoyl-CoA desaturase genes: role in cellular metabolism and preadipocyte differentiation. *Biochemical and Biophysical Research Communications* 266, 1–4.

Kruk, Z.A., Malau-Aduli, A.E.O., Thomson, A.M., Siebert, B.D., Pitchford, W.S., Bottema, C.D.K. 1997a. Are ß-carotene and fatty acid composition in cattle related? In: *Breeding: Responding to client needs*. P. Arthur, N. Fogarty, A. Gilmour and S. Mortimer (eds). *Proceedings Association for the Advancement of Animal Breeding & Genetics*, Dubbo, Australia, 12, 278–282.

Kruk, Z.A., Malau-Aduli, A.E.O., Thomson, A.M., Siebert, B.D., Pitchford, W.S., Bottema, C.D.K. 1997b. Do breed and season affect ß-carotene and the fatty acid composition of muscle phospholipids in cattle? In: *Vitality of Meat* John Bass (ed.). *Proceedings of the 43rd International Congress on Meat Science and Technology*, Auckland, New Zealand, C32, 314–315.

Kruk, Z.A., Malau-Aduli, A.E.O., Pitchford, W.S., Bottema, C.D.K. 1998. Genetics of fat colour in beef cattle. *Proceedings 6th World Congress on Genetics Applied to Livestock Production*, Armidale, Australia, 23, 121–124.

Kruk, Z.A., Siebert, B.D., Pitchford, W.S., Davis, J., Harper, G.S. and Bottema, C.D.K. 2004. Effects of vitamin A on growth performance and carcass quality in steers. *Livestock Science* 119, 12–21.

Laliotis, G., Bizelis, I., Rogdakis, E., 2010. Comparative approach of the de novo fatty acid synthesis (lipogenesis) between ruminant and non ruminant mammalian species: from biochemical level to the main regulatory lipogenic genes. *Current Genomics* 11, 168.

Li, C., Aldai, N., Vinsky, M., Dugan, M. E. R., McAllister, T. A. 2012. Association analyses of single nucleotide polymorphisms in bovine stearoyl-CoA desaturase and fatty acid synthase genes with fatty acid composition in commercial cross-bred beef steers. *Animal Genetics* 43, 93–97.

Li, C., Vinsky, M., Aldai, N., Osoro, K., Dugan, M.E.R. 2010. Associations of a bovine SCD SNP with fatty acid composition in concentrate-finished yearling bulls of different muscular hypertrophy genotypes. *Proceedings of 9th World Congress on Genetics Applied to Livestock Production (WCGALP)*, Leipzig, Germany, 505.

Lock, A.L., Harvatine, K.J., Drackley, J.K., Bauman, D.E., 2006. Concepts in fat and fatty acid digestion in ruminants. *Proceedings Intermountain Nutrition Conference* pp. 85–100.

Lorenz, S., Buettner, A., Ender, K., Nürnberg, G., Papstein, H.-J., Schieberle, P., Nürnberg, K., 2002. Influence of keeping system on the fatty acid composition in the longissimus muscle of bulls and odorants formed after pressure-cooking. *Eur Food Res Technol* 214, 112–118.

Lourenco, M., Ramos-Morales, E., Wallace, R.J., 2010. The role of microbes in rumen lipolysis and biohydrogenation and their manipulation. *Animal* 4, 1008–1023.

Malau-Aduli, A.E.O., Siebert, B.D., Pitchford, W.S., Bottema, C.D.K. 1995. Genetic variation in the fatty acid composition of cattle fat and muscle In: W.S. Pitchford (Editor). *Breeding for Quality & Profit. Proceedings Australian Association of Animal Breeding and Genetics* 11, 554–557.

Malau-Aduli, A.E.O., Siebert, B.D., Bottema, C.D.K., Pitchford, W.S. 1996. Genetic comparison of the fatty acid composition of intramuscular and adipose tissue lipids from early and late maturing beef cattle. In: *Partnerships for Sustainable Livestock Production and Human Welfare* K. Watanabe (ed.). Proceedings 8th Animal Science Congress, Asian-Australasian Association of Animal Production Tokyo, Japan, 2, 2–33.

Malau-Aduli, A.E.O., Siebert, B.D., Bottema, C.D.K., Pitchford, W.S. 1997a. A comparison of the fatty acid composition of triacylglycerols in adipose tissue from Limousin and Jersey cattle. *Australian Journal of Agricultural Research* 48 (5), 715–722.

Malau-Aduli, A.E.O., Siebert, B.D., Bottema, C.D.K., Pitchford, W.S. 1997b. Genotype and site differences in the fatty acid composition of muscle phospholipids in cattle. In: *Breeding: Responding to Client Needs*, P. Author, N. Fogarty, A. Gilmour and S. Mortimer (eds). *Proceedings Association for the Advancement of Animal Breeding & Genetics*, Dubbo, Australia, 12, 580–584.

Malau-Aduli, A., Siebert, B.D., Bottema, C., Pitchford, W.S., 1998a. Breed comparison of the fatty acid composition of muscle phospholipids in Jersey and Limousin cattle. *Journal of Animal Science* 76, 766–773.

Malau-Aduli, A.E.O., Siebert, B.D., Bottema, C.D.K., Pitchford, W.S 1998b. Mode of inheritance of triacylglycerol fatty acids in beef adipose tissue. *Journal of Animal Science* 76 (Suppl 1), 153.

Malau-Aduli, A.E.O., Siebert, B.D., Bottema, C.D.K., Deland, M.P.B., Pitchford, W.S. 1998c. Heritabilities of triacylglycerol fatty acids from the adipose tissue of beef cattle at weaning and slaughter. *Proceedings 6th World Congress on Genetics Applied to Livestock Production*, Armidale, Australia, 25, 181–184.

Malau-Aduli, A.E.O., Siebert, B.D., Bottema, C.D.K., Pitchford, W.S. 1998d. Genetic and phenotypic correlations between triacylglycerol fatty acids at weaning and slaughter in beef cattle. In: *JAM van Arendok* (ed.) 49th *Annual Meeting of the European Association for Animal Production*, Warsaw, Poland, Volume 48, Series 4, Section GC 3.5, 16.

Malau-Aduli, A.E.O., Edriss, M.A., Siebert, B.D., Bottema, C.D.K., Pitchford, W.S., 2000a. Breed differences and genetic parameters for melting point, marbling score and fatty acid composition of lot-fed cattle. *Journal of Animal Physiology and Animal Nutrition* 83, 95–105.

Malau-Aduli, A.E.O., Siebert, B.D., Bottema, C.D.K., Pitchford, W.S. 2000b. Breed differences and heterosis in triacylglycerol fatty acid composition of bovine adipose tissue. *Journal of Animal Physiology and Animal Nutrition* 83 (2), 106–112.

Malau-Aduli, A.E.O., Edriss, M.A., Siebert, B.D., Bottema, C.D.K., Deland, M.P.B., Pitchford, W.S. 2000c. Estimates of genetic parameters for triacylglycerol fatty acids in beef cattle at weaning and slaughter. *Journal of Animal Physiology and Animal Nutrition* 83 (4–5), 169–180.

Malau-Aduli, A.E.O., Siebert, B.D., Bottema, C.D.K., Pitchford, W.S. 2000d. Heterosis, sex and breed differences in the fatty acid composition of muscle phospholipids in beef cattle. *Journal of Animal Physiology and Animal Nutrition* 83 (3), 113–120.

Malau-Aduli, A.E.O., Holman, B. 2010. Genetics-nutrition interactions influencing wool spinning fineness in Australian crossbred sheep. *Journal of Animal Science* 88 (E-Suppl 2), 469.

Malau-Aduli, A.E.O., Bignell, C.W., McCulloch, R., Kijas, J.W., Nichols, P.D. 2011. Genetic association of delta-six fatty acid desaturase single nucleotide polymorphic molecular marker and muscle long chain omega-3 fatty acids in Australian lamb. In: Stefaan De Smet (ed.). Global challenges to production, processing and consumption of meat. *Proceedings of the 57th International Congress of Meat Science and Technology*, 7–12 August 2011, Ghent, Belgium, 57, 126.

Malau-Aduli, A.E.O., Nightingale, E., McEvoy, P., Eve, J.U., John, A.J., Hobbins, A.A., et al. 2012a. Teaching Animal Science and Genetics to Australian university undergraduates to enhance inquiry-based student learning and research with sheep: Growth and conformation traits in crossbred prime lambs. *British Journal of Educational Research* 2(1), 59–76.

Malau-Aduli, A.E.O., Holman, B.W.B., Lane, P.A. 2012b. Influence of sire breed, protein supplementation and gender on wool spinning fineness in first-cross Merino sheep. World Academy of Science, *Engineering and Technology Scientia Special Journal* 67, 1029–1036.

Malau-Aduli, A.E.O., Akuoch, J.D.D. 2012. Sire genetics, protein supplementation and gender effects on wool comfort factor in Australian crossbred sheep. *American Journal of Experimental Agriculture* 2(1), 31–46.

Malau-Aduli, A.E.O., Lane, P.A. 2012. Synergistic nexus between research-led teaching and inquiry-based student learning in Animal Sciences: Sharing the University of Tasmania experience. *Proceedings of the Teaching and Learning in the Animal Sciences Conference: Challenging Old Assumptions and Break New Ground for the 21st Century*, 20–22 June 2012, The Lowell Center, University of Wisconsin–Madison, USA, pp 72.

Mandell, I., Buchanan-Smith, J., Holub, B., Campbell, C., 1997. Effects of fish meal in beef cattle diets on growth performance, carcass characteristics, and fatty acid composition of longissimus muscle. *Journal of Animal Science* 75, 910–919.

McLachlan, J.C. 2006. The relationship between assessment and learning. *Medical Education* 40, 716–717.

Mezöszentgyörgyi, D., Husvéth, F., Lengyel, A., Szegleti, C., Komlósi, I., 2001. Genotype-related variations in subcutaneous fat composition in sheep. *Animal Science* 72, 607–612.

Nute, G.R., Richardson, R.I., Wood, J.D., Hughes, S.I., Wilkinson, R.G., Cooper, S.L., Sinclair, L.A., 2007. Effect of dietary oil source on the flavour and the colour and lipid stability of lamb meat. *Meat Science* 77, 547–555.

Ohsaki, H., Tanaka, A., Hoashi, S., Sasazaki, S., Oyama, K., Taniguchi, M., Mukai, F., Mannen, H. 2009. Effect of SCD and SREBP genotypes on fatty acid composition in adipose tissue of Japanese Black cattle herds. *Animal Science Journal* 80, 225–232.

Ordovas, J.M. 2008. Genotype–phenotype associations: Modulation by diet and obesity. *Obesity* 16 (Suppl 3), S40–S46.

Ordovas, J.M., Corella, D. 2005. Genetic variation and lipid metabolism: Modulation by dietary factors. *Curr Cardiol Rep* 7, 480–486.

Ordovas, J.M., Corella, D. 2004a. Genes, diet and plasma lipids: the evidence from observational studies. *World Rev Nutr Diet* 93, 41–76.

Ordovas, J.M., Corella, D. 2004b. Nutritional genomics. *Annual Review of Genomics and Human Genetics* 5, 71–118.

Orrù, L., Cifuni, G.F., Piasentier, E., Corazzin, M., Bovolenta, S., Moioli, B. 2011 Association analyses of single nucleotide polymorphisms in the LEP and SCD1 genes on the fatty acid profile of muscle fat in Simmental bulls. *Meat Science* 87, 344–348.

O'Quinn, P.R., Nelssen, J.L., Goodband, R.D., Tokach, M.D., 2000. Conjugated linoleic acid. *Animal Health Research Reviews* 1, 35–46.

Perry, D., Nicholls, P.J., Thompson, J.M., 1998. The effect of sire breed on the melting point and fatty acid composition of subcutaneous fat in steers. *Journal of Animal Science* 76, 87–95.

Pitchford, W.S., Deland, M.P.B., Siebert, B.D., Malau-Aduli, A.E.O., Bottema, C.D.K. 2002. Genetic variation in fatness and fatty acid composition of crossbred cattle. *Journal of Animal Science* 80 (11), 2825–2832.

Pitchford, W.S., Deland, M.P.B., Siebert, B.D., Malau-Aduli, A.E.O., Bottema, C.D.K. 2001. Breed differences and genetic parameters for fat traits of crossbred cattle, In: Neville Jopson (ed.). *Biotechnology. Proceedings Association for the Advancement of Animal Breeding and Genetics Queenstown,* New Zealand, 14, 477–480.

Raes, K., De Smet, S., Demeyer, D., 2004. Effect of dietary fatty acids on incorporation of long chain polyunsaturated fatty acids and conjugated linoleic acid in lamb, beef and pork meat: a review. *Animal Feed Science and Technology* 113, 199–221.

Ryan, S.M., Unruh, J.A., Corrigan, M.E., Drouillard, J.S., Seyfert, M., 2007. Effects of concentrate level on carcass traits of Boer crossbred goats. *Small Ruminant Research* 73, 67–76.

Schennink, A., Heck, J.M., Bovenhuis, H., Visker, M.H.P.W., Van Valenberg, H.J.F., Van Arendonk, J.A.M. 2008. Milk fatty acid unsaturation: genetic parameters and effects of stearoyl-CoA desaturase (SCD1) and acyl CoA: diacylglycerol acyltransferase 1 (DGAT1). *Journal of Dairy Science* 91, 2135–2143.

Schmidely, R., Glasser, F., Doreau, M., Sauvant, D., 2008. Digestion of fatty acids in ruminants: a meta-analysis of flows and variation factors. 1. Total fatty acids. *Animal* 2, 677–690.

Scollan, N.D., Choi, N.J., Kurt, E., Fisher, A.V., Enser, M., Wood, J.D., 2001a. Manipulating the fatty acid composition of muscle and adipose tissue in beef cattle. *British Journal of Nutrition* 85, 115–124.

Scollan, N.D., Dhanoa, M.S., Choi, N.J., Maeng, W.J., Enser, M., Wood, J.D., 2001b. Biohydrogenation and digestion of long chain fatty acids in steers fed on different sources of lipid. *Journal of Agricultural Science* 136, 345–355.

Shingfield, K.J., Ahvenjarvi, S., Toivonen, V., Arola, A., Nurmela, K.V.V., Huhtanen, P., Griinari, J.M., 2003. Effect of dietary fish oil on biohydrogenation of fatty acids and milk fatty acid content in cows. *Animal Science* 77, 165–179.

Siebert, B.D., Malau-Aduli, A.E.O., Bottema, C.D.K., Deland, M.P.B., Pitchford, W.S. 1998. Genetic variation between crossbred weaner calves in triacylglycerol fatty acid composition. *Proceedings 6th World Congress on Genetics Applied to Livestock Production,* Armidale, Australia, 25, 177–180.

Siebert, B.D., Pitchford, W.S., Malau-Aduli, A.E.O., Deland, M.P.B., Bottema, C.D.K. 1999. Breed and sire effects on fatty acid composition of beef fat, In: P. Vercoe, N. Adams and D. Masters (eds). *Rising to the Challenge – Breeding for the 21st Century Customer. Proceedings Association for the Advancement of Animal Breeding and Genetics,* Mandurah, Western Australia, 13, 389–392.

Siebert, B.D., Pitchford, W.S., Kuchel, H., Kruk, Z.A., & Bottema, C.D.K. 2000. The effect of beta-carotene on desaturation of ruminant fat. Asian-Aus. *Journal of Animal Science* 13, 185–188.

Siebert, B.D., Kruk, Z.A., Davis, J., Pitchford, W.S., Harper, G.S. and Bottema, C.D.K. 2006. Effect of low vitamin A status on fat deposition and fatty acid desaturation in beef cattle. *Lipids* 41, 365–370.

Taniguchi, M., Utsugi, T., Oyama, K., Mannen, H., Kobayashi, M., Tanabe, Y., et al. 2004. Genotype of stearoyl-CoA desaturase is associated with fatty acid composition in Japanese Black cattle. *Mammalian Genome* 15, 142–8.

Van Nevel, C.J., Demeyer, D.I., 1996. Influence of pH on lipolysis and biohydrogenation of soybean oil by rumen contents *in vitro*. *Reproduction Nutrition Development* 36, 53–63.

Watkins, P.J., Kearney, G., Rose, G., Allen, D., Ball, A.J., Pethick, D.W., Warner, R.D. 2012. Effect of branched-chain fatty acids, 3-methylindole and 4-methylphenol on consumer sensory scores of grilled lamb meat. *Meat Science* Doi: http://dx.doi.org/10.1016/j.meatsci.2012.08.011

Webb, E.C., O'Neill, H.A., 2008. The animal fat paradox and meat quality. *Meat Science* 80, 28–36.

Wong, E., Johnson, C.B., Nixon, L.N., 1975. The contribution of 4-methyloctanoic (hircinoic) acid to mutton and goat meat flavour. *New Zealand Journal of Agricultural Research* 18, 261–266.

Wood, J.D., Enser, M., Fisher, A.V., Nute, G.R., Sheard, P.R., Richardson, R.I., et al., 2008. Fat deposition, fatty acid composition and meat quality: A review. *Meat Science* 78, 343–358.

Wood, J.D., Richardson, R.I., Nute, G.R., Fisher, A.V., Campo, M.M., Kasapidou, E., et al. 2004. Effects of fatty acids on meat quality: A review. *Meat Science* 66, 21–32.

Zhang, S., Knight, T.J., Reecy, J.M., Beitz, D.C. 2008. DNA polymorphisms in bovine fatty acid synthase are associated with beef fatty acid composition. *Animal Genetics* 39, 62–70.

Zhou, G., Yang, A., Tume, R., 1993. A relationship between bovine fat colour and fatty acid composition. *Meat Science* 35, 205–212.

21 Nutritional Epigenomics

Congjun Li

Bovine Functional Genomics Laboratory, Agricultural Research Service, United States Department of Agriculture, MI, USA[1]

Introduction

The discoveries and technological advances during the last decade have taken the epigenomics field to a completely new level. The modern version of **epigenomics** includes molecular mechanisms that influence the phenotypic outcome of a gene or genome, in the absence of changes to the underlying DNA sequence. The focus of this chapter is the nutrient-specific modulation of genetic regulation, especially gene expression. As this area continues to unfold, it is becoming increasingly evident that a host of genomic interrelationships with diet exist that encompass the broad topic of **nutrigenomics**, defined as the interaction between nutrition and an individual's genome. The concept of nutrigenomics, mechanisms of epigenomic regulation such as DNA methylation and histone modification, and the recent discoveries in this field are introduced in this chapter. The specific implication of epigenomics in animal science and the impact on farm animal production are discussed.

Epigenomics involves studying the phenomenon of changes in regulating gene expression and phenotype that do not depend on changes in gene sequences and establishing links between gene expression and phenotype. Epigenomics is genomics that goes beyond DNA sequences (Katsnelson, 2010). Although **epigenomics** refers to the study of global changes across the entire genome, **epigenetics** involves studying single genes or groups of genes. In other words, cells' information is inherited by the next generation through genetic and epigenetic routes. Genetic information is encoded in the DNA sequence while epigenetic information is defined by DNA modification (DNA methylation) and chromatin modifications (methylation, phosphorylation, acetylation, and ubiquitination of histone cores). The study of epigenomics/epigenetics explores heritable, reversible modifications of DNA and chromatin that do not change primary nucleotide sequences or a set of chemical modifications not encoded within DNA, which coordinate how and when genes are expressed (Katsnelson, 2010). Epigenomics demonstrates the changes in regulating gene activities that act without, or independently of, changes in gene sequences. Many life phenomena can be explained by epigenomics or epigenetics. Several types of epigenetic mechanisms are involved in regulating gene expression in many biological processes. These mechanisms, such as DNA methylation and histone posttranslational modifications, have been recognized for decades, and they are intricately interconnected with each other. Some of these processes, such as the formation of microRNA, have only recently been discovered. Genomic imprinting, gene silencing, X chromosome inactivation, position effects, reprogramming, and the progress of carcinogenesis are all known epigenetic processes. By definition, in addition to DNA methylation and histone posttranslational modifications, RNA splicing, RNA editing, microRNA, and prions can be included as epigenetic mechanisms for regulating genes (Table 21.1).

These regulatory mechanisms for modulating gene function are multifaceted and complex. This type of regulation determines up- or down-regulation and the scope of gene responses to the activation of different signaling pathways. Epigenetic mechanisms also contribute to stable, cell-type-specific patterns of gene activities (silencing or activation) (Woodcock, 2004). This is a key property of living systems of robustness, the ability to maintain phenotypic stability in the face of diverse perturbations arising from environmental changes (Stelling et al., 2004). **Nutrigenomics** is the intriguing topic of nutrition and epigenetics, concentrating especially on nutrient-epigenetic regulation in animals. Nutrigenomics addresses the interaction between dietary components and genetic functions and the effect of nutrients on gene expression, and provides a basis for understanding the biological activity of dietary components. Research in epigenomics, especially nutrigenomics, is still in its infancy in farm animals. As this field continues to expand, it is becoming increasingly evident that a host of genomic interrelationships with aspects of the external environment, such as temperature or nutrient availability, is present (Table 21.2).

[1]Mention of trade names or commercial products in this article is solely for the purpose of providing specific information and does not imply recommendation or endorsement by the U.S. Department of Agriculture. The USDA is an equal opportunity provider and employer.

Molecular and Quantitative Animal Genetics, First Edition. Edited by Hasan Khatib.

Table 21.1 Mechanisms of epigenomic regulations.

Epigenomic Regulations
1. Histone posttranslational modifications (PTMs):
acetylation
methylation
phosphorylation
ubiquitination, and so on.
2. DNA methylation
3. Genomic imprinting
4. miRNA
5. Prion proteins
6. and so on.

Table 21.2 Factors affecting epigenomic regulation.

Nutrition (dietary factors)
Environmental agents
Radiation exposure
Infectious agents
Immunological factors
Genetic factors
Toxic agents
Mutagens

For years, scientists have understood that biological fate is not completely controlled by the DNA sequence and that genome sequences play only a partial role in determining an individual's biological characterization (phenotype). Cells, either free-living or part of a multi-cellular organism, must be able to rapidly respond to changes in their external environment, such as temperature or nutrient availability, to exploit and survive in changing conditions. The function of DNA is not as fixed as previously thought. The interaction between genes and the environment plays a crucial role in determining human and animal resistance to various types of stress. Various environmental and nutritional factors are known to result in changes in phenotype in many organisms, described as phenotypic **plasticity**. The recognition that nutrient availability has the capacity to modulate the molecular mechanisms underlying an organism's physiological functions has prompted a revolution in the field of animal and human nutrition (Mutch et al., 2005).

Epigenomic machinery and gene regulation

Genome, chromatin, and epigenomics

The genome of a female Hereford cow has been sequenced by the **Bovine Genome Sequencing and Analysis Consortium** (Elsik et al., 2009), a team of researchers led by the National Institutes of Health and the US Department of Agriculture. This genome is one of the largest ever sequenced. The results, published in the journal *Science* on April 24, 2009 (*Science* podcast, 2009), are likely to have a major impact on livestock breeding. The sequences were obtained by more than 300 scientists in 25 countries after six years of effort. The bovine genome is about 3 Gb (3 billion base pairs). It contains approximately 22,000 genes of which 14,000 are common to all mammalian species. Bovines share 80% of their genes with humans, and cows are much more similar to humans than rodents. Cows also share about 1000 genes with dogs and rodents that have not been identified in humans. The number of sequenced species is increasing at a staggering rate due to the development of the technology to rapidly and cheaply sequence full genomes.

However, sequencing of whole genomes has not yet answered all questions of biology. For example, in cattle, as well as all mammals, about 200 different types of cells have different morphology and perform different functions. Yet these cells all share the same genome set. The astonishing diversity in phenotypic plasticity regarding morphology and functions is determined by the different **epigenomes**. This diversity is characterized by cell-specific patterns of gene expression controlled by regulatory sites in the genome and the "epigenomic regulatory program" that we have just started to learn about. Pluripotent stem cells including embryonic stem cells and induced pluripotent stem cells use a complex network of genetic and epigenetic pathways to maintain a delicate balance between self-renewal and multi-lineage differentiation.

The bovine genome is composed of approximately 3 billion nucleotides and, if stretched in a single, continuous strand, would span almost 2 m. The organization of this entity within the tiny space of the nucleus is an essential challenge for every eukaryotic cell (Taube and Barton, 2006). Adding to the complexity, several important biological functions such as DNA replication, DNA repair, and transcription move along the DNA in a precisely regulated manner. These functions are all involved with very complicated functional machineries made up of many enzymes and proteins. In other words, in addition to genomic sequences, there are some highly dynamic programs of DNA-associated biological processes such as DNA replication, initiated by binding of DNA replication complexes and activation of DNA replication enzymes, transcription with binding of transcription factors, or the position of the nucleosomes. These activities start with binding of functional proteins such as transcription factors, DNA replication complexes, and so on, to the DNA regulatory sites. Access to these sites is coordinated through **chromatin**. Dark staining nuclear material was observed by early microscopists and called chromatin, which later proved to be the structural answer to the compact organization of genome DNA. Chromatin structure and its place in regulated gene expression, cellular homeostasis, and loss of function in disease form ever-expanding and fundamental interests for scientists in the field of biology.

Chromatin, the functional platform of the genome, is constituted of **nucleosomes**, complexes of DNA, RNA, and proteins (Figure 21.1). Chromatin is probably the most complex molecular ensemble in the cell. The most important genome architecture element is the basic repeats of nucleosomes in which DNA wraps around histone proteins. Multiple histones wrap into a 30 nm filament consisting of nucleosome arrays. The octamer complex of the histone is made up of two molecules of each histone (H2A, H2B, H3, and H4). These octamer complexes are also called **core histones**.

The amino-terminal portion of the core histone proteins contains a flexible and highly basic tail region. This region is conserved across various species and is subject to various posttranslational modifications. The linker histone H1 further condenses chromatin by binding to DNA between the nucleosome core particles. Depending on the states of the chromatin structure (euchromatin or heterochromatin), the proposed structure of the 30 nm chromatin filament for the DNA repeat length per nucleosome ranges from 177 to 207 bp (about 147 bp of DNA wrapping around the core histone and 20 to 50 bp of linker DNA connecting the nucleosomes) (Ball, 2003; Lee and Orr-Weaver, 2001; Taube and Barton, 2006; Robinson and Rhodes, 2006; Wong et al., 2007). In contrast with DNA, chromatin is not stationary but highly dynamic, mainly through posttranslational modifications of histones. Generally, chromatin can be classified as condensed, transcriptionally silent heterochromatin or less-condensed, transcriptionally active euchromatin. Most genome DNA (telomeres, pericentric regions, and an area rich in repetitive sequences) is believed to be packed into heterochromatin. There is also looping of large stretches of chromatin from a chromosome to generate the local secondary structure poised for transcription (Gilbert et al., 2004). The dynamic nature of the chromatin anticipates different conformational forms existing in the nucleus at different cell proliferation status, too. More importantly, the chromatin structure is influenced by the modification of DNA or histones that encompass the chromatins.

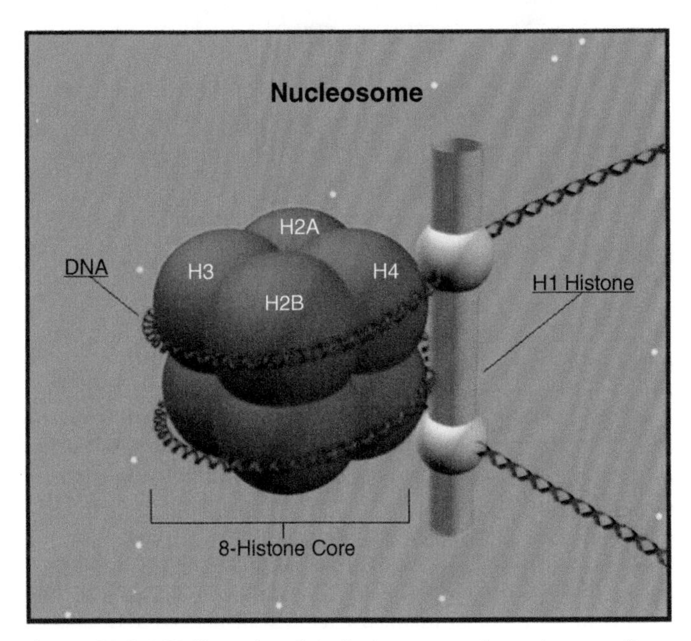

Figure 21.1 A 3D illustration of the basic structure of a nucleosome. (from http://en.wikipedia.org/wiki/File:Nucleosome.jpg).

Nutrients and methylation of nucleosome

Chromatin is not only a platform for storing genetic information but also regulates transcriptional processes based on its modification on both components: DNA and histones. Increasing evidence suggests the existence of extensive crosstalk among epigenetic pathways that modify DNA, histones, and nucleosomes (Figure 21.2). Mapping the higher-order chromatin structure and chromatin–nuclear matrix interactions provide important insights into the three-dimensional organization of the genome and a framework in which the existing genomic data of epigenetic regulation can be integrated to discover new rules of gene regulation.

In the mammalian genome, DNA methylation is an epigenetic mechanism involving the transfer of a methyl group onto the C5 position of the cytosine to form 5-methylcytosine. DNA methylation occurs mostly within the context of a CpG dinucleotide. Genome regions that contain a high frequency of CpG sites are called CpG islands or CG islands. The "p" in CpG refers to the phosphodiester bond between the cytosine and the guanine, which indicates that the C and the G are next to each other in sequence. DNA methylation regulates gene expression by recruiting proteins involved in gene repression or by inhibiting the

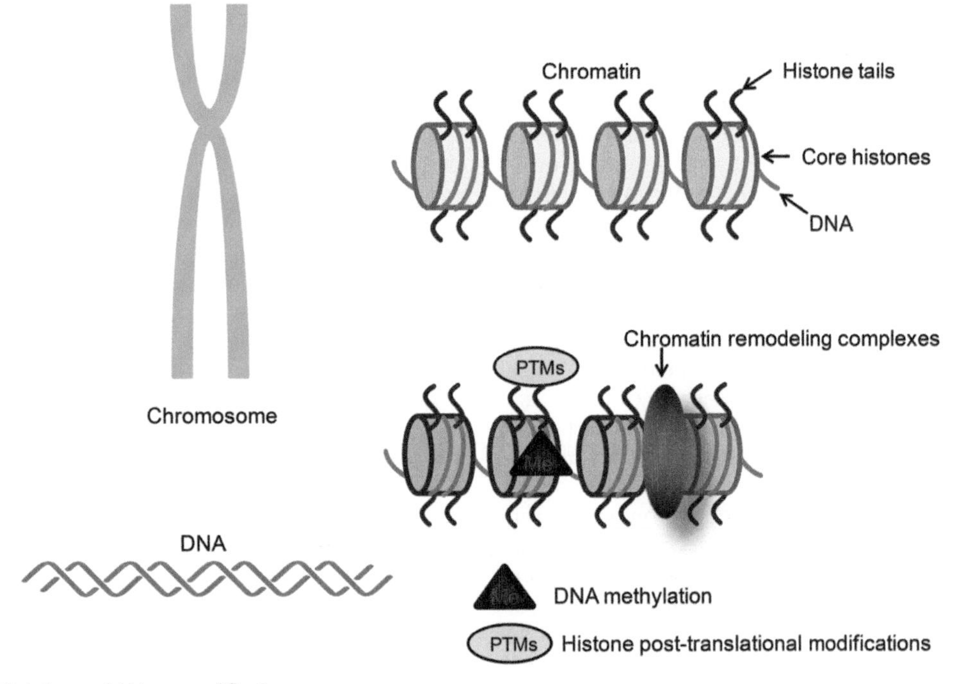

Figure 21.2 DNA methylation and histone modification.

binding of transcription factor(s) to DNA. During development, the pattern of DNA methylation in the genome changes as a result of a dynamic process involving *de novo* DNA methylation and demethylation. As a consequence, differentiated cells develop a stable and unique DNA methylation pattern that regulates tissue-specific gene transcription. DNA methylation has many basic functions. Methylation within gene regulatory elements such as promoters, enhancers, insulators, and repressors generally suppresses the function of the gene. Methylation within gene-deficient regions, such as in pericentromeric heterochromatin, is critical for maintaining the structure and integrity of the chromosome (Friso and Choi, 2002). DNA methylation is also a vital mechanism for normal embryonic development. Methylation of the mammalian genome endures remarkable changes during early development and seems to be an integral mechanism for differentiating and forming various tissues and organs. Starting after fertilization, DNA methylation patterns undergo establishment, reestablishment, and maintenance. Those changes are called "epigenetic reprogramming." These modifications are important for normal embryo and placental developments. DNA methylation patterns also go through establishment, maintenance, erasure, and reestablishment throughout life and are passed to the next generation (see Chapter 24). In mammals, epigenetic reprogramming in germ cells and during preimplantation, especially its effects on imprinting genes, predominantly establishes developmental stages. In humans, however, a decreased level of genomic DNA methylation is a common factor in tumorigenesis and usually appears in the early stage of carcinogenesis and precedes the mutation and deletion events that occur later in the development of cancer. However, hypomethylation may contribute to genomic instability, structural changes in chromosomes, and enhancement of gene activities.

Histones are post translationally modified by various enzymes, including methylation, primarily at their flexible N-termini. Histone methylation can take place in lysine and arginine residues. These modifications play an important role in modulating chromatin structure and thus regulate gene expression, DNA repair, DNA replication, and many other chromatin-associated biological processes (Kouzarides, 2007).

Nutrients modifying DNA and histone methylation

Many dietary components can modulate DNA and histone methylation. DNA methylation and histone methylation are *S*-Adenosyl methionine (ademetionine, AdoMet, SAM, SAMe, SAM-e: Figure 21.3), a common cosubstrate involved in methyl group transfers dependent.

ATP is the cofactor most used for methylation enzymes, and SAM is the second most used cofactor. The methyl groups required for establishing and maintaining DNA methylation and histone methylation come from sole dietary methyl donors in association with specific enzymes and cofactors (Ulrey et al., 2005). The conveying of a methyl group to the fifth carbon of cytosine within CpG dinucleotides is catalyzed by **DNA methyltransferases** (DNMTs). SAM is the sole final methyl donor produced by **one-carbon metabolism** (one-carbon metabolism is composed of the cytosolic and mitochondrial folate cycles and the methionine cycle). SAM is also the universal methyl donor in all cell types. **SAM** is not only critical for DNA methylation but is also very important in nucleic acid metabolism and histone methylation. Therefore, it is obvious that any nutrients that are coenzymes of one-carbon metabolism, dietary methyl donor

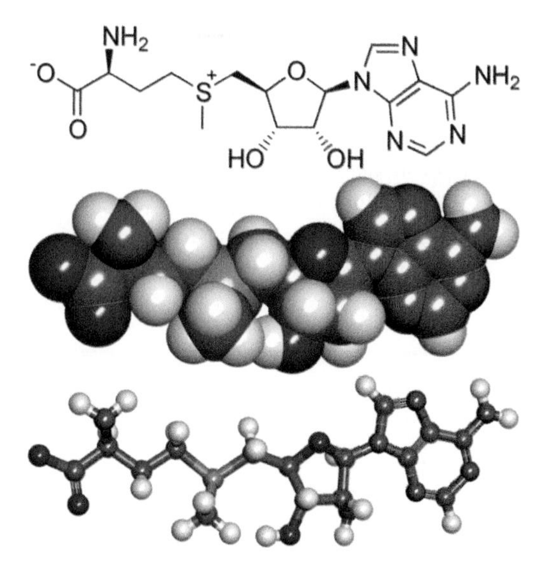

Figure 21.3 *S*-Adenosyl methionine (ademetionine, AdoMet, SAM, SAMe, SAM-e) (Modified from http://en.wikipedia.org/wiki/File:S-adenosylmethionine.png).

nutrients, nutrients that can modify one-carbon metabolism, and any nutrients that may modify the activity of DNMTs affect DNA and histone methylation. Dietary folate is a classic example of nutrients whose deficiency has profound effects on DNA methylation.

Folates exist in blood plasma most commonly as polyglutamated methyl-tetrahydrofolate (methyl-THF). After transport into cytoplasm by reduced folate carrier (RFC), methyl-THF plays a role as a methyl group donor for producing tetrahydrofolate (THF) and a precursor for homocysteine conversion to methionine, which is catalyzed by methionine synthase. Vitamin B_{12} is required as a cofactor in this process. Methionine is converted to SAM by methionine adenosyltransferase. Up to half of the daily intake of methionine is converted to SAM, and the majority of all subsequent methylation reactions take place in the liver. Folates are the major determinant of DNA and histone methylation because the folate-derived methyl group is a major methyl donor for homocysteine remethylation and therefore determines the status of **SAM**.

Numerous studies have also demonstrated that changes in DNA methylation level can come as the consequences of additional dietary factors, such as alcohol consumption, which can affect the bioavailability of SAM. The direct effects of alcohol on folate metabolism primarily are due to inhibition of the enzyme methionine synthase, resulting in a decreased level of downstream products methionine and SAM and an increased level of precursors, homocysteine and S-adenosylhomocysteine (SAH). SAH also functions as an inhibitor of SAM-dependent methyltransferases.

The effects of folate deficiency on DNA methylation are multifarious. The effects are cell type and target organ dependent. They are also gene and site-specific (Jhaveri et al., 2001). Animal studies have indicated that prenatal feeding of a methyl-supplemented diet can increase DNA methylation and decrease the expression of genes in offspring. Limited folate supply in human results in increased levels of homocysteine and reduced DNA methylation. However, the *de novo* establishment and maintenance of DNA and histone methylation require complicated metabolic pathways and interact with many other essential biochemical reactions needed for cellular activities for survival, proliferation, and differentiation. The importance of biological complexity has been realized recently. The principle of one gene leads to one protein leads to one metabolite is too simplified for biological systems. The development of next-generation DNA sequencing to study DNA methylation across the genome will help us understand the significance of methyl donor production and insufficiency and unravel the mechanisms underlying disease risk associated with deficiencies in these pathways.

The effects of maternal nutrition on fetuses/offspring

Epidemiological and animal studies have shown that adverse environments or suboptimal nutritional conditions during pregnancy can alter the physiology of offspring and increase their predisposition to many diseases in adult life. There is growing evidence that maternal diet during stages of pregnancy can induce physiological and genetic changes in fetal tissues in different species. This phenomenon is referred to as **developmental programming**. The nutrition of the pregnant mother can modify epigenetic marks, such as DNA methylation, and gene expression in the offspring can be altered due to the modification of epigenetic marks. One of the best studied examples of epigenetically sensitive genes via maternal diet is the Agouti viable yellow (Avy) locus in the mouse. Methyl supplements in the diet of pregnant mice increased the methylation level of the Agouti gene, and consequently led to changes in the coat color (from brown to yellow) of the offspring. While "viable yellow" mice are larger, obese, hyperinsulinemic, more susceptible to cancer, and, on average, shorter lived than their non-yellow siblings, pseudoagouti phenotype with minimal ectopic expression) mice are lean, healthy, and longer lived than their yellow siblings (Cooney et al., 2002; Wolff et al., 1998). These studies have clearly established that maternal diet has a transgenerational effect on the offspring through epigenetic modifications.

Expression of imprinted gene can be influenced by maternal nutrition was found not only in mouse and rat, but also reported recently in farm animals such as cow and sheep. Imprinted genes play a significant role in fetal growth and development. They are expressed in a parent of-origin-specific allele manner so that so that alleles inherited from one parent are expressed, whereas alleles inherited from the other parent are silenced. These genes have roles in embryonic survival and growth, obesity and insulin resistance, spermatogenesis defects, energy regulation, motor and learning dysfunction, deafness, weight loss, placental growth, tumorigenesis, apoptosis, fertility, and maternal behavior. Epigenetic modifications, such as DNA methylation and acetylation, regulate expression of imprinted genes. The different maternal diets alter the expression of DNA methyltransferases (DNMTs), thereby changing the DNA methylation status, is believed as the underlying mechanism.

Another striking example of "Nature versus Nurture" is the honeybee society, the queen/worker developmental divide. The honeybee queen and workers are genetically identical, however, queen and workers are different in their anatomical, behavioral and physiological characters, as well as the longevity of the queen. The magic formula responsible for the making of a queen is now known as "Royal jelly," which is concentrated with mixture of proteins, essential amino acids, unusual lipids, vitamins, and other compounds. The "nurse" workers produce the royal jelly with their head glands. The queen larvae get fed in large quantities and over extended periods of royal jelly. The phenotypic alternations that lie at the core of honeybee social organization imply the existence of epigenetic controls in honeybee and providing a link between environmental factors and phenotypic differences.

The discoveries of a family of highly conserved DNA cytosine-5-methyltransferases (DNMTs) in honeybees suggest the epigenetic integration of environmental and genomic signals in honeybees. Recent evidence also indicates silencing DNA methylation in young larvae mimics the effects of nutrition on early developmental processes that determine the reproductive fate of honeybee females. Mapping of entire methylomes of the brains of honeybee queens and workers also presented a complete picture of the transcriptional programs governed by differential DNA methylation (Lyko et al., 2010). The epigenome-based findings created an astonishing opportunity for testing the hypothesis that DNA methylation in honeybees is involved in a nutritionally driven phenomenon of queen development.

Nutrients and histone modification

Chromatin remodeling and histone modification

Chromatin is composed of DNA and various modified histones and non-histone proteins, which have an impact on cell differentiation, gene regulation, and other key cellular processes. Histones are subject to a wide variety of posttranslational modifications such as lysine acetylation, lysine and arginine methylation, serine and threonine phosphorylation, lysine ubiquitination, sumoylation and ADP ribosylation, and proline isomerization. These modifications occur primarily within the histone N-terminal tails protruding from the surface of the nucleosome and on the globular core region. Histone modifications, especially acetylated histones and nucleosomes, represent a type of epigenetic tag within chromatin.

Histone acetylation and deacetylation

The histones are acetylated and deacetylated on lysine residues in the N-terminal tail and on the surface of the nucleosome core as part of gene regulation. These reactions are typically catalyzed by enzymes with **histone acetyltransferase** (**HAT**) or **histone deacetylase** (**HDAC**) activity. The source of the acetyl group in histone acetylation is acetyl-Coenzyme A, and in histone deacetylation, the acetyl group is transferred to Coenzyme A (Figure 21.4).

HATs are enzymes that acetylate conserved lysine amino acids on histone proteins by transferring an acetyl group from acetyl-CoA to form ε-*N*-acetyllysine. Because this reversible process of histone acetylation occurs at the ε–amino group of lysine residues, HATs are also named lysine acetyltransferases (KATs). Histone acetyltransferases can also acetylate non-histone proteins, such as transcription factors and nuclear receptors to facilitate gene expression. Major HAT families include the GNAT superfamily, the MYST superfamily, and p300/CBP. All the GNAT superfamily members share structural and sequence similarity to Gcn5 (general control of nuclear-5) protein. The most relevant HATs of this family are Gcn5, PCAF, Hat1, Elp3, and Hpa2. The MYST family was named after its founding members: MOZ, Ybf2/Sas3, Sas2, and Tip60. P300/CBP are often referred to as a single entity because the two proteins are structural and

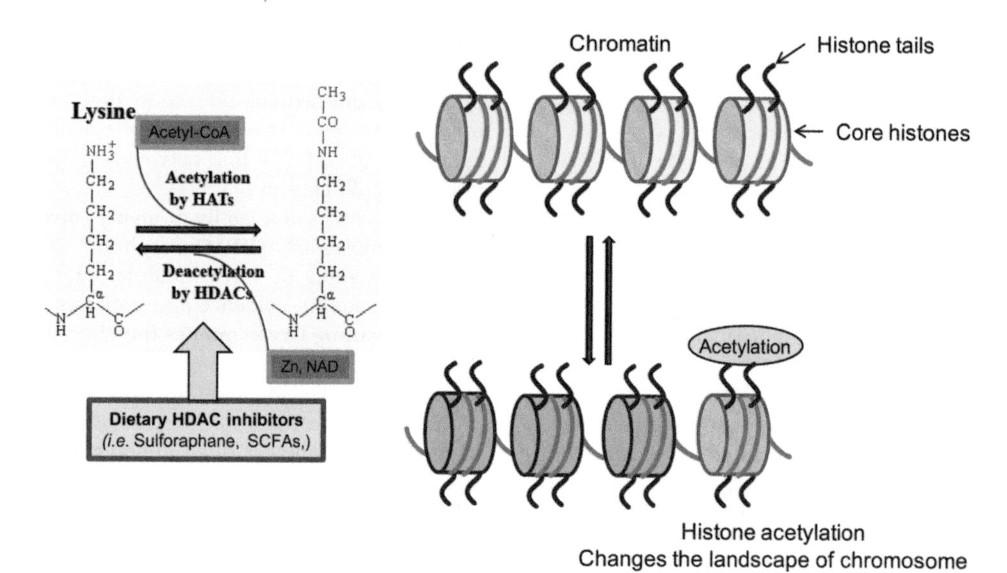

Figure 21.4 Dynamic state of histone acetylation.

functional homologs. Their function has been shown to be interchangeable.

HDACs are a class of enzymes that remove acetyl groups from an ε-N-acetyl lysine amino acid on a histone opposite the action of HATs. HDAC proteins are now also called lysine deacetylases (KDACs). KDAC is more precise for the description of these enzymes' function rather than their target because KDACs also target non-histone proteins. HDAC proteins are classified in four groups based on function and DNA sequence similarity. The first two groups are considered "classical" HDACs whose activities are inhibited by trichostatin A (TSA), whereas the third group is a family of NAD+-dependent proteins not affected by TSA. The fourth group is an atypical category of its own, based solely on DNA sequence similarity to the others.

Histone modifications are proposed to affect chromosome function through at least two distinct mechanisms. The first mechanism suggests modifications may alter the electrostatic charge of the histone resulting in a structural change in histones or their binding to DNA. The earlier model of action underlying the acetylated histone was called the charge neutralization model. Acetylation removes the positive charge on the histones, thus decreasing the interaction of the N-termini of histones with the negatively charged phosphate groups of DNA. As a consequence, the condensed chromatin is transformed into a more relaxed structure associated with greater levels of gene transcription. This relaxation can be reversed by HDAC activity. Relaxed, transcriptionally active DNA is referred to as **euchromatin**. More condensed (tightly packed) DNA is referred to as **heterochromatin**. Condensation can be brought about by processes including deacetylation and methylation; the action of methylation is indirect and has no effect upon the charge. The concept of this model is based on the fact that transcriptionally active genes are correlated with the rapid turnover of histone acetylation. This requires that the HATs and HDACs must act continuously on the affected histone tail. This model also indicates that histone acetylation plays an important role in regulating gene expression based on the evidence that acetylation can decompact nucleosome arrays. Hyperacetylated chromatin is transcriptionally active, and hypoacetylated chromatin is silent (*cis*-modifying effects). This model has been challenged by recent research developments. The second mechanism proposes that these modifications are binding sites for protein recognition modules, such as the bromodomains or chromodomains, that recognize acetylated lysines or methylated lysine, respectively (*trans*-modifying effects). Therefore, the histone posttranslational modifications provide a binding surface for proteins to associate with chromatin and regulate DNA-templated processes such as transcription and DNA replication and repair. This model has been characterized in the most detail. Functional protein recruitment is defined with functional consequences, which means it may activate or repress the outcome on gene expression. Although the links between histone posttranslational modifications (especially acetylation) and chromatin structure and cell cycle progression, transcription, DNA replication, and overall chromosome functions are very clear, the modulation of genome expression as a consequence of chromatin structural changes is much more complex.

Dietary factors and histone deacetylase inhibition

Histone (lysine) acetyltransferases (HATs/KATs) and deacetylases (HDACs) maintain the steady-state balance of histone acetylation. Several dietary factors can inhibit HDAC activity (Figure 21.5). These nutrients include butyrate (formed during microbial fermentation of dietary fiber in the gastrointestinal tract of mammalian species), diallyl disulfide (in garlic and other *Allium* vegetables), and sulforaphane (in cruciferous vegetables). The anticancer effects of HDAC inhibitors (HDACi) are well established. The links between histone posttranslational modifications and chromatin structure and cell cycle progression, DNA replication, and overall chromosome functions are very clear. The modulation of genome expression as a consequence of chromatin structural changes is likely a basic mechanism.

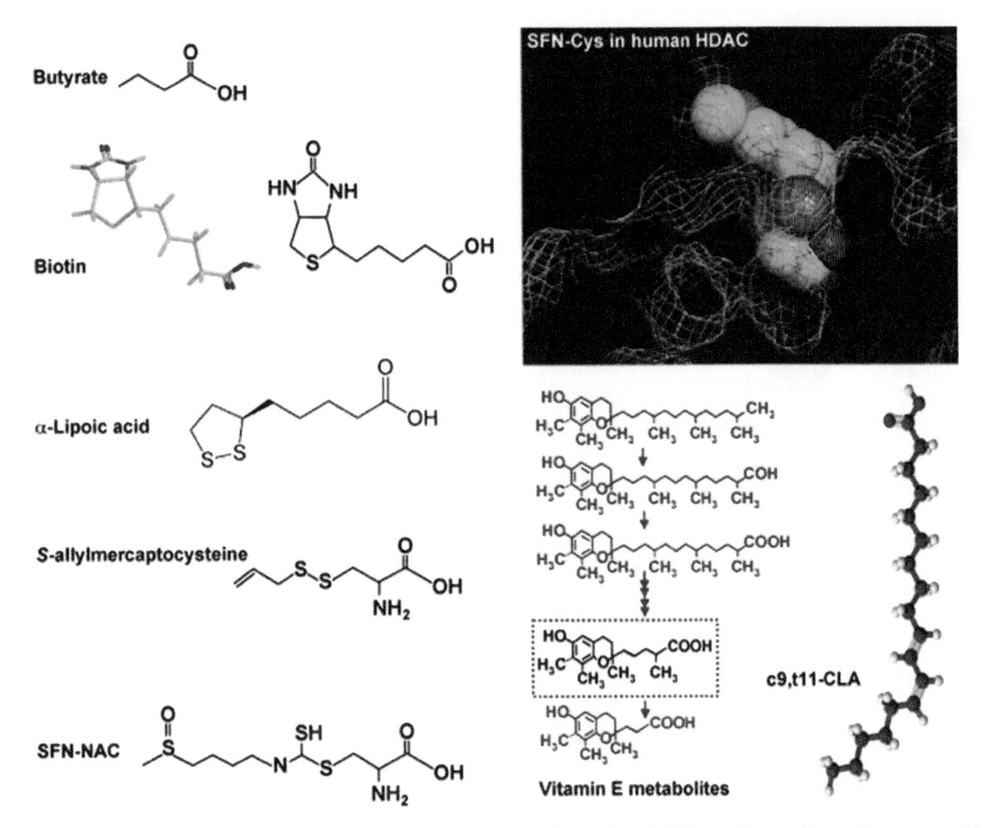

Figure 21.5 Dietary HDAC inhibitors (from Dashwood RH and Ho E, Dietary histone deacetylase inhibitors: from cells to mice to man, *Seminars in Cancer Biology* 7(5): 363–369, 2007. Reprinted with permission from Roderick H. Dashwood, Emily Ho. Dietary histone deacetylase inhibitors: From cells to mice to man. *Seminars in Cancer Biology*, vol. 17, issue 5. Copyright © Elseiver, 2007.

Nutrients and epigenetics in bovine cells: One definitive example of the nutrient-epigenetic-phenotype relationship

Research in epigenomics, especially nutrigenomics, is still in its infancy in farm animals. A definitive example of the nutrient-epigenetic-phenotype relationship can be perceived by examining volatile fatty acids (VFAs, i.e., acetate, propionate, and butyrate), also referred to as short-chain fatty acids (SCFAs), and the regulation of gene expression induced by SCFAs. Short-chain fatty acids, especially butyrate, participate in metabolism as nutrients and as histone deacetylase inhibitors (HDACi) regulating the "epigenomic code" in bovines. HDACi are also an emerging class of novel anti-cancer drugs that cause growth arrest, differentiation, and apoptosis of tumor cells in humans. SCFAs are formed during microbial fermentation of dietary fiber in the gastrointestinal tract of mammalian species and then are directly absorbed at the site of production. These compounds contribute up to 70% of the energy requirements of ruminant species. It is well established that rumen papillae development is driven by volatile fatty acids, with butyrate being particularly powerful. Rates of SCFA production and absorption in ruminants, which has been calculated as about 5 mol/kg dry matter intake, are much higher than in other animals and in humans. The major SCFAs in either the rumen or the large intestine are acetate, propionate, and butyrate, and they are produced in ratios varying from about 75:15:10 to 40:40:20. The concentrations of SCFAs in the rumen are highly variable, and the total amount present usually fluctuates between 60 and 150 mM. In sheep, butyrate concentrations in the digestive tract and blood are usually between 0.5 and 13 mM (Bergman, 1990). Acetate and propionate have relatively higher concentrations as they have a prominent position in providing energy for ruminant metabolism. Butyrate is low in relative concentration but appears to be involved in metabolic processes beyond its role as a nutrient with high effectiveness as an inhibitor of histone deacetylases (HDACs). Butyrate involvement has been determined in cell differentiation, proliferation, motility, and, in particular, induction of cell cycle arrest and apoptosis. These biological effects on the cell cycle and apoptosis have undergone intensive investigation, with the aim of developing butyrate as a therapeutic agent for cancer treatment. SCFAs are common and important nutrients in cattle. Therefore, understanding their biological importance, beyond their use as a simple energy supply, is important. This knowledge could lead to improvements in the efficiency of production of food animals.

In addition to their nutritional value, SCFAs, especially butyrate, modulate cell differentiation, proliferation, and motility, and induce cell cycle arrest and apoptosis. The therapeutic potential of butyrate for cancer has been intensively studied (Myzak and Dashwood, 2006). The cell cycle regulatory effects of butyrate at the cellular and molecular levels in normal bovine cells have also been studied (Figure 21.6). The principle biochemical change in cells treated with butyrate and other histone deacetylase (HDAC) inhibitors is the global hyperacetylation of histones. The links between histone posttranslational

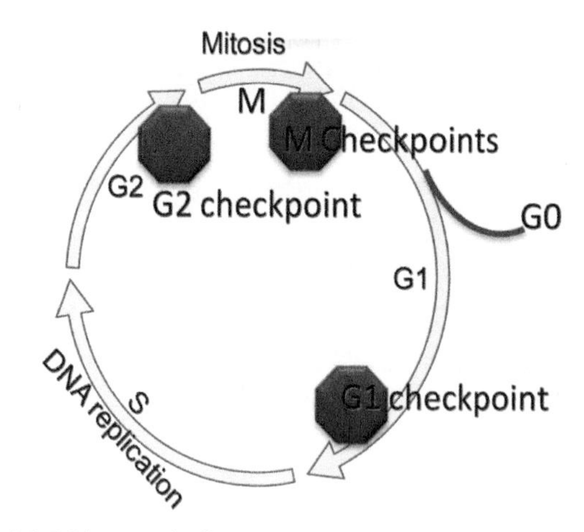

Figure 21.6 Diagram of cell cycle and cell cycle checkpoints.

modifications and chromatin structure and cell cycle progression, DNA replication, and overall chromosome stability become very clear (Wolffe and Guschin, 2000a,b).

The modulation of genome expression as a consequence of chromatin structural changes is likely a mechanism with a major role in determining tissue responses. Data from *in vitro* experiments on bovine cells (MDBK) show that as a direct result of the hyperacetylation of histones induced by butyrate treatment at physiological concentrations (2.5 to 10 mM), cultured bovine cells are arrested in the early G1 phase and DNA synthesis is eliminated. At a relatively high concentration (10 mM), butyrate also induces apoptosis in an established bovine MDBK cell line. In primary cultures of isolated ruminal epithelial cells, DNA replication is also inhibited by butyrate treatment. In addition, butyrate may also alter histone methylation as a histone deacetylase inhibitor, suggesting interplay between histone acetylation and methylation. When bovine cells (MDBK) were treated with 10 mM butyrate for 24 h, cell morphology became distorted. Cells with large vacuoles and ragged membranes, lacking distinct intracellular organelles and increasing spaces between cells were readily visible and recurrent. Flow cytometry analysis of cell population profiles for DNA content and BrdU labeling also confirmed that the cells were arrested at the G1 and G1/S boundary. The incorporation of the BrdU label suggested that DNA synthesis was blocked by butyrate treatment. Western blotting also confirmed that butyrate induced hyperacetylation of histones.

Histone acetylation and gene regulation underlying the mechanisms of short-chain fatty acid effects on cellular functions

Acetylation of histone tails is essential for diverse cellular processes, such as DNA replication and cell cycle progression. Butyrate-induced histone hyperacetylation, however, has divergent activities, including the induction of cell cycle arrest, gene expression, and apoptosis (Figure 21.7).

The development of **next-generation sequencing** (NGS) (Chapter 17) has provided novel tools for expression profiling and genome analysis (Kahvejian et al., 2008; Ozsolak and Milos, 2011; Shendure and Ji, 2008). A study using next-generation sequencing technology provided a more complete characterization of the RNA transcripts of MDBK cells (Wu et al., 2012). The study focused on the comparison between a control group of cells (without butyrate treatment) and cells treated with 10 mM butyrate for 24 h. The samples were deep-sequenced, with an average of more than 67 million reads per sample, and the results were used to estimate the differences induced by butyrate treatment. The NGS results showed very reliable and detailed profiling of the changes in gene expression induced by butyrate in a normal bovine cell line. The phenomenal number of genes fell within a broad range of functional categories providing a very detailed molecular basis for the butyrate-induced biological effects. **Transcriptome** (the set of all RNA molecules, including mRNA, rRNA, tRNA, and other non-coding RNA produced in one or a population of cells) characterization of bovine cells using RNA-sequencing (RNA-seq) identified transcriptional control mechanisms by butyrate.

Genome-wide ChIP-seq mapping and analysis reveal butyrate-induced chromatin modification

Chromatin-immunoprecipitation-DNA sequencing (ChIP-seq) is used to analyze protein interactions with DNA. ChIP-seq combines ChIP with massively parallel DNA sequencing to identify the binding sites of DNA-associated proteins such as histone and transcription factors). The HDAC inhibition activity of butyrate makes it a great inducer of the hyperacetylation of histone in cells. Discovering how the epigenomic landscape is modified by butyrate-induced histone acetylation is a critical step in the path to understanding how this nutrient affects specific transcriptome changes at the mechanistic level. Utilizing next-generation sequencing technology, combined with ChIP technology, histone modification (acetylation) induced by butyrate and large-scale mapping of the epigenomic landscape of normal histone H3 and acetylated histone H3K9 and H3K27 has been completed recently. The analysis indicated that the distribution of histone H3, acetyl-H3K9, and acetyl-H3K27 correlated with transcription activity induced by butyrate. Analysis of the consensus sequences bound to H3, acetyl-H3K9, and acetyl-H3K27 reveals several consensus sequences (motifs) from each ChIP-seq data set. **Sequence motifs** are short, recurring patterns in DNA that are presumed to have a biological function. The motifs usually indicate sequence-specific binding sites for proteins such as nucleases and transcription factors, and in our case, one important chromatin component, histone H3. Computational analysis of the ChIP-seq data makes it possible to derive genome-wide binding patterns for the histone H3 core, acetyl-H3K9, and acetyl-H3K27. Those results reveal that butyrate-induced acetylation of H3K9 and H3K27 changes the sequence-based binding preference of histone H3. The differences in the binding motifs for acetyl-H3K9 and acetyl-H3K27 indicate that histone modification (acetylation) at various lysine sites changes the histone H3 binding preferences either independently or cooperatively. In either situation, butyrate-induced acetylation of histone plays a role in changing the chromatin structure, defining genetic regulatory networks, and interpreting the regulatory program of individual genes. In addition to the variation of the binding preferences of acetylated histone H3 as the result of acetylation, a high degree of conservation in histone binding is also evidently presented. This may present a key property of living systems of robustness, for example, the ability to maintain phenotypic stability in the face of diverse perturbations arising from environmental changes. The recent progress in understanding epigenetics at the molecular level is remarkable; however, epigenetics is closely related to robustness as both lie between genotypes and

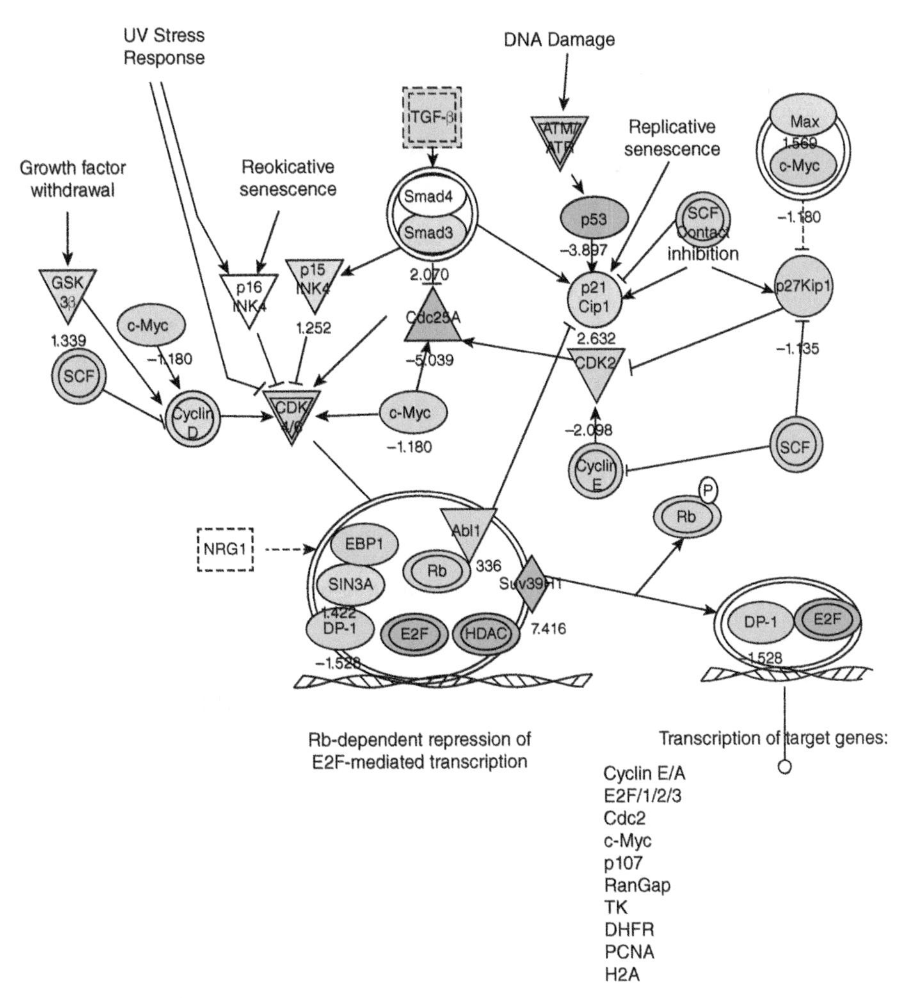

Figure 21.7 Regulation of the cell cycle: The G1/S checkpoint control and effects of butyrate treatment. The color indicates the expression level of the genes (red indicating up-regulated genes and green indicating down-regulated genes). Reproduced with permission from Wu S, Li RW, Li W, Li CJ. Transcriptome characterization by RNA-seq unravels the mechanisms of butyrate-induced epigenomic regulation in bovine cells. *PLoS One*. 2012;7(5):e36940.

phenotypes. More understanding of the molecular level of robustness and epigenetics is needed. The mapping of epigenomic landscape modified by butyrate-induced histone acetylation was a crucial starting point for an in-depth evaluation of the mechanisms involved in bovine rumen epithelial epigenomic regulation.

The bacterial community composition of the bovine rumen, ruminal functions, and environmental impacts: Nutrigenomics of the host-microbiome interactions

The rumen is a complex microbial ecosystem and has played a critical role in sustainable agriculture throughout human civilization. Microorganisms in the rumen, such as bacteria, archaea, protozoa, and fungi, perform essential fermentation, including converting plant fiber into small molecules, such as volatile fatty acids (VFA) and vitamins, to produce meat, milk, and wool for human consumption, thus influencing the host's nutrition. In addition, rumen microorganisms modulate host immunity and enhance host resistance to invading pathogens. Furthermore, certain bacteria detoxify naturally occurring compounds in the diet that are harmful to the host. Microorganisms in the rumen are highly responsive to changes in diet, host genetics, and host physiology as well as geographic and environmental factors. The balance between the complex community of the gut **microbiome** (the totality of microbes), food nutrients, and intestinal genomic and physiological environment is increasingly recognized as a major contributor to human and animal health and disease. Due to the high-throughput methodologies, the complexity, evolution with age, environment, and individual nature of the gut **microflora** (the bacterial population in the digestion track) have been more thoroughly investigated.

Microbiomics is the molecular fingerprint of the wide microbial diversity in the human and animal alimentary tract (microbiome); **metagenomics** uses the next-generation sequencing technique to profile microbial DNA; two technologies that provide insight into host–microbe communications mechanisms.

Nutritional systems biology assembles all *-omics* information in nutrients and biological (human or animal samples) computational databases, analyzes, and develops interactive networks in an overview level of the whole organism through the development of quantitative mathematical models of biological network structure and interactions.

The capacity of the rumen microbiota to produce volatile fatty acids (VFAs) has important implications in animal well-being and production, as well as environmental impacts. In rumen, hydrogen is produced during the anaerobic fermentation of nutrients. This hydrogen can be used during the synthesis of volatile fatty acids (VFAs) and microbial protein synthesis. Ruminal methanogens reduce carbon dioxide to methane (CH_4) and eliminate the excess hydrogen from NADH. Methane production in ruminants is an energetically (2–15% of ingested gross energy) wasteful process, since a portion of the animal's feed, which is converted to methane, is eructed as gas. Furthermore, emission of methane, the greenhouse gas, in the environment contribute to global warming by trapping outgoing terrestrial infrared radiation.

Temporal changes in the rumen microbiota in response to butyrate infusion were investigated using pyrosequencing of the 16S rRNA gene. Twenty-one phyla were identified in the rumen microbiota of dairy cows. The rumen microbiota harbored 54.5 ± 6.1 genera (mean ± SD) and 127.3 ± 4.4 operational taxonomic units (OTUs), respectively. However, the core microbiome comprises 26 genera and 82 OTUs. Butyrate infusion altered molar percentages of three major VFAs. Butyrate perturbation had a profound impact on the rumen microbial composition. A 72-h infusion led to a significant change in the number of sequence reads derived from four phyla, including the two most abundant phyla, Bacteroidetes and Firmicutes. As many as 19 genera and 43 OTUs were significantly impacted by butyrate infusion. Elevated butyrate levels in the rumen seemingly had a stimulating effect on butyrate-producing bacteria populations. The resilience of the rumen microbial ecosystem was evident as the abundance of the microorganisms returned to their predisturbed status after infusion withdrawal. The findings provide insight into the perturbation dynamics of the rumen microbial ecosystem and should guide efforts in formulating optimal uses of probiotic bacteria treating human diseases. A solid understanding of the major components of rumen microbial ecosystems and their interactions is a prerequisite to successful ruminal manipulation, such as increasing efficiency of fiber digestion and reducing ruminal methanogenesis.

Summary

The phenotypic characterization of an animal can be changed by modifying epigenetic mechanisms such as histone posttranslational modification, microRNA, and other mechanisms. The multiple layers of regulatory control of gene expression provide many paths in which cells can control their responses to external stimuli. An explosion of research efforts in recent years has begun to uncover common molecular mechanisms underlying epigenetic phenomena. With the rapid development of biotechnologies, epigenomic approaches will help us characterize genome-wide epigenetic markers that are targets for dietary regulation. The discovery and characterization of these novel epigenetic markers and epigenetic mechanisms will facilitate understanding of how dietary factors modulate these epigenetic regulatory mechanisms. This area will provide a great research opportunity and a better understanding of the role of dietary components in changing epigenetic patterns and have an important impact on functional genomic research in bovines and in the farm animal industry.

In the meantime, we have to realize that epigenetics and epigenomics are much more complicated and more dynamic than previously thought. Epigenomics research and even developing maps of the epigenomic markers in farm animals increasingly require standardized platforms and procedures. Quick and efficient detection of epigenomic markers and better understanding of the factors that induce changes in these markers require great effort. Substantial international collaboration is essential to accomplish this goal. The fast development in this field has attracted strong interest in exploiting epigenetic phenomena. In the near future, we should expect to see the epigenetic landscape take shape in animal science.

References

Ball, G.F.M., Pantothenic acid | physiology, in *Encyclopedia of Food Sciences and Nutrition*, 2nd edn, Editor-in-Chief: C. Benjamin, 2003, Academic Press: Oxford. p. 4339–4345.

Bergman, E.N., Energy contributions of volatile fatty acids from the gastrointestinal tract in various species. *Physiol Rev*, 1990. 70(2): p. 567–90.

Bovine Genome Sequencing and Analysis Consortium, Elsik, C.G., R.L. Tellam, K.C. Worley, R.A. Gibbs, et al., The genome sequence of taurine cattle: a window to ruminant biology and evolution. *Science*, 2009. 324(5926): 522–528.

Cooney, C.A., A.A. Dave, and G.L. Wolff, Maternal methyl supplements in mice affect epigenetic variation and DNA methylation of offspring. *J Nutr*, 2002. 132(8 Suppl): 2393S–2400S.

Friso, S. and S.W. Choi, Gene-nutrient interactions and DNA methylation. *J Nutr*, 2002. 132(8 Suppl): 2382S–2387S.

Gilbert, N., S. Boyle, H. Fiegler, K. Woodfine, N.P. Carter, and W.A. Bickmore., Chromatin architecture of the human genome: gene-rich domains are enriched in open chromatin fibers. *Cell*, 2004. 118(5): 555–566.

Jhaveri, M.S., C. Wagner, and J.B. Trepel, Impact of extracellular folate levels on global gene expression. *Molecular Pharmacology*, 2001. 60(6): 1288–1295.

Kahvejian, A., J. Quackenbush, and J.F. Thompson, What would you do if you could sequence everything? *Nature Biotechnology*, 2008. 26(10): 1125–1133.

Katsnelson, A., Genomics goes beyond DNA sequence. *Nature*, 2010. 465(7295): 145.

Kouzarides, T., Chromatin modifications and their function. *Cell*, 2007. 128(4): 693–705.

Lee, J.Y. and T.L. Orr-Weaver, Chromatin, in *Encyclopedia of Genetics*, Editors-in-Chief: B. Sydney and H.M. Jeffrey. 2001, Academic Press: New York. p. 340–343.

Lyko, F., S. Foret, R. Kucharski, S. Wolf, C. Falckenhayn, and R. Maleszka, The honey bee epigenomes: differential methylation of brain DNA in queens and workers. *PLoS Biol*, 2010. 8(11): p. e1000506.

Mutch, D.M., W. Wahli, and G. Williamson, Nutrigenomics and nutrigenetics: the emerging faces of nutrition. *FASEB Journal: Official Publication of the Federation of American Societies for Experimental Biology*, 2005. 19(12): 1602–1616.

Myzak, M.C. and R.H. Dashwood, Histone deacetylases as targets for dietary cancer preventive agents: lessons learned with butyrate, diallyl disulfide, and sulforaphane. *Curr Drug Targets*, 2006. 7(4): 443–452.

Ozsolak, F. and P.M. Milos, RNA sequencing: advances, challenges and opportunities. *Nature Reviews Genetics*, 2011. 12(2): 87–98.

Robinson, P.J.J. and D. Rhodes, Structure of the "30 nm" chromatin fibre: A key role for the linker histone. *Current Opinion in Structural Biology*, 2006. 16(3): 336–343.

Science Podcast, 04/24/09. *Science*, 2009. 324(5926): p. 537.

Shendure, J. and H. Ji, Next-generation DNA sequencing. *Nature Biotechnology*, 2008. 26(10): 1135–1145.

Stelling, J., U. Sauer, Z. Szallasi, F.J. Doyle, and J. Doyle. Robustness of cellular functions. *Cell*, 2004. 118(6): 675–685.

Taube, J.H. and M.C. Barton, Chromatin and Regulation of Gene Expression. In *Gene Expression and Regulation*, J. Ma (ed.) 2006, Springer New York. p. 95–109.

Ulrey, C.L., L. Liu, L.G. Andrews, and T.O. Tollefsbol, The impact of metabolism on DNA methylation. *Human Molecular Genetics*, 2005. 14 Spec No 1: R139–147.

Wolff, G.L., R.l. Kodell, S.R. Moore, and C.A. Cooney., Maternal epigenetics and methyl supplements affect agouti gene expression in Avy/a mice. *FASEB J*, 1998. 12(11): 949–957.

Wolffe, A.P. and D. Guschin, Review: chromatin structural features and targets that regulate transcription. *J Struct Biol*, 2000a. 129(2–3): 102–122.

Wolffe, A.P. and D. Guschin, Review: chromatin structural features and targets that regulate transcription. *Journal of Structural Biology*, 2000b. 129(2–3): 102–122.

Wong, H., J.M. Victor, and J. Mozziconacci, An all-atom model of the chromatin fiber containing linker histones reveals a versatile structure tuned by the nucleosomal repeat length. *PloS One*, 2007. 2(9): e877.

Woodcock, C.L., Chromatin: physical organization, in *Encyclopedia of Biological Chemistry*, Editors-in-Chief: J.L. William and M.D. Lane. 2004, Elsevier: New York. p. 464–468.

Wu, S., R.W. Li, W. Li, and C.J. Li, Transcriptome characterization by RNA-seq unravels the mechanisms of butyrate-induced epigenomic regulation in bovine cells. *PLoS One*, 2012. 7(5): e36940.

Review questions

1. How does nutrigenomics relate to epigenomics?

2. What are the differences between genetics and epigenetics.

3. What are some epigenetic mechanisms?

4. What are pluripotent stem cells and what is their importance?

5. What is the nucleosome and why is it important?

6. What is chromatin?

7. Describe the functions mechanisms of DNA methylation.

8. What are the first and second most used cofactors of methylation?

9. What are HATs?

10. Describe the two mechanisms that affect chromosome function.

11. What are the butyrate concentrations in bovine and sheep?

12. When butyrate is at a high concentration level what does it induce?

13. Describe HDACs. What can be used to inhibit HDACs?

14. What technologies can be used to map the epigenomic landscape of histones?

Section 4

Genetics of Embryo Development and Fertility

22 Genomics of Sex Determination and Dosage Compensation

Jenifer Cruickshank and Christopher H. Chandler
Department of Biological Sciences, State University of New York at Oswego, NY, USA

Genotypic sex determination (GSD)

Among the species that most people are familiar with, **genotypic sex determination**, or GSD, is the most prevalent type of sex-determining system. In species with GSD, the sex of a developing organism is determined primarily by its genotype – in other words, the alleles that it carries at one or more sex-determining loci (in contrast to species with environmental sex determination, discussed later in this chapter). Not all GSD mechanisms are alike, however. You have probably heard of the X and Y sex chromosomes shared by most mammals; individuals with an XX genotype (two copies of the X chromosome) are female, while those with an XY genotype are male. But in other species, such as birds and snakes, males are **homogametic** (i.e., carry two copies of the same sex chromosome), and females are the **heterogametic** sex (with two different kinds of sex chromosomes). In still other species, there are no distinct sex chromosomes; instead, sex is determined by ploidy level, by the cumulative effects of many loci across the genome, or even by cytoplasmically inherited parasites.

XX/XY systems

XX/XY systems, in which females are XX and males are XY, have evolved independently numerous times in many different groups of animals, and even in some plants. In most of these cases, the X and the Y chromosome share a common ancestor. However, in some species, as time progressed, recombination near the male-determining gene on the Y chromosome slowed down. As a result, mutations accumulated on the Y chromosome, and as the Y chromosome diverged from the X, the non-recombining region began to spread. Different species are at different stages of this process. In placental mammals, for example, most of the genes on the Y chromosome have been lost, and only a small region of homology between the X and the Y chromosomes, called the **pseudoautosomal region** or PAR, remains.

Why does an XX genotype lead to females, while an XY genotype leads to males? In placental mammals, the gene *Sry*, located on the Y chromosome, plays a critical role (see Figure 22.1). Mutations that knock out *Sry* result in female development even in XY individuals, while XX mice carrying an extra copy of *Sry* on an autosome develop as males (although they are sterile, because additional Y-linked genes are needed for successful sperm production). *Sry* encodes a transcription factor, which probably activates expression of *Sox9*, triggering the bipotential gonad to differentiate into a testis and suppressing the genes in the female-determining pathway (see Figure 22.1). On the other hand, in the absence of *Sry*, a different set of genes – *Wnt4*, *Rspo1*, and *Foxl2*, among others – is activated. Besides triggering ovarian differentiation, expression of these genes suppresses expression of the male genes. Thus, proper sexual differentiation in mammals is maintained by mutual repression between male and female genes, ensuring that development proceeds along only one branch of this pathway. Finally, once the gonad has differentiated into either a testis or an ovary, it begins producing sex steroid hormones, signaling the rest of the body to develop as either a male or female. Recent work, however, suggests that sexual development in the rest of the body is not entirely dependent on these hormones. Instead, there is evidence that some cells may "know" their sex even in the absence of estrogens or testosterone. For instance, male and female embryos in some mammals including mice, cattle, and pigs differ in certain ways (such as size, glucose metabolism, protein metabolism, and gene expression patterns) even before the gonads have differentiated into either testes or ovaries.

Because the X and Y chromosomes differ so drastically in size, structure, and gene content, there are large differences in gene dosage for sex-linked genes between males and females. As a result, mammals have evolved a complex dosage compensation mechanism that equalizes the expression levels of X-linked genes between the sexes, by almost completely inactivating one of the X chromosome copies in females (see the section on Dosage Compensation later in this chapter).

ZZ/ZW systems

In some other animals, the female is heterogametic, so the sex chromosomes are called Z and W to distinguish them from the X and Y. Like XX/XY systems, ZZ/ZW sex chromosomes have

Molecular and Quantitative Animal Genetics, First Edition. Edited by Hasan Khatib.

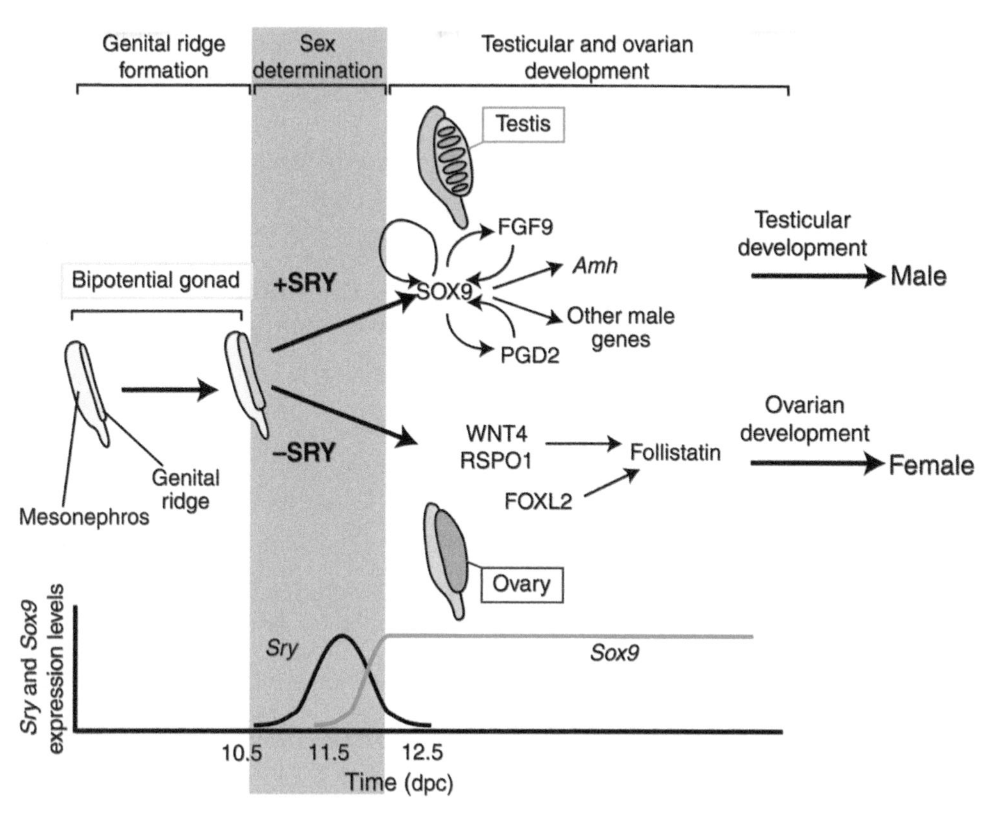

Figure 22.1 Overview of genes involved in sex determination in mammals, specifically the mouse. Reprint of Figure 1 from Kashimada and Koopman 2010, *Development*, 137:3921–3930.

evolved independently in numerous groups of organisms, including Lepidoptera (butterflies and moths), snakes, and birds. The Z and W chromosomes of many species show many parallels to X and Y chromosomes, but there are some important differences, too. Birds are probably the most well-studied ZZ/ZW group, with most of our knowledge of avian sex determination coming from experiments using the chicken as a model.

The latest evidence suggests that sex determination in birds is not controlled by a dominant female-determining gene on the W chromosome, the way the mammalian *Sry* does. Instead, avian sex determination is probably initiated by the dosage of one or more Z-linked genes. A good candidate for a dosage-sensitive "master" regulator of sex determination in birds is *DMRT1*. *DMRT1* is a transcription factor belonging to a family of genes containing a motif known as the DM domain. *DMRT1* is also found in mammals and other vertebrates, where it also plays a key role in sexual development. For example, mutations in *DMRT1* are associated with intersexual gonad development in humans, and duplicated copies of *dmrt1* seem to play key regulatory roles in sex determination in a number of vertebrate species, including the medaka fish (*Oryzias latipes*) and the African clawed frog (*Xenopus laevis*). Likewise, related genes sharing a DM domain are also involved in sexual development even in very distantly related invertebrates, including worms and flies. These observations, combined with the fact that *DMRT1* is found on the avian Z chromosome, provide good reasons to suspect that this gene may play an important role in sex determination in birds. Experimental evidence confirms these suspicions: experimental knockdown of DMRT1 expression in ZZ birds (which would normally develop as males) leads to feminization of the gonads.

However, this is an incomplete picture of avian sex determination. First, a number of lines of evidence suggest that avian sex determination is at least partially cell autonomous. Recall that the traditional view of sex determination in mammals involves sex determination in the gonads, which then signal the rest of the body to differentiate as either male or female via sex hormones. In birds, however, a number of **gynandromorphs** – individuals composed of a mixture of ZZ and ZW cells, displaying a male phenotype on one side of the body and a female phenotype on the other – have been studied (see Figure 22.2). If somatic sex differentiation were directed entirely by sex hormones, which circulate throughout the entire body, then the occurrence of these gynandromorphs would be impossible; their mere existence implies that at least some cells in the body "know" their sex regardless of the signals coming from the gonads. In addition, cell transplant experiments between chick embryos show that male somatic gonad cells transplanted into a female gonad retain their male identity, and vice versa, implying that these donor cells "know" their own genotypic sex, even when they are surrounded by cells of the opposite sex. Moreover, there are differences in gene expression between male and female chicken embryos even before the gonads have differentiated into testis or ovary. Besides that, *DMRT1* is not expressed in somatic tissues outside the urogenital system, even in tissues that still develop sexual dimorphism in an apparently cell-autonomous manner. Some other sex-specific signal or signals must therefore be responsible for these differences, but we do not yet know what those signals are. There is still much work to do before the genetics of sex determination in birds have been fully deciphered.

Figure 22.2 A gynandormorph chicken. The left half of this bird displays male characteristics, while its right half displays female characteristics. This "split" phenotype provides evidence that sexual identity is cell autonomous, instead of being controlled entirely by sex hormones circulating throughout the bird's entire body. Reprint of Figure 1 from Zhao et al. 2010, *Nature*, 464:237–242. Reprinted by permission from Macmillan Publishers Ltd: *Nature*: Zhao et al. 2010, *Nature*, 464:237–242, copyright 2010.

Finally, like the X and Y in mammals, the Z and W chromosomes of most birds are highly differentiated, with the W chromosome containing few genes, lots of repetitive elements, and being largely heterochromatic. Birds, therefore, share a similar problem with mammals: males and females differ in the copy number, and therefore dosages, of Z-linked genes. However, birds seem to lack any sort of widespread dosage compensation mechanism to equalize expression levels of Z-linked genes between the sexes, the way X-chromosome inactivation does in mammals. Instead, dosage compensation, if it truly exists at all in birds, is variable, acting on a gene-by-gene basis.

Functional genomics and evolution of sex chromosomes

Because of their unusual inheritance patterns, sex chromosomes in most species have evolved unique characteristics. Two common themes in sex chromosome evolution are sex-specific adaptation and chromosomal decay.

As previously mentioned, the Y chromosome of mammals and W chromosome of birds are similar in many ways. Both contain few genes relative to their X and Z counterparts. Moreover, because each of these chromosomes is only found in a single sex, they are freed from any sort of conflicting selective pressures between the sexes. For example, the Y chromosome is free to accumulate mutations that benefit males, even if these same mutations are harmful to females. Today, of the few genes present on the mammalian Y chromosome, many are expressed only in the testis, with roles in male fertility. Similarly, mutations benefitting females can collect on the W chromosome. One interesting study looked at gene expression in breeds of chicken that had experienced artificial selection on female-related traits, such as egg-laying, for a century or more, comparing them to breeds selected for male functions, like aggressive fighting ability or ornamental plumage. This study found that many of the genes on the W chromosome were up-regulated in the female-selected breeds relative to male-selected breeds, supporting the idea that sex chromosomes can play an important role in sex-specific adaptation.

The Y and W chromosomes in many species can also "decay" over evolutionary time as a result of their altered recombination patterns. The X chromosome recombines fully in female mammals, as does the Z chromosome in male birds, and as a result, selection can effectively filter out deleterious mutations while maintaining beneficial ones. Most of the Y and W chromosomes, on the other hand, do not cross over during meiosis. This problem is further amplified because these chromosomes are only found in one of the two sexes, and then only in a single copy, meaning that the Y and W chromosomes have a smaller effective population size than the X and Z chromosomes. As a result, these chromosomes are much more subject to the forces of genetic drift, leading to their evolutionary decay, as mutations knocking out non-essential genes on the Y chromosome become fixed.

Other systems

Though only a few are traditionally considered agriculturally important, some species have no true sex chromosomes at all. The most prominent example is insects in the order Hymenoptera, which includes bees, wasps, and ants. This group exhibits **haplodiploidy**, meaning that ploidy level determines sex: unfertilized eggs, which are haploid, become males, while fertilized eggs, which are diploid, become females.

However, the mechanism by which ploidy level determines sex does vary among sub-groups. For instance, some species exhibit complementary sex determination. In these cases, ploidy level *per se* does not directly determine sex, but rather heterozygosity at a highly polymorphic locus does. In honeybees, heterozygosity at the *csd* locus leads to female development. Because there are so many alleles of this locus in most populations, diploids are nearly always heterozygous, and thus female. Inbreeding, however, can lead to diploid homozygotes, which develop as males, as do hemizygotes. Following on the vertebrate theme of gene duplications leading to changes in sex determination, the *csd* locus evolved in honeybees from a duplicated copy of *tra*, another gene with a conserved role in sex determination across insects.

There are a number of other fascinating examples of organisms with bizarre sex determination systems. The common pillbug *Armadillidium vulgare* has homomorphic Z and W sex chromosomes. Some populations of this species are infected with *Wolbachia*, an intracellular bacterial parasite (of which different strains also infect many other arthropods). Because this parasite is transmitted by mothers to their offspring (but not with 100% efficiency), males are essentially a "dead end" for *Wolbachia*. This bacterial lineage has therefore evolved the remarkable ability to manipulate its hosts' reproduction for its own advantage. Pillbugs that are infected with *Wolbachia*, for example, develop as females regardless of their sex chromosome make-up! In some populations, therefore, the W chromosome has disappeared almost entirely, and sex is determined not genetically, but by infection status: individuals who inherit the infection from their mother become females, while the "lucky" few who do not become males.

Environmental sex determination (ESD)

In other species, environmental factors can also play a role in sex determination. These species are said to exhibit **environmental sex determination (ESD)**. The most common form of ESD is **temperature-dependent sex determination (TSD)**. In TSD species, the incubation temperature that an egg or embryo experiences determines what sex it becomes. In painted turtles (*Chrysemys picta*), cooler incubation temperatures are more likely to produce males, while warmer temperatures are more likely to produce females. However in other species, the reverse is true: higher temperatures may lead to males and cooler conditions to females. In yet others, such as the American snapping turtle (*Chelydra serpentina*), females develop with high and low incubation temperatures and males with those in the moderate range. TSD is most prevalent in reptiles: all crocodiles and alligators, most turtles, and some lizards exhibit TSD. Additionally, TSD is also found in some fishes, such as the Atlantic silverside (*Menidia menidia*).

In some invertebrates, different environmental cues can play a role in sex determination. For example, in mermithid nematodes, which parasitize mosquito larvae, the host's size and the number of other nematodes it is infected with determine sex. In larger hosts with relatively few parasites, the nematodes are more likely to become females, whereas smaller hosts carrying a dense parasite load are more likely to produce male nematodes.

Exactly how these environmental factors lead to male or female development is still somewhat of a mystery. Reptiles with TSD do share many of the same genes involved in sex determination in mammals and birds, such as *Dmrt1* and *Sox9*. Some of these genes are expressed in a temperature-dependent way, even in species that have recently evolved GSD from TSD. However, the mechanisms underlying this thermal sensitivity have yet to be elucidated.

In reality, GSD and ESD are probably not truly distinct mechanisms. Instead, they probably represent the endpoints along a continuum. According to this view, sex is determined entirely by genotype at one end of the continuum, entirely by the environment at the other end, and by some combination of genetics and the environment in between (like most other traits). For example, in some amphibians and reptiles with GSD, extreme temperatures during incubation or development can cause sex reversal. Likewise, in some species with temperature-dependent sex determination, including reptiles and fish, there is genetic variation that can influence an individual's likelihood of developing as a male or a female at a given temperature.

Dosage compensation in mammals: X chromosome inactivation

Random X inactivation

Female mammals have two X chromosomes per cell. As a means of **dosage compensation**, so that the activity of the X chromosome in the homogametic sex (female mammals) is comparable to that in the heterogametic sex (male mammals), one X chromosome is inactivated. This silencing of one of the X chromosomes – an epigenetic event – occurs early in development as embryonic stem cells differentiate and lose their pluripotency. Whether it is the maternally-derived or paternally-derived X chromosome that is in inactivated in a particular cell is random.

Figure 22.3 A tortoiseshell cat showing orange and black patches of fur indicative of heterozygosity at the X-linked O gene. (Photo courtesy of Emily Thompson, SUNY Oswego.)

The epigenetic modifications to the inactive X chromosome (the Xi) are carried or reestablished through successive rounds of cell division, such that the descendants of a given cell will maintain the same X inactivation, whether it is the maternally- or paternally-derived X. The classic example of this phenomenon is tortoiseshell and calico cats (see Figure 22.3). The O coat color gene with two alleles – orange and black – is on the X chromosome in cats. Males are therefore hemizygous for either the orange or the black allele and will be either black or orange (depending on the rest of their coat color gene complement) but not both. Tortoiseshell (female!) cats are heterozygous at the O locus and will have black patches and orange patches of fur depending on which X chromosome – the one carrying the orange allele or the one carrying the black allele – is silenced in the lineage of cells that became that patch. In addition to the black and orange patches, calico cats also have white patches resulting from their nonrecessive genotype for the spotting gene. Of course there are many other coat color and pattern genes, and an individual's full coat color and pattern genotype with its particular dominance and epistasis relationships will determine its appearance.

In eutherian mammals, the process of **X chromosome inactivation** involves transcription of the ***Xist*** (Xi specific transcript) gene, found in a region of the X chromosome known as the X-inactivation center. *Xist* transcripts remain in RNA form, and multiple alternative splice variants form a heterogeneous population of RNAs. *Xist* RNAs physically coat the territory of the to-be-silenced X, and in doing so recruit other gene silencing factors (such as those that result in histone methylation) that all together result in transcriptional silencing of the Xi. Expression of the *Xist* gene on the Xi is upregulated, and on the active X chromosome (Xa) it is repressed (see Figure 22.4).

Xist RNA is thought to initiate and maintain X chromosome inactivation via the recruitment of chromatin modifying factors,

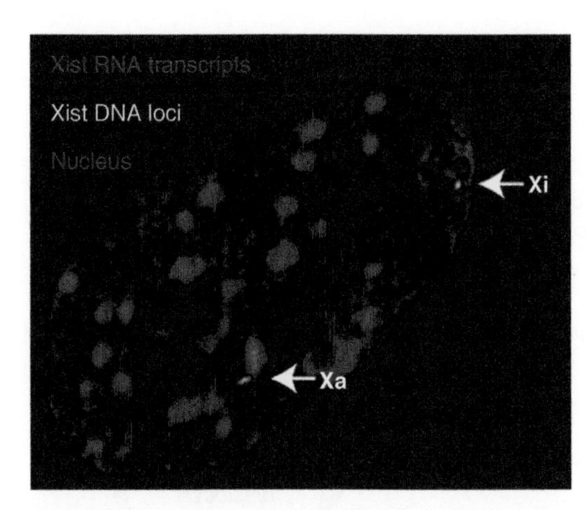

Figure 22.4 X chromosomes identified by fluorescent labeling of the *Xist* gene. Flourescently labeled *Xist* RNA is visible in the vicinity of the inactive X chromosome (Xi), while the acive X chromosome (Xa) has no *Xist* RNA near it. Image by B Reinius and C Shi from Wikimedia Commons (http://en.wikipedia.org/wiki/File:XistRNADNAFISH.jpg), modified from Reinius et al. *BMC Genomics* 2010, 11:614.

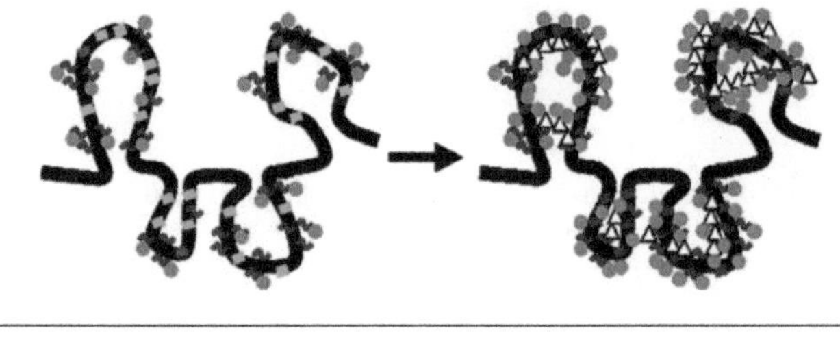

Key
- X-linked gene on
- X-linked partially repressed
- X-linked gene off
- Xist RNA
- Chromosomal DNA loops
- Nucleosomal repressor complex
- Δ Scaffolding complex

Figure 22.5 X chromosome silencing mechanisms. Reproduced from Figure 3B from Brockdorff 2011 *Development* 138, 5057–5065.

such as proteins that interact with histones (see Figure 22.5). Other involved factors are noncoding RNAs that are important in the initiation of X chromosome inactivation. Others, such as Line-1 (long interspersed nuclear element, which is a retrotransposon and highly repetitive element in genomes) appear to be involved in Xi chromosome territory organization. *Xist*-coated regions of the chromosome tend to be rich in repetitive sequences as are often found in gene-poor areas, in introns, and between genes. Compartmentalization within the nucleus also seems critical to establishment and maintenance of a silenced X.

The X-inactivation center also harbors *Tsix*, a noncoding gene transcribed in the antisense orientation to *Xist*. The future Xa has active *Tsix* expression that represses *Xist* expression. On the future Xi, *Tsix* is repressed, thereby allowing expression of *Xist* and its downstream silencing effects. Other regulators of the initiation and maintenance of X chromosome inactivation have yet to be identified. Female cells devote resources to producing *Xist* and other X chromosome inactivation components that male cells do not, and this discrepancy itself may also contribute to phenotypic differences between the sexes. *Xist* may therefore be considered a primary sex-determining gene and have additional, as yet undiscovered roles.

Imprinted X chromosome inactivation

X chromosome inactivation actually occurs twice in the developing mammalian embryo. Prior to the random inactivation event described previously, an earlier, imprinted inactivation of the paternal X chromosome occurs. This inactivation of the paternal X begins in earnest after the two-cell stage of embryo development, and continues to the early blastocyst stage, where 80% of genes (at least in mice) on the paternally-derived X chromosome are silenced. *Xist* RNA is expressed solely from the paternal X chromosome at this stage. The paternal X chromosome is reactivated in the late blastocyst period, at least in the inner cell mass, which goes on to form the embryo proper. Subsequently in those cells, *Xist* is randomly expressed from either the maternally- or paternally-derived X. However, in the cells of the trophectoderm (which goes on to form the extra-embryonic tissues), the imprinted inactivation remains, where the paternal X chromosome stays silenced, and the maternal X remains active (see Figure 22.6).

Unlike eutherian mammals, marsupials have no *Xist* gene, although the inactivated X chromosome bears many of the same epigenetic features (histone methylation patterns, etc.) as silenced X's in placental mammals. Interestingly, in female marsupials, X inactivation is imprinted, such that the paternally-derived X chromosome is universally silenced and the maternally-derived X chromosome is active in all somatic cells. Therefore, given this consistent pattern of X inactivation, one would not expect to find a calico kangaroo.

Activity patterns of sex chromosomes during gametogenesis

Much of what is known about the state of sex chromosome chromatin and expression of sex chromosome genes results from studies in mice, where extensive observations of sex chromosome activity have been made. X and Y chromatin states and gene expression levels differ markedly between males and females as well as across different time points in gamete and embryo development and in adult somatic cells.

During oogenesis (the process of producing an ovum), the previously silenced X chromosome is reactivated as part of overall genome-wide reprogramming events and reversal of other epigenetic changes. Both X chromosomes remain transcriptionally active through meiosis. The details of oogenic X chromosome reactivation are yet to be revealed, although it is expected that the mechanisms are similar to the reactivation of the paternal X chromosome in early embryonic development. Reactivation of both the randomly-inactivated X chromosome that occurs during female germ cell development and of the imprinted paternal X during early embryogenesis involves the loss of *Xist* RNAs. Presumably, other epigenetic marks are also stripped from the reactivating chromosome.

In male primordial germ cells – those cells that are the early progenitors of sperm cells – both X and Y are transcriptionally active and remain so through the period of mitotic arrest (from late embryogenesis until shortly after birth in mice). In the early postnatal days, prospermatogonia resume mitosis and begin spermatogenesis, which involves multiple mitoses followed by meiosis and spermiogenesis (the process of spermatids transforming into mature spermatozoa). The X and Y chromosomes are active through the rounds of mitosis until pachytene of meiosis where

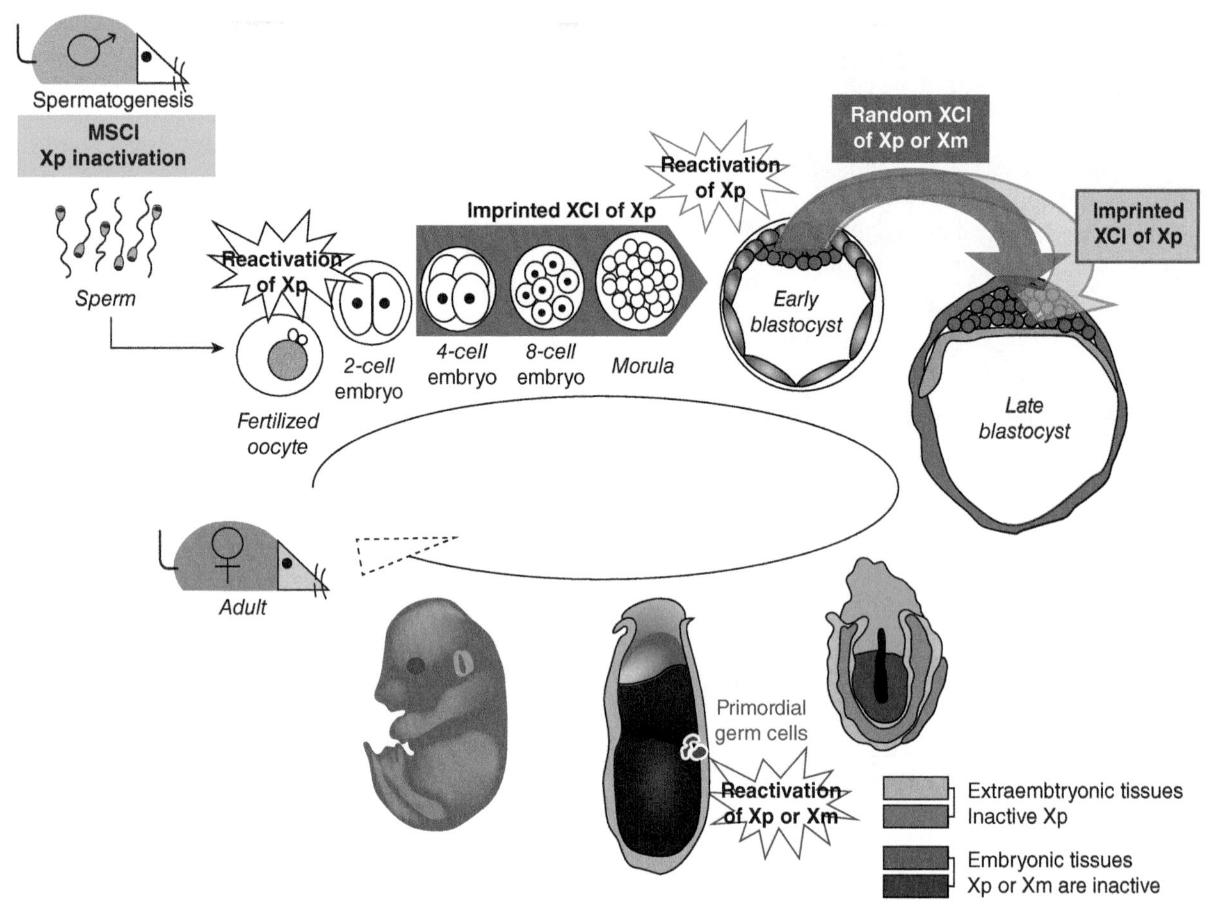

Figure 22.6 X inactivation and reactivation in embryonic and gametic development. Reproduced from Figure 4 in Céline Morey, Philip Avner. Genetics and epigenetics of the X chromosome. *Annals of the New York Academy of Sciences*, pp. E18–E33. © John Wiley & Sons, 2011.

they both undergo silencing, a process termed **meiotic sex chromosome inactivation** (MSCI; see Figure 22.6). The X and Y chromosomes are condensed together in a separate domain in prophase I of meiosis referred to as the sex body or XY body. During MSCI, there is evidence for some gene expression from the X chromosome, but it is far less than that of the autosomes. It has been proposed that the inactivation that occurs in MSCI is simply carried over through all of male meiosis and into embryonic development in marsupials, given their imprinted X inactivation in adult cells. However, at least in the opossum (*Monodelphis domestica*), X chromosomes silenced by MSCI are reactivated after meiosis in spermiogenesis, just as occurs in mice, then the universal paternal X inactivation occurs subsequently in early embryonic development in female embryos.

The trigger for MSCI seems to be a lack of pairing between the X and Y chromosomes during meiosis. Where autosomes synapse fully with each other, the X and Y remain unpaired, which seems to result in a repression of transcription. Further evidence of this phenomenon has been found in XYY humans and mice where the Y chromosomes remain active during meiosis, presumably due to their homologous pairing with each other. Likewise in female birds (ZW), the only partly synapsing Z and W chromosomes in female gametogenesis are also silenced.

MSCI and its misapplication in individuals with abnormal sex chromosome complements may explain infertility in X0 females and in XXY and XYY males. In X0 females, the X chromosome remains unpaired during meiosis and therefore undergoes MSCI and likely causes the oocyte loss observed in these individuals. In XXY and XYY males, the inappropriate expression from the X and Y chromosomes due to a lack of silencing, respectively, seems to result in germ cell arrest and meiotic failure.

Escape from X inactivation

X inactivation is not absolute; there are some genes that go unsilenced. The pseudoautosomal region on the X and Y chromosomes retain homology and pairing, and genes that lie in this region escape X inactivation, thereby leading to biallelic expression in both male and female somatic cells. Some additional genes not located in the pseudoautosomal region also escape X inactivation, although there seems to be little congruency across species in terms of how many and which genes remain active on the silenced X chromosome. Escape from inactivation also seems to be inconsistent across individuals and even across tissues within an individual. The activity of some escape genes may change with development and aging: being silenced initially and having transcription from both copies later. X inactivation escapees appear to vary from species to species with regard to the overall number of escapee genes and the position of the nonsilenced genes along the chromosome.

Notably, biallelic expression of X-linked genes does not seem to result in markedly higher overall expression of the X-escapees. It has been suggested that an active Xi allele may be partially silenced by repressive modification of adjacent loci or that the active Xi allele may lack the modifications that would normally lead to upregulation of the Xa allele. So-called insulator proteins that bind to insulator sequences may assist in keeping escape genes active. These insulator sequences have been found in regions between silenced sections of the Xi and genes that remain active.

Xi-linked genes that escape inactivation, particularly those outside the pseudoautosomal region contribute to sex-based differences between male and female somatic cells. For example, during MSCI, protein-coding genes are silenced, but many X-linked microRNAs – especially those thought to be engaged in male meiosis – escape MSCI. Where there is an expression difference, there is the potential for a difference in physiological phenotype.

Abnormalities in chromosomal sex

Nondisjunction of chromosomes during meiosis can produce gametes with an abnormal sex chromosome number. If these gametes successfully form a zygote, an embryo with an abnormal sex chromosome complement can be formed. XXY individuals develop as male but are sterile, likely due to the lack of MSCI as noted earlier. X0 individuals develop as female, but they, too, are sterile likely as a result of a lack of MSCI. The germ cells in so-affected individuals degenerate, which negatively affects the normal function of the gonad that can affect downstream sexual differentiation. This is most apparent in humans with Klinefelter syndrome (XXY) or Turner syndrome (X0).

XXY cats have been identified, usually because they were tortoiseshell or calico males. Very few XXY dogs have been documented as there is no obvious phenotype to indicate the chromosomal abnormality. There have been a few reports of feline and canine X monosomy with phenotypes of abnormal ovarian cycling and small stature. XXX cats and dogs have also been documented, most with abnormal estrous cycles. Chromosomal abnormalities have been found in livestock as well, with affected individuals showing testicular hypoplasia (XXY cattle, sheep, and pig) or ovarian hypoplasia (XO and XXX horse and XXX cattle).

Sex reversal

Sex reversal syndromes result from a discord between chromosomal sex and gonadal sex. Sex reversed XY individuals often develop as true hermaphrodites with ovotestes (gonads comprising both ovarian and testicular tissue). In some cases, female reproductive organs are present in addition to male organs. One such case was a cat with an unknown causative mutation (it had a normal, single *SRY* gene) that presented with a normal-appearing penis and scrotum, an intact uterus, epididymides, and ovotestes.

Sex reversed XX individuals have a normal karyotype yet develop testes, or in some cases ovotestes. In some instances, XX sex reversal arises from a translocation of the *Sry* gene to an autosome or in the case of one mutant mouse, a duplication of *Sry* was transferred to the X chromosome. Other cases of XX sex reversal are *Sry*-negative, that is, affected individuals do not have a stray *Sry* gene. *SRY*-negative XX sex reversal has been identified in the American cocker spaniel breed of dogs (among others), where it is inherited in a sex-limited autosomal recessive fashion.

Pseudohermaphrodites have concordant chromosomal and gonadal sex, but one or more features of the internal or external genitalia is that of the opposite sex. These disorders are due to abnormal hormone production or reception, and are not considered true sex reversal disorders (see Box 22.1).

Box 22.1

Polled, hermaphrodite goats

In the late nineteenth century, an intersexed phenotype appearing in conjunction with a polled (hornless) phenotype in goats was documented. Later analysis indicated that the hornlessness was autosomal dominant (PP and Pp animals are polled and pp individuals have horns: Figure 22.7). Further investigations revealed that the polled hermaphrodite goats were in fact XX, PP individuals, and – following the discovery of *SRY* in 1990 – that the disorder was *SRY* negative.

More recent studies identified the mutation responsible for the so-called polled/intersex syndrome: an 11.7-kb deletion on chromosome 1. The deleted DNA region (called PIS) does not contain any coding sequence, but does serve as a regulatory region for at least three genes, *PISRT1* (PIS regulated transcript number 1), *PFOXic* (promoter FOXL2 inverse complementary), and *FOXL2* (forkhead box L2). *PISRT1* and *PFOXic* are non-coding RNAs that appear to regulate expression of *FOXL2*, which encodes a transcription factor.

In normal female gonads, *PISRT1*, *PFOXic*, and *FOXL2* are expressed from early ovarian formation until adulthood; however, expression of all three are absent in PIS–/– gonads. *FOXL2* appears to participate in switching on the female pathway, including regulating female steroidogenesis, and in silencing male-specific genes in the developing XX gonad. So when the regulatory region (PIS) is missing, ovarian *FOXL2* does not get "turned on", leading to XX goats that do not develop as normal females.

In the horn buds in both females and males, these three genes are not expressed in PIS+/+ wild type goats, but are expressed in PIS+/– heterozygotes and PIS–/– goats homozygous for the mutation. Presumably, the misexpression of *FOXL2* in the horn buds is due to the partial or complete absence of the PIS region which ultimately causes misregulation of FOXL2 target genes that are directly responsible for horn growth.

Figure 22.7 Horned, PIS$^{+/+}$ goat (*left*) and polled, PIS$^{-/-}$ goat (*right*). Reproduced with by permission from Macmillan Publishers Ltd: *Nature Genetics*. Photo Courtesy of Eric Pailhoux.

Summary

Sex determination – the process by which an organism "decides" to become a male or a female –is a critical step in development. Whether it is initially triggered by chromosome composition or by some environmental cue, sex determination is governed by the subsequent activation or repression of key genes that set the course for further sexual development. Genotypic sex determination results from individuals inheriting a particular combination of sex chromosomes, such as the XY system in mammals and the ZW system in birds. The cells of female mammals, being the homogametic sex, undergo inactivation of one of the X chromosomes as a form of dosage compensation of the X-linked genes. This inactivation results from production of *Xist* RNA from the X chromosome that will be silenced. Sex chromosome silencing also plays roles in other developmental processes such as gametogenesis. With environmental sex determination, factors such as temperature or population density can determine sex in the developing organisms.

Further reading

Arnold AP. 2012. The end of gonad-centric sex determination in mammals. *Trends Genet.* 28: 55–61.

Bachtrog D, Kirkpatrick M, Mank JE, McDaniel SF, Pires JC, Rice WR, Valenzuela N. 2011. Are all sex chromosomes created equal? *Trends Genet.* 27:350–357.

Basu R, Zhang L-F. 2011. X Chromosome inactivation: A *Silence That Needs to be Broken. Genesis* 49:821–834.

Bellot DW, Skaletsky H, Pyntikova T, Mardis ER, Graves T, et al. 2010. Convergent evolution of chicken Z and human X chromosomes by expansion and gene acquisition. *Nature.* 466: 612–6.

Bergero R, Charlesworth D. 2009. The evolution of restricted recombination in sex chromosomes. *Trends Ecol Evol.* 24:94–102.

Berletch JB, Yang F, Disteche CM. 2010. Escape from X inactivation in mice and humans. *Genome Biology* 11:213.

Bermejo-Alvarez P, Rizos D, Rath D, Lonergan P, Gutierrez-Adan A. 2010. Sex determines the expression level of one third of the actively expressed genes bovine blastocysts. *PNAS* 107(8):3394–3399.

Blackmore MS, Charnov EL. 1989. Adaptive variation in sex determination in a nematode. *Am Nat.* 134:817–23.

Brockdorff N. 2011. Chromosome silencing mechanisms in X-chromosome inactivation: unknown unknowns. *Development* 138:5057–5065.

Charlesworth D, Charlesworth B, Marais G. 2005. Steps in the evolution of heteromorphic sex chromosomes. *Heredity* 95:118–128.

Chow JC, Heard E. 2010. Nuclear Organization and Dosage Compensation. *Cold Spring Harb Perspect Biol* 2:a000604.

Chue J, Smith CA. 2011. Sex determination and sexual differentiation in the avian model. *FEBS Journal.* 278:1027–1034.

Conover DO, Kynard BE. 1981. Environmental sex determination: interaction of temperature and genotype in a fish. *Science.* 213:577–579.

Conover DO, Van Voorhees DA. 1990. Evolution of a balanced sex ratio by frequency-dependent selection in a fish. *Science.* 250:1556–1558.

Cordaux R, Bouchon D, Grève P. 2011. The impact of endosymbionts on the evolution of host sex-determination mechanisms. *Trends Genet.* 27:332–341.

de Vries M, Vosters S, Merkx G, D'Hauwers K, Vansink DG, Ramos L, de Boer P. 2012. Human Male Meiotic Sex Chromosome Inactivation. *PLoS One* 7(2):e31485.

Eggers S, Sinclair A. 2012. Mammalian sex determination – insights from humans and mice. *Chromosome Res.* 20:215–238.

Ellegren H. 2011. Sex chromosome evolution: recent progress and the influence of male and female heterogamety. *Nat Rev Genet.* 12:157–166.

Ewert MA, Lang JW, Nelson CE. 2005. Geographic Variation in the Pattern of Temperature-dependent Sex Determination in the American Snapping Turtle (*Chelydra Serpentina*). *J Zoology* 265(1):81–95.

Gao Y, Hyttel P, Hall VJ. 2011. Dynamic Changes in Epigenetic Marks and Gene Expression During Porcine Epiblast Specification. *Cellular Reprogramming* 13(4):345–360.

Hasselmann M, Gempe T, Schiøtt M, Nunes-Silva CG, Otte M, Beye M. 2008. Evidence for the evolutionary nascence of a novel sex determination pathway in honeybees. *Nature.* 454:519–522.

Heard E, Turner J. 2011. Function of the Sex Chromosomes in Mammalian Fertility. *Cold Spring Harb Perspect Biol* 3:a002675.

Janzen FJ, Phillips PC. 2006. Exploring the evolution of environmental sex determination, especially in reptiles. *J Evol Biol.* 19:1775–1784.

Kashimada K, Koopman P. 2010. *Sry:* the master switch in mammalian sex determination. *Development.* 137:3921–3930.

Ledig S, Hiort O, Wünsch L, Wieacker P. 2012. Partial deletion of DMRT1 causes 46,XY ovotesticular disorder of sexual development. *Eur J Encorinol.* 167:119–124.

Livernois AM, Graves JAM, Waters PD. 2012. The origin and evolution of vertebrate sex chromosomes and dosage compensation. *Heredity.* 108:50–58.

Mahadevaiah SK, Royo H, VandeBerg JL, McCarrey JR, Mackay S, Turner JMA. 2009. Key Features of the X Inactivation Process Are Conserved between Marsupials and Eutherians. *Current Biology* 19:1478–1484.

Mank JE, Ellegren H. 2009. All dosage compensation is local: Gene-by-gene regulation of sex-biased expression on the chicken Z chromosome. *Heredity.* 102:312–320.

Matson CK, Murph MW, Sarver AL, Griswold MD, Bardwell VJ, Zarkower D. 2011. DMRT1 prevents female reprogramming in the postnatal mammalian testis. *Nature.* 476:101–104.

Matsuda M, Nagahama Y, Shinomiya A, et al. 2002. *DMY* is a Y-specific DM-domain gene required for male development in the medaka fish. *Nature.* 417:559–563.

Matsuda M, Shinomiya A, Kinoshita M, et al. 2007. *DMY* gene induces male development in genetically female (XX) medaka fish. *Proc Natl Acad Sci USA.* 104:3865–3870.

Meyers-Wallen VN. 2006. Sex chromosomes, sexual development, and sex reversal, in *The Dog and Its Genome,* Ostrander EA, Giger U, Lindblad-Toh K (eds). Cold Spring Harbor Laboratory Press. Cold Spring Harbor, New York.

Meyers-Wallen VN. 2012. Gonadal and sex differentiation abnormalities of dogs and cats. *Sex Dev.* 6(1–3):46–60.

Moghadam HK, Pointer MA, Wright AE, Berlin S, Mank JE. 2012. W chromosome expression responds to female-specific selection. *Proc Natl Acad Sci USA.* 109:8207–8211.

Morey C, Avner P. 2011. Genetics and epigenetics of the X chromosome. 2010. *Ann N Y Acad Sci.* 1214:E18–33

News and Views: Mutant of the *Month.* 2007. *Nature Genetics* 39:585.

Okamoto I, Otte AP, Allis CD, Reinberg D, Heard E. 2004. Epigenetic dynamics of imprinted X inactivation during early mouse development. *Science* 303(5658):644–649.

Pannetier M, Elzaiat M, Thépot D, Pailhoux E. 2012. Telling the story of XX *Sex Reversal* in the *Goat*: Highlighting *the Sex-Crossroad in Domestic Mammals. Sex Dev.* 6:33–45.

Patrat C, Okamoto I, Diabangouaya P, Vialon V, Le Baccon P, Chow J, Heard E. 2009. Dynamic changes in paternal X-chromosome activity during imprinted X-chromosome inactivation in mice. *PNAS* 106(13):5198–5203.

Reinius B, Shi C, Hengshuo L, Sandhu KS, Radomska KJ, Rosen GD, Lu L, Kullander K, Williams RW, Jazin E. 2010. Female-biased expression of long non-coding RNAs in domains that escape X-inactivation in mouse. *BMC Genomics* 11:614.

Rhen T, Schroeder A, Sakata JT, Huang V, Crews D. 2010. Segregating variation for temperature-dependent sex determination in a lizard. *Heredity.* 106:649–60.

Rigaud T, Juchualt P, Mocquard J-P. 1997. The evolution of sex determination in isopod crustaceans. *Bioessays.* 19:409–416.
Rosnina Y, Jainudeen MR, Hafez ESE. 2000. Genetics of Reproductive Failure in *Reproduction in Farm Animals* 7th edn, Hafez ESE, Hafez B (eds). Lippincott Williams & Wilkins.
Sarre SD, Georges A, Quin A. 2004. The ends of a continuum: genetic and temperature-dependent sex determination in reptiles. *Bioessays.* 26:639–645.
Schlafer DH, Valentin B, Fahnestock G, Froenicke L, Grahn RA, Lyons LA, and Meyers-Wallen VN. 2011. A case of *SRY*-positive 38,XY true hermaphroditism (XY Sex Reversal) in a cat. *Vet. Path.* 48(4):817–822.
Sekido R, Levell-Badge R. 2009. Sex determination and SRY: down to a wink and a nudge? *Trends Genet.* 25:19–29.
Smith CA, Roeszler KN, Ohnesort T, Cummins DM, Farlie PG, Doran TJ, Sinclair AH. 2009. The avian Z-linked gene *DMRT1* is required for male sex determination in the chicken. *Nature.* 461:267–271.
Tattermusch A, Brockdorff N. 2011. A scaffold for X chromosome inactivation. *Human Genetics* 130:247–253.
Valenzuela N. 2008. Evolution of the gene network underlying gonadogenesis in turtles with temperature-dependent and genotypic sex determination. *Integr Comp Biol.* 48:476–485.
Valenzuela N. 2008. Relic thermosensitive gene expression in a turtle with genotypic sex determination in a turtle with genotypic sex determination. *Evolution.* 62:234–240.
Vandeputte M, Dupont-Nivet M, Chavanne H, Chatain B. 2007. A polygenic hypothesis for sex determination in the European sea bass *Dicentrarchus labrax. Genetics.* 176:1049–1057.
Yoshimoto S, Ikeda N, Izutsu Y, Shiba T, Takamatsu N, Ito M. 2010. Opposite roles of *DMRT1* and its W-linked paralogue, *DM-W*, in sexual dimorphism of *Xenopus laevis:* implications of a ZZ/ZW-type sex-determining system. *Development.* 137:2519–2526.
Zhao D, McBride D, Nandi S, McQueen HA, McGrew MJ, Hocking PM, et al. 2010. Somatic sex identity is cell autonomous in the chicken. *Nature.* 464:237–242.

Review questions

1. What kind of protein does *Sry* encode?

2. The traditional view of vertebrate sex determination is that the sex chromosomes cause the gonads to differentiate into either testes or ovaries, and that the gonads then use sex steroid hormones to "tell" the rest of the body what sex it is. What evidence is there that this model is incomplete, at least in some animals?

3. How does avian sex determination differ from mammalian sex determination? How does avian dosage compensation differ from mammalian dosage compensation?

4. How are differences between the X and the Y chromosomes (or between the Z and W chromosomes) thought to evolve?

5. What role has gene duplication played in the evolution of sex determination mechanisms?

6. How could inbreeding lead to the development of diploid males in bee species with complementary sex determination?

7. How do you think climate change might impact species with environmental sex determination?

8. In female mammals, which X chromosome copy (maternal or paternal) is inactivated?

9. What abnormality would a tortoiseshell male cat likely be harboring? Why?

10. How does the Xist gene initiate X chromosome inactivation?

11. Why is meiotic sex chromosome inactivation (MSCI) thought to lead to infertility in X0 and XXY individuals?

12. Why does it seem that sex determination and differentiation abnormalities are better studied in cats and dogs than in other animals?

23

Functional Genomics of Mammalian Gametes and Preimplantation Embryos

Şule Doğan,[1] Aruna Govindaraju,[1] Elizabeth A. Crate,[1,2] and Erdoğan Memili[1]

[1]Department of Animal and Dairy Sciences, Mississippi State University, MS, USA
[2]Department of Biology, New College of Florida, FL, USA

Introduction

Early mammalian development is a fascinating process where haploid male and female gametes fuse to form a single diploid cell, known as a **zygote**. This eventually proliferates into billions of cells to form the organism. Early gametogenesis and embryogenesis set the stage for subsequent development during which not only vital phenotypes are formed, but also certain diseases and abnormalities afflict the organism. Functional genomics, the study of gene products, is important because it helps to better understand the effects of genotype, environment, development, and so on, on molecular phenotypes. For this reason, the study of gamete and embryo development and functional genomics approach are justified. This chapter provides an overview of gamete and embryo development, including the most commonly employed functional genomics methods followed by brief descriptions of studies employing these methods. The model organism and the phenotype emphasized are bovine and fertility, respectively. By using functional genomics approaches on these basic processes of gametogenesis and embryogenesis will allow us to improve fertility of mammals as well as humans.

Gamete and embryo development

The roots of mammalian development begin with genesis of the male and female gametes, spermatozoa and oocytes, respectively, followed by fertilization and embryogenesis (Figures 23.1 and 23.2; reformatted from Memili et al., 2012). **Spermatogenesis** begins with formation of primordial germ cells, or PGCs, that originate as epiblast cells in the extraembryonic mesoderm. The PGCs migrate to the gonads and they are sequestered into the seminiferous tubules, and then germ cells enter mitotic arrest. Following the birth of the male, these cells locate to the basement membrane and become spermatogonial stem cells, or SCCs, where they will remain in that stage until puberty. The species specific spermatogenic cycle begins with the surge of gonadotropins around the pubescent period (Phillips et al., 2010). Through mitotic cell divisions, SCCs develop into primary spermatocytes, which then divide through Meiosis I into two equivalent secondary spermatocytes. These two secondary spermatocytes both undergo Meiosis II, which results in round spermatids that are the direct precursor to the mature spermatozoon. Through spermiation, spermatids loose organelles and some of the cytoplasm and acquire a self-propelling tail, and become spermatozoa. The final step is the maturation and formation of spermatozoa in the epididymis.

Oogenesis is the process of ovum differentiation during which the ovum develops a very rich and complex cytoplasm containing stores of cytoplasmic enzymes and organelles required for maintenance of metabolism and support of early embryogenesis (Memili and First, 2000). The production begins with the germ cells that differentiate into oogonium. These oogonia are bound together by cellular bridges that ensure that they divide in sync and divide by mitosis into a limited number of egg precursor cells called primary oocytes. Shortly before birth the primary oocytes of female fetuses undergo prophase I of the first meiotic division, where they arrest until puberty. Following the onset of puberty, the meiosis continues for a pair of primary oocytes. The division of the cells during telophase results in a secondary oocyte, which contains a majority of the cytoplasm and cellular constituents, and a polar body that is much smaller and unequal in volume. The second meiosis takes place in a similar way resulting again in an unequal cytokinesis, with the majority of the secondary oocyte being contained by the ovum and a secondary polar body holding a partial nucleus. This process of oogenesis is continued until the onset of menopause.

Fertilization is the fusion of the spermatozoon and the egg that results in the formation of a new organism, the zygote (Memili and First, 2000). The sperm and ovum meet within the female fallopian tube following the movement of the sperm through obstacles such as vaginal fluids with low pH, macrophages, and the cervix. During the process to reach the ovum the sperm undergoes capacitation, which is noted by an increase in calcium in the sperm cell. This hyperactivates the cell and prepares it for fertilization, and certain conformation changes including an efflux of cholesterol and lipid raft production. Both of

Molecular and Quantitative Animal Genetics, First Edition. Edited by Hasan Khatib.

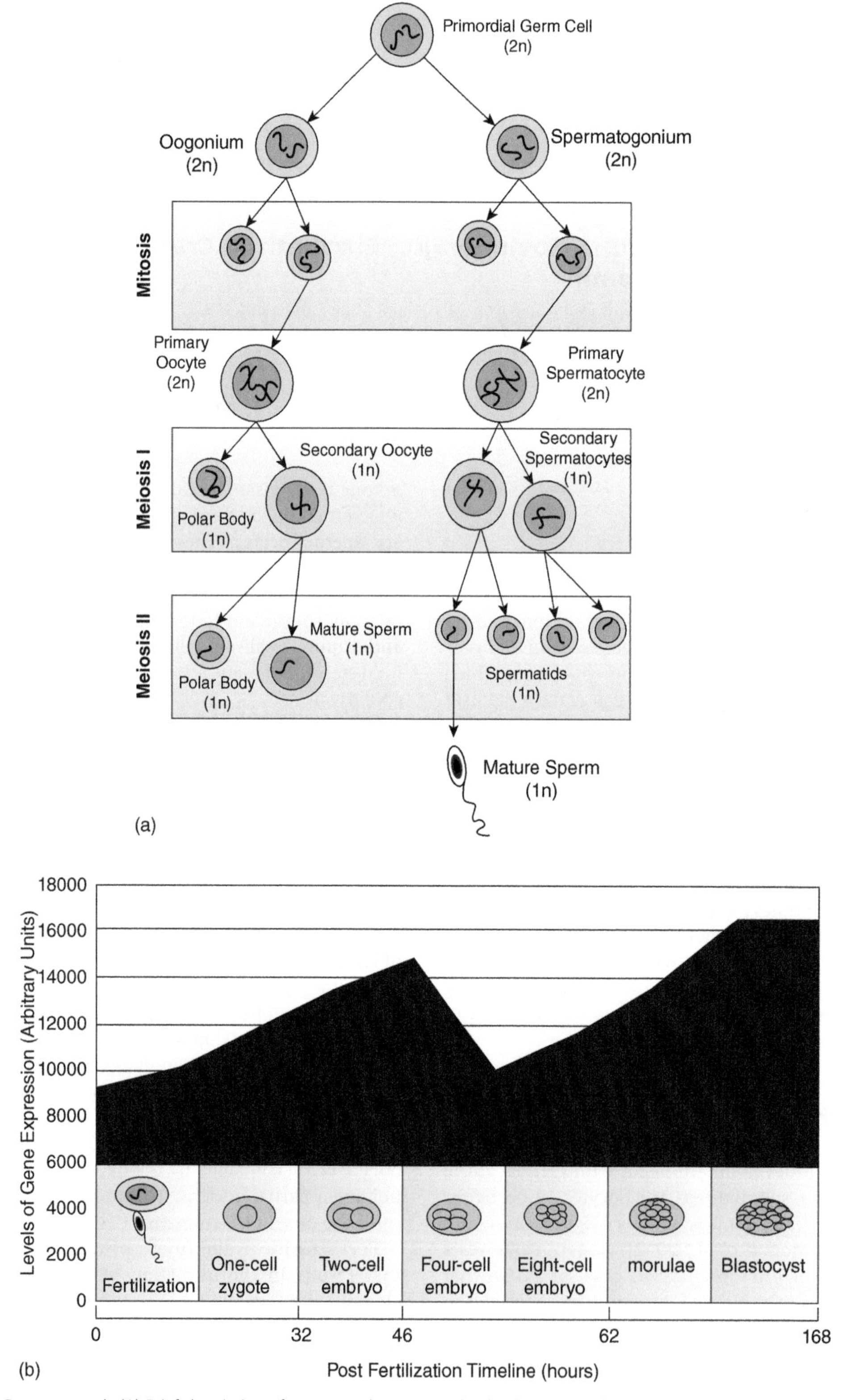

Figure 23.1a and b Gametogenesis (A) Brief description of oocyte and sperm production in ovary and testis, respectively. Formation of haploid germ cells from the diploid precursory cells is illustrated in this figure. Embryogenesis (B) following fertilization, zygote formation and beyond in bovine are demonstrated in this figure, including the gene expression profile (adapted from Memili and First, 2000 and Kues et al., 2008).

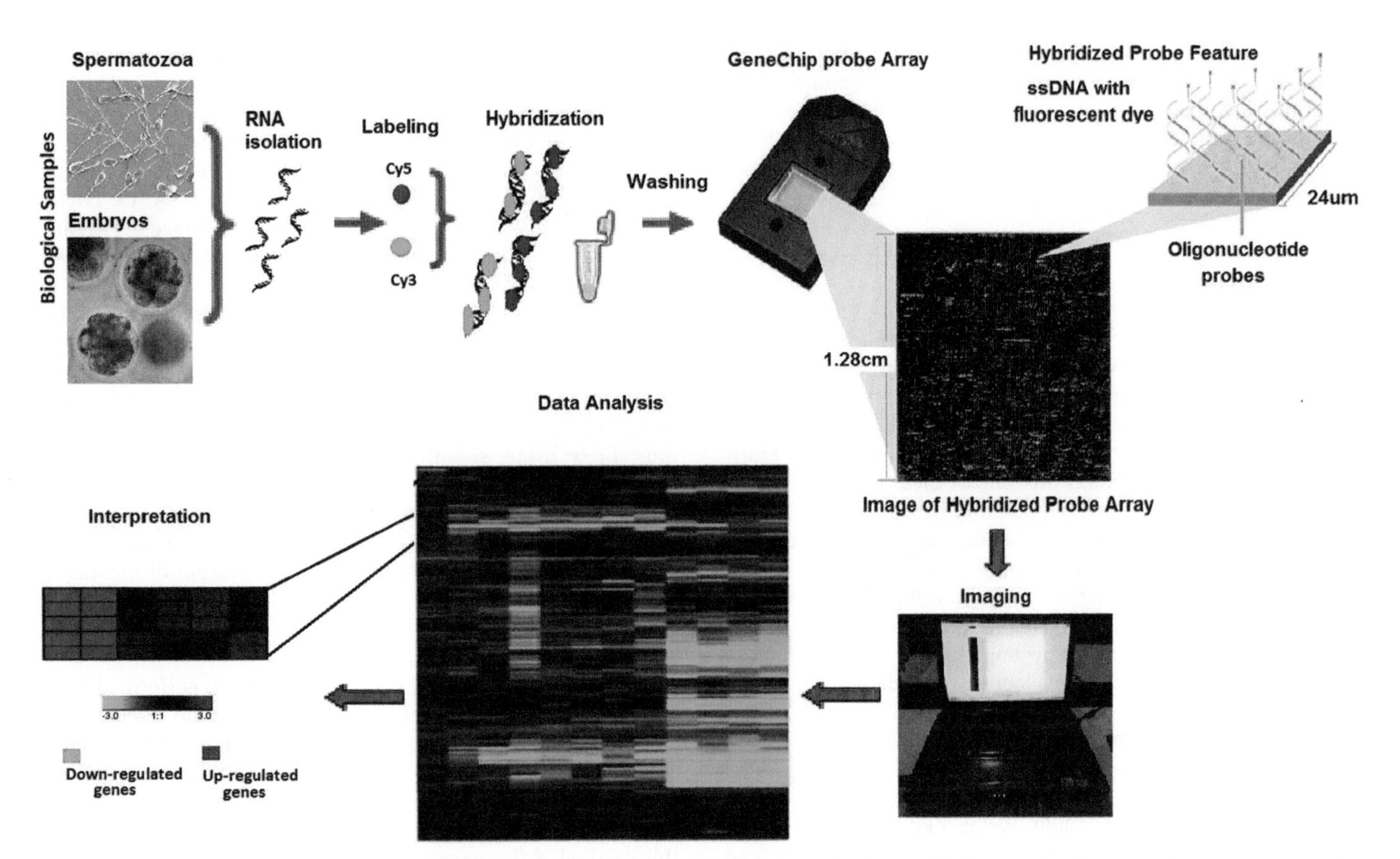

Figure 23.2 Microarray workflow. The RNAs are extracted from a variety of biological samples in which bovine spermatozoa and embryos are represented as an example here. Following RNA isolation, these transcripts are labeled using fluorescent dyes such as Cy5and Cy3, and then they are hybridized with specific probes in the array such as cDNA probes or oligonucleotides. Following washes, single stranded, fluorescently labeled DNA targets are bound by oligonucleotide probes of the array surface. Each probe features are made up millions of copies of a specific probes, leading up to 200.000 different probes for the genes of interest. After hybridization, these microarray chips are processed using specific imaging systems that are precisely analyzed using software such as Significance Analysis of Microarrays (SAM). Based on these results, differentially expressed genes are determined in biological samples among the individuals or treatments. Briefly, on the SAM plot, a non-zero fold change parameter is selected to display the upregulated and the downregulated significant genes with red and green, respectively. Then, a heat map from all processed data using hierarchical clustering may be generated by taking advantage of such software, GeneTraffic UNO (Iobion Informatics LLC, www.iobion.com). Besides, for multiple comparisons of the data, one-way analysis of variance (ANOVA) may be used as well. The chip picture is obtained from www.affymetrix.com.

these changes increase the likelihood of the sperm positively interacting with the oocyte and initiating the fertilization process. Prior to fertilization two further obstacles meet the sperm: the cumulus cells and the zona pellucida. The outermost structure surrounding the oocyte is called the cumulus-cell oocyte complex, a matrix-like enclosement that the sperm must penetrate through the use of its contained enzymes. The sperm then undergoes an acrosomal reaction, in which the sperm head loses the acrosome to reveal enzymes that can then penetrate the ovum. Entry of only one spermatozoon is ensured through the changes in the egg zona pellucida to block polyspermy.

The development of an **embryo** begins with the maternal and paternal genome fusion and is followed by precise and coordinated cell divisions that occur within the first few hours of fertilization known as the cleavage stage (Figure 23.1a and 23.1b). Groups of cells called blastomeres proliferate within the embryo eventually resulting in 32-cell morulae. The time required for cell cleavage is variable, with the first cell cleavage taking around 18–36 h to complete and the second and third about 8–12 h each. The morulae contracts and hollows to form a blastocyst that is composed of two cell groups the trophectoderm and the inner cell mass. The bovine blastocyst contains around 100 cells and will expand until the zona pellucida is lost and then elongates several centimeters before implanting in the uterus.

While large supplies of maternally inherited components support the newly fertilized embryo, it is thought that diverse sets of paternal molecules may be transferred into the egg during fertilization and that they might play roles in early embryogenesis (Govindaraju et al., 2012). Even following zygotic gene expression, the maternally derived factors still act to regulate the embryo. Activation of the embryonic genome appears to occur in wave-like patterns where diverse sets of genes are expressed in embryos and in blastocysts (Misirlioglu et al., 2006; Kues et al., 2008). During embryonic development, the genomes of both gametes are differentially marked through the genomic imprinting and they need to be reprogrammed via epigenetic marks in each generation. While genome refers only the genes, epigenome consists of both the genes and the marks that are responsible for the gene regulation. These epigenetic marks help maintain the memories of the specific genes such as imprinted genes expressed in embryo by only one of two parental alleles, remaining the other allele as silent based on the gender of the offspring. The main pathway to maintain this epigenetic remodeling in the gametes is DNA metyhlation during which methylation of DNA is erased and then rebuilt in germ cell development (reviewed by Rodriguez-Osorio et al., 2012).

Transcriptomics

During the 1900s, DNA was thought to determine a cell's ultimate fate; however, discovery of mRNA has had a major influence on the DNA-centric concept, leading to innovations in modern molecular biology. While genomics is termed as the determination of entire DNA sequence in an organism, **transcriptomics** is an expression profiling of these DNA sequences, leading to the RNA-centric world from DNA-centric concept. Transcriptome profiling allowed scientists to focus on the functions of the gene rather than on its presence, resulting new field of **functional genomics** at the end of the 1990s. Transcriptomics have also been used for identification of, not only coding RNAs, but also non-coding RNAs, providing clues to gene regulation, function, and interaction in development and disease.

Transcriptomics of messenger RNAs

We know that the oocyte has large amounts of transcripts being translated during oocyte maturation and early embryonic development (Memili and First, 2000). We also know that spermatozoa contain a diverse set of transcripts including mRNAs and microRNAs (Feugang et al., 2010; Govindaraju et al., 2012); however, functions, if any, of these transcripts during gamete and embryo development are not yet fully known. Transcripts are expressed in unique patterns in the developing embryo and hence, are vital for embryonic development (Memili et al., 1998; Misirlioglu et al., 2006). Therefore, study of transcripts in germ cells and in preimplantation embryos is important for better understanding the basic processes regulating development and for advancement in biotechnology.

Several methods have been developed to identify transcripts in biological samples such as expressed sequence tags (EST), Norther -blot, serial analysis of gene expression (SAGE) (Yamamoto et al., 2001), tiling arrays, RT-PCRs, microarrays, and RNA-sequencing. The Northern blotting method requires not only size separation of RNAs in an agarose gel, but also transfer of RNAs onto a membrane following hybridization of radiolabeled probes. The EST is a tag-based sequencing method that is now an open source to the public in GenBank (www.ncbi.nlm.nih.gov/genbank/). The SAGE approach involves use of 10–14bp tags obtained by cloning. This enables identification of transcripts from both known and unknown genes (www.sagenet.org; reviewed by (Robert, 2008), while the tiling array is another kind of microarray hybridization technique. Although tiling arrays are considered to be a subtype of microarrays, they are generally used to detect the entire genome or contigs of the genome. For example, the applications of these tiling array platforms are Chip-chip (for protein-protein interaction), MeDIP-Chip (for locations of Methylated DNA) and transcriptome mapping (for detection of novel transcript) (Reviewed by Liu, 2007). We will focus on here three commonly used transcriptomics approaches in the study of gametes and embryos; qRT-PCR, microarrays and RNA-sequencing.

Quantitative real time reverse-transcription polymerase chain reaction (qRT-PCR)

Referring to central dogma, reverse transcription is to produce complementary DNA (cDNA) synthesis of either total or messenger RNAs. Polymerase chain reaction (PCR) is a common method for *in vitro* amplification of either DNA or cDNA. Reverse- transcription PCR (RT-PCR) takes advantage of PCR and incorporates this into reverse transcription approach to be able to amplify transcripts. Quantitative real time PCR (qPCR) is designed to quantify amplifications using either DNA or cDNA, which is also known as Real-time PCR. Moreover, Quantitative Real Time reverse-transcription PCR (qRT-PCR) automatically amplifies cDNA using single or double-step amplifications where the quantity of products (amplicon) are simultaneously monitored using specific software supplied with the qPCR machine from a company such as Roche, Agilent, and so on. In a conventional PCR, the amplified DNA is quantified by running in a gel. QRT-PCR method is based on a specific or non-specific detection chemistry that allows the quantification of the amplified product.

Methodology

Following isolation of RNA, synthesis of cDNA is performed using reverse-transcription reaction. For the template preparation, quantity and quality of the RNA and cDNA can be measured using NanoDrop spectrophotometer and Agilent Bioanalyzer, respectively. The former requires templates in nanograms compared to the latter method that can detect picograms of RNAs. In this method, a fluorescence stain is used to monitor the quantification of these amplicons. Three fluorescence-based approaches performed at the moment are scorpion, SYBR-green and Taq-man. While SYBR-Green chemistry that binds to double strand (ds) DNA can detect all products, including non-specific amplicons, Taq-man is designed based on a fluorogenic 5′ nuclease chemistry that uses probes to identify the specific amplicons. This allows Taq-man to only hybridize the probes with the specific target, leading to the fluorescent signal without the background and false positive signals. By using Taq-man probes tagged with two different reporter dyes, two different target genes can be detected in the same reaction. However, the Taq-man probes should be separately designed for each target, which is not required in SYBR-Green system (www.appliedbiosystems.com). Primers for qRT-PCR are generally designed based on spanning an exon-exon boundary method wherein one of the primers skips an intron and targets the junction of two adjacent exons. This results in amplification of cDNA (i.e., RNA) and genomic (g) DNA into two different sizes, providing a good control for presence of genomic DNA contamination in the original RNA samples. The following link provides information on primer design: (www.premierbiosoft.com/primerdesign/index.html).

Data analysis

Real-time PCR contains exponential, linear and plateau amplification steps during the process. At the exponential step, cycle threshold (Ct) is calculated for each sample based on the threshold line, and these Ct values are used for the data analysis later. Absolute and relative quantification approaches can be used to analyze the qPCR data. The former represents the last abundance or copy number of the genes of interest in the sample while the latter compares the differentially expressed genes among groups or treatments (www.appliedbiosystems.com).

Several statistical approaches including SAS codes can be incorporated in to analysis of qPCR data (Yuan et al., 2006). The relative expression software (REST©) (www.gene-quantification.de/rest.html) is useful and publicly available data analysis software for analyses of qPCR results. This software has been exploited to analyze differentially expressed transcripts in different tissues and cell types such as rat liver (Pfaffl et al., 2002) or bovine matured oocytes, zygotes (Uzun et al., 2009) and embryos (Misirlioglu et al., 2006). In addition to these, Genex and Biogazelle are other quantification software for analyses of qPCR results (http://genex.gene-quantification.info; www.biogazelle.com). PowerNest software is also used to design an exponential qPCR experiment based on a pilot data, and it is currently free to the public (www.powernest.net).

Advantages and disadvantages

QRT-PCR is the most suitable methods currently available for detection of transcripts in biological samples and for quantification of differentially expressed genes among treatments or individuals. It is still preferred over some of the other methods due to its flexibility, cost-effectiveness and time-saving without a requirement for a gel running step. The most significant limitation of this method is that it targets one gene at a time; thus, it is not an independent transcriptome profiling method with a high-throughput data. However, this method is still being used in some experiments as a validation, and/or preliminary method. It is also an ideal method for a study where a panel of genes is being analyzed.

Applications

Several studies on sperm mRNA transcriptome profiling have been conducted by taking advantage of qRT-PCR approach. As an example, Das et al. (2010) obtained spermatozoa from varying stallion sperm samples to isolate transcripts from fresh, frozen, and extended ejaculates, and epididymis. Using qRT-PCR, the authors demonstrated that RNA isolation using TRIzol reagent (Invitrogen) was more efficient than the spin-column method (commercially available such as at Qiagen). They concluded that sperm RNA extraction should be carefully enhanced by avoiding any DNA contamination (Das et al., 2010). In another study, Lambard et al. (2004) isolated different populations of human sperm mRNAs via discontinuous density gradients by sorting spermatozoa based on their motility from the same semen. According to their results, protamine1 (*prm1*) transcripts in motile spermatozoa locating low density gradient were significantly higher compared to those in high density fractions. The authors then concluded that examining sperm transcriptomics using RT-PCR might be a diagnostic tool for prediction of fertilization success (Lambard et al., 2004). In addition to transcriptomics profiling of the sperm, detection of paternally expressed genes in blastocysts was also done using qRT-PCR approach. For example, one of the paternally expressed genes, *Nnat* was found to be absent in buffalo zygotes generated via activation of intra cytoplasmic sperm injection (ICSI), illuminating failure of male genome activation (Chankitisakul et al., 2012).

Quantitative PCR tool has also been incorporated into the functional genomics- and computational biology-based studies. For example, the high mobility group nucleosomal 3a (*hmgn3a*) and SWI/SNF related, matrix associated, actin dependent regulator of chromatin, subfamily a-like1 (*smarcal1*) transcripts, contributing to embryonic gene activation (EGA) in bovine embryos were examined using the qPCR approach followed by comparative functional genomics among mammals (Uzun et al., 2009). Additionally, Rodriguez-Osorio et al., (2010) determined the abundance of DNA methyltransferases (*dnmt*) transcripts in bovine oocytes and preimplantation embryos by using qPCR method.

In many studies, qPCR has been used as a validation method for both microarray and RNA-seq data. For example, microarray results obtained from bovine matured oocytes and embryos were validated and confirmed by qPCR experiments (Misirlioglu et al., 2006). Gilbert et al. (2007) performed qPCR to validate sperm transcriptome data generated using microarrays. In another study, the microaarray results obtained from the *in vivo*- versus *in vitro*-generated bovine blastocysts were validated using for a panel of six genes (*cyp51a1, fads1, tdgf1, habp2, apoa2,* and *slc12a2*) (Clemente et al., 2011). The validation studies typically focus on conserved "patterns" of gene expression rather than the same levels in microarray versus qPCR because of the differing natures of these two approaches.

Microarrays

The history of the microarrays started with a pioneer, Affymetrix Chip® technology in late 1980s. **Microarrays** initially it started out with a simple glass chip array. More complicated ones have been manufactured with a solid surface, containing known nucleotides and cDNA probes. Briefly, fluorescent tagged cDNA are synthesized from isolated mRNA and then they are incubated with the chip where hybridization takes place depending on their complementarities of probes embedded on the microarrays. Thus, the expressions of specific genes are quantified by detecting these fluorescent signals. Using these microarrays, more than 20,000 scientific papers have shed lights onto roles of mRNA transcripts in mammals (www.affymetrix.com). A microarray workflow is displayed in Figure 23.2.

According to Robert (2008), microarrays can be classified into three groups; cDNA microarrays with a few hundred base pairs on the glass slides, the oligonucleotides with the 40–70bp of ssDNA and lastly the Affymetrix Chip system with 25bp of ssDNA on the quartz-coated surface. In addition, another microarray platform from Arrayit produces comparative genomic hybridization (CGH) microarrays using oligos, cDNAs, and BACs (bacterial artificial chromosome) (http://arrayit.com). Also, OneMicroarray (www.genomics.agilent.com) and NimbleGen (www.nimblegen.com) microarray platforms are offered by Agilent technology and Roche, respectively. In addition to these, Illumina has recently come up with an advanced microarray platform, Omni family of microarrays that contains 5 million variants per sample (www.illumina.com). Microarrays containing few numbers of expression tags (ESTs) such as ~934 are also called macroarrays. These were used in a bovine embryo transcriptomics study (Thelie et al., 2009). Macroarrays are not appreciated compared to microarrays because of their limited effectiveness and sensitivities.

Methodology

Microarrays are designed in such a way that cDNA probes or oligonucleotides are embedded in to the chips, and can measure transcriptome levels of several thousands of genes in a biological sample. Their methodology relies on the DNA hybridization of transcripts followed by serial amplifications, including sample preparation and labeling. Once the samples are labeled with two specific dyes (Cy3 and Cy5), imaging is performed using microscopy (see Figure 23.2 for detailed microarray workflow). Since microarrays deal with a limited amount of a starting material, an additional amplification step can be included after RNA extraction. Two major techniques of this amplification are linear and exponential amplifications. The former is based on *in vitro* transcription (IVT) by T7 RNA polymerase, leading to the amplification of antisense (aRNA) RNAs. In the latter amplification PCR is used to generate double-stranded cDNAs. Combination of these two amplification techniques in a study showed that the linear amplification method was more reproducible, cost-effective and compatible with both sense and antisense arrays over the IVT method (Laurell et al., 2007). For example, this IVT method was performed in a study where bovine oocyte and embryos were used (Thelie et al., 2009).

Data analysis

Significance Analysis of Microarrays (SAM) was developed to analyze microarray results, and is still being used by several research studies in the life sciences. This software enables the calculations per gene scored above threshold, using permutations. It provides an estimated ratio of gene scores that are mainly identified by chance, which is called false discovery rate (FDR) (www-stat.stanford.edu/~tibs/SAM/). For instance, microarray data of transcripts from Human lymphoblastoid cell lines were analyzed using SAM. In the same study, three comparative methods that were used to determine the differential gene expression were SAM, fold change and pairwise fold change (Tusher et al., 2001). In addition to SAM, another bioinformatics tool to analyze the microarray data is Gene Expression Profile Analysis Suite (GEPAS), which is publicly available (www.gepas.org and/or www.babelomics.org). This software allows scientists not only to analyze their data but also to interpret them by performing advanced functional profiling such as gene annotation and clustering, and so on. As an example, Garrido et al. (2009) analyzed a data set of human sperm transcriptomics by using GEPAS (Garrido et al., 2009).

Advantages and disadvantages

Following the qRT-PCR technology, which is limited to "one gene at a time," microarray technique had become a high-throughput method by the late 1990s. Although microarrays are now becoming outdated, they are still acceptable approaches for transcriptomics today. Since these array platforms are generally designed to be used in human and mouse samples, lack of species-specific oligonucleotide probes is an impediment of this technique. However, this issue might be overcome by designing commercially available custom platforms and/or by improving preparation of the home-made cDNA libraries. For example, Yang et al. (2009) developed a home-made cDNA library of a swine sperm-specific library using non-redundant 5′end complementary DNA. In another study, transcriptome profiling of microarray was completed using bovine embryo-specific probes (Kepkova et al., 2011). Following the completion of genome sequencing of several livestock such as bovine, porcine, and equine, and so on, the companies have started producing the species-specific microarrays that contain larger numbers of probes that cover a majority of the genomes. For example, Affymetrix offers new microarray strips (model and applied research organism's gene 1.1st) array strips to be used for several livestock animals (www.affymetrix.com). However, this renovation does not change the fact that microarray platforms are a design based upon a known reference genome and hence, identify a limited number of transcripts (30,000–40,000) with their embedded cDNA probes or oligonucleotides.

Applications

Microarrays have been exploited in several studies by using different mammalian tissue types such as cell lines (Tusher et al., 2001) and/or germ cells, including spermatozoa (Garrido et al., 2009), oocytes and embryos (Misirlioglu et al., 2006; Clemente et al., 2011). One of the first studies conducted by taking advantage of microarrays in matured bovine oocytes and embryos was published in 2006. This revealed a remarkable knowledge to better understand specific embryonic transcripts, likely playing important roles in embryonic genome activation (EGA) (Misirlioglu et al., 2006).

Somatic cells nuclear transfer (SCNT), a method to generate cloned embryos in mammals, was utilized in a study conducted by Smith et al. (2005). In this study, the expressions of reprogrammed genes in cloned bovine embryos were shown to be significantly higher by exploiting microarrays. Rodriguez-Osorio et al.(2009) studied the extent of molecular reprogramming in bovine embryos derived by *in vitro* fertilization (IVF) where four generations of successive cloning by chromatin transfer were used in SCNT approach. The results of this study revealed that although the SCNT embryos had gone through significant levels of reprogramming, there were "cumulative errors" in epigenetic reprogramming in animal cloning (Rodriguez-Osorio et al., 2009). Bovine Affymetrix microarrays also were carried out to identify changes of transcriptome at the initiation of the blastocyst elongation window, that is, from Day 7 to Day 13 in the *in vitro* and *in vivo*- derived blastocysts. According to the results of this study, 50 and 288 genes were differentially expressed in the *in vivo*-derived blastocysts on day 7 and day 13, respectively compared to their *in vitro* counterparts (Clemente et al., 2011).

In addition to embryo transcriptome profiling, sperm mRNAs have been determined by utilizing microarrays in different studies. In the first example of these studies, sperm mRNAs from testes and ejaculates of normal fertile men were detected using microarrays, suggesting that testis-specific infertility can be diagnosed by investigating mRNA expression in spermatozoa (Ostermeier et al., 2002). Later, the same group showed that a total of 228 unique sperm transcripts were identified, and hence speculated that some of these specific mRNAs might be associated with regulation of male fertility (Ostermeier et al., 2005). Moreover, mRNA repertoires of sperm from fertile and infertile males were investigated to determine differentially expressed mRNAs among mammals. Garrido et al. (2009) demonstrated significant differences in sperm transcriptome between infertile and fertile men. Transcripts of genes were found to be highly expressed in fertile men's spermatozoa as compared to their infertile counterparts and to be related to spermatogenesis, sperm motility, histone modification, DNA repair, and oxidative stress regulation. Sperm transcriptome profiles of idiopathic infertile men were shown to be different than that of fertile men (Montjean et al., 2012). Besides, in a bovine sperm transcriptome study conducted using microarray it was shown that the mRNAs found in spermatids were functionally different than those in spermatozoa. Yang et al. (2010a) demonstrated seasonal changes in swine sperm triscriptome using sperm-specific oligonucleotide microarray. In summary, the scientists suggested that sperm quality that is an important factor affecting male fertility can be diagnosed by transcriptome profiling of sperm mRNAs (Gilbert et al., 2007).

Transcriptomics of small non coding RNAs (sncRNA)

Apart from protein coding mRNAs, transcription of genomes also produces RNA molecules that are not translated into any proteins. These are precisely called **non-coding RNAs**. While some of these transcripts are long (long-non coding RNAs) such as *Xist*, some are short (short non coding RNAs or sncRNAs) such as **micro RNAs** (miRNA), **small interfering RNAs** (siRNAs), and **piwi interacting RNAs** (piRNAs). Although they do not code for proteins, they have a wide range of vital functions in almost all tissues of the animal. This section will focus on the current methods for the sncRNAs and functions of some of the sncRNAs illuminated by using these methods. Because the methods for the study of mRNAs are similar to those used for the study of sncRNAs, here we will only include the differences that must be kept in mind when studying sncRNAs.

Micro RNAs are only 21 to 23 nt long and are synthesized as nascent miRNA transcript, pri-miRNA in the nucleus which is processed by the enzyme Drosha to form pre-miRNA. Pre-miRNA is exported to the cytoplasm by a protein called exportin 5, where it undergoes cleavage by RNAse III enzyme, Dicer to form mature miRNA. One of the strands of mature miRNA is then incorporated into RNA induced silencing complex (RISC) which is directed towards the cleavage of mRNA containing complementary sequence of the incorporated miRNA strand. The miRNAs and the other sncRNAs are often called regulatory RNAs that control gene expression at transcriptional or post transcriptional levels through profound mechanisms to silence genes and alter cellular processes (Figure 23.3). Since the identification of the first miRNA lin-4 which is essential for postembryonic developmental events in *C. elegans* the search for novel sncRNAs and their roles has expanded (reviewed by Memili et al., 2012).

Small interfering RNAs (siRNAs) are double-stranded transcripts that are 20–25 nucleotides in length. They are notable for their actions on repressing gene expression through the RNA interference pathway. In addition to endogeneous siRNA, synthetic siRNAs developed in the laboratory are shown to induce RNAi in mammalian cell lines (Elbashir et al., 2001). This has heightened the scope of using siRNA in gene therapeutics and functional genomics studies to knock down a particular gene of interest.

PIWI-interacting RNAs, or piRNAs, are the most recently discovered small non-coding RNAs, and they are known to play a crucial role in germline stability through interaction with piwi proteins. These RNAs were first described as a class of longer RNAs that acted to silence repetitive components. They are typically 24–34 nt in length and contain a phosphorylated 5′ end and a 2′-O-methyl configuration that is similar to the structure of siRNAs (Stefani and Slack, 2008). They were first called repeat-associated siRNAs (rasiRNAs), because they were small RNAs being complementary to repetitive transposable elements. However, their names have been changed because these rasiRNAs co-purify piwi proteins. There are two pathways that have been explored as the possible piRNA biogenesis method: the primary pathway where an initial group of piRNAs combat the transposable elements and the ping-pong cycle pathway during which the piRNA population is refined through the expression of target active transposons (Siomi et al., 2011).

Quantitative real-time reverse transcription polymerase chain reaction (qRT-PCR)

As aforementioned, qRT-PCR is an improvised method where quantification is performed while the DNA is amplified, thus eliminating the post-amplification contamination and providing accurate measurement. We are often interested in quantifying sncRNA expression because presence of sncRNAs reflects the

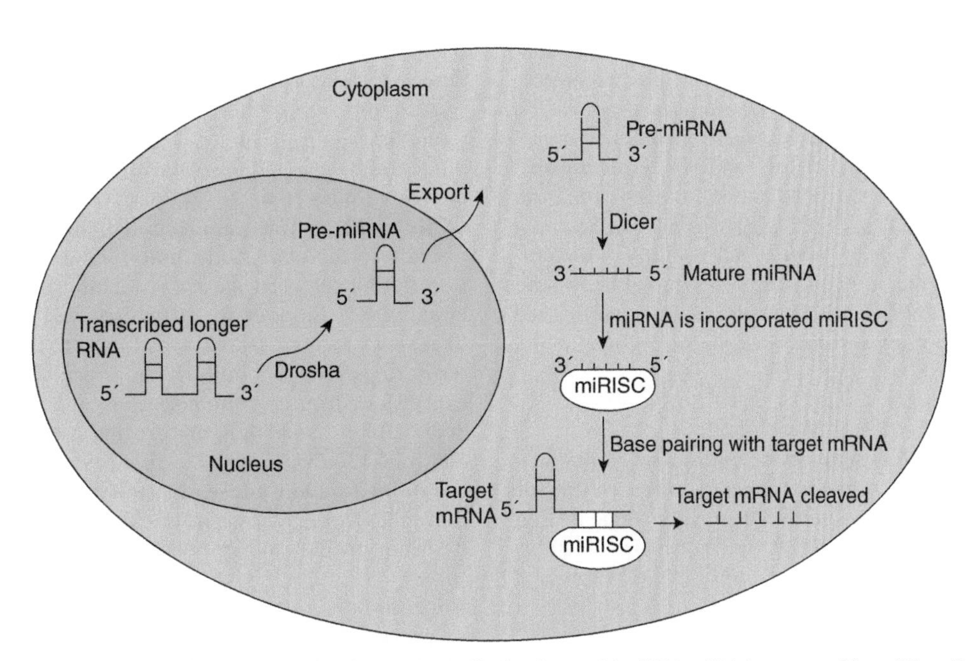

Figure 23.3 MicroRNA biogenesis and function. Initially miRNA is transcribed as longer Pri-miRNA, which is processed by a RNase III enzyme, Drosha to form Pre-miRNA. Pre-miRNA is exported to cytoplasm by a protein called exportin. Inside cytoplasm, as a result of Dicer processing, mature miRNA forms. One of the strands of miRNA gets incorporated into RNA-induced silencing complex (RISC) to form the final unit for mRNA translational repression.

role of their target mRNAs. However, it is challenging to quantify sncRNAs using PCR. For example the miRNA precursor has a hairpin structure which makes it difficult for primers to bind to. Furthermore, mature miRNAs are very small, almost the size of a standard PCR primer (Schmittgen et al., 2008).

Methodology

Prior knowledge of sequence information about the precursor and mature miRNA is required to design the primers. Formation of isoforms among the miRNA is a major challenge in primer designing. It is often the precursor miRNA that is chosen for the amplification rather than mature miRNA because its extremely small size renders it to be a poor choice. Primers are designed with a maximum possible T_m difference between forward and reverse primers of $\leq 2°C$ and a primer length between 18 and 24 nts. T_m of 55–59°C is considered ideal for the primers, however due to size constraints; some are designed with a T_m that is below 55°C. The primers should not have any 3′ GC clamps, and a minimum amplicon size should be 55 bp (Schmittgen et al., 2008). For the miRNA primer design, miScript Primer Assays from Qiagen offer both ready to use and custom design primers (www.qiagen.com/products/miscriptprimerassays). The two commonly used methods for qRT-PCR to detect expression of mature miRNA levels are stem-loop RT primers method and a modified oligo(dT) technique. The former technique is based on the reverse synthesis of cDNA of specific miRNAs using stem-loop RT primers. In the modified oligo(dT) technique, all transcripts are reverse transcribed, which allows detection of both target miRNA and normalizing mRNA with the same RT reaction. The two methods were found to be similar; however, using stem-loop RT is more time consuming as compared to the modified oligo(dT) method (Fiedler et al., 2010). Since mature miRNAs do not have Poly (A) tails, nondirect RT approach requires a poly (A)-tailing of miRNAs before RT reaction. In addition, TaqMan® MicroRNA Assays (Applied Biosystems) are designed based on the direct-RT method using a miRNA-specific stem-loop RT primer. This Stem-loop RT based TaqMan® assay has become a common approach for both detection and validation of miRNAs because of its sensitivity and specificity (www.invitrogen.com). The reaction protocol is similar as for mRNA, where cycle number at which the reaction crossed an arbitrarily-placed threshold (CT) is determined and the relative amount of each miRNA to an internal control, generally 18S rRNA was determined using the equation $2 - \Delta CT$ where $\Delta CT = (CT_{miRNA} - CT_{18S\ rRNA})$ (Livak and Schmittgen, 2001). Pre-miRNA can be synthesized using *in vitro* transcription using primers that are annealed to ~50 bp upstream and downstream of the specific miRNA hairpin. This synthesized miRNA can be used to draw a standard curve (Schmittgen et al., 2008).

Data analysis

Accuracy of the qRT-PCR experiment relies on data normalization and analysis. Multiple reference genes are used widely for normalizing the data. The mean miRNA expression value is used as a new method for miRNA qRT-PCR data normalization. This mean normalization also reduces the noise of overall expressed miRNA genes (Mestdagh et al., 2009).

Advantages and disadvantages

QRT-PCR is highly sensitive and is still the gold standard method to verify microarray data because of its specificity. The qPCR-arrays make this approach possible for high-throughput miRNA expression profiling (Chen et al., 2009). However, for miRNA high-throughput qRT-PCR can give variable results in quantifying DNA. Compared to microarray and Northern blot results, qRT-PCR results may fail to give perfect correlation in miRNA profiling (Morin et al., 2008). This is because the method relies on primers to be designed individually for detecting each of the reference miRNA sequences.

Applications

MicroRNAs play essential roles in cell differentiation and tissue morphogenesis, and in disease. There are specific miRNAs for specific developmental stages of mammalian development. For example, using the qRT-PCR method, several miRNAs including miR-302b, miR-302b, miR-302c, and miR-373 were shown to be expressed exclusively in human embryonic stem cells (Suh et al., 2004). RNAi experiments using siRNAs have been designed to gain knowledge on the functions of genes in gamete and embryo development. For example, when bovine importin alpha8 (KPNA7) is knocked down in embryos, the number of embryos developing to the blastocyst stage was decreased. This has helped assign the role of KPNA7 gene product in transport of essential nuclear proteins required for early embryogenesis (Tejomurtula et al., 2009). Tang et al., using real time RT PCR, showed that miRNAs expressed during oogenesis can shape the gene expression profile of the mature oocyte (Tang et al., 2007). Regulation of miRNAs was identified in the embryonic stem cells (ESCs) during trophectodermal differentiation. In the same study, miR-290, miR-302, and miR-17~92 were found to be highly expressed in embryonic stem cells. While the expression of miR-34b was similar between the eight-cell and blastocyst, miR-182 was down-regulated between the two-cell and four-cell embryos. In addition, miR-103 was detected to be similar in the zygote and four-cell embryos (Viswanathan et al., 2009). The expression of miRNAs specific to bovine cumulus cells was detected during late oogenesis. Additionally, a total of 1798 novel miRNA sequences were identified and 64 of which were found in the miRBase database. The expression of let-7b, let-7i, and miR-106a were detected using Taqman assay and only the expression of miR106a was found to be different in the follicles with varying size. According to this result, let-7 and let-7i targeted to MYC mRNA, while miR-106a targeted WEE1A mRNA in cumulus oocyte complex (Miles et al., 2012).Sperm miRNA such as miR-34c out of six miRNAs was demonstrated in zygotes but not in the oocytes. In the same study, miR-34c was detected to have function during cell cleavage via regulation of Bcl2 expression (Liu et al., 2012). Bjork et al., (2010) showed that miR-18 targeting heat shock factor 2 (HSF2) was found to be abundant in mouse testis, suggesting that miR-18 has a regulatory function on HSF2 expression during spermatogenesis and gametogenesis (Bjork et al., 2010). In addition, mature and precursor forms of miR-21 and miR-130a were found with a linear increase during bovine oocyte maturation and early embryo development, suggesting that miRNAs might have a function during maternal embryonic transition (MET) (Mondou et al., 2012).

Microarrays

As previously mentioned, a microarray is a chip based method where one can see and measure the expression of large number of genes at the same time. Axiom® miRNA target site genotyping arrays are designed to detect SNPs, miRNAs, and their target sites (http://media.affymetrix.com/support/technical/brochures/axiom_mirna_flyer.pdf).

Methodology

This method exploits the hybridization between miRNA and the DNA from which it is transcribed. In a typical miRNA microarray, instead of mRNA expression, miRNA is being measured in the chip. Briefly, first miRNAs are oxidized by sodium periodate, which converts 3′ terminal adjacent hydroxyl groups into dialdehyde. This dialdehyde groups are allowed to react with biotin-X-hydrazide through a condensation reaction resulting in biotinylated miRNA. Next, oligonucleotide probes complementary to miRNAs are immobilized on glass slides which are preincubated with streptavidin, and washed. Then, the hybridization is allowed to take place. Quantum dots are labeled on the captured miRNAs through specific interaction of streptavidin and biotin. Signals are detected with a laser confocal scanner (Liang et al., 2005). Axioms are created to identify known and novel SNPs, indels in gene regions of miRNA promoters, miRNA precursor and miRNA stem-loop regions.

Data analysis

In order to eliminate the experimental variations and to make the biological variations more meaningful, data should be normalized first and then must be analyzed. Image processing, which relays the actual intensity of each spot, is also important. Many image processing methods have been developed to distinguish data from background noise (Allison et al., 2006).

Advantages and disadvantages

Microarray is a convenient way to compare the expression in two or more sets of different samples at the same time. However, the prior sequence knowledge of all genes is necessary. Although cross species hybridization is possible, the chances of missing the species specific information are high.

Applications

According to a recent study, normal miRNA function was identified to be down-regulated during oocyte development, which was related to lack of *Dgcr8* that is necessary for the miRNA pathway (Ma et al., 2010). In another study, miRNAs in bovine cloned embryos (Day 17) during elongation were determined using microarray approach, followed by validation of the data via qRT-PCR. This study concluded that reprogramming of miRNAs might occur in cloned bovine embryos (Castro et al., 2010). In addition, miR-21 targeting *Reck* gene was detected to be highly expressed miRNA during implantation in mouse. Although miR-21 was not found during pseudopregnancy and under delayed implantation, it was abundant in the subluminal stromal cells at implantation sites on day 5 of pregnancy (Hu et al., 2008).

Deregulated expression of some miRNA is associated with cancer. A miRNA microarray experiment has identified the miRNAs associated with chemotherapy response in ovarian cancer. In this study, let-7i expression was found to be significantly reduced in chemotherapy-resistant patients compared to those who responded to chemotherapy. Reduced let-7i expression is responsible for increased resistance of ovarian and breast cancer cells to the chemotherapy drug, *cis*-platinum (Yang et al., 2008). In another study, miRNA microarray was employed to compare miRNA expression profiles of testis tissues from immature and mature rhesus monkeys (Yan et al., 2009). Nelson et al. (2004) developed microarray based high throughput expression profiling for miRNAs. They have used a technique called RNA-primed, array-based Klenow enzyme assay because it involves Klenow fragment of DNA polymerase I. Later on, Liu et al. (2008) described a protocol for miRNA microarray including miRNA oligo probe design, array fabrication and miRNA target preparation (by reverse transcription), target-probe hybridization on array, signal detection and data analysis.

A number of recent studies have shed lights onto roles of miRNAs in germ cell development. For example, a miRNA expression profiling assay where all expressed miRNAs were isolated and cloned has revealed that 5% of total miRNAs are exclusively expressed in testis (Ro et al., 2007). Using real time PCR based 220-plex miRNA expression profiling method, miR-17–92 and miR-290–295 clusters are shown to be highly expressed during cell cycle progression and throughout the development of primordial germ cells (Hayashi et al., 2008). Deletion of miRNA processing enzyme, *Dicer1* in primordial germ cells has resulted in spermatogenic defects leading to subfertility (Maatouk et al., 2008).

The miRNA regulation is not only limited to gamete formation, but is also important for early embryogenesis. Maternal miRNAs are necessary for embryogenesis as the lack of maternal miRNAs has resulted in a decrease in the average number of progeny produced following fertilization by wild-type males (Suh et al., 2010). MicroRNA profiling using qPCR studies among sperm, egg, and zygotes showed the expression of miR-34c in sperm and zygotes, but not in oocytes. Inhibiting miR-34c has resulted in the suppression of first cleavage division (Liu et al., 2012).

RNA sequencing

RNA sequencing is a highly sensitive method that gives accurate expression levels of a number of genes without prior knowledge of the genes whose transcripts are studied.

Methodology

This method is based on the deep sequencing technique as explained in the previous section. Sequencing of small RNA involves first the small RNA library preparation. To do so, small RNAs are isolated and purified, and then the RNA pool is ligated with 3′ and 5′ adapters with T4 ligase. These RNAs are converted to single stranded cDNA by a suitable reverse transcriptase enzyme and the resulting cDNAs are PCR amplified to proceed for sequencing. Each unique sequence is counted and then the repeats are added to respective unique read (Morin et al., 2008).

Data analysis

Data analysis depends on accurate mapping of sequencing reads to corresponding reference genomes or else *de novo* assembly should be efficient. Once the mapping is completed, the alignment can be explored on a genome browser such as UCSC Genome Browser or the Integrative Genomics Viewer. Quantitative expressions for annotated genes are provided by the total number of reads mapping to the coordinates of each of the annotated genes. It should also be noted that the total number of reads for a given transcript is roughly proportional to both expression level and the length of the transcript. That means, a long transcript will have more reads mapping to it as compared to a short gene of equal expression level. Because of this reason, measuring short transcripts including sncRNAs will always be over shadowed by long transcripts in the same sample (Costa et al., 2010).

Therefore, in regards to analyzing sncRNA, separation of small RNAs from total RNA is preferred.

Advantages and disadvantages
RNA sequencing facilitates identification of novel sncRNAs since the technique does not require existing genomic data. This is particularly useful when there is no well-established sequence knowledge of an organism under study. It also gives absolute quantification of the transcripts. The disadvantages of this method are unlike RT-PCR. It is time-consuming and data analysis of huge reads of transcriptome is tedious.

Commonly used computational tools for sncRNAs
There is an exclusive database for miRNA called miRBase. It is an online repository for miRNA nomenclature, sequence data, annotation, and target prediction (Griffiths-Jones et al., 2008). The deepBase is a computational tool which was developed to perform the comprehensive annotation and discovery of small RNAs from the huge transcriptomic data output from RNA-seq. The deepBase contains deep sequencing data from 185 small RNA libraries from varieties of tissue and cell lines of seven organisms: *Homo sapiens, Mus musculus, Gallus gallus, Ciona intestinalis, Drosophila melanogaster, Caenhorhabditis elegans,* and *Arabidopsis thaliana*. It also allows visualization of deep sequencing data from multiple technological platforms (Yang et al., 2010b). A database called TarBase is a manually curated collection of experimentally tested miRNA targets in human, mouse, fruit fly, worm, and zebra fish (Sethupathy et al., 2006).

Applications
Identification and characterization of sncRNAs expressed in various stages of gamete formation and zygote development will allow us to perform functional studies. Silencing of a gene by miRNA and siRNA uncovers the functional role of a gene. MicroRNAs and siRNAs use the same RNA-processing complex to direct silencing and hence till date functional genomics studies that rely upon this gene silencing mechanisms. The RNA interference method was employed to find gene functions during oocyte development and differentiation. A transgenic mouse was developed expressing a transgene dsRNA against the gene *Mos*, coding for a serine-threonine kinase that functions upstream of MAP kinase. Mos accumulates during oocyte growth and is not translated until resumption of meiosis. Those transgenic mice exhibited decreased mitogen-activated protein (MAP) kinase activity, and the metaphase II eggs exhibited spontaneous parthenogenetic activation (Stein et al., 2003).

Expression profiles for 191 previously annotated miRNAs, 13 novel miRNAs, and 56 candidate miRNAs were identified by deep sequencing of small RNA libraries in human embryonic stem cells (Bar et al., 2008). Using this approach in bovine kidney cell line, Glazov et al. (2009) identified 219 out of 356 known bovine miRNAs and also 115 corresponding miRNA sequences. They also reported five new bovine orthologs of known mammalian miRNAs and discovered 268 new bovine miRNAs that are not identified in other mammalian genomes. They found seven new bovine mirtron gene candidates (type of microRNAs that are located in the introns of the mRNA encoding genes), 10 small nucleolar RNA (snoRNA) loci with possible miRNA-like function. Deep sequencing of small RNAs in mouse oocytes has revealed the presence of miRNAs, piRNAs, and also endosiRNAs (Tam et al., 2008; Watanabe et al., 2008). Homeostasis of siRNAs and miRNAs in mouse oocytes were shown to be regulated by Dicer and Argonaute2 protein (Ago2) to control retrotransposon expression and mRNA expression (Kaneda et al., 2009) where real-time PCR was used for miRNA profiling and cDNA microarray analysis for identifying genes regulated by Dicer and Argonaute2 proteins. Krawetz et al., (2011) reported a survey of miRNAs in human spermatozoa demonstrating the distributions of nc-RNAs as 7% miRNAs, 17% piRNAs, and 65% repeat-associated small RNAs.

Question
An RNAseq experiment in embryo has shown the abundance of particular RNA transcripts. If the sequence of its corresponding gene has no reported annotation, how will you perform functional studies of that gene?

Proteomics

Once translated from their transcripts, proteins travel to their destinations in the cell and they perform functions vital to cell and organismal physiology. Expression of proteins reflects the molecular happenings in organisms related to development, disease, environment, and so on. This section will review a set of diverse methods for the study of proteins in gametes and embryos that help illuminate developmental mechanisms. Figure 23.4 provides a flow chart of the methods for the study of proteins. Extensive review of proteomics and its applications in animal biology are available in a recent book on proteomics (Eckersall and Whitfield, 2011).

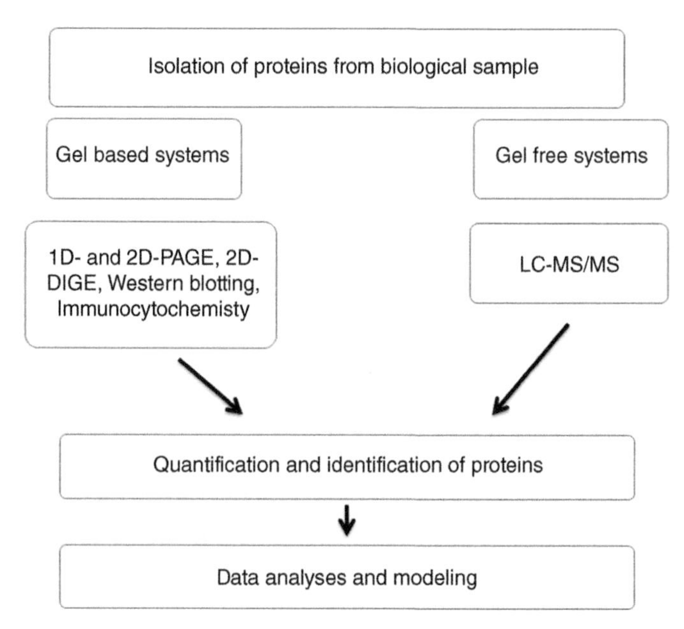

Figure 23.4 Protein methods. Proteins are macromolecules with vital functions in cells during development. Meaningful data can be generated only through a well-designed study with appropriate proteomics approaches-which may include high-throughput methods such as MS and 2D-DIGE or reductionist methods such as western blotting and immunocytochemistry. Often, two approaches are used to validate the results. Quantifying proteins and analyzing the data are essential for developing models to illuminate molecular and cellular underpinnings of developmental mechanisms.

Polyacrylamide gel electrophoresis (PAGE), two-dimensional page (2D-PAGE), and two-dimensional differential in gel electrophoresis (2D-DIGE)

These methods have been vital in advancing protein science and hence, can be applied to diverse study areas in animal biology.

Methodology

The first step of proteomics studies is sample preparation during which proteins are isolated from the biological materials via lysis in appropriate buffer and then fractionated according to their molecular weights in polyacrylamide gels. Protein extraction should be performed using a suitable buffer for different target proteins such as membrane, cytoplasmic or nuclear proteins. In 2D-PAGE, proteins are first separated according to their molecular weights and then isoelectric points. Proteins are then stained with comassie blue, silver stain, or with other dyes. Depending on the protein mixture and size of the gel, it is possible to visualize hundreds of protein spots in 2D-PAGE.

The more advanced gel based proteomics approach is 2D-DIGE enabling detection of differentially expressed proteins from two different samples in one gel (Unlü et al., 1997). This is accomplished by labeling the two samples with fluorescent CyDyes (Cy2, Cy3, and Cy5), and the mixture of the two samples with another dye so that the mixture will serve as control. The protein mixtures are first fractionated according to their molecular weights and then based on their isoelectric points. The gels are then scanned by a robotic system which takes the pictures of the gels with precise coordinates of florescent dyed protein spots. The spot intensities are determined by the software of the imaging system, and spots of interests are extracted out by the robot for protein identification using mass spectrometry.

Data analysis

Signal intensities or densities of the protein spots can be quantified if the proteins were labeled using radiolabeled isotopes. The other stains, that is, comassie blue or silver stained protein spots may not be reliably quantified. Analyses of the differentially expressed proteins in 2D-PAGE and 2D-DIGE are determined most confidently by the scanner, which is a part of the robotic system. The protein spots can be easily tracked down within and across the gels and their densities can be measured using the software within the imaging system.

Advantages and disadvantages

Although fractionation of proteins using one dimensional gels provide a general picture of the proteins present in a particular cell type, the separation of these proteins is just based on the molecular weights and that too many proteins can be located in a small range of the gel. Thus, unless linked with additional approaches such as immunoblotting, and so on, one dimensional gels have inherently limited resolution of proteins. Two dimensional gels provide better separation and once coupled with an automated system, the image analyses are more reliable. However, protein spots picked by the robot might still contain mixtures of proteins rather than a single protein. Furthermore, 2D-DIGE is a labor intensive method, requiring a scanner included the software for the protein spots. Finally, large numbers of cells are needed to generate significant data: This is usually a hindrance for research involving oocytes and embryos whose availability is often limited.

Applications

Proteomics methods and their applications in reproductive biology have been extensively reviewed by Wright et al. (2012), Arnold and Frohlich (2011). Using 2D-PAGE and radiolabeling approaches, Susor et al. (2007) detected 120–240 proteins during maturation of GV stage oocytes to MII stage. Eighty and 23 differentially expressed proteins in sperm from both Angus and Holstein bulls of varying fertility were recently identified using 2D-DIGE, respectively (Memili, unpublished data). Expression of the proteins identified was validated using western blotting. The proteins provide potential markers for male fertility where they can be used to predict sperm viability and thus bull fertility.

Mass spectrometry

Although several kinds of mass spectrometry (MS) methods have been developed, their core principles remain similar. Using this technology, vast array of proteins from a mixture can be identified in one experiment, providing a global view of proteome and their predicted functions in cell.

Methodology

The first step is preparation of protein samples, which influences the amounts and quality of proteins to be studied. Proteins are isolated/fractionated and enriched if desired and then analyzed using MS. Protein spots cut out from 2D-PAGE or 2D-DIGE are digested and then identified on the basis of protein mass to charge ratio using MS, a method called **peptide mass fingerprint**.

Proteins from mixtures can be identified using peptide separation by liquid chromatography where peptides are separated according to their hydrophobicity or charge state using reverse-phase high performance chromatography (RF-HPLC) or cation exchange chromatography, respectively.

Liquid chromatography-mass spectrometry (LC-MS)

This method provides a separation of proteins or peptides through LC followed by tandem mass spectrometry, that is, LC-MS/MS. That is, the proteins are first digested with a sequence-specific endopeptidase such as trypsin, and then they are separated by liquid chromatography. The last step of separation is done through a reverse phase chromatography such as nano-LC (Frohlich and Arnold, 2006).

Following fractionation of proteins through LC, the resultant peptides are then processed in MS where peptides are ionized before the mass analyzer. There are two ionization methods, electrospray ionization (EIS) and matrix-assisted laser desorption ionization (MALDI).

Quantification of proteins and data analysis

Quantification of proteins in LC-MS/MS experiments is based on quantification of individual peptides. This creates a challenge because many proteins share common peptides. Thus, an improved quantification can be achieved through monitoring peptides proteotypic for the corresponding isoforms. Precursor ion monitoring is a method where signal intensity of each distinct precursor ion with a match is used to determine amounts of this peptide in the proteins analyzed (Arnold and Frohlich, 2011). Spectral counting is another protein quantification method where relative abundance of each peptide in a protein is determined by using the number of MS/MS spectra recorded by the MS machine for each peptide (Carvalho et al., 2008).

To quantify proteins effectively, covalent labeling can be introduced to proteins or peptides with chemically identical tags that differ in their isotope composition (i.e., ^{12}C vs ^{13}C). In a typical experiment, control and test samples can be labeled with different isotopes, combined after the labeling and fractionated and then analyzed through MS. This enables reliable determination of abundance of each peptide in the samples. Besides, Gygi et al. (1999) introduced isotope-coded affinity tag (ICAT) to introduce an isotope tag into proteins. Only limited number of peptides can be labeled since in this method only the cyctein residues can be isotope tagged. This limitation can be overcome by using ICPL method (Smith et al., 2005) in which primary amino acid groups are labeled. In isobaric tag for relative and absolute quantification (iTRAQ) method, the tag is introduced at the level of peptides, and signal intensities generated during MS/MS step are used to quantify fragments (Choe et al., 2007; Ross et al., 2004).

An analysis of the mass spectrometry data is currently tedious and challenging. This is partly because of the large amounts of data generated and lack of knowledge in complete proteome of organisms as well as incomplete annotation of mammalian proteins. Thus, software packages and statistical methods included in the proteomics methods should be diligently applied in the data analyses.

Advantages and disadvantages

A major advantage of mass spectrometry is that large amounts of data can be produced in a single experiment. Challenges include misleading information if not enough repeats are not performed and the time consuming aspect of data analysis, which requires considerable expertise in computational biology. In addition, large sample requirements (e.g., for the study of oocytes or germ cells) can be considered a limitation.

Applications

While the exact numbers of proteins are not yet defined for each mammal, we now know several essential proteins in the developing gametes. For example, using mass spectrometry, Peddinti et al. (2008) obtained 125 proteins differentially expressed in sperm from bulls with different fertility. While some of these proteins were known, some were predicted. These proteins can potentially be used to predict bull fertility.

Western blotting

Also known as immunoblotting, this common method has long been used to detect expression of proteins in diverse sets of cells, tissues, or fluids of various species. Over the years, improvements have been made in the sensitivity and feasibility of the method by generating more specific primary antibodies; however, challenges and limitations still exist.

Methodology

Cells or tissues are lyzed with a buffer containing protease inhibitors to avoid protein degradation. Following lysis step, appropriate amounts of proteins are first boiled in a buffer containing SDS (Sodium dodecyl sulfate) and then chilled on ice for unraveling and denaturizing proteins, respectively. Afterwards, protein extracts are mixed with a loading buffer to trace the protein extract while they are size separated in the gel. The proteins from the test samples, positive and negative controls and size markers are then separated either in 1D or 2D-PAGE as described in the previous section. The size fractionated proteins on the acrylamide gel are then transferred to membranes such as nitrocellulose and that are pre-equilibrated in transfer buffer using dry, semi-dry or wet transfer methods. Several important factors such as the duration and voltage need to be empirically determined to obtain optimum transfer of the proteins on to the membranes. Efficiency of transfer can be determined by staining the membranes using Ponceau dye that enables visualization of the proteins. If the membrane is stained, the dye needs to be washed off and then the nonspecific sites in the membrane are blocked using blocking buffer containing a bovine serum albumin (BSA), casein or dry milk solutions. Next, the membranes are incubated with primary antibodies that will bind to their protein targets, washed with washing buffer and then incubated with a labeled secondary antibody that will bind to the primary antibody on the membrane. Following another series of washes, the membrane is incubated with substrates that result in fluorescent or other color emissions. The pictures of protein bands can then be taken using X-ray or UV-based systems to determine the intensities of protein signals.

Data analysis

Because of the nature of western blotting method, the data generated are not quantitative. That is, the data are semi-quantitative at best. Typically, ratios of signal intensities of proteins of interest to those from an internal control, usually housekeeping gene proteins, are presented as protein expression profiles (www.bio-rad.com). The experiments are repeated at least three times, using independently isolated proteins in each repeat. Depending of the sample size, that is, proteins from sperm of 20 bulls, and the technical and biological repeats, the data can then be analyzed using classical statistical approaches such as the t-test. However, because the cells are pooled, that is, proteins are isolated from 100 oocytes or embryos, expression profiles of the proteins of interest in individual oocytes or embryos are not known using this method even with the use of advanced statistical tools.

Advantages and disadvantages

This method enables scientists to detect expression of proteins for which antibodies exist. In addition, posttranslational modifications of the proteins can be determined using antibodies against the modified proteins (Walsh, 2006). The disadvantages are that: (1) this approach is limited to detection of total proteins in lyzed cells and hence, may not determine cellular locations of proteins unless proteins are isolated from specific cell fractions or organelles, (2) the expression levels determined are semi-quantitative, (3) prior knowledge of the proteins of interest is required and antibodies might not be available for all proteins and that the specificity of the antibodies might be variable, (4) this is a low throughput method where one or very few proteins at a time can be studied.

Applications

Western blotting had been a classical method to detect protein expression in reproductive tissues as well as many other types. For example, Nakamura et al. (2010) has demonstrated expression of Hexokinase 1, Phosphofructokinase M, and Gluthatione S-Transferase mu class 5 in fibrous sheath of mouse sperm using western blotting and immunocytochemistry. Expression dynamics of Classical and non-classical major histocompatibility complex class I (MHC-1) protein have been demonstrated in bovine oocytes and developing embryos using western blotting (Doyle et al., 2009).

Immunocytochemistry

Through the use of antibodies against proteins of interests, immunocytochemistry makes it possible to study proteins in the cell thereby providing information about their locations in the cells. By measuring the location of proteins, this technique can also be semiquantitative.

Methodology

Typically, the cells are attached onto a coverslip treated with polylysine to stabilize the cells. However, embryos can be studied without fixation, and tissues often need to be thin sections before the next steps. The cells are then permeabilized usually with detergents such as 0.1% Triton X-100 to allow the antibodies to reach their target proteins in various destinations inside the cell. The cells are then incubated with blocking buffer and then with an antibody or antibodies against proteins of interest followed by washes and incubation with secondary antibodies with fluorescent labels. Following the final washes, the cells or tissues are visualized under a fluorescent microscope. Depending on the channels available in the microscope, several proteins can be studied. Either epiflourescence or confocal microscope is needed in order to generate quantitative data on protein expression levels because this microscopy enables collection of protein expression in the whole cell such as embryo using a z-series approach. At the end, the signals can be pseudo colored and can be superimposed with data on light microscopy of the cell image and nuclear staining, supporting the "seeing is believing" theory.

Data analysis

Quantitative data on signal intensities of proteins in a cell can be generated using z-series of immunostained samples under confocal microscopy. Otherwise, the images of samples taken using fluorescent microscope alone provide qualitative assessment of signal intensities. Depending upon juxtaposition of the proteins, the image contains an intense background signals, masking the actual florescent signal coming from the protein of interest. Software provided with the microscope is needed to determine spatial locations and amounts of proteins in cells. In addition, temporal location of the protein can be determined using time lapse microscopy system.

Advantages and disadvantages

The main advantage of immunocytochemistry is that cellular locations as well as quantitative amounts of proteins can be assessed thereby enabling the study of relationships among protein locations, expression, and cell morphology. The drawbacks are that: (1) relatively small numbers of proteins can be studied at one time, (2) signal intensities of some proteins especially the low expressed ones can be challenging especially if the background signals are high, (3) antibodies are not available against all of the proteins in an organism and that antibodies might not be specific enough, and (4) quantitative proteomics and analyses of relationship among cellular locations, levels and cell shapes require expensive settings such as confocal microscopes and sophisticated software.

Applications

Because location of proteins in the cell is directly related to the functions of the proteins and cell physiology, immunocytochemistry has been an instrumental tool for both discovery and hypothesis driven research endeavors. This method has also been used to validate expression of proteins that are detected using molecular biology approaches such as western blotting and mass spectrometry. Associations between histone deacetylation and large-scale chromatin remodeling in mouse GV stage oocytes was demonstrated using immunocytochemistry (De La Fuente et al., 2004). For example, Wu et al. (2007) has elegantly demonstrated expression and functions of PAWP, a sperm-specific WW domain-binding protein, in sperm and zygotes using this method.

Question

You have determined 50 differentially expressed proteins in bovine MII oocytes versus blastocysts using 2D-DIGE. To your surprise, once you identified the differentially expressed protein spots using Masspectrometry, 20 out of the 50 differentially expressed protein spots turned out to be the same proteins, although they were of different molecular weights in the 2D-DIGE gels. Provide an explanation that involves molecular biology of proteins for this puzzle.

Systems biology

At the end of 1990s, functional genomics were defined as a field of molecular biology to analyze the functions and interactions of the biological samples such as DNA, RNA, and proteins. Last but not least, advanced use of genomics was led to another phenomenon in the 2000s, which is systems biology. Although systems biology is fundamentally based on biological sciences, it is a multi-disciplinary field in which bioinformatics and computational biology are mostly constructed as core elements (see Figure 23.5 for the workflow of system biology-based experiments). In this following section, we will summarize the basics of system biology and of how it is being incorporated in to studies on gametes biology.

Methodology

In addition to experimental data, in system biology, public sources such as NCBI can be used to obtain gene expression data including other research articles to interpret the results and to draw a conclusion from them. Cell physiology involves interactions of many macromolecules as well as small molecules within the cell. Thus, a systems biology approach provides a comprehensive understanding of molecular networks and regulation of functions in the cells and organisms. First, either experimental or computational data are obtained from the genomic, transcriptomic, and proteomic studies or the databases. Second, a suitable computational tool is used to compute the data followed data analysis using a statistical-based approach. Last, a network, an interactome and/or a pathway is drawn using the input, shedding light onto novel roles of these transcripts and proteins in mammalian development.

Applications

Several studies have used system biology accompanying tools in advanced computational biology in mouse (Cloonan et al., 2008) and bovine (Mamo et al., 2011). For instance, transcriptome of mouse embryonic stem cells was identified using RNA-seq technology combined with the bioinformatics tools, ending up with a related pathway using IPA (Cloonan et al., 2008). In a microarray study validated by qPCR, corresponding proteins of bovine

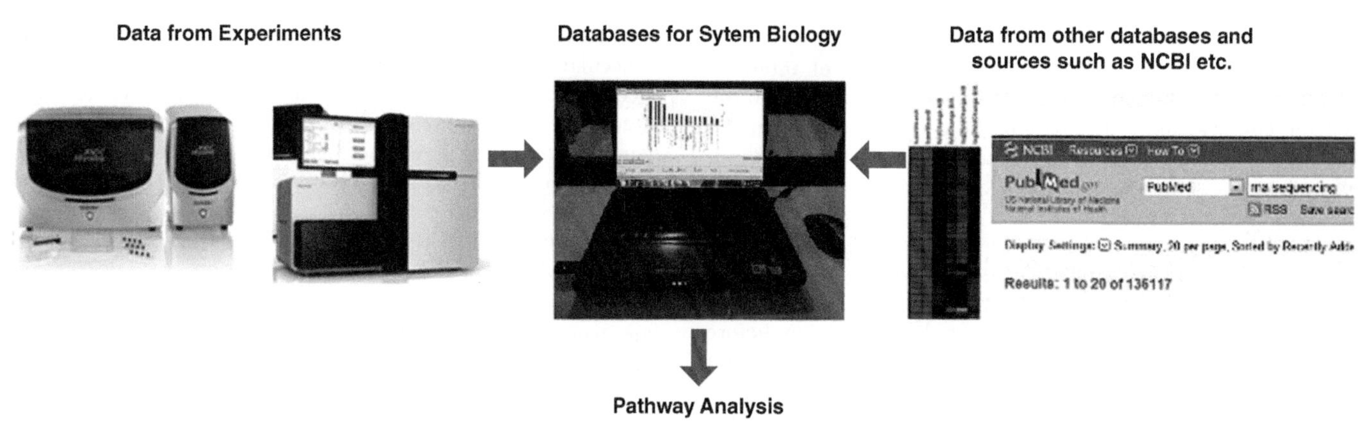

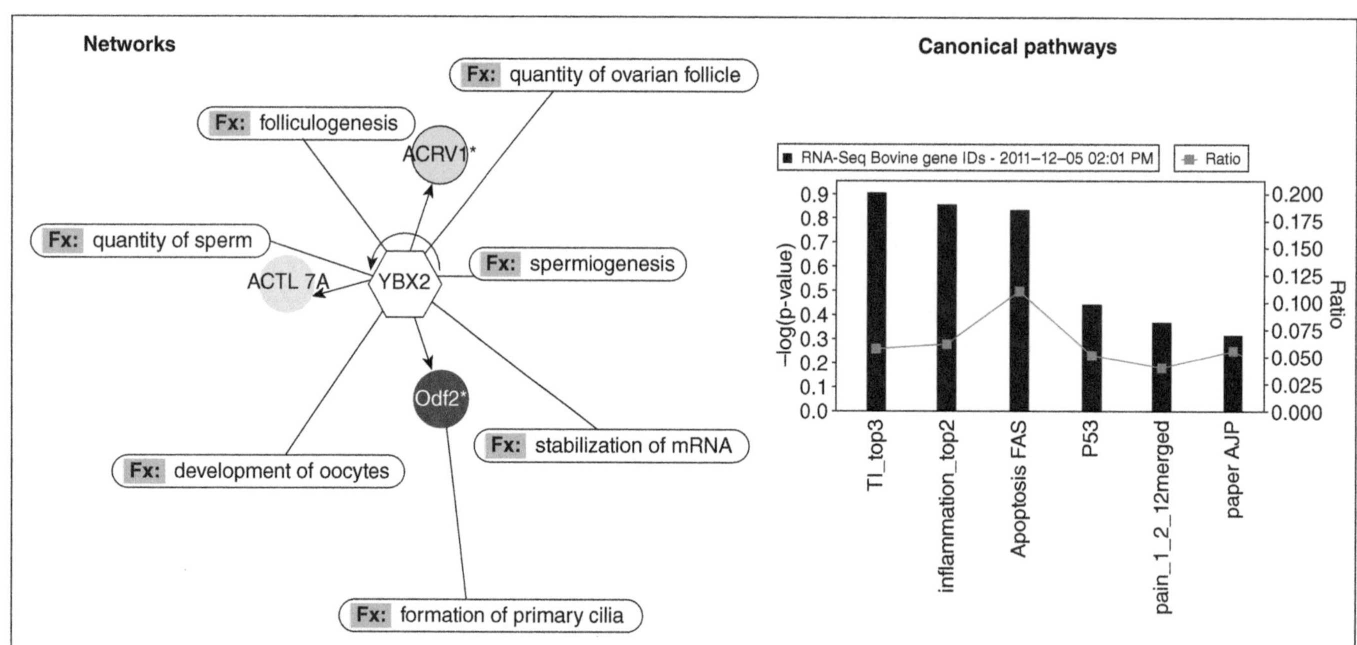

Figure 23.5 Systems Biology. A network and a canonical pathways chart are displayed here as two examples of how system biology works using Ingenuity pathway analysis software (IPA). Note that these two interactomes are obtained from two different data set followed by sperm proteomics and embryo RNA-seq. experiments. The network on the left side of the figure is generated from the sperm microarray data whereas the canonical pathways are created from bovine embryo RNA-seq data using IPA software (unpublished results). Up- and down-regulated genes are represented as red and green marks, respectively in the network revealing the interactions between the proteins of interest. On the right side, canonical pathways are represented as blue bar charts containing the genes of interest with the ratio of their log values (yellow plot). In addition to experimental data, public sources can also be used to obtain the gene annotations and their expression data etc. such as NCBI including other research articles. In the figure, "RNA sequencing" is searched at NCBI –PubMed to obtain the research articles containing the data previously published by others. Note that all RNA-seq data might not be obtained from NCBI if the researchers are not published them in the research articles. In that case, these data might be separately requested from the researchers if applicable (see the related equipment in the following links: www.illumina.com, www.affymetrix.com www.eppendorf.com, and www.ingenuity.com).

embryos were obtained to determine interactions among the data (Kepkova et al., 2011). Another recent study showed several differentially expressed gene products between Day 7 and Day 13 bovine embryos using bioinformatics followed by RNA-seq based transcriptomics followed by incorporation of the resultant information into pathways (Clemente et al., 2011).

Data analysis

Some of the commonly used computational tools for genomics, transcriptomics, and proteomics are as following:

- BLAST: Basic Local Alignment Search Tool (www.ncbi.nlm.nih.gov/tools/) multiple sequence alignment for transcripts and nucleotides.
- UniProt (www.uniprot.org) multiple sequence alignment proteomics.
- COBALT (www.ncbi.nlm.nih.gov/tools/cobalt) multiple sequence alignment for proteomics.
- DAVID (http://david.abcc.ncifcrf.gov/) Gene annotation, conversion, and clustering.
- AgBase (www.agbase.msstate.edu/) Gene annotation, conversion, and clustering.
- Ensembl/BioMart (www.ensembl.org) Gene annotation, conversion, and clustering.
- CDTree (www.ncbi.nlm.nih.gov/Structure/cdtree/cdtree.shtml) Classification of proteins sequences, 3D-structures and conserved domains tools.

- GO (www.geneontology.org/) Gene annotation, functions.
- String (http://string-db.org/) Interactomes for proteomics.
- PATIKA (www.patika.org) Pathway analysis and interactomes.
- KEGG (www.genome.jp/kegg/kegg2.html) Gene annotation, conversion, and clustering, pathway analysis and interactomes.
- REACTOME (www.reactome.org/ReaactomeGWT/entrypoint.html) Pathway analysis and interactomes.
- I2D; Interologous interaction database (http://ophid.utoronto.ca/ophidv2.201/) Interactomes.
- Ingenuity pathway analysis tool (www.ingenuity.com/) Pathway analysis and interactomes tools (licensed).
- MetaCore (www.genego.com) Pathway analysis and interactomes tools (licensed).
- dbSNP (www.ncbi.nlm.nih.gov/snp/) Genomics-based SNP.
- PID; Pathway Interaction Database (http://pid.nci.nih.gov/) Pathway analysis and interactomes – human only.
- Pathway Commons (www.pathwaycommons.org/pc/) (Cerami et al., 2011).
- Cyclone (http://nemo-cyclone.sourceforge.net/) Organism-specific pathway/genome databases (Le Fevre et al., 2007).
- SIGEN@E (www.sigenae.org/) Animal Species Expressed sequence tags (EST) Contigs Access through Ensembl and BioMart.

Conclusion

In conclusion, studying gene products such as transcripts and proteins has shed light onto the modern biology in many fronts including gene-environment interactions, molecular and cellular basis of disease, genesis of gametes and embryos, and so on. It is important to note that, while more advanced methods for the detection of transcripts and proteins are being developed, efforts are also underway for better determining how these molecules function. That is, a number of sophisticated methods involving experiments at the bench and computational biology are being developed for studying transcriptome and proteome. In this review, we have extensively summarized current functional genomics approaches being used in gamete biology and embryo development as well.

Further reading

Anders S, and Huber W. 2010. Differential expression analysis for sequence count data. *Genome Biology* 11:R106.

Auer PL, and Doerge RW. 2010. Statistical design and analysis of RNA sequencing data. *Genetics* 185:405–416.

Cloonan N, Forrest AR, Kolle G, Gardiner BB, Faulkner GJ, Brown MK, et al. 2008. Stem cell transcriptome profiling via massive-scale mRNA sequencing. *Nature Methods* 5:613–619.

Cumbie JS, Kimbrel JA, Di Y, Schafer DW, Wilhelm LJ, Fox SE, et al. 2011. GENE-counter: a computational pipeline for the analysis of RNA-Seq data for gene expression differences. *PloS One* 6:e25279.

De La Fuente R, Viveiros MM, Burns KH, Adashi EY, Matzuk MM, and Eppig JJ. 2004. Major chromatin remodeling in the germinal vesicle (GV) of mammalian oocyte is dispenable for global transcriptional silencing but required for centromeric heterochromatin function. *Developmental Biology* 275:447–458.

Driver AM, Penagaricano F, Huang W, Ahmad KR, Hackbart KS, Witlbank MC, Khatib H. 2012. RNA-Seq analysis uncovers transcriptomic variations between morphologically similar *in vivo*- and *in vitro*-derived bovine blastocysts. *BMC Genomics* 13:118.

Garber M, Grabherr MG, Guttman M, and Trapnell C. 2011. Computational methods for transcriptome annotation and quantification using RNA-seq. *Nature Methods* 8:469–477.

Goecks J, Nekrutenko A, Taylor J, and Galaxy T. 2010. Galaxy: a comprehensive approach for supporting accessible, reproducible, and transparent computational research in the life sciences. *Genome Biology* 11:R86.

Huang W, and Khatib H. 2010. Comparison of transcriptomic landscapes of bovine embryos using RNA-Seq. *BMC Genomics* 11:711.

Marioni JC, Mason CE, Mane SM, Stephens M, and Gilad Y. 2008. RNA-seq: an assessment of technical reproducibility and comparison with gene expression arrays. *Genome Research* 18:1509–1517.

Ozsolak F, Platt AR, Jones DR, Reifenberger JG, Sass LE, McInerney P, Thompson JF, Bowers J, Jarosz M, and Milos PM. 2009. Direct RNA sequencing. *Nature* 461:814–818.

Perry GH, Melsted P, Marioni JC, Wang Y, Bainer R, Pickrell JK, Michelini K, Zehr S, Yoder AD, Stephens M, Pritchard JK, and Gilad Y. 2012. Comparative RNA sequencing reveals substantial genetic variation in endangered primates. *Genome Research* 22(4):602–610.

Ross PJ and Chitwood JL. 2011. 139 Transcriptome analysis of single bovine embryos by RNA-Seq. *Reproduction, Fertility, and Development* 24:182.

Weber AP, Weber KL, Carr K, Wilkerson C, and Ohlrogge JB. 2007. Sampling the Arabidopsis transcriptome with massively parallel pyrosequencing. *Plant Physiology* 144:32–42.

Wright PC, Noirel J, Ow SY, and Fazeli A. 2012. A review of current proteomics technologies with a survey on their widespred use in reproductive biology investigations. *Theriogenology* 77:738–765.

References

Allison DB, Cui X, Page GP, and Sabripour M. 2006. Microarray data analysis: from disarray to consolidation and consensus. *Nat Rev Genet* 7:55–65.

Arnold GJ and Frohlich T. 2011. Dynamic proteome signatures in gametes, embryos and their maternal environment. *Reproduction, Fertility and Development.* 23:81–91.

Bar M, Wyman SK, Fritz BR, Qi J, Garg KS, Parkin RK, et al. 2008. MicroRNA discovery and profiling in human embryonic stem cells by deep sequencing of small RNA libraries. *Stem Cells* 26:2496–2505.

Bjork JK, Sandqvist A, Elsing AN, Kotaja N, and Sistonen L. 2010. miR-18, a member of Oncomir-1, targets heat shock transcription factor 2 in spermatogenesis. *Development* 137: 3177–3184.

Carvalho PC, Hewel J, Barbosa VC, Yates JR III. 2008. Identifying differences in protein expression levels by spectral counting and feature selection. *Genet Mol Res* 7:342–356.

Castro FO, Sharbati S, Rodriguez-Alvarez LL, Cox JF, Hultschig C, and Einspanier R. 2010. MicroRNA expression profiling of elongated cloned and *in vitro*-fertilized bovine embryos. *Theriogenology* 73:71–85.

Cerami EG, Gross BE, Demir E, Rodchenkov I, Babur O, Anwar N, et al. 2011. Pathway Commons, a web resource for biological pathway data. *Nucleic Acids Research* 39:D685–690.

Chankitisakul V, Tharasanit T, Phutikanit N, Tasripoo K, Nagai T, and Techakumphu M. 2012.

Lacking expression of paternally-expressed gene confirms the failure of syngamy after intracytoplasmic sperm injection in swamp buffalo (*Bubalus bubalis*). *Theriogenology* 77: 1415–1424.

Chen Y, Gelfond JA, McManus LM, and Shireman PK. 2009. Reproducibility of quantitative RT-PCR array in miRNA expression profiling and comparison with microarray analysis. *BMC Genomics* 10:407.

Choe L, D'Ascenzo M, Relkin NR, Pappin D, Ross P, Williamson B, et al. 2007. Eight-plex quantification of changes in serebrospinal fluid protein expression in subjects undergoing intraveneous immunoglobulin treatment for Alzheimer's disease. *Proteomics* 7:3651–3660.

Clemente M, Lopez-Vidriero I, O'Gaora P, Mehta JP, Forde N, Gutierrez-Adan A, et al. 2011. Transcriptome changes at the initiation of elongation in the bovine conceptus. *Biology of Reproduction* 85:285–295.

Costa V, Angelini C, De Feis I, and Ciccodicola A. 2010. Uncovering the complexity of transcriptomes with RNA-Seq. *J Biomed Biotechnol* 2010:853916.

Das PJ, Paria N, Gustafson-Seabury A, Vishnoi M, Chaki SP, Love CC, et al. 2010. Total RNA isolation from stallion sperm and testis biopsies. *Theriogenology* 74:1099–1106, 1106e1091–1092.

Doyle J, Ellis SA, O'Gorman GM, Donoso IMA, Lonergan P, and Fair T. 2009. Classical and non-classical Major Histocompatibility Complex class I gene expression in *in vitro* derived bovine embryos. *Journal of Reproductive Immunology.* 82(1):48–56.

Eckersall DP and Whitfield PD. 2011. *Methods in Animal Proteomics.* John Wiley & Sons, Inc., Hoboken.

Elbashir SM, Harborth J, Lendeckel W, Yalcin A, Weber K, and Tuschl T. 2001. Duplexes of 21-nucleotide RNAs mediate RNA interference in cultured mammalian cells. *Nature* 411:494–498.

Feugang JM, Rodriguez-Osorio N, Kaya A, Wang H, Page G, Ostermeier GC, Topper EK, Memili E. 2010. Transcriptome analysis of bull spermatozoa: implications for male fertility. *Reproductive Biomedicine Online* 21(3):312–324.

Fiedler SD, Carletti MZ, and Christenson LK. 2010. Quantitative RT-PCR methods for mature microRNA expression analysis. *Methods in Molecular Biology* 630:49–64.

Frohlich T and Arnold GJ. 2006. Proteome research based on modern liquid chromatography-tandem mass spectrometry: separation, identification and quantification. *J Neural Transm* 113:973–994.

Garrido N, Martinez-Conejero JA, Jauregui J, Horcajadas JA, Simon C, Remohi J, and Meseguer M. 2009. Microarray analysis in sperm from fertile and infertile men without basic sperm analysis abnormalities reveals a significantly different transcriptome. *Fertility and Sterility* 91:1307–1310.

Gilbert I, Bissonnette N, Boissonneault G, Vallee M, and Robert C. 2007. A molecular analysis of the population of mRNA in bovine spermatozoa. *Reproduction* 133:1073–1086.

Glazov EA, Kongsuwan K, Assavalapsakul W, Horwood PF, Mitter N, and Mahony TJ. 2009. Repertoire of bovine miRNA and miRNA-like small regulatory RNAs expressed upon viral infection. *PLoS One* 4(7): e6349.

Griffiths-Jones S, Saini HK, van Dongen S, and Enright AJ. 2008. miRBase: tools for microRNA genomics. *Nucleic Acids Res* 36:D154–158.

Govindaraju A, Uzun A, Robertson L, Atli MO, Kaya A, Topper E, et al. 2012 Dynamics of microRNAs in bull spermatozoa. *Reproductive Biology and Endocrinology* 10:82.

Gygi SP, Rist B, Gerber SA, Turecek F, Gelb MH, Aebersold R. 1999. Quantitative analysis of complex protein mixtures using isotope-coded affinity tags. *Nat Biotechnol.* 17(10):994–999.

Hayashi K, Chuva de Sousa Lopes SM, Kaneda M, Tang F, Hajkova P, Lao K, et al. 2008. MicroRNA biogenesis is required for mouse primordial germ cell development and spermatogenesis. *PLoS One* 3:e1738.

Hu SJ, Ren G, Liu JL, Zhao ZA, Yu YS, Su RW, et al. 2008. MicroRNA expression and regulation in mouse uterus during embryo implantation. *The Journal of Biological Chemistry* 283: 23473–23484.

Kaneda M, Tang F, O'Carroll D, Lao K, and Surani MA. 2009. Essential role for Argonaute2 protein in mouse oogenesis. *Epigenetics Chromatin* 2:9.

Kepkova KV, Vodicka P, Toralova T, Lopatarova M, Cech S, Dolezel R, et al. 2011. Transcriptomic analysis of *in vivo* and *in vitro* produced bovine embryos revealed a developmental change in cullin 1 expression during maternal-to-embryonictransition. *Theriogenology* 75:1582–1595.

Krawetz SA, Kruger A, Lalancette C, Tagett R, Anton E, Draghici S, and Diamond MP. 2011. A survey of small RNAs in human sperm. *Hum Reprod* 26:3401–3412.

Kues WA, Sudheer S, Herrmann D, Carnwath JW, Havlicek V, Besenfelder U, et al. 2008. Genome-wide expression profiling reveals distinct clusters of transcriptional regulation during bovine preimplantation development *in vivo*. *Proc Natl Acad Sci USA* 105:19768–19773.

Lambard S, Galeraud-Denis I, Martin G, Levy R, Chocat A, and Carreau S. 2004. Analysis and significance of mRNA in human ejaculated sperm from normozoospermic donors: relationship to sperm motility and capacitation. *Molecular Human Reproduction* 10:535–541.

Laurell C, Wirta V, Nilsson P, and Lundeberg J. 2007. Comparative analysis of a 3′ end tag PCR and a linear RNA amplification approach for microarray analysis. *Journal of Biotechnology* 127:638–646.

Le Fevre F, Smidtas S, and Schachter V. 2007. Cyclone: java-based querying and computing with Pathway/Genome databases. *Bioinformatics* 23:1299–1300.

Liang RQ, Li W, Li Y, Tan CY, Li JX, Jin YX, and Ruan KC. 2005. An oligonucleotide microarray for microRNA expression analysis based on labeling RNA with quantum dot and nanogold probe. *Nucleic Acids Res* 33:e17.

Liu CG, Calin GA, Volinia S, and Croce CM. 2008. MicroRNA expression profiling using microarrays. *Nat Protoc* 3:563–578.

Liu XS. 2007. Getting started in tiling microarray analysis. *PLoS Computational Biology* 3:1842–1844.

Liu WM, Pang RT, Chiu PC, Wong BP, Lao K, Lee KF, and Yeung WS. 2012. Sperm-borne microRNA-34c is required for the first cleavage division in mouse. *Proc Natl Acad Sci U S A* 109:490–494.

Livak KJ, and Schmittgen TD. 2001. Analysis of relative gene expression data using real-time quantitative PCR and the 2(-Delta Delta C(T)) Method. *Methods* 25:402–408.

Ma J, Flemr M, Stein P, Berninger P, Malik R, Zavolan M, Svoboda P, and Schultz RM. 2010. MicroRNA activity is suppressed in mouse oocytes. *Curr Biol* 20(3):265–270.

Maatouk DM, Loveland KL, McManus MT, Moore K, and Harfe BD. 2008. Dicer1 is required for differentiation of the mouse male germline. *Biol Reprod* 79:696–703.

Mamo S, Mehta JP, McGettigan P, Fair T, Spencer TE, Bazer FW, and Lonergan P. 2011. RNA sequencing reveals novel gene clusters in bovine conceptuses associated with maternal recognition of pregnancy and implantation. *Biology of Reproduction* 85:1143–1151.

Memili E, Dominko T and First NL. 1998. Onset of transcription in bovine oocytes, 2-, 4- and 8-cell embryos. *Molecular Reproduction and Development* 51:36–41.

Memili E, and First NL. 2000. Regulation of gene expression during early bovine embryogenesis: A review of timing and mechanism as compared to other species. *Zygote* 8:87–96.

Memili E, Dogan S, Rodriguez-Osorio N, Wang X, de Olivera RV, Mason MC, et al. 2012. Makings of the best spermatozoa: Molecular determinants of high fertility. In *Male Fertility*. Bashamboo A. (ed.), InTech, Open Access Publisher.

Mestdagh P, Van Vlierberghe P, De Weer A, Muth D, Westermann F, Speleman F, and Vandesompele J. 2009. A novel and universal method for microRNA RT-qPCR data normalization. *Genome Biol* 10:R64.

Miles JR, McDaneld TG, Wiedmann RT, Cushman RA, Echternkamp SE, Vallet JL, and Smith TP. 2012. MicroRNA expression profile in bovine cumulus-oocyte complexes: possible role of let-7 and miR-106a in the development of bovine oocytes. *Animal Reproduction Science* 130:16–26.

Misirlioglu M, Page GP, Sagirkaya H, Kaya A, Parrish JJ, First NL, and Memili E. 2006. Dynamics of global transcriptome in bovine matured oocytes and preimplantation embryos. *PNAS* 103:18905–18910.

Mondou E, Dufort I, Gohin M, Fournier E, and Sirard MA. 2012. Analysis of microRNAs and their precursors in bovine early embryonic development. *Molecular human reproduction* 18:425–434.

Montjean D, De La Grange P, Gentien D, Rapinat A, Belloc S, Cohen-Bacrie P, et al. 2012. Sperm transcriptome profiling in oligozoospermia. *Journal of Assisted Reproduction and Genetics* 29:3–10.

Morin RD, O'Connor MD, Griffith M, Kuchenbauer F, Delaney A, Prabhu A, et al. 2008. Application of massively parallel sequencing to microRNA profiling and discovery in human embryonic stem cells. *Genome Res* 18:610–621.

Nakamura N, Mori C, and Eddy EM. 2010. Molecular complex of three testis-specific izozymes associated with the mouse sperm fibrous sheath: Hexokinase 1, Phosphofructokinase M, and Gluthatione S-Transferase mu class 5. *Biology of Reproduction* 82:504–515.

Nelson PT, Baldwin DA, Scearce LM, Oberholtzer JC, Tobias JW, and Mourelatos Z. 2004. Microarray-based, high-throughput gene expression profiling of microRNAs. *Nat Methods* 1:155–161.

Ostermeier GC, Dix DJ, Miller D, Khatri P, and Krawetz SA. 2002. Spermatozoal RNA profiles of normal fertile men. *Lancet* 360:772–777.

Ostermeier GC, Goodrich RJ, Diamond MP, Dix DJ, and Krawetz SA. 2005. Toward using stable spermatozoal RNAs for prognostic assessment of male factor fertility. *Fertility and Sterility* 83:1687–1694.

Peddinti D, Nanduri B, Kaya A, Feugang JM, Burgess SC, and Memili E. 2008. Comprehensive proteomic analysis of bovine spermatozoa of varying fertility rates and identification of biomarkers associated with fertility. *BMC Syst Biol* 2:19.

Pfaffl MW, Horgan GW, and Dempfle L. 2002. Relative expression software tool (REST) for group-wise comparison and statistical analysis of relative expression results in real-time PCR. *Nucleic Acids Research* 30:e36.

Phillips BT, Gassei K, and Orwig KE. 2010. Spermatogonial stem cell regulation and spermatogenesis. *Philos Trans R Soc Lond B Biol Sci* 365:1663–1678.

Ro S, Park C, Sanders KM, McCarrey JR, and Yan W. 2007. Cloning and expression profiling of testis-expressed microRNAs. *Dev Biol* 311:592–602.

Robert C. 2008. Challenges of functional genomics applied to farm animal gametes and prehatching embryos. *Theriogenology* 70:1277–1287.

Rodriguez-Osorio N, Dogan S, and Memili E. 2012. *Epigenetics of Mammalian Gamete and Embryo Development, in Livestock Epigenetics* Vol. 1. Wiley-Blackwell, Oxford.

Rodriguez-Osorio N, Wang H, Rupinski J, Bridges SM and Memili E. 2010 Comparative functional genomics of mammalian DNA methyltransferases. *Reproductive Biomedicine Online* 20:243–255.

Rodriguez-Osorio N, Wang Z, Kasinathan P, Page GP, Robl JM,and Memili E. 2009. Transcriptional reprogramming of gene expression in bovine somatic cell chromatin transfer embryos. *BMC Genomics* 10:190.

Ross PL, Huang YN, Marchese JN, Williamson B, Parker K, Hattan S, et al. 2004. Multiplexed protein quantification in Saccharomyces cerevisiae using amine-reactive isobaric tagging reagents. *Mol Cell Proteomics* 3:1154–1169.

Schmittgen TD, Lee EJ, Jiang J, Sarkar A, Yang L, Elton TS, and Chen C. 2008. Real-time PCR quantification of precursor and mature micro-RNA. *Methods* 44:31–38.

Sethupathy P, Corda B, and Hatzigeorgiou AG. 2006. TarBase: A comprehensive database of experimentally supported animal microRNA targets. *RNA* 12:192–197.

Siomi M, Sato K, Pezic D, and Aravin A. 2011. PIWI-interacting small RNAs: the vanguard of genome defense. *Nature Reviews Molecular Cell Biology* 12:246–258.

Smith SL, Everts RE, Tian XC, Du F, Sung LY, Rodriguez-Zas SL, et al. 2005. Global gene expression profiles reveal significant nuclear reprogramming by the blastocyst stage after cloning. *Proc Natl Acad Sci* 102:17582–17587.

Stefani G, and Slack F. 2008. Small non-coding RNAs in animal development. *Molecular Cell Biology* 9:219–230.

Stein P, Svoboda P, and Schultz RM. 2003. Transgenic RNAi in mouse oocytes: a simple and fast approach to study gene function. *Dev Biol* 256:187–193.

Suh MR, Lee Y, Kim JY, Kim SK, Moon SH, Lee JY, et al., 2004. Human embryonic stem cells express a unique set of microRNAs. *Dev Biol* 270:488–498.

Suh N, Baehner L, Moltzahn F, Melton C, Shenoy A, Chen J, and Blelloch R. 2010. MicroRNA function is globally suppressed in mouse oocytes and early embryos. *Curr Biol* 20:271–277.

Susor A, Ellederova Z, Jelinkova L, Halada P, Kavan D, Kubelka M, Kovarova H. 2007. Proteomic analyses of porcine oocytes during *in vitro* maturation reveals essential role for the ubiquitine C-terminal hydrolase-L1. *Reproduction* 134:559–568.

Tam OH, Aravin AA, Stein P, Girard A, Murchison EP, Cheloufi S, et al. 2008. Pseudogene-derived small interfering RNAs regulate gene expression in mouse oocytes. *Nature* 453: 534–538.

Tang F, Kaneda M, O'Carroll D, Hajkova P, Barton SC, Sun YA, et al. 2007. Maternal microRNAs are essential for mouse zygotic development. *Genes & Development* 21:644–648.

Tejomurtula J, Lee KB, Tripurani SK, Smith GW, and Yao J. 2009. Role of importin alpha8, a new member of the importin alpha family of nuclear transport proteins, in early embryonic development in cattle. *Biol Reprod* 81:333–342.

Thelie A, Papillier P, Perreau C, Uzbekova S, Hennequet-Antier C, and Dalbies-Tran R. 2009. Regulation of bovine oocyte-specific transcripts during *in vitro* oocyte maturation and after maternal-embryonic transition analyzed using a transcriptomic approach. *Molecular Reproduction and Development* 76:773–782.

Tusher VG, Tibshirani R, and Chu G. 2001. Significance analysis of microarrays applied to the ionizing radiation response. *Proceedings of the National Academy of Sciences of the United States of America* 98:5116–5121.

Unlü M, Morgan ME, and Minden JS. 1997. Difference gel electrophoresis: a single gel method for detecting changes in protein extracts. *Electrophoresis* 18(11):2071–2077.

Uzun A, Rodriguez-Osorio N, Kaya A, Wang H, Parrish JJ, Ilyin VA, and Memili E. 2009. Functional genomics of HMGN3a and SMARCAL1 in early mammalian embryogenesis. *BMC Genomics* 10:183.

Viswanathan SR, Mermel CH, Lu J, Lu CW, Golub TR, and Daley GQ. 2009. microRNA expression during trophectoderm specification. *PloS One* 4:e6143.

Walsh CT. 2006. *Posttranslational Modifications of Proteins. Expanding Nature's Inventory.* ISBN 0-9747077-3-2. Roberts and Company Publishers.

Watanabe T, Totoki Y, Toyoda A, Kaneda M, Kuramochi-Miyagawa S, Obata Y, et al. 2008. Endogenous siRNAs from naturally formed dsRNAs regulate transcripts in mouse oocytes. *Nature* 453:539–543.

Wu AT, Sutovsky P, Manandhar G, Xu W, Katayama M, Day BN, et al. 2007. PAWP, a sperm-specific WW domain-binding protein, promotes meiotic resumption and pronuclear development during fertilization. *J Biol Chem* 282:12164–12175.

Yan N, Lu Y, Sun H, Qiu W, Tao D, Liu Y, et al. 2009. Microarray profiling of microRNAs expressed in testis tissues of developing primates. *J Assist Reprod Genet* 26:179–186.

Yang CC, Lin YS, Hsu CC, Wu SC, Lin EC, and Cheng WT. 2009. Identification and sequencing of remnant messenger RNAs found in domestic swine (Sus scrofa) fresh ejaculated spermatozoa. *Animal Reproduction Science* 113:143–155.

Yang CC, Lin YS, Hsu CC, Tsai MH, Wu SC, and Cheng WT. 2010a. Seasonal effect on sperm messenger RNA profile of domestic swine (Sus Scrofa). *Animal Reproduction Science* 119: 76–84.

Yang JH, Shao P, Zhou H, Chen YQ, and Qu LH. 2010b. deepBase: a database for deeply annotating and mining deep sequencing data. *Nucleic Acids Res* 38: D123–130.

Yang N, Kaur S, Volinia S, Greshock J, Lassus H, Hasegawa K, et al. 2008. MicroRNA microarray identifies Let-7i as a novel biomarker and therapeutic target in human epithelial ovarian cancer. *Cancer Res* 68:10307–10314.

Yamamoto M, Wakatsuki T, Hada A, and Ryo, A. 2001. Use of serial analysis of gene expression (SAGE) technology. *Journal of Immunological Methods* 250:45–66.

Yuan JS, Reed A, Chen F, and Stewart CN, Jr. 2006. Statistical analysis of real-time PCR data. *BMC Bioinformatics* 7:85.

24 The Genetics of *In Vitro* Produced Embryos

Ashley Driver
Division of Human Genetics, Cincinnati Children's Hospital, OH, USA

In vitro production: from livestock to humans

In vitro embryo production (**IVP**) is the process that involves manipulation of both oocyte and sperm to create embryos within a laboratory environment. During the *in vitro* process, gametes are fused and the subsequent embryos are cultured and transferred back into a recipient. Depending on the species, the purpose for utilizing IVP can vary from infertility (humans) to genetic selection and improvement (livestock). In terms of livestock species such as cattle and pigs, IVP allows for the selection of oocytes and sperm with proven genetic merit and as such creates progeny with desired genes and subsequent improved production traits.

Use of IVP is also a valuable tool in research allowing for isolation of the embryo for analysis of its internal genetic framework. During the pre-implantation period of embryo development, numerous genetic hurdles must be passed before implantation and subsequent development can occur. As this is the same time period when embryos are manipulated during IVP, researchers have been able to isolate and determine key genetic components underlying these developmental milestones.

One of the significant events occurring post-fertilization is cellular division. During this time individual cells of the embryo, known as blastomeres, undergo cytokinesis followed by asynchronous cellular divisions (Fujimori et al., 2009). Around the eight-cell stage mammalian embryos face a major genetic hurdle known as **Embryonic Genome Activation** (**EGA**). At this stage in development, the initial maternal transcripts driving early embryonic development are degraded while embryo specific transcripts are activated (Figure 24.1).

Once the EGA is successfully completed, the embryo will continue with cellular divisions until it reaches a transition point. It's at this time when the individual cells in the embryo compact and the cells gain apical/basal polarity. Prior to this, all cells in the embryo were **totipotent**; that is, with the potential to transform into any cell type in the body including placental cells. However, this transition period will allow some cells to become **pluripotent**, with a more specific fate. These cells will only be able to transition into one of the three major germ cell layers (ectoderm, endoderm, and mesoderm) and do not form placental tissues. Potency of the cells in the embryo is critical to mammalian development to help distinguish two major embryonic cell types: the inner cell mass (which is the precursor to the fetus) and the fluid containing trophectoderm (precursor to placental cells). Once these two cell populations are created, the embryo is deemed a blastocyst, and is preparing to hatch and subsequently implant into the uterus.

Genetically, numerous markers have been established as key factors for this transition to blastocyst stage. For example, a study in mouse embryos showed stage specific expression of the genes *Nanog*, *Klf1*, and *Myc* (de Vries et al., 2008). All three genes appeared to be activated after the eight-cell stage in mice with the last two only being present in morula and blastocyst stage embryos. This suggests potential roles specific to this transition period late in pre-implantation embryo development. By understanding genes involved during this critical event, researchers can then begin to understand how genetic abnormalities may cause failure in development.

In addition to the transcriptomic dynamics of the embryo, there are critical changes occurring at the DNA level. In placental mammals, chemical groups can bind to the DNA structure to alter gene function. This is known as **epigenetics** (*epi* translates to "outer" in Greek). In most cells the expression of a gene is by both alleles known as biallelic expression. However, by placement of a chemical group onto an allele one copy of a gene may be silenced causing expression of only one of two parental alleles; termed **monoallelic expression**. Thus, expression of this gene is reduced. When this phenomena occurs due to methyl groups binding to the DNA, it is called genomic **imprinting** (see Chapter 21). In the embryo, this is of specific interest as there are numerous changes to the methyl patterns along the genome from the one cell stage to the blastocyst stage (**Figure 24.2**). However, as these changes are occurring, specific genes that receive monoallelic parent-specific methylation marks are protected and the methylation pattern is maintained. Should methylation patterns be perturbed during this time, subsequent changes in gene expression and function could result.

Molecular and Quantitative Animal Genetics, First Edition. Edited by Hasan Khatib.

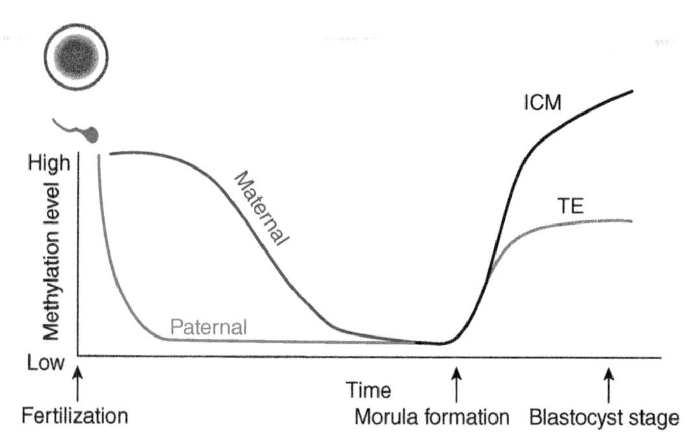

Figure 24.1 Embryonic genome activation. During mammalian preimplantation development maternal RNA transcripts (upper line) drive initial development. By the 8–16 cell stage these transcripts are degraded and embryo specific transcripts (lower curve) are activated.

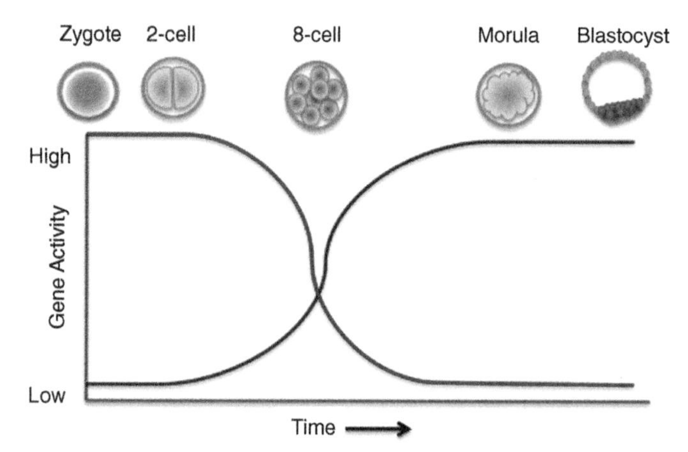

Figure 24.2 Patterns of methylation during bovine pre-implantation embryo development. When oocyte and sperm fuse at fertilization there is a rapid de-methylation of the parental genomes as the cells of the embryo divides. When the cells undergo compaction there is a transition to differentiation. It's at this time that the two differing cell types in the developing blastocyst undergo differential levels of remethylation. After the blastocyst stage is obtained, the embryo can implant into the uterus. ICM = inner cell mass and TE = trophectoderm of the blastocyst embryo.

Table 24.1 Large-scale expression studies profiling mammalian pre-implantation embryo development.

Year	Authors	Species
2000	Ko et al.	Mouse
2003	Lonergan et al.	Bovine
2004	Hamatani et al.	Mouse
2004	Tanaka et al.	Mouse
2005	Wang et al.	Mouse
2005	Whitworth et al.	Porcine
2006	Jeong et al.	Mouse
2006	Hamatani et al.	Mouse
2006	Mamo et al.	Bovine
2006	Mişirlioglu et al.	Bovine
2008	Kues et al.	Bovine
2009	Smith et al.	Mouse
2009	Zhang et al.	Human
2010	Bauer et al.	Porcine
2011	Beilby et al.	Ovine
2012	Dorji et al.	Bovine

Unlocking developmentally important genes in the pre-implantation embryo

Although IVP is useful in research, it has numerous clinical concerns including reduced embryo development and low pregnancy rates. The most recent human IVP data in the United States from the Center of Disease Control (CDC) shows the highest live birth rate (from IVP) is 41.5% for women less than 35 years of age (www.cdc.gov/art). In cattle, outcomes are similar with only approximately 45% of IVP embryos resulting in pregnancy (Farin et al., 2006). Being that the IVP process is time consuming and costly, it is important to understand why these success rates appear so low. One way to study this is through translation of the genetic circuitry driving the individual embryo's development. Due to ethical constraints, research in human embryos is limited and as such animal models have become increasingly popular. Interestingly, recent findings suggest that during the pre-implantation stage, cattle have the highest genetic similarity to human embryos and therefore serve as a strong animal model (Berg et al., 2011). Thus, cattle embryo studies to determine developmentally important genes have gained increasing attention.

Numerous studies have presented the stage specific expression of genes in the bovine pre-implantation embryo as well as in other species allowing for a comparative approach for the study of development (Table 24.1).

However, there is difficulty in understanding what genes are essential for development to occur and which genes may be less critical. To detect the level of importance of genes, studies have been directed towards comparing developed embryos with those showing abnormal or arrested characteristics. Within these studies, comparisons can be made at the genomic and transcriptomic level to translate potential errors in the genetic circuitry driving development. For example, an *in vitro* fertilization system has been utilized for the discovery of genes and pathways involved in the transition from morula to blastocyst stage.

One example of this system is that used by the lab of Khatib et al. which compares differences between bovine embryos developed through the pre-implantation period (to a day 8 blastocyst) and embryos that arrest between days 5 and 8 of development (degenerates). In order to accomplish this, putative zygotes are cultured for 5 days in culture. At this point in time, the bovine embryo should have 16–32 cells and show evidence of coalescence (cells loose individual boundaries and form a tight ball) deeming it a morula (Figure 24.3a). Embryos with these physical qualities are selected and further cultured to day 8 post-fertilization. By this time the embryo should have developed a fluid filled cavity, deeming it a "blastocyst" (Figure 24.3b). Those embryos that formed into a morula by day 5 of culture, but fail to form a fluid cavity were deemed "degenerates" and are shown in Figure 24.3(c).

To establish genes that have potential roles in the transition from morula to blastocyst, degenerates (Figure 24.3c) were com-

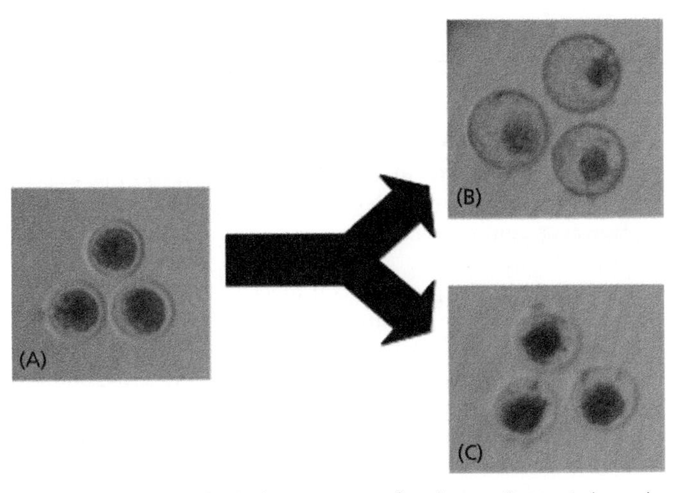

Figure 24.3 Morphological assessment of embryos. Compacted morula (A) that were cultured until day 8 of development and either showed signs of blastocoele formation (B) or degeneration (C).

pared to blastocysts (Figure 24.3b). Using this system, a total of 67 genes were differentially expressed between the two populations (Huang et al., 2010). A later study by Zhang et al. (2011) reported 17 heat shock protein genes showing differential expression ranging from 1.5- to 7.6-fold. Studies such as these provide key insight to genes that could be further assayed and potentially used as developmental markers.

IVP: potential source of genetic alteration?

Although IVP serves as a valuable source for isolating internal genetic mechanisms in the embryo, there is also controversy that the process itself may present alterations in genetic expression. Numerous studies in mouse and cattle embryos have presented evidence of transcriptomic differences between IVP and *in vivo* (within animal) production. A recent study by Driver et al. (2012) performed an in-depth comparison of bovine blastocyst embryos produced by IVP and *in vivo* techniques using RNA-sequencing (see Chapter 17). Over 17,000 genes were detected, with 793 genes showing differential expression between IVP and *in vivo* techniques (FDR < 0.05). In addition, 4800 genes showed evidence of alternative splicing (**AS**). The phenomena of AS occurs in eukaryotes, and is a post transcriptional process whereby introns and certain exons are removed and the remaining exons are assembled into a final RNA transcript. Removing and assembling certain exons together can generate various splice variants. If a single gene has numerous splice variants produced, it can have numerous different functions as the mRNA may code for different proteins. Interestingly, the aforementioned study found 873 genes displaying differential alternative splicing between the two pools suggesting that embryos were processing RNA differently due to the differing production environment (Driver et al., 2012). Evidence for alternative splicing in bovine embryos was also shown by Zhang et al. (2011) where multiple splice variants were found for four heat shock protein genes. These genes are critical in the stress response due to environmental stressors and as such may be indicators of adaptation of the embryo to different growth and culture conditions. Therefore, environmental influence on splice variants may give rise to the differing developmental capacity of embryos.

While prior studies have utilized the whole pre-implantation embryo, there is a continuing effort to understand potential impacts on the two major cell types of the blastocyst: the inner cell mass (ICM) and trophoblast cells. The rationale behind this study design is that a better understanding could be obtained as to how certain manipulations may affect the specific cellular types of the embryo such as the placental tissue (derived from trophoblast cells) or embryo proper (derived from inner cell mass). In a study of mouse embryos, the ICM and trophoblast cell populations were isolated in embryos produced by IVP or *in vivo* production to make comparisons at the transcriptomic level. Trophoblast cell populations from IVP embryos had 108 differentially expressed genes in comparison to trophoblast from *in vivo* embryos. When metabolic pathways were analyzed based on this data, researchers found significant reductions in expression of genes involved in placentation, which may give insight to the seemingly low pregnancy rates in IVP embryos. Similarly, analysis was carried out on the isolated ICM, which is gives rise to the fetus. Comparison of gene expression between *in vivo* and IVP ICM cells detected 310 differentially expressed transcripts. These differences in gene expression can translate into dramatic differences regarding which biological pathways are actively driving the development of the embryo.

In addition to alterations in RNA processing and structure, scientists have also reported changes at the epigenetic level between *in vivo* and *in vitro* embryo. A study done on porcine embryos showed decreasing methylation levels in IVP embryos from the eight-cell to blastocyst stages (Deshmukh et al., 2011). In contrast, methylation levels would be expected to increase, as shown in Figure 24.2. Thus, this suggests that the IVP embryos may not be recovering proper methylation levels during growth from the eight-cell stage onward. In cattle, differential methylation between *in vivo* and *in vitro* derived embryonic liver tissue was found, suggesting aberration due to the IVP process (Hiendleder et al., 2006). In terms of development, abnormal methylation patterns have been linked to overgrowth disorders in livestock such as large offspring syndrome. Similarly in humans, controversial evidence has suggested potential "imprinting syndromes" in children produced by IVP methods. In humans, Beckwith–Weidemann Syndrome (**BWS**) has been associated with abnormalities in methylation of the *KCNQ1OT* gene (Gicquel et al., 2003). However, further studies need to be pursued to determine how IVP may influence the methylation level. Further discussion of epigenetics including a case of couple with a BWS child is presented in **Supplementary Videos 1** and **2**.

PGD: genetic screening and human embryos

The integration of genetics and embryology has evolved from a research to clinical setting with the potential to determine an embryo's predisposition to specific disorders. **Pre-implantation genetic diagnosis** (**PGD**) is the process of screening embryos for chromosomal abnormalities and genetic mutations. If individuals have a known history of specific genetic disorders, they can have embryos produced *in vitro* and screened with PGD. There are two main techniques used for PGD: **Fluorescent *in-situ* hybridization** (**FISH**) and **polymerase chain reaction** (**PCR**). Three days post-fertilization human embryos are biopsied and a single cell is screened. DNA target-specific probes are used to determine potential chromosomal imbalance due to translocation or could be used for the selection of female embryos (if a known X-linked disorder is present). In addition to FISH, the use of PCR allows for screening of monogenic diseases with known mutations. For example, PCR has shown to be effective in screening embryos for Gaucher disease type I (Altarescu et al., 2011). Individuals with this disorder have mutations in the gene *GBA*, which codes for an enzyme that breaks down a fatty substance (glucocerebroside) into a sugar (glucose) and a fat molecule. Individuals with Gaucher disease may have a build-up of glucocerebroside in their liver, bone marrow, spleen, lungs, and, in rare cases, the brain. In the most severe cases, affected individuals may incur medical problems (including seizures and brain damage) at infancy and fail to live past the age of two. By way of biopsy and PCR, selection of embryos allowed for a healthy child unaffected by Gaucher disease type I to be born to a couple where the female was homozygous for a disease related mutation and the male was a carrier. Thus, the technology provides the opportunity for couples with infertility or pre-existing medical conditions to have the assurance that their child may not have to face a given disease.

Screening the embryo: to infinity and beyond?

Although the use of PGD can be beneficial in selecting embryos predisposed to detrimental disease, it also serves as a point of controversy. Since some disorders are gender specific (i.e., X-linked disorders), use of PGD could result in gender selection. In addition, development of embryos subject to cell biopsy is reduced as one study reported a 21.7% pregnancy rate across 57 participating clinics in Europe (Harper et al., 2010). The purpose of PGD can range beyond selection for the embryo's direct benefit. There are documented cases of "savior siblings," where embryos are selected *in vitro* to serve as a donor to a living sibling in need. In these cases, embryos are created through IVP and screened for human leukocyte antigen (**HLA**), which is used to determine potential donors for transplants. If a family has a disease-affected child, PGD provides the opportunity to create a donor to help rescue the affected child. There are reports of PGD being used for HLA typing as one couple utilized the technology to create a child that was able to be a hematopoietic stem cell transplant to a sibling.

In addition to these controversies, the use of genetic markers for screening embryos for non-disease related phenotypes has gained attention. With the expanse of information on the human genome, markers for physical traits have been discovered. For example, there are numerous markers related to eye color that have been established. In fact, genetic models are now being created to use a combination of single nucleotide polymorphisms (SNPs) to predict both hair and eye color. A study presented in 2011 reported the use of six SNPs to predict eye color in individuals and seven SNPs to predict skin pigmentation (Spichenok et al., 2011). Another study showed that selection could be based on genetic predictors such as SNPs in known genes with roles in human hair pigmentation. Using only 13 SNPs from 11 genes with known roles in hair pigmentation they could predict red, black, and blonde hair with 90% accuracy (Branicki et al., 2011). Brown hair color could be predicted with 80% accuracy (Branicki et al., 2011). Novel predictive tools such as these may potentially become more available to integrate into techniques such as PGD. "**Designer babies**" has become the term coined for embryos that would be selected based on physical traits, and has continued to grow in controversy. Numerous questions have been raised regarding the ethics of this process, including how far the selection process could go. In 2006, a clinic in California offered services such as eye color selection (**see Supplementary Video 3**). However, controversial responses caused a halt in services one week after starting.

As a result of the growing controversy, PGD is restricted in numerous countries. In Europe, the Human Fertilization and Embryology Act (HFEA) was passed to create oversight of IVP clinics and those performing PGD (Knoppers and Isasi, 2004). In the United States, although there is not a formal oversight committee, efforts are being made to create more clear guidelines for PGD use (Hudson, 2006). Nonetheless, certain countries have still outlawed the use of PGD including Germany, Switzerland, and Austria (Knoppers and Isasi, 2004).

Embryogenetics: what's next?

Overall, genetics has become a valuable tool in understanding the biological framework of early embryonic development. Efforts are continually made to understand how our manipulations of the embryo may have downstream developmental consequences. Additionally, the development of more high-throughput genetic techniques will allow for more streamline fine-scale screenings of embryos *in vitro*. There is a continual need for further research in this area to better understand how genetics drives the success of the early mammalian embryo.

Summary

Overall, *in vitro* production of mammalian embryos has provided a unique opportunity to study and benefit multiple species. Continued research will allow for further translation of the genetic machinery driving development and provide avenues to improve on current embryo manipulation techniques. In addition, the growing use of genetic techniques will provide an ever-growing topic of controversy regarding the limits to which we should apply these methods to embryology. Nonetheless, genetics has provided numerous opportunities to better interpret and understand the embryo and will continue to do so in the future.

Key terms

Beckwith–Weidemann Syndrome (BWS): Congenital birth defect that results in symptoms such as overgrowth (tongue, body and organ size) and childhood cancer.

"Designer Babies": Slang given for embryos that are selected for non-disease related phenotypes such as eye or hair color.

Embryonic Genome Activation (EGA): The event where maternal RNA transcripts are degraded and embryo specific transcripts are activated.

Epigenetics: The study of heritable characteristics altering gene expression that do not alter the DNA sequence. In Greek, *epi-* means outer.

Fluorescent *In-situ* Hybridization (FISH): A technique that detects target-specific DNA sequences on chromosomes.

Imprinting: Parent-specific monoallelic expression of genes.

***In vitro* embryo production (IVP):** Manipulation of gametes to create and mature embryos in a laboratory setting.

Pluripotent: The potential to transform into one of the three major germ cell layers (ectoderm, endoderm, and mesoderm). Cells with this potency cannot transform into placental cells.

Polymerase Chain Reaction (PCR): A procedure that amplifies a few copies of DNA using sequence specific primers.

Pre-implantation Genetic Diagnosis: A procedure done on embryos when parents have a pre-existing condition to screen for unaffected offspring through biopsy of one cell (blastomere).

Totipotent: With the potential to transform into any cell type in the body including cells of the placenta.

References

Altarescu, G., P. Renbaum, T. Eldar-Geva, I. Varshower, B. Brooks, R. Beeri, et al. 2011. Preimplantation genetic diagnosis (PGD) for a treatable disorder: Gaucher disease type 1 as a model. *Blood Cells, Mol Dis* 46(1):15–18.

Bauer, B. K., S. C. Isom, L. D. Spate, K. M. Whitworth, W. G. Spollen, S. M. Blake, et al., 2010. Transcriptional profiling by deep sequencing identifies differences in mRNA transcript abundance in *in vivo*-derived versus *in vitro*-cultured porcine blastocyst stage embryos. *Biol Reprod* 83(5):791–798.

Beilby, K. H., S. P. de Graaf, G. Evans, W. M. Maxwell, S. Wilkening, C. Wrenzycki, and C. G. Grupen. 2011. Quantitative mRNA expression in ovine blastocysts produced from X- and Y-chromosome bearing sperm, both *in vitro* and *in vivo*. *Theriogenology* 76(3):471–481.

Berg, D. K., C. S. Smith, D. J. Pearton, D. N. Wells, R. Broadhurst, M. Donnison, and P. L. Pfeffer. 2011. Trophectoderm lineage determination in cattle. *Dev Cell* 20(2):244–255.

Branicki, W., F. Liu, K. van Duijn, J. Draus-Barini, E. Pospiech, S. Walsh, et al. 2011. Model-based prediction of human hair color using DNA variants. *Human Gen* 129(4):443–454.

de Vries, W. N., A. V. Evsikov, L. J. Brogan, C. P. Anderson, J. H. Graber, B. B. Knowles, and D. Solter. 2008. Reprogramming and differentiation in mammals: motifs and mechanisms. *Cold Spring Harbor Symposia on Quantitative Biology* 73:33–38.

Deshmukh, R. S., O. Ostrup, E. Ostrup, M. Vejlsted, H. Niemann, A. Lucas-Hahn, B. et al. 2011. DNA methylation in porcine preimplantation embryos developed *in vivo* and produced by *in vitro* fertilization, parthenogenetic activation and somatic cell nuclear transfer. *Epigenetics: Official Journal of the DNA Methylation Society* 6(2):177–187.

Dorji, O. Y, K. Miyoshi, and M. Yoshida. 2012. Gene expression profile differences in embryos derived from prepubertal and adult Japanese Black cattle during *in vitro* development. *Reprod Fertil Dev* 24:370–381.

Driver, A. M., F. Penagaricano, W. Huang, K. R. Ahmad, K. S. Hackbart, M. C. Wiltbank, and H. Khatib. 2012. RNA-Seq analysis uncovers transcriptomic variations between morphologically similar *in vivo*- and *in vitro*-derived bovine blastocysts. *BMC Genom* 13:118.

Farin, P. W., J. A. Piedrahita, and C. E. Farin. 2006. Errors in development of fetuses and placentas from *in vitro*-produced bovine embryos. *Theriogenology* 65(1):178–191.

Fujimori, T., Y. Kurotaki, K. Komatsu, and Y. Nabeshima. 2009. Morphological organization of the mouse preimplantation embryo. *Reprod Sci* 16(2):171–177.

Gicquel, C., V. Gaston, J. Mandelbaum, J. P. Siffroi, A. Flahault, and Y. Le Bouc. 2003. *In vitro* fertilization may increase the risk of Beckwith-Wiedemann syndrome related to the abnormal imprinting of the KCN1OT gene. *Am J Hum Gen* 72(5):1338–1341.

Hamatani, T., M. Carter, A. Sharov, and M. S. H. Ko. 2004. Dynamics of global gene expression during mouse preimplantation development. *Dev Cell* 6:117–131.

Hamatani, T., M. S. H. Ko, M. Yamada, N. Kuji, Y. Mizusawa, M. Shoji, et al. 2006. Global gene expression profiling of preimplantation embryos. *Hum Cell* 19:98–117.

Harper, J. C., E. Coonen, M. De Rycke, G. Harton, C. Moutou, T. Pehlivan, et al. 2010. ESHRE PGD Consortium data collection X: cycles from January to December 2007 with pregnancy follow-up to October 2008. *Hum Reprod* 25(11):2685–2707.

Hiendleder, S., M. Wirtz, C. Mund, M. Klempt, H. D. Reichenbach, M. Stojkovic, M. et al. 2006. Tissue-specific effects of *in vitro* fertilization procedures on genomic cytosine methylation levels in overgrown and normal sized bovine fetuses. *Biol Reprod* 75(1):17–23.

Huang, W., B. S. Yandell, and H. Khatib. 2010. Transcriptomic profiling of bovine IVF embryos revealed candidate genes and pathways involved in early embryonic development. *BMC Genom* 11:23.

Hudson, K. L. 2006. Preimplantation genetic diagnosis: public policy and public attitudes. *Fertility Steril* 85(6):1638–1645.

Jeong, H. J., H. J. Kim, S. H. Lee, K. Kwack, S. Y. Ahn, Y. J. Choi, et al. 2006. Gene expression profiling of the pre-implantation mouse embryo by microarray analysis: comparison of the two-cell stage and two-cell block. *Theriogenology* 66:785–766.

Knoppers, B. M. and R. M. Isasi. 2004. Regulatory approaches to reproductive genetic testing. *Hum Reprod* 19(12):2695–2701.

Ko, M. S.H., J. R. Kitchen, X. Wang, T. A. Threat, X. Wang, A. Hasegawa, et al. 2000. Large-scale cDNA analysis reveals phased gene expression patterns during preimplantation mouse development. *Development* 127:1737–1749.

Kues, W. A., S. Sudheer, D. Herrmann, J. W. Carnwath, V. Havlicek, U. Besenfelder, et al. 2008. Genome-wide expression profiling reveals distinct clusters of transcriptional regulation during bovine preimplantation development *in vivo*. *Proc Natl Acad Sci* 105: 19768–19773.

Lonergan, P., D. Rizos, A. Gutiérrez-Adán, P. M. Moreira, B. Pintado, J. de la Fuente, and M. P. Boland. 2003. Temporal divergence in the patterns of messenger RNA expression in bovine embryos cultured from the zygote to blastocyst stage *in vitro* or *in vivo*. *Biol Reprod* 69:1424–1431.

Mamo, S., C. A. Sargent, N. A. Affara, D. Tesfaye, N. El-Halawany, K. Wimmers, et al. 2006. Transcript profiles of some developmentally important genes detected in bovine oocytes and *in vitro*-produed blastocysts using RNA amplification and cDNA microarrays. *Reprod Domest Anim* 41:527–534.

Misirlioglu, M., G. P. Page, H. Sagirkaya, A. Kaya, J. J. Parrish, N. L. First, and E. Memili. 2006. Dynamics of global transcriptome in bovine matured oocytes and preimplantation embryos. *Proc Natl Acad Sci U S A* 103(50): 18905–18910.

Smith, S.L., R. E. Everts, L. Y. Sung, F. Du, R. L. Page, B. Henderson, et al. 2009. Gene expression profiling of single bovine embryos uncovers significant effects of *in vitro* maturation, fertilization, and culture. *Mol Reprod Dev* 76: 38–47.

Spichenok, O., Z. M. Budimlija, A. A. Mitchell, A. Jenny, L. Kovacevic, D. Marjanovic, et al. 2011. Prediction of eye and skin color in diverse populations using seven SNPs. Forensic science international. *Genetics* 5(5):472–478.

Tanaka, T. S. and M. S. Ko. 2004. A global view of gene expression in the preimplantation embryo: morula versus blastocyst. *Eur J Obstet Gynecol Reprod Biol* 115(Suppl 1):S85–91.

Wang, S., C. A. Cowan, H. Chipperfield, and R. D. Powers. 2005. Gene expression in the preimplantation embryo: *in-vitro* developmental changes. *Reprod Biomed Online* 10:607–616.
Whitworth, K. M., C. Agca, J. G. Kim, R. V. Patel, G. K. Springer, N. J. Bivens, et al. 2005. Transcriptional profiling of pig embryogenesis by using a 15-K member unigene set specific for pig reproductive tissues and embryos. *Biol Reprod* 72:1437–1451.
Zhang, B., F. Penagaricano, A. Driver, H. Chen, and H. Khatib. 2011. Differential expression of heat shock protein genes and their splice variants in bovine preimplantation embryos. *J Dairy Sci* 94(8):4174–4182.
Zhang, P., M. Zucchelli, S. Bruce, F. Hambiliki, A. Stavreus-Evers, L. Levkov, et al. 2009. Transcriptome profiling of human preimplantation development. *PLoS One* 4:e7844.

Review questions

1. What is the main purpose for the use of IVP in livestock species? Could you create an example of what traits may be targeted? How does this differ from humans?

2. What genetic hurdles does the embryo face during preimplantation development?

3. Draw the methylation profile of gametes and the embryo from fertilization to blastocyst stage. What is a possible consequence of abnormal methylation during this time period?

4. What is alternative splicing? Why would an embryo utilize alternative splicing during development?

5. List some ways in which IVP may affect the embryo at the transcriptomic and genomic level. Can you think of any other comparisons you could make? Why?

6. What are the two main techniques used for PGD? Can you think of some limitations to the PGD other than those mentioned? Are there other genetic techniques you would use?

Supplementary videos

Supplementary Video 1: BBC Segment-1
www.youtube.com/watch?v=iUyEmr1oTTE&feature=related
Supplementary Video 2: BBC Segment 2
www.youtube.com/watch?v=d9SB0teMQo0&feature=related
Supplementary Video 3: Designer Babies-Ethical?
http://www.cbsnews.com/news/designer-babies-ethical/

Section 5

Genetics of Animal Health and Biotechnology

25 Understanding the Major Histocompatibility Complex and Immunoglobulin Genes

Michael G. Gonda
Animal Science Department, South Dakota State University, SD, USA

Introduction

When I was a student, trying to understand the genetics of the Major Histocompatibility Complex and immunoglobulins seemed overwhelming. Animals need to synthesize millions of different immunoglobulins to effectively respond to the diversity of antigens encountered throughout the animal's lifetime. Yet, most animal genomes harbor only about 20,000–25,000 genes, most of which are not involved with immunoglobulin production. How can so many different immunoglobulins be synthesized by such a limited pool of genes in the genome? In contrast, an individual animal can express only two Major Histocompatibility receptors (i.e., the two alleles at each Major Histocompatibility locus) and these receptors must somehow recognize a large number of foreign peptides for an individual to mount an effective adaptive immune response. How can these receptors recognize the diversity of peptides that an animal will encounter in its environment?

To understand immunogenetics, we must be able to answer these questions. This chapter attempts to explain the basic genetic mechanisms for how immunoglobulins are produced and how the Major Histocompatibility Complex is able to recognize the diverse set of foreign peptides that a population may be exposed to in its environment. These mechanisms are not identical in all animal species and any attempt to generalize about these mechanisms inevitably results in an oversimplification of how animals respond to pathogens in their environment. In fact, one criticism of most introductory immunology textbooks is their implicit assumption that what is known in humans and rodent species is equally applicable to livestock and companion animals. Throughout this chapter, I emphasize differences among livestock and companion animals, but my goal is not to elaborate on these differences extensively or to comprehensively discuss the entire state of knowledge of this topic. My aim is to ensure that this chapter is accessible and relevant both to advanced undergraduate students on an introductory animal genetics course and to graduate students who need an introduction to this topic.

This chapter is written for advanced undergraduate and graduate students of animal breeding and genetics. By necessity, this chapter does assume some familiarity with basic genetics terminology. It will also be helpful to understand basic immunology and we have included a section in this chapter titled "Overview of the immune system," which is intended to provide a review of basic immunology. However, this chapter will be most useful if you have been exposed to the field of immunology before reading this chapter. Reading the first couple of chapters in a basic immunology textbook should suffice for students who have not been taught immunology (see References section).

Overview of the immune system

An overview of the immune system can be found in Figure 25.1. To illustrate the function of the immune system, we will examine what happens when an Angus heifer becomes infected with a bacterial pathogen. While being vaccinated, this heifer cuts her leg on a sharp part of the squeeze chute. The wound was superficial, but the cut allowed the pathogen to invade the heifer's body.

Normally, this bacterial pathogen wouldn't have been able to even enter the heifer's leg because of her first line of defense against pathogens, called the **anatomical barrier**. The anatomical barrier consists of the skin and mucous membranes. The skin acts as a protective covering, preventing entry of undesirable microorganisms into the body. Besides acting as a mechanical barrier, skin pH ranges from 3 to 5, reducing the proliferation of many microorganisms that come into contact with the skin. Mucous membranes line the gastrointestinal, respiratory, and urogenital tracts. Mucous traps microorganisms before they have an opportunity to invade the animal and mucous-trapped microorganisms are swept away by cilia, which are small hairs in epithelial cells lining these tracts. Before disregarding the importance of these anatomical barriers, consider that the vast majority of potential pathogens never enter an animal's body because of the presence of skin and mucous membranes. However, in this

Molecular and Quantitative Animal Genetics, First Edition. Edited by Hasan Khatib.

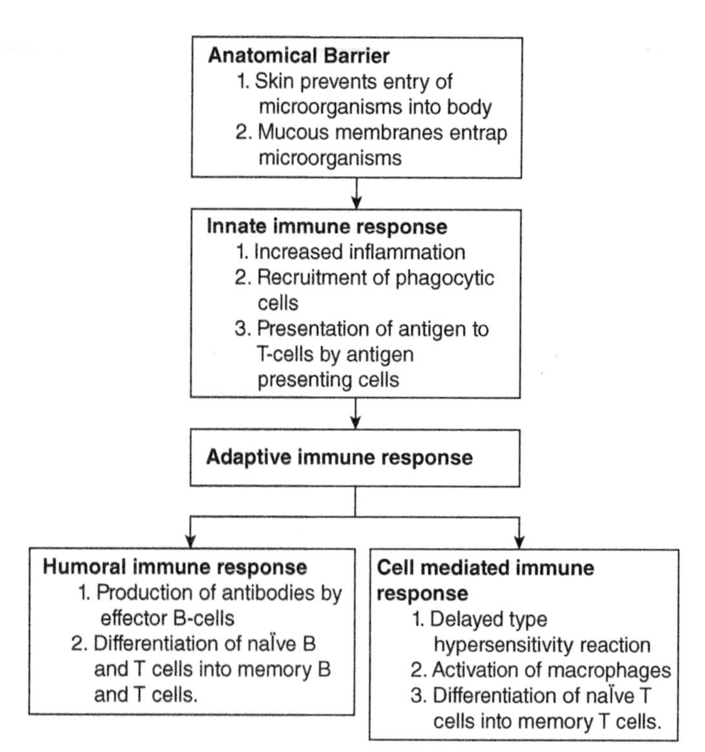

Figure 25.1 Overview of immune response in mammals. Animals first attempt to prevent entry of pathogens into the body by anatomical barriers. If pathogens enter the body, the innate ("non-specific") immune system reacts to the pathogen within hours. The adaptive ("specific") immune response takes about two weeks to develop, but creates memory cells that can "remember" the pathogen in the future and respond faster if the animal encounters the pathogen a second time. Some pathogens elicit a stronger humoral immune response than a cell-mediated response, and other pathogens elicit a stronger cell-mediated immune response. However, both adaptive immune responses are elicited by pathogens.

circumstance, the heifer cut her leg, allowing the pathogen to bypass the anatomical barrier and enter the heifer's body.

After permeating the anatomical barrier, the immune response can be divided into the **innate immune response** and the **adaptive immune response**. The innate immune response is not specific to a particular pathogen (i.e., is "non-specific"), while the adaptive immune response is tailored to the invading microorganism (i.e., is "specific"). The bacterial species will encounter the heifer's innate immune response first. The heifer's innate immune system responds by increasing inflammation near the wound site and attracting phagocytic cells to the site of infection. Inflammation raises the temperature of the body near the wound site and increases capillary permeability near the wound. This increased permeability facilitates migration of phagocytic cells from the blood to the wound site. Most of these phagocytic cells are neutrophils, but a few dendritic cells and macrophages also migrate to the wound site. All phagocytic cells are able to digest and kill microorganisms, but only the dendritic cells and macrophages can present peptides derived from the pathogen to T-cells. Cells that can perform this function are called **antigen presenting cells**.

Most pathogens are eliminated by the innate immune system. The innate immune response can begin working shortly after an infection develops. However, for some pathogens, the innate immune response is insufficient for eliminating an infection. In this circumstance, the adaptive immune response is normally activated. Let's assume the bacterial pathogen is able to survive the innate immune response long enough for an adaptive immune response to develop. At this point, the bacteria have started to replicate in the heifer's body, resulting in numerous bacterial cell copies throughout the heifer's tissue near the wound site. An antigen presenting cell takes up one of these newly replicated bacterial cells, digests the bacteria into numerous small peptide fragments, and then presents one of these peptides on the surface of its plasma membrane. These peptides are bound on the surface of antigen presenting cells by a **MHC class II** receptor. Had the pathogen been a virus or intracellular bacteria, then the peptide may also have been bound to an **MHC class I** receptor. Unlike MHC class II receptors, MHC class I receptors are present on the surface of most cells in the body. These MHC receptors are important for triggering the adaptive immune response. How the MHC receptors differentiate between self and foreign peptides to activate the adaptive immune response is a major focus of this chapter.

In this story, however, the bacterial pathogen has been taken up by an antigen presenting cell. The antigen presenting cell migrates to a nearby lymph node where peptides from the bacterial cell are presented to **T-helper cells**. The T-helper cells secrete cytokines that in turn activate the **humoral immune response**, **cell-mediated immune response**, or both. The humoral and cell-mediated immune responses both comprise the adaptive immune response. The cell-mediated immune response further increases inflammation and activation of macrophages and is most effective against viruses and intracellular bacteria. The humoral immune response activates B-cells, specifically those B-cells with immunoglobulins that have already successfully bound to the pathogen. Activation of these B-cells further increases B-cell proliferation and antibody production. Each B-cell has a distinct immunoglobulin structure that can only bind a very specific site on the surface of a pathogen (i.e., an **epitope**). The vast majority of B-cells do not have immunoglobulins that can bind to this bacterial pathogen, but the few B-cells that can bind to this pathogen will be further activated by T-helper cells. Immunoglobulins can trap pathogens, recruit phagocytic cells to sites where the pathogen has invaded, and activate an innate immune defense called the **complement system**.

Had bacterial-derived peptides also been presented on MHC class I receptors, then these receptors would activate **T-cytotoxic cells**. T-cytotoxic cells scan the surfaces of host cells looking for signs of host cell infection. In normal, uninfected cells, these MHC class I receptors are bound to peptides that are normally produced by the cell. The T-cytotoxic cells are normally not activated when a "self" protein is bound to an MHC class I receptor. However, when foreign peptides are complexed to MHC class I receptors, the T-cytotoxic cells can recognize the infection and releases proteins that cause the infected cell to die by apoptosis.

Activation of the adaptive immune response also produces **memory cells**. Memory cells are long-lived B- and T-cells that can respond much quicker to the invading pathogen than other B- and T-cells. If the heifer is exposed to this same bacterial species again, then these memory cells can generate a much faster and stronger immune response during the second and later exposures.

Given all these different weapons available to combat this bacterial species and other pathogens, it would appear that the heifer will be well protected from infection. However, if the immune system perfectly eliminated all pathogens, then we would have no disease and vaccines would be unnecessary. Although many microorganisms are eventually eliminated by the host immune system, pathogens have evolved the ability to survive and adapt to immune responses. Over time, selective pressure on pathogens by host immune systems has selected for pathogens that can successfully evade immune responses. In turn, natural selection has favored animals that can survive and reproduce in the presence of pathogens. This "evolutionary dance" between pathogens and hosts is still occurring today and has shaped how immune systems function and the ways in which pathogens evade these immune systems.

The major histocompatibility complex loci

The **Major Histocompatibility Complex** (**MHC**) loci are necessary for generating an adaptive immune response. Three classes of MHC receptors are found in animals: MHC class I, II, and III. The MHC class I and II receptors are responsible for presentation of peptides to T-cells. The MHC class III loci encode for complement and other molecules important for generation of an effective immune response, but do not present antigens to T-cells. We will not discuss MHC class III loci further.

MHC class I loci

The **MHC class I** loci can be further subdivided into "classical" (type Ia) and "non-classical" (types Ib, Ic, and Id) loci. The **non-classical MHC class Ib**, **Ic**, and **Id** loci are lowly polymorphic and have a much more limited range of peptides that can bind to their receptors. Some non-classical loci (e.g., type Ic) do not bind peptides. The **classical MHC class Ia** receptors are much more polymorphic, bind a much larger range of peptides, and are primarily responsible for presentation of peptides to T-cytotoxic cells. Most nucleated cells have MHC class I classical receptors.

Inside cells, older proteins are continually being replaced by newer proteins. As this turnover happens, the older proteins are degraded by proteasomes into peptide fragments. These peptide fragments are transported to the lumen of the endoplasmic reticulum, where some peptides are complexed with MHC class I type Ia receptors. This peptide-receptor complex is then transported to the surface of the plasma membrane. In uninfected cells, these peptides are only derived from host cell proteins and do not bind strongly to receptors on the T-cytotoxic cells. However, in viral or bacterially infected cells, some of these peptides may be derived from the viral or bacterial invader. The T-cytotoxic cells harbor receptors that have higher affinity for this foreign peptide-MHC complex, which activates the T-cytotoxic cells. The T-cytotoxic cells subsequently produce cytokines and other proteins that trigger apoptosis in the infected host cell.

The classical MHC class I receptors can bind a wide range of peptides. However, not every peptide can bind to every MHC class Ia receptor. Each receptor can bind only a specific set of peptides that will fit inside the groove of the MHC class Ia receptor. The peptide needs to be 8–10 amino acids in length. The peptides that can bind to a receptor depend on the shape and amino acid sequence of the MHC receptor. Thus, two different MHC class Ia receptors with different amino acid sequences (i.e., different alleles) may bind a different set of peptide fragments. The end result is that some individuals may be able to bind a peptide from one microorganism and present the peptide on its cell surface, while other individuals may not harbor an MHC class Ia receptor that can bind a peptide from the same microorganism.

Structurally the MHC class I receptor consists of one α-polypeptide chain and a molecule of β-microglobulin. Usually, three MHC class I loci are present in the genomes of most animals (A, B, and C), but this number varies among species.

MHC class II loci

The **MHC class II** receptors are expressed on the surface of antigen presenting cells, which classically include dendritic cells, macrophages, and B-cells. Some animal species also express MHC class II receptors on other cell types, including resting T-cells. Their function is similar to MHC class I receptors except that peptides are presented to T-helper cells instead of T-cytotoxic cells. As bacterial cells and other extracellular, foreign proteins are taken up by antigen presenting cells, they are transferred to endosomal compartments within the cell that degrade the proteins into small peptides. These peptide fragments are then complexed with MHC class II receptors and expressed on the surface of the antigen presenting cells. T-helper cells harbor receptors that recognize foreign peptides complexed with MHC class II receptors and this recognition by T-helper cells stimulates the adaptive immune response.

Like the MHC class I receptors, each MHC class II receptor allele can bind to a different repertoire of antigenic peptides. Thus, one allele may be able to bind a peptide from a microorganism, but a different allele may not be able to bind peptides from that same microorganism. Structurally the MHC class II receptor consists of two polypeptide chains (α and β) and does not include β-microglobulin. Usually six MHC class II loci are present in the genome of mammalian species (DP-α &-β, DQ-α &-β, and DR-α &-β), but the actual number of loci depends on the species.

Genetic diversity at the MHC loci

The MHC is a good example of **heterozygote advantage** and may be one cause of heterosis for disease susceptibility in livestock species (Figure 25.2). An animal that is heterozygous at an MHC locus more likely has higher fitness than an animal that is homozygous at the same MHC locus. The heterozygous animal can present a larger range of peptides to T-helper cells and thus has a higher probability of generating an adaptive immune response to a pathogen.

The diversity of MHC loci and alleles in species can likely be explained by evolutionary pressure to maintain sufficient genetic diversity at the MHC receptor loci. If a population was only exposed to one or a few pathogens over time, evolution should favor selection of MHC loci and alleles that bind most strongly to antigens derived from these pathogens. A population would likely have a higher frequency of a small number of MHC alleles that best protect individuals from the small number of pathogens in their environment. However, most populations are exposed to a large, diverse number of pathogens. An MHC allele that can bind peptides from one pathogen may not be able to bind peptides from other pathogens. It becomes increasingly likely that an MHC allele that confers increased resistance to one pathogen may actually confer increased susceptibility to other pathogens.

Because a single MHC allele cannot protect a population from all pathogens that a population may be exposed to, evolution would favor maintenance of genetic diversity at the MHC loci.

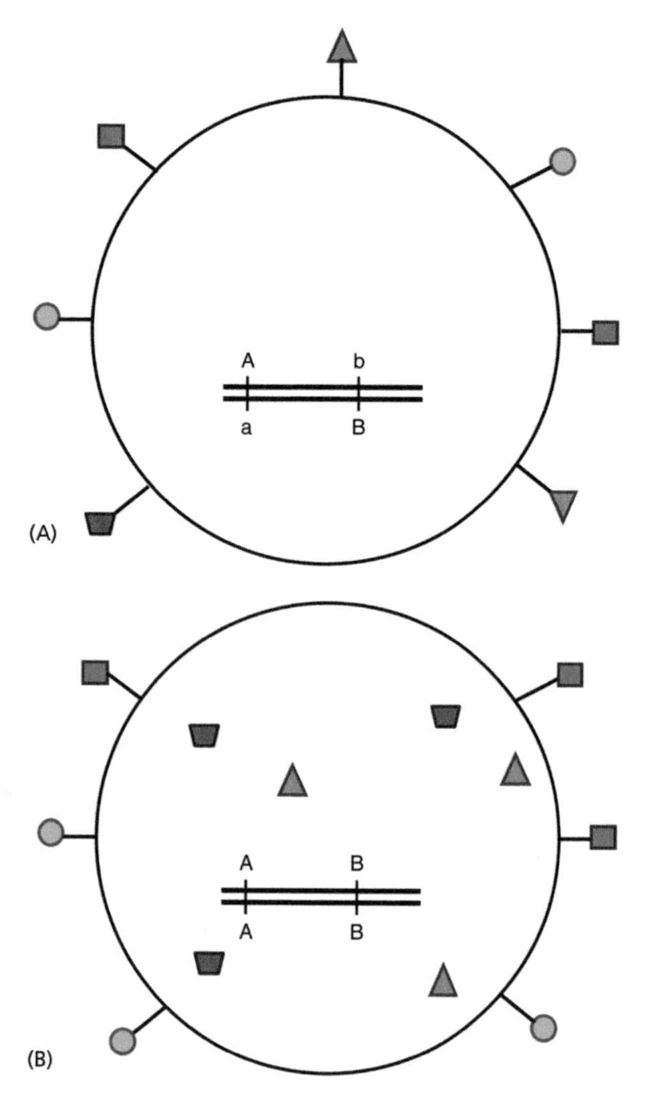

Figure 25.2 The importance of heterozygote advantage at the Major Histocompatibility Loci. (**A**) This individual is heterozygous at these two MHC loci (Aa and Bb). As a consequence, each MHC allele can present on its surface a wider range of peptides derived from foreign bodies. The lines extending from the surface of the cell are the MHC receptors. The colored shapes (circle, square, triangle, and trapezoid) are different peptides derived from a foreign body. (**B**) This individual is homozygous at these two MHC loci (AA and BB). The individual is not able to present on its cell surface all of the peptides that the individual in diagram **A** can. As a result, a pathogen is more likely to escape detection in this individual than in the individual in diagram **A**. In this diagram, the peptides derived from foreign bodies that remain inside the cell cannot bind to MHC receptors and thus cannot be presented on the surface of the cell.

In this scenario, the welfare of the population is favored over the welfare of any specific individual. Within a population, some individuals will not harbor an allele that can strongly bind peptides from a pathogen. If these individuals are exposed to the pathogen, they are more likely to succumb to the infection. However, if at least some individuals in the population harbor MHC alleles that confer resistance to the pathogen, then the population can continue to survive even if some individuals within the population do not.

The MHC alleles have been associated with resistance and susceptibility to many infectious diseases in livestock, companion animals, rodents, and humans. These loci are important determinants of disease susceptibility. However, their effects on disease susceptibility and resistance should not be overstated. Many other loci contribute towards genetic susceptibility to disease and non-genetic factors also play a significant role. Susceptibility to disease is clearly a polygenic trait.

From an animal breeding standpoint, the long-term consequences of genetic selection for MHC alleles are unknown. Selection for an allele associated with resistance to a specific, economically important disease may or may not result in a healthier population in the long term. If the allele substitution effect on resistance to the economically important disease is large, then selection should result in a population that is more resistant to this disease. However, selection would also result in less genetic diversity at the MHC loci. This loss of genetic diversity could have profound implications on the health of the population in the future.

Immunoglobulin loci

Immunoglobulin function

Immunoglobulins are a class of receptor produced by B-cells that are either found on the surface of B-cells or secreted into extracellular fluid. Immunoglobulins are the effector molecules of the humoral immune response. Unlike MHC receptors, immunoglobulin receptors are highly specific for a particular epitope on an **antigen**. An immunoglobulin bound to the surface of a B-cell membrane is called a **B-cell receptor** or **antigen receptor**. Immunoglobulins secreted into extracellular fluid are called **antibodies**. The terms B-cell receptor and antibody are not interchangeable, but both B-cell receptors and antibodies are immunoglobulins. Binding of antigen to a B-cell receptor, along with T-helper cell activation, can elicit a humoral immune response. Upon activation of a B-cell, the B-cell begins to proliferate and differentiate into **effector cells** and **memory cells**. Through this process, only B-cells harboring immunoglobulins that can bind to the pathogen will be activated. The effector cells produce antibodies, which have the following functions for neutralizing pathogens.

1. **Agglutination and neutralization of antigen** – Antibodies bind to pathogens in extracellular fluid, preventing the pathogens from invading host cells.
2. **Opsonization** – Phagocytic cells harbor **F_c receptors** on their surfaces that bind to the constant regions of some antibodies. Through these receptors, phagocytic cells can bind to antibody-bound pathogen, facilitating phagocytosis.
3. **Activation of the complement pathway** – Some antibodies can activate the complement pathway when antibodies are bound to the surface of a microorganism. The complement pathway is a component of the innate immune response and, when activated, leads to lysis of the microorganism.

Immunoglobulin classes and structures

Immunoglobulins have similar structures regardless of immunoglobulin class. An immunoglobulin molecule consists of two heavy chain polypeptides and two light chain polypeptides. The heavy chains have a heavier molecular weight and are larger than the light chains. Each heavy chain and each light chain are

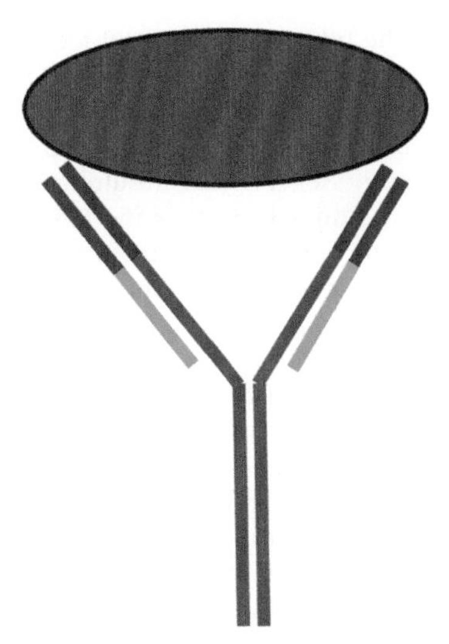

Figure 25.3 Interaction between an antigen and an antibody molecule. The variable regions (red) of the light and heavy antibody chains interact with the antigen (purple oval). The constant (D, J, and C) regions of the light (green) and heavy (blue) antibody chains do not bind to antigen. Hence, the variable domains have evolved to be more polymorphic than the other antibody domains.

structurally identical. The two heavy chains are linked together by a disulfide bond, and each light chain is linked by additional disulfide bonds to a heavy chain. The light chains have three domains: V (variable), J (joining), and C (constant). The heavy chains have four domains: V, D, J, and C. The antibodies bind to antigen at the V antibody domains and thus these antibody regions are the most variable among B-cell immunoglobulins (Figure 25.3). Each domain is encoded by its own locus. Additionally, the light chains have two loci: kappa and lambda.

The class of an immunoglobulin molecule is determined by the C domain of the heavy chain. Five immunoglobulin classes are present in most livestock and companion animal species.

1. *IgM* – This class is found on the surface of **naïve B-cells**. IgM antibodies are produced following development of the **primary immune response**, which is the adaptive immune response that occurs following the first exposure to an antigen. Although IgM antibodies are produced during the **secondary immune response**, this immunoglobulin is not the most prevalent antibody class produced in response to the antigen. IgM forms a pentameric ring when secreted from B-cells and thus, because of its large size as a secreted molecule, rarely escapes from the bloodstream into surrounding tissues.

2. *IgD* – The IgD class, like IgM, is found on the surface of naïve B cells. However, IgD antibodies are not produced during either the primary or secondary immune responses. The IgD class is much more plastic than the other immunoglobulin classes and does not appear to participate significantly during immune responses. Its role in immunology is unknown.

3. *IgG* – This immunoglobulin class is not present on naïve B cells, but IgG antibodies are produced during the primary immune response. During the secondary immune response, IgG is the predominant antibody produced against the antigen. IgG is the smallest immunoglobulin class and thus can most easily escape from the bloodstream to sites of tissue inflammation.

4. *IgA* – Although present in serum, IgA is predominantly localized to mucosal surfaces (e.g., gastrointestinal tract, respiratory tract, urogenital tract) and secretions (e.g., colostrum, milk, saliva, tears). The primary purpose of IgA is to act as a sentry near potential points of entry into the body. The IgA antibodies can bind to and neutralize pathogens before they can enter into tissues. IgA cannot participate in opsonization or complement activation. In its secreted form, IgA is found as a dimer. Interestingly, species vary in the amount of IgA found in some secretions; for example, IgA is more frequently found in colostrum in horses and pigs than ruminants, dogs, and cats.

5. *IgE* – This immunoglobulin class primarily functions to defend the animal against parasitic worms. IgE is also associated with allergic reactions. This antibody can trigger inflammation by binding to receptors on basophils and mast cells. This binding releases inflammatory molecules.

Within each class, immunoglobulins can be further subdivided into subclasses. For example, cattle have three IgG subclasses (IgG_1, IgG_2, and IgG_3) while dogs have four IgG subclasses (IgG_1, IgG_2, IgG_3, and IgG_4). The number of subclasses often varies among species. Immunoglobulins within a subclass will also sometimes have slightly different functions; for example, bovine IgG_2 can participate in antigen agglutination while bovine IgG_1 cannot.

Immunologists use different terminology to describe polymorphisms within the immunoglobulin genes. Next, are the definitions of terms you may encounter when studying immunogenetics.

- **Allotype** – Polymorphisms can be found within the immunoglobulin loci. These within-species differences in nucleotide sequence within the immunoglobulin loci are called **allotypes**. If an allotype was associated with increased disease resistance, then similar to other polymorphisms, the allotype could be included in a genetic selection program.
- **Idiotype** – To recognize foreign invaders, animals need to produce numerous, diverse immunoglobulins, each with different amino acid sequences in its variable region that can recognize a different epitope. These variants are created during and after B-cell maturation. Germ line DNA in sperm and egg cells does not harbor these **idiotypes** and thus, from an animal breeding standpoint, idiotypes cannot be included in an animal breeding program.

Although idiotypes are not useful to animal breeders, in a sense these idiotypes are selected for when an animal is exposed to a pathogen. Each B-cell has its own unique immunoglobulin idiotype. When a pathogen infects an animal, only B-cells with immunoglobulin idiotypes that can bind to the pathogen will be "selected" for by the animal. These B-cells will begin to proliferate, drastically increasing their numbers in response to the pathogen.

Generating immunoglobulin diversity

No single method for producing immunoglobulin diversity (or idiotypes) exists. This diversity is generated by a combination of immunoglobulin gene rearrangement, nucleotide insertion and deletion, receptor editing, gene conversion, and somatic hypermutation. To complicate matters further, the importance of each of these methods for generating immunoglobulin diversity varies among species. For example, gene conversion plays a relatively large role in generating immunoglobulin diversity in ruminants, but in horses, gene rearrangement plays a more significant role than gene conversion. Somatic hypermutation only plays a role after B-cells are exposed to antigen.

Immunoglobulin gene rearrangements

Immunoglobulin molecules are encoded for in the genome by three loci for light chains and four loci for heavy chains. The light chain loci encode for the V, J, and C immunoglobulin light chain domains and the heavy chain loci encode for the V, D, J, and C immunoglobulin heavy chain domains. Although each chain consists of these domains, the genome harbors multiple copies of each of these loci. For example, at the immunoglobulin heavy chain loci, pigs harbor 20 V loci, 2 D loci, and 1 J locus. Only one of each V, D, J, and C locus is transcribed, however. As the B-cell matures, these loci are rearranged so that only one V, D, J, and C locus is expressed per immunoglobulin chain.

The loci that encode each of the light chain domains and the heavy chain domains can be found clustered together in the same location in the genome. For example, all of the loci that encode the equine immunoglobulin heavy chain domains are found together in a cluster on chromosome 24. Within these clusters, all of the V loci are found at the 5′ end of the cluster, followed by the D loci (for heavy chain loci), the J loci, and then finally the C loci. All livestock and companion animal species harbor at least one heavy chain C domain corresponding to each of the immunoglobulin classes in the following order: M, D, G, E, and A. (One possible exception is the cat, which may not harbor a constant domain for IgD.) The constant region included in the immunoglobulin heavy chain determines the immunoglobulin class produced by the B-cell.

Gene rearrangement occurs in all B-cells during B-cell maturation in the bone marrow. At the light chain loci (Figure 25.4), first a randomly selected V locus is recombined with a randomly selected J locus. The intervening V and J loci are subsequently spliced from the DNA. At this point, the V-J loci are still separated from the C loci by an intron. The V-J and C loci are then transcribed into one full-length messenger RNA molecule. The intron is spliced from the messenger RNA, resulting in a mature messenger RNA that has linked together the V-J-C loci. This messenger RNA molecule is then translated and processed to form the mature immunoglobulin light chain.

Heavy chain gene rearrangement (Figure 25.5) occurs similarly to rearrangement of the light chain genes, except that the V-D loci are recombined first. Once the V-D loci are recombined, a second recombination event occurs in the DNA leading to the joining of the V-D-J loci. The V-D-J and C loci are subsequently transcribed into one full-length messenger RNA molecule and messenger RNA processing and translation proceeds similarly to the light chains. The enzymes primarily responsible for recombination at the immunoglobulin loci are (1) recombinase and (2) DNA repair enzymes. Recombinase splices out the intervening DNA between the loci that are to be combined and the DNA repair enzymes join the two loci together. A mutation in a DNA repair enzyme in horses causes severe combined immunodeficiency disorder (SCID), which leads to lack of immunoglobulin production.

Gene rearrangement leads to the generation of different immunoglobulin glycoproteins, which each will bind a different epitope. For example, horses harbor 25 lambda light chain V loci, four lambda light chain J loci, at least sevenheavy chain V loci, 10 heavy chain D loci, and five heavy chain J loci. Thus, theoretically horses can generate 35,000 different immunoglobulin struc-

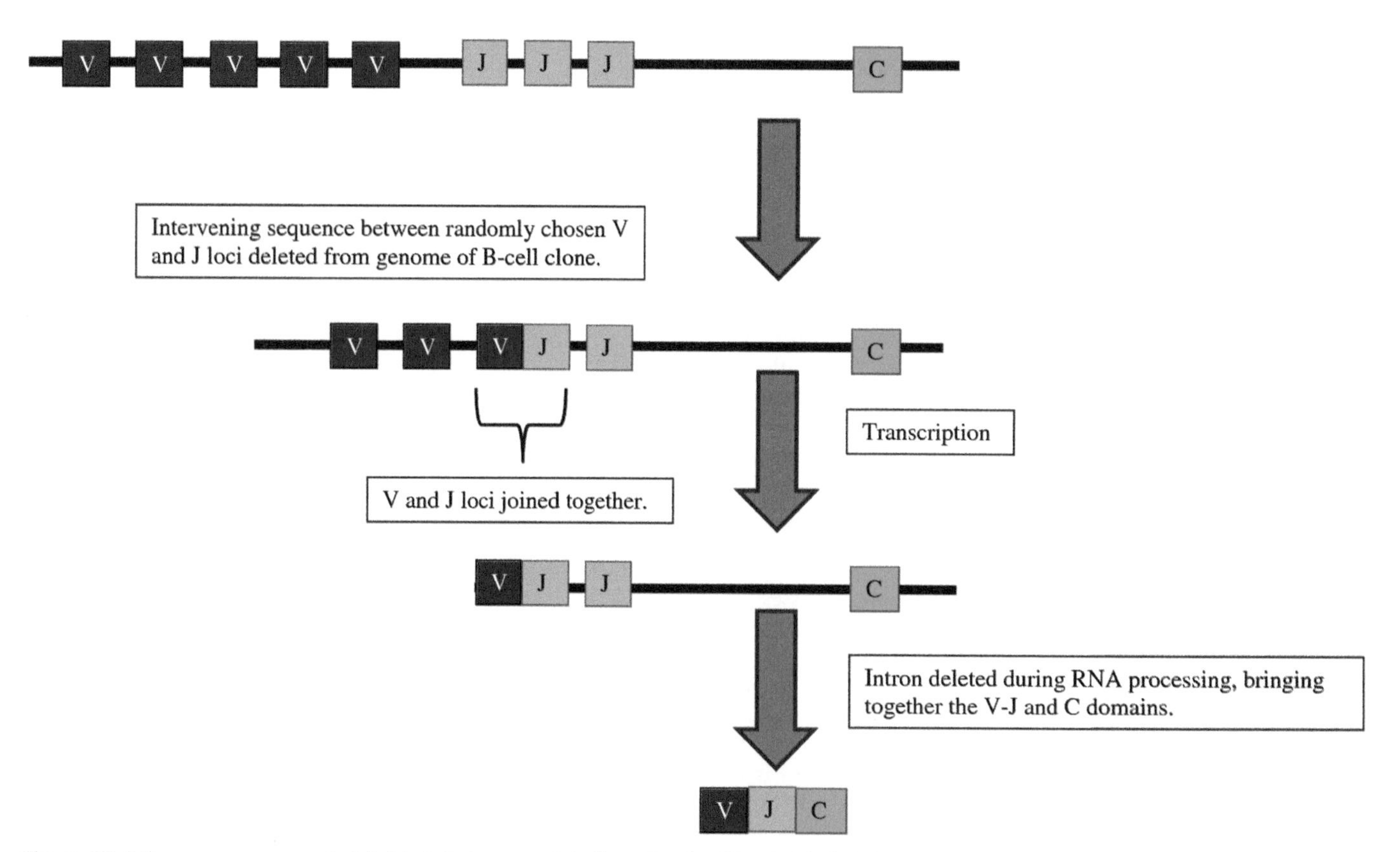

Figure 25.4 Gene rearrangement of light chain immunoglobulin molecules. The V and J loci that are combined are chosen randomly. Adapted from Fig. 5.4 *Kuby Immunology*, 4th edn by R.A. Goldsby, T.J. Kindt, and B.A. Osborne, W.H. Freeman & Co., New York, NY.

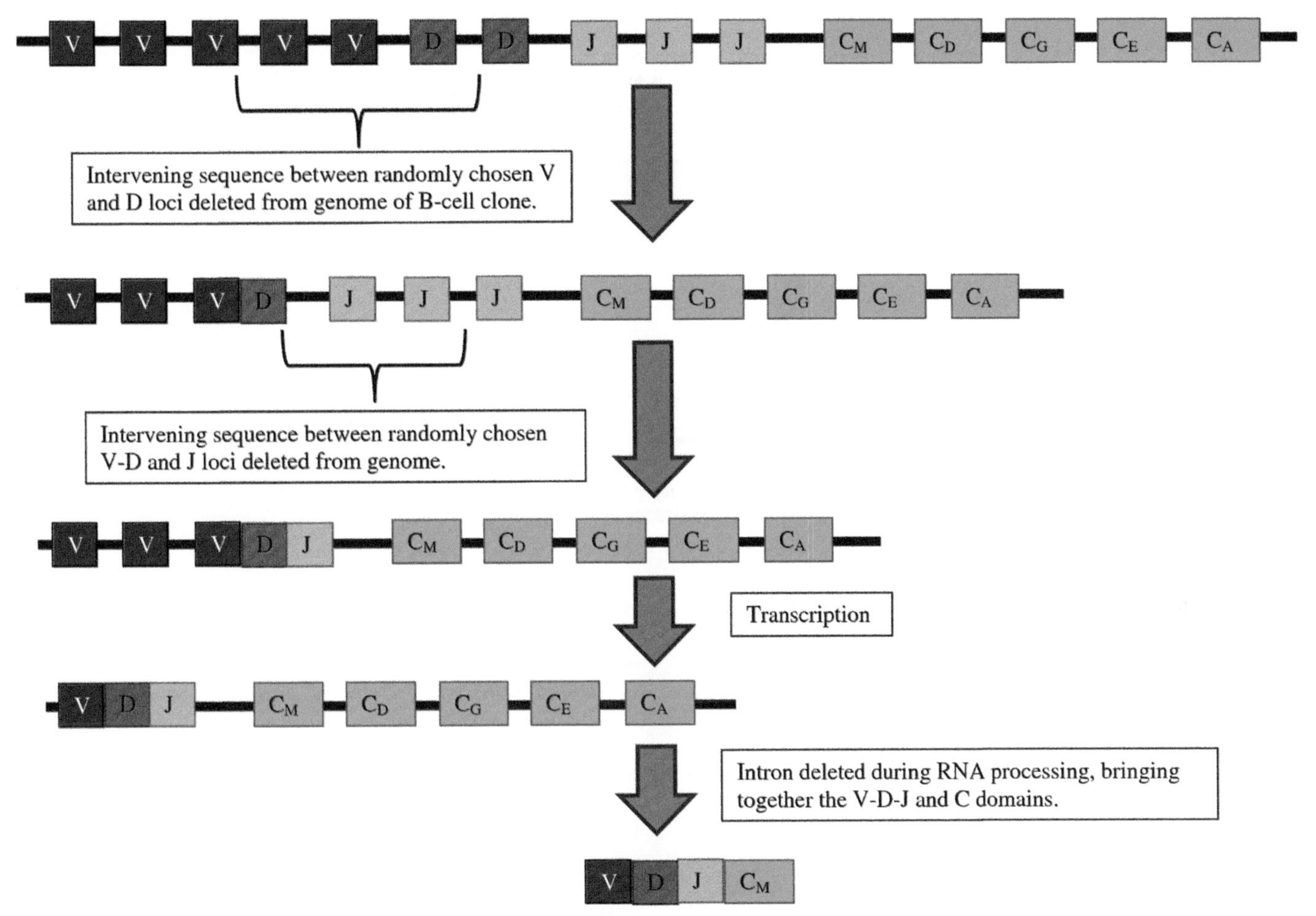

Figure 25.5 Gene rearrangement of heavy chain immunoglobulins. During class switching, the C_M or C_D constant domains are replaced with the new constant domain (either C_G, C_E, or C_A). If, for example, the class switches to C_A, then all of the 5′ proximal constant domains are deleted along with the rest of the intron during RNA processing. Adapted from Fig. 5.5, *Kuby Immunology*, 4th edn by R.A. Goldsby, T.J. Kindt, and B.A. Osborne, W.H. Freeman & Co., New York, NY.

tures from gene rearrangements alone, and this number does not consider the different loci present in the kappa light chain.

Class switching

The heavy chain constant domain locus determines the immunoglobulin class that the B-cell produces. When an immature B-cell matures into a naïve B-cell (that is, a B-cell that has not been exposed to foreign antigens), the B-cell produces only IgM and IgD immunoglobulins. In naïve B cells, therefore, the C domains that are integrated into the mature immunoglobulin are either M or D. When a naïve B-cell encounters antigen, **class switching** will occur. The B-cell will switch its constant region domain from M and D to either G, E, or A. The class selected by the B-cell is determined by **cytokines** released into the environment by the B-cell and other immune cells. Cytokine expression is in turn affected by the antigen that the immune cells are exposed to and the environment surrounding the immune cells.

Nucleotide insertions and deletions

When gene rearrangement occurs, oftentimes nucleotides are inserted or deleted at the ends of these loci before the loci are spliced together. These insertions and deletions alter the amino acid sequence of the immunoglobulin and often also change the translation reading frame. Changing the reading frame often causes the introduction of premature stop codons and immunoglobulin proteins that are non-functional. As expected, two-thirds of these insertions and deletions result in a **non-productive rearrangement** and a non-viable immunoglobulin chain.

As stated earlier, the genome harbors two clusters of light chain loci called kappa and lambda. During gene rearrangement, B-cells randomly choose to incorporate either the kappa or lambda light chains into their immunoglobulin molecule. If an insertion or deletion causes a non-productive rearrangement, then the B-cell attempts to use the other allele at the same light chain cluster. If this rearrangement still results in a non-productive rearrangement, then the B-cell attempts to use the alternative light chain locus alleles. If, after four attempts (both alleles at the kappa and lambda loci) the B-cell still does not produce a productive light chain rearrangement, then the B-cell undergoes apoptosis.

Gene conversion

Gene rearrangements are a significant source of immunoglobulin diversity, especially in species such as humans and rodents that harbor hundreds of heavy chain V loci. However, livestock species

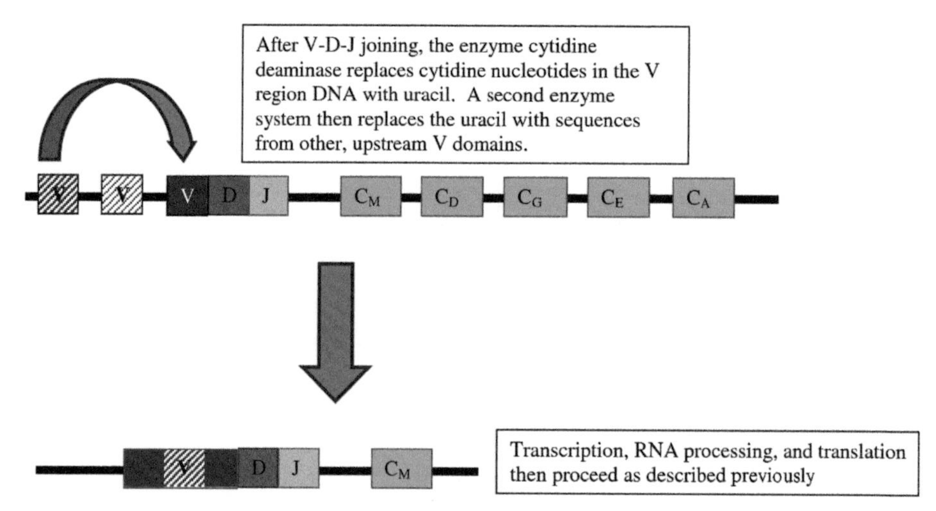

Figure 25.6 Gene conversion. This figure shows gene conversion in the heavy chain after V-D-J joining, but gene conversion also occurs in the light chains. Adapted from Fig. 17.10, *Veterinary Immunology*, 9th edn by I.R. Tizard, "How Antigen-Binding Receptors Are Made," p. 183, Copyright Elsevier, 2013.

often don't harbor comparatively many functional heavy chain V loci (~7–20). For many livestock species, especially ruminants, **gene conversion** is a more important source of immunoglobulin diversity (Figure 25.6). After V-(D)-J rearrangement, an enzyme called cytidine deaminase randomly replaces cytidine nucleotides in the selected V locus with uracil. Uracil is not a deoxynucleotide and thus is recognized by the cell as a mistake in the DNA sequence. Repair enzymes remove the uracils. After uracil is removed, DNA sequences from both functional and non-functional V loci upstream of the selected V loci are inserted into the gap left from the removal of the uracils. In this manner, V loci that are **pseudogenes** can still contribute towards generation of immunoglobulin diversity even if the loci themselves are not functional.

Receptor editing

We previously discussed how naïve B-cells can switch their heavy chain constant loci in response to antigen. After antigen exposure, naïve B-cells can also change their V or J locus, too. Although the intervening loci between the selected V and J loci have been deleted from the DNA, the V and J loci that are upstream and downstream from the selected loci are still present in the B-cell's genome. After exposure to antigen, the B-cell can still switch to using one of these inactive V or J loci. Further, B-cell genomes still harbor loci for both kappa and lambda light chains. In response to antigen exposure, the B-cell can switch the light chain locus that is used to produce immunoglobulin. Naïve B-cells often edit their receptors to increase affinity between their immunoglobulins and the antigens. Receptor editing is also used when a B-cell immunoglobulin can react with self-antigens.

Somatic hypermutation

When B-cell receptors bind to antigen, a signal transduction reaction is triggered that results in up-regulation of cytidine deaminase, the same enzyme that initiates gene conversion. As described for gene conversion, cytidine deaminase replaces cytidine nucleotides in the immunoglobulin loci at random with uracil nucleotides. Repair enzymes then replace the uracils with either thymidine or a random deoxynucleotide, depending upon the specific repair enzyme used. This change in nucleotide sequence can result in a change in the sequence of the immunoglobulin protein, which may in turn increase or decrease the affinity of the B-cell receptor for the antigen.

B-cell receptors with stronger affinities for antigens trigger higher rates of B-cell proliferation. Thus, B-cells with immunoglobulins that bind more strongly to antigen will expand at a faster rate, resulting in relatively more B-cells that bind most strongly to the antigen. Through this process, the immune system selects for B-cells that can produce antibodies that bind strongest to the antigen and thus can eliminate the foreign invader most effectively. Although this process has no direct relevance to animal breeding, it does mimic selection on a cellular level!

Summary

- The MHC receptors on the surface of cells bind foreign peptides from bacteria, viruses, and other pathogens and present these peptides to immune cells. These receptors are necessary for activation of the adaptive immune response.
- Three types of MHC loci exist. The MHC Class I loci primarily bind peptides derived from viruses and intracellular bacteria. The MHC Class II loci primarily bind peptides derived from extracellular bacteria. The MHC Class III loci do not bind peptides, but instead are a class of proteins important for activation of the immune response.
- Each MHC receptor allele encodes for a protein that can bind a large number of peptides. However, a single MHC receptor allele cannot bind all possible peptides derived from pathogens that could be present in the environment.
- The MHC loci are therefore highly polymorphic. Although any single individual in a population may not harbor an MHC receptor allele that can respond to a specific pathogen, the population as a whole likely will harbor MHC alleles that can trigger an immune response to the pathogen.
- Immunoglobulins are receptors that are the effectors of the humoral immune response, which is one arm of the adaptive immune response in animals.

• A population of B-cells in an animal must be able to produce immunoglobulins that can recognize all of the diverse types of pathogens that may be encountered in the environment. Literally millions of different types of immunoglobulins (i.e., idiotypes) are required for an animal to fulfill this function. These idiotypes are created from only hundreds of genes in livestock.
• This idiotype diversity is produced through several genetic mechanisms including gene rearrangements, gene conversion, nucleotide insertion and deletion, receptor editing, and somatic mutation.

Key terms

Adaptive immune response – The branch of the immune response that is specific to the foreign antigen in the body. The immune response is "tailored" to the foreign antigen.

Agglutination – Clumping together of antigen by antibodies.

Allotype – A polymorphism in an immunoglobulin locus.

Anatomical barrier – Skin and mucous membranes; structures that prevent entry of microorganisms into the body.

Antibody – A secreted immunoglobulin molecule that is part of the humoral immune response. Facilitates pathogen neutralization and sometimes oposonization and complement activation.

Antigen – A substance that can elicit an immune response.

Antigen presenting cell – Macrophages, dendritic cells, B-cells, and in some species resting T-cells. These cells have the ability to present peptide fragments from antigens complexed with MHC II receptors on their cell surface.

B-cell receptor – An immunoglobulin that is attached to a B-cell membrane and acts as a receptor for B-cell activation.

Cell-mediated immune response – A branch of the adaptive immune response that causes a delayed-type hypersensitivity reaction and increases macrophage activity. Often elicited in response to viral or intracellular bacterial infection.

Complement – A series of soluble proteins that are part of the innate immune response. Can target a microorganism for elimination and lyse microorganisms.

Classical Major Histocompatibility Complex I Loci – Highly polymorphic loci that encode for receptors that bind intracellular peptides and present these peptides on the surface of most nucleated cells.

Class switching – When a B-cell switches the class of immunoglobulins that is being produced. Specifically, the heavy chain constant region of the immunoglobulin locus is changed, resulting in a class switch.

Cytotoxic T-cell – An adaptive immune cell that can recognize foreign peptides complexed to MHC class I receptors on the surface of cells and can then cause the infected cell to die by apoptosis.

Effector cell – A B- or T-cell that has differentiated from a naïve cell into a cell that is producing an immune response (e.g., antibody production, cytokines) to an antigen.

Epitope – A site on the surface of an antigen that can be recognized by an immunoglobulin.

Fc receptor – A receptor on the surface of macrophages that can bind antibody. Increases contact between antibody-bound antigens and phagocytic cells.

Gene conversion – Occurs when DNA sequences from both functional and non-functional V loci upstream of the selected V locus are inserted into the selected V locus. Gene conversion is an important source of immunoglobulin diversity in some species (e.g., ruminants).

Helper T-cell – An adaptive immune cell that produces cytokines that direct the immune system to produce either an adaptive or cell-mediated response.

Heterozygote advantage – Individuals who harbor two different alleles at MHC loci will be able to bind a wider range of foreign peptides than individuals who harbor only a single allele (i.e., homozygotes). Heterozygotes will be able to mount an effective immune response against a wider range of pathogens.

Humoral immune response – A branch of the adaptive immune response that produces antibodies to protect against pathogens.

Idiotype – Differences in the amino acid sequence of immunoglobulin molecules that change the binding properties of the immunoglobulins. Many of the different idiotypes among immunoglobulins are formed by gene rearrangements, nucleotide insertions and deletions, and gene conversion and are not encoded for in the genomic DNA.

Immunoglobulin – A glycoprotein that is produced by B-cells and that can bind only very specific epitopes on the surface of antigens. Can be present in both membrane-bound and secreted forms.

Innate immune response – A branch of the immune response that is non-specific, i.e., is not tailored to a specific pathogen.

Major histocompatibility complex I – See classical and non-classical MHC I.

Major histocompatibility complex II – Receptors that can bind peptides derived from exogenous antigen and present these peptides on the surface of antigen presenting cells.

Major histocompatibility complex III – Loci that encode for complement proteins, cytokines, and other proteins with immunological function and that are found proximal to the MHC class I and II loci.

Memory cells – After stimulation of an adaptive immune response, some B- and T-cells differentiate into memory cells, which are long-lived cells that allow an animal to respond stronger and faster to re-exposure to the same pathogen.

Naïve B- and T-cells – B- and T-cells that have not encountered antigen, but that are mature lymphocytes.

Non-classical Major Histocompatibility Complex I Loci – Lowly polymorphic loci proximal to the classical MHC class I loci that function in the immune response. Some non-classical MHC I loci appear to encode receptors similar to classical MHC class I loci, while others do not appear to have similar functions.

Non-productive rearrangement – A gene rearrangement at a cluster of immunoglobulin loci that does not result in a functional immunoglobulin heavy or light chain.

Opsonization – An antibody function that increases contact between antibody-bound antigen and phagocytic cells.

Primary immune response – The immune response of an animal after its first exposure to an antigen.

Pseudogene – A locus in the genome that has features similar to a functional gene, but does not produce a functional protein.

Receptor editing – The ability of a B-cell to switch different V and J domains in an immunoglobulin gene after exposure to antigen. Also, the ability of a B-cell to switch from a kappa to a lambda light chain and vice versa.

Secondary immune response – The immune response of an animal after its second exposure to an antigen.

Somatic hypermutation – Increased mutation rate in variable domains of immunoglobulin molecules that can change the affinity of the immunoglobulin molecule to an antigen.

Further reading and references

Kindt, T.J., B.A. Osborne, and R.A. Goldsby. 2006. *Kuby Immunology*. 6th edn. W.H. Freeman & Co., New York, NY.

Tizard, I.R. 2013. *Veterinary Immunology*. 9th edn. Elsevier Inc., St. Louis, MO.

Review questions

1. Normally, most receptors are highly specific for a particular ligand, for example, the growth hormone receptor binds to growth hormone ligand. The MHC and B-cell receptors need to bind to a much more diverse set of molecules. Compare and contrast how B-cell receptors and the MHC receptors are able to bind to the diverse number of microorganisms that mammalian cells will need to respond to in their environment.

2. The MHC III loci have diverse, distinct functions that are different from the MHC I and II loci. Why do you think these loci were labeled as part of the Major Histocompatibility Complex?

3. What are the potential consequences of selecting for a single MHC allele in a population?

4. Can animal breeders incorporate idiotypes into a livestock selection program? Why or why not?

5. What is the difference between receptor editing and class switching?

26 Livestock and Companion Animal Genetics: Genetics of Infectious Disease Susceptibility

Michael G. Gonda
Animal Science Department, South Dakota State University, SD, USA

Introduction

In the previous chapter, we discussed how receptor diversity is generated and maintained in livestock and companion animal populations so that these populations can recognize and respond to the diversity of pathogens that they will encounter in their environment. It has been demonstrated that polymorphisms in these loci – especially the Major Histocompatibility Complex (MHC) loci – are associated with susceptibility to disease. However, one should not ignore the reality that many other loci are important for mounting an effective immune response. Susceptibility to disease is a **polygenic trait**, which means that a large number of **polymorphisms** at many different loci, plus environmental factors, contribute to whether an individual is susceptible or resistant to disease. Although we have discovered polymorphisms that have a "major" effect on disease susceptibility, it is rare to find a polymorphism that is the sole genetic determinant of whether an animal is susceptible or resistant to disease.

Before continuing on with this discussion, we need to define what we mean by "disease." Diseases can be placed into one of four general categories:

1. Simply-inherited genetic defects – Sometimes considered to be a disease, a single (or, rarely, a few) genetic mutation can cause a large, deleterious change in the phenotype of an animal. Usually, the genetic mutation causes abnormal growth and development of the animal. The environment has little to no effect on expression of these genetic defects. An example of a simply-inherited genetic defect would be Arthrogryposis Multiplex (AM) in Angus cattle. This defect is controlled by a single genetic locus and causes abnormal fetal growth, leading to physical deformities and death shortly after parturition.

2. Infectious disease – These diseases are caused by a foreign microorganism for example, *Mycobacterium paratuberculosis* (the cause of Johne's disease in cattle). We normally think of this class of diseases first when considering disease in livestock and companion animals.

3. Metabolic disease – These diseases are usually caused by feeding diets that are not matched well to the physiological and environmental state of the animal. Examples include ketosis, milk fever, and displaced abomasums in cattle. Metabolic diseases are usually polygenic.

4. Autoimmune disease – Autoimmune diseases occur when the animal's immune system recognizes "self" antigens as foreign or "non-self." An example of an autoimmune disease is systemic lupus erythematosus in dogs. Sometimes autoimmune disease can be triggered by an infection, and generally autoimmune diseases are polygenic.

For the purposes of this chapter, our definition of disease will be limited to infectious diseases.

Why is studying the genetics of disease susceptibility important?

First, understanding the genetics of disease susceptibility could lead to the development of estimates of genetic merit for disease susceptibility in animals. These estimates of genetic merit are usually called expected progeny differences (EPDs). The dairy industry uses the terminology "predicted transmitting ability" (PTA), but PTAs are equivalent to EPDs. These estimates of genetic merit can be calculated from phenotypic records of disease incidence in relatives and progeny, genomic information, or both. We can then use these estimates of genetic merit to select for individuals that are less genetically susceptible to certain diseases. Alternatively, these estimates of genetic merit could be used to manage animals differently based on the risk that an animal has for being susceptible to a disease. For example, perhaps one may wish to keep animals that are genetically more susceptible to disease in a separate location, barn, or pasture. These estimates of genetic merit are already available for some diseases for example, subclinical mastitis in dairy cattle.

Second, livestock and companion animals can be good biomedical models for human diseases. Historically, most basic

Molecular and Quantitative Animal Genetics, First Edition. Edited by Hasan Khatib.

research on the immune system has been completed on rodent species, especially mice but also rats and rabbits. Several biomedical models of disease have also been created in livestock and companion animals. The pig is the most extensively used livestock species for this purpose; pigs have been used as biomedical models for infections important to human health, vaccine development, and development of the immune system. Cattle have also been used to improve human health. Hematech Inc. (Sioux Falls, SD) has replaced bovine immunoglobulin loci with human immunoglobulin loci to create cattle that can produce large quantities of human immunoglobulins. These human immunoglobulins can be used to treat an array of human health problems, including infectious and autoimmune diseases.

Third, understanding the genetics of the immune response could lead to the development of more effective vaccines and therapeutics.

Present applications of genetic selection tools for predicting disease susceptibility

First, we cannot overemphasize the fact that disease resistance/susceptibility is a polygenic trait in most cases. We do not expect to find a single locus or gene that completely determines whether an animal will be susceptible or resistant to a disease. Instead, it is the effects of hundreds of genotypes within an individual that determines an animal's genetic predisposition for being susceptible to disease. As an analogy, let's assume that the loci that affect disease susceptibility are represented by 100 marbles. A blue marble is a favorable genotype for disease resistance/susceptibility and a red marble is an unfavorable genotype for disease resistance/susceptibility. An animal's genetic predisposition for disease resistance/susceptibility is then determined by counting the number of blue and red marbles. Animals with more blue marbles are more resistant to the disease and animals with more red marbles are less resistant to the disease. If we change the color of one marble from blue to red (or vice versa), we wouldn't expect much change in the animal's genetic predisposition for disease resistance/susceptibility. Similarly, we should not expect polymorphisms at one locus to have a large effect on disease resistance/susceptibility.

This analogy made some simplifying assumptions, including that each locus (or marble) contributed equally to expression of the phenotype and that the effects of each locus interact additively. However, the analogy effectively illustrates the principle of polygenic variation. Also note that although we often measure disease resistance/susceptibility as a binary ("sick" or "healthy") trait, true disease resistance/susceptibility is in reality continuously distributed. Disease resistance/susceptibility is therefore a threshold trait. Animals are often not completely resistant or completely susceptible to a disease. As an example, consider two animals, "Barney" and "Bert." Barney is more susceptible to a disease than Bert. When Barney and Bert are exposed to low pathogen loads, neither becomes sick. When both animals are exposed to a moderate pathogen load, Barney becomes sick and dies while Bert will become sick but recovers. However, when Barney and Bert are exposed to very high pathogen loads, both animals become sick and die. Neither animal is completely susceptible or resistant to the pathogen, but Bert is more resistant to the pathogen than Barney.

In addition to the action of a large number of loci on a trait, the animal's environment often has a significant effect on trait expression. We can think of quite a few non-genetic factors that determine how an animal responds to a pathogen: exposure to the pathogen, stress levels in the animal, nutrition, vaccination protocols, herd management, and so on. Much of the variation we observe for a trait, even within a contemporary group, can still be explained by non-genetic factors. Good genetics does not negate the importance of nutrition, management, and veterinary practices. An animal's genetic potential for disease resistance/susceptibility may never be realized if the animal is not cared for properly.

Despite these hurdles, geneticists have developed tools for genetic selection for increased disease resistance. In most cases, these genetic tools have been developed with the aide of an **indicator trait** for disease resistance/susceptibility. An indicator trait is a trait that by itself is not economically important but is genetically correlated with a trait that is economically important, in this example disease resistance/susceptibility. New genetic tools for selecting animals with increased disease resistance are in the pipeline, and thus this chapter will likely be out of date by the time this textbook is published.

This summary of tools for genetic selection for disease resistance does not include traits related to productive life or longevity. Disease susceptibility is certainly a contributing factor towards longevity of an animal, along with other considerations such as fertility, behavior, feed intake, and production traits. However, we have limited our discussion to only those traits which are primarily concerned with infectious disease susceptibility. Further, several livestock species (e.g., horses, beef cattle) are not represented because genetic tools for selecting for disease resistance are not yet available in these species.

Dairy cattle

Perhaps the oldest and most well-known selection tool for disease susceptibility/resistance is the development of estimates of genetic merit for mastitis susceptibility. Mastitis is an intramammary infection that can be caused by a large number of diverse pathogens. This disease is most economically important in dairy cattle, although certainly mastitis occurs in other species, including humans. Mastitis increases costs for producers through increased number of veterinary treatments, labor, and milking time. Mastitis can also lead to premature culling and reduced milk quality. Unlike some other diseases, eradication and vaccination programs will likely be ineffective because of the diversity of pathogens that can cause mastitis in dairy cattle. Genetic selection was considered to be a tool that could be useful in reducing the frequency of mastitis in dairy herds.

A common theme throughout this chapter will be the difficulty of defining the optimal way for measuring disease resistance/susceptibility. Mastitis has been defined differently for the purposes of genetic selection depending upon management and information recording practices. In Scandinavia, the incidence of treatments for clinical mastitis on individual cows are recorded and used directly for estimating genetic merit. Because these countries record mastitis treatment records, incidence of clinical mastitis can be used directly as the phenotype for estimating genetic merit for this trait. Further, disease incidence is recorded directly by veterinarians because antibiotics in these countries can only be prescribed by licensed veterinarians. This characteristic of dairy management practices in Scandinavian countries allows for increased confidence in the data collected on clinical mastitis incidence. Genetic selection for decreased clinical mastitis based on these treatment records has been shown to be effective in increasing resistance to mastitis.

Most breed associations in other countries, however, do not routinely record disease incidence data, and when this information is recorded, the quality of the data is not as good as found in Scandinavia. Dairy breeders in these countries must rely on other phenotypes for estimating genetic merit for mastitis susceptibility/resistance. In some countries, such as the USA, somatic cell counts are routinely recorded. A somatic cell count is the number of somatic cells in milk. Most of these somatic cells are involved with the immune response to pathogens in the udder. The predominant somatic cell found in milk is the neutrophil, which is a phagocytic cell that is part of the animal's innate immune response. When an udder infection occurs, neutrophils and other white blood cells transit to the udder and multiply to attempt to control the infection. Thus, somatic cell counts are a useful indicator trait for mastitis susceptibility in cattle.

Somatic cell counts are transformed into somatic cell scores prior to their use in genetic evaluations. The somatic cell score has several properties that are more amenable for genetic analysis, the most important being that somatic cell scores are normally distributed. Somatic cell counts (SCC) are transformed into somatic cell scores (SCS) as follows:

$$SCS = \log_2\left(\frac{SCC}{100{,}000}\right) + 3 \tag{26.1}$$

An indicator trait is a trait that by itself is not economically important but is genetically correlated with an economically important trait. An indicator trait must meet three criteria: (1) the indicator trait (e.g., somatic cell score) must be genetically correlated with the trait of interest (e.g., mastitis susceptibility), (2) the trait of interest must be heritable, and (3) the trait of interest must be economically important or have value to the breeder. Milk somatic cell scores are genetically correlated with incidence of mastitis in cattle. Although genetic correlation (and heritability) estimates will vary across populations and environments, the genetic correlation between somatic cell scores and clinical mastitis susceptibility averages around 0.70. Keeping in mind that genetic correlations can vary from -1 to $+1$, this correlation is relatively high. In words, a positive genetic correlation between somatic cell score and clinical mastitis susceptibility means that cows with higher somatic cell scores would more likely have clinical mastitis.

Clinical mastitis is also heritable, which meets our second criteria for an indicator trait. Like most disease phenotypes, clinical mastitis susceptibility is lowly heritable. Most heritability estimates for this trait range from 0.02 to 0.10, depending on the population, environment, and statistical model used to estimate heritability. The heritability of somatic cell score is higher, averaging about 0.15 when considering average test day somatic cell scores across lactation. Finally, as we have established previously, clinical mastitis susceptibility is an economically important trait in dairy cattle. Thus, somatic cell scores are a good indicator trait for incidence of clinical mastitis and are commonly used to estimate genetic merit for mastitis susceptibility in dairy cattle breeds.

Like many livestock species, dairy breed associations publish economic selection indices which are intended to allow producers to rank animals based on only a single number: the economic value of the animal (see Chapter 9). Ideally, economic selection indices take into account all economically important traits and weight each trait according to its economic value to the producer. Economic indices published by a breed association must by necessity be generic in nature and not be tailored to a specific dairy operation. The economic weight assigned to each trait therefore assumes average costs and incomes for all dairy producers who use the index. Three commonly employed selection indices in US dairy cattle are net merit ($NM), cheese merit ($CM), and fluid merit ($FM). Net merit is designed for most dairy producers and is based on the average price of milk across the USA. Fluid merit is designed for producers who do not receive premiums for protein while cheese merit is designed for producers who sell milk directly to a cheese plant.

Because mastitis susceptibility is assumed to be equally important to producers regardless of the price of milk, the relative weight placed on somatic cell score ranges from -7 to -9% for each of the three indices. The relative percentage emphasis is negative because lower somatic cell scores are more desirable. Recall that lower somatic cell scores are associated with decreased susceptibility to mastitis.

Swine

Susceptibility to most infectious diseases is a polygenic trait. However, in swine, susceptibility to enterotoxigenic *Escherichia coli* is a simply-inherited trait, controlled by only one locus. Enterotoxigenic *E. coli* with F4 fimbriae invades the pig by attaching its fimbriae to receptors on the surface of intestinal cells in the pig's gastrointestinal tract. In some pigs, the receptor that the fimbriae attaches to is not present on the surface of intestinal cells. Because the receptor is not present, enterotoxigenic *E. coli* cannot attach to the intestinal cells and colonize the gastrointestinal tract. The presence or absence of this receptor is determined by the genotype of a single locus in the pig's genome.

Unlike many simply-inherited genetic conditions, susceptibility to enterotoxigenic *E. coli* is a completely dominant trait. Only individuals who do not harbor a receptor allele that can bind to the *E. coli* fimbriae are protected from disease. Heterozygous individuals are as susceptible to infection as individuals who are homozygous for the receptor allele that can bind to the *E. coli* F4 fimbriae.

The gene that encodes for susceptibility to enterotoxigenic *E. coli* infection has not yet been discovered. The locus encoding susceptibility/resistance to this trait has been localized to a small region on pig chromosome 13. For many years, the most promising candidate gene for this trait was Mucin-4 (Muc-4) because (1) a polymorphism in this gene was in strong linkage disequilibrium with susceptibility/resistance to enterotoxigenic *E. coli* adhesion and (2) the gene encoded for a sialoglycoprotein known to bind *E. coli* F4 fimbriae. Recently, however, researchers found that the region of the swine genome in strongest linkage disequilibrium with susceptibility/resistance to *E. coli* infection does not include Muc-4. Instead, these researchers proposed that HEG homolog 1 (Heg-1) and Integrin beta-5 (Itgb-5) were more likely responsible for susceptibility/resistance to this infection. The fact that the causative mutation underlying this simply-inherited phenotype has still not been identified underscores the difficulty in identifying causative mutations for complex phenotypes affected by hundreds of loci and environmental factors.

A genetic test for susceptibility to enterotoxigenic *E. coli* does exist despite the absence of a known causative mutation for this trait. The most common polymorphism used for this genetic test is found in Muc-4. As mentioned previously, although this polymorphism is probably not the causative mutation for susceptibility/resistance to *E. coli* infection, the Muc-4 polymorphism is in strong linkage disequilibrium with this phenotype. After genetic testing, selection for resistance to *E. coli* colonization is relatively straightforward: only breed individuals that do not harbor an allele for susceptibility to *E. coli*. The frequency of the susceptibility allele can be quickly reduced to zero in only a single generation. Of course, most infectious diseases are polygenic and genetic selection will be much less straightforward for polygenic diseases.

Sheep

Scrapie is an infectious disease of sheep that is caused by an abnormally folded **prion** protein. The disease attacks the nervous system and causes a diversity of clinical signs, including a "hopping gait," uncoordinated movement, abnormal behavior, and pruritus. No treatment for scrapie is available and the disease can be transmitted by contact with infected sheep, particularly by contact between a lamb and its ewe. Similar diseases exist in other species, including humans (Creutzfeldt–Jakob Disease) and deer (chronic wasting disease).

Like enterotoxigenic *E. coli* in pigs, susceptibility to scrapie is determined by alleles at a single locus: the prion gene. Three polymorphisms in the prion gene are known to affect susceptibility to scrapie. These polymorphisms are found within codons 136, 154, and 171 in the prion gene. The variant with the largest effect on scrapie susceptibility is found at codon 171. Two alleles, each encoding a different amino acid, are found at codon 171: arginine (R) and glutamine (Q). Individuals who are homozygous RR at codon 171 are highly resistant to scrapie infection. Only one RR individual has ever been reported to harbor a scrapie infection. Individuals who are heterozygous QR at codon 171 are also highly resistant to scrapie, but not as resistant as individuals homozygous for the R allele. Individuals who are homozygous QQ at codon 171 are susceptible to scrapie infection; a high percentage of individuals infected with scrapie are QQ at codon 171.

Alleles at codon 136 also affect susceptibility to scrapie, although codon 136 has a larger effect on scrapie susceptibility in Europe than in the USA. This geographical difference in scrapie susceptibility may be due to strain differences in the abnormal prion protein between Europe and the USA. At codon 136, two alleles are present: alanine (A) and valine (V). Individuals who are homozygous AA are more resistant to scrapie infection than individuals who are homozygous VV. Although AA individuals are more resistant to scrapie infection than VV individuals, this resistance is not complete. Individuals with an AA genotype at codon 136 can become infected with scrapie.

The effect of alleles at codon 154 is more ambiguous. Two alleles are present at codon 154: arginine (R) and histidine (H). However, the effect of each genotype on scrapie susceptibility is unclear. In some studies, individuals who are homozygous RR are more resistant than individuals who are HH, but in other studies, the opposite is true. Although this codon is classically considered to determine scrapie susceptibility in sheep, it is becoming increasingly more likely that the codon 154 alleles are in linkage disequilibrium with the causative mutations for scrapie susceptibility.

Genetic testing is available for scrapie susceptibility in sheep. Because the prion gene has a large effect on scrapie susceptibility, selecting sheep that are homozygous for resistance alleles (especially at codon 171) should significantly decrease the frequency of scrapie in sheep flocks. One caveat is that, although the prion gene has a large effect on scrapie susceptibility, it is possible that other loci with smaller effects may modify resistance/susceptibility to this disease. These modifier loci may explain cases where individuals homozygous for the resistance genotype at codon 136 can still become infected with scrapie.

A misunderstanding often arises when discussing the meaning of genetic susceptibility to scrapie. Genetic susceptibility to scrapie is not the same as being infected with scrapie. An entire flock could be genetically susceptible to scrapie; however, individuals in the flock may never become infected with scrapie if they are never exposed to the abnormal prion protein. A scrapie infection requires both genetic susceptibility to scrapie and exposure to the scrapie-causing prion protein.

Chickens

Marek's disease is a lymphoproliferative disease caused by herpes viruses. Four different forms of Marek's disease exist: cutaneous, neural, ocular, and visceral. The clinical signs of disease are highly variable and depend on the form of Marek's disease. The disease is spread through contact with dust particles and dander that are harboring the virus. Another route of transmission is contact with infected chickens. No treatment is available for Marek's disease; however, a vaccine is available. Unfortunately, the vaccine is not completely effective at controlling the disease. Besides vaccination and good sanitation practices, selective breeding can be used to reduce incidence of Marek's disease. Marek's disease susceptibility is heritable and is a polygenic trait in chickens.

It has been recognized that Marek's disease susceptibility (as measured by mortality to the disease) was a heritable trait as early as the 1960s. Heritability estimates for this trait range from about 0.12 to 0.72 and appear to be highly dependent upon the population studied and other environmental factors, including Marek's disease strain. Several studies have shown that genetic selection for Marek's disease susceptibility and resistance is possible. Genetic lines of chickens with high and low mortalities after exposure to Marek's disease have been created. Since at least the early 1970s, the inclusion of Marek's disease susceptibility/resistance has been considered for selection indices in the poultry industry. Because of the economic losses that would result from a Marek's disease outbreak, many poultry breeders are incorporating Marek's disease susceptibility/resistance as part of their selective breeding programs.

These selective breeding programs often included DNA marker information at an earlier time point than other livestock species. The MHC B alleles were known to be associated with Marek's disease; the B 21 allele at this locus increases resistance to this disease. Numerous **quantitative trait loci** (QTL) have also been discovered that affect resistance to Marek's disease. Because of the vertical integration of the poultry industry and proprietary nature of selective breeding programs, it is difficult to gauge the extent of inclusion of molecular markers in breeding programs for Marek's disease resistance/susceptibility. It is clear that resistance to this disease is being included in layer selection programs, however.

Current research into genetic selection for livestock health

It is not helpful to animal genetics students to include all of the current research into the genetics of animal health traits. The examples given in this chapter were selected because of their importance to animal health and increasing attention given to the disease by animal genetics researchers. The reasons why animal geneticists have focused on these diseases as well as the challenges faced by geneticists to estimate breeding values for resistance/susceptibility to these diseases are emphasized. Although key discoveries will be mentioned, an exhaustive list of all of the heritability estimates and discovered QTL will be avoided.

Johne's disease

Johne's disease is an infectious disease of cattle that is caused by *Mycobacterium paratuberculosis*. Most calves that develop Johne's disease appear to become infected with the causative bacterium early in life. *M. paratuberculosis* infects the small intestine of cattle and clinical signs of the disease usually do not present itself until about 4 or 5 years of age. Clinical signs of the disease include rapid weight loss, decreased appetite, diarrhea, and decreased milk production. Subclinical *M. paratuberculosis* infection often will go undetected in dairy herds and results in decreased milk yields.

The disease presents challenges to veterinarians and dairy producers. First, detecting cattle that are infected with *M. paratuberculosis* is difficult. Cattle often do not begin producing antibodies to *M. paratuberculosis* until about their second or third lactation, which decreases the effectiveness of antibody testing for *M. paratuberculosis* at young ages. Fecal culture of *M. paratuberculosis* has been used to diagnose *M. paratuberculosis* infection, but *M. paratuberculosis* shedding in feces is intermittent. Thus, the available diagnostic tests for *M. paratuberculosis* infection do not detect all of the infected animals in a herd. Second, a good vaccine for *M. paratuberculosis* does not exist. To control an outbreak of Johne's disease in a dairy herd, producers need to continually test their herd with the available diagnostic tests for the disease and cull all animals that are positive for infection by the diagnostic test. This test and cull approach will take many years to eliminate *M. paratuberculosis* from a dairy herd, but is the only option available to dairy producers. For this reason, animal geneticists have been interested in testing whether selective breeding could be used to reduce the incidence of Johne's disease.

Susceptibility to *M. paratuberculosis* is heritable. The heritability of this trait ranges from about 0.05–0.15 depending on the population tested, disease prevalence, and other environmental factors. Several QTL have also been identified for susceptibility to this infection. Interestingly, QTL are not always consistent across studies. This lack of consistency is not uncommon when discovering QTL for economically important traits in livestock, and may be the result of false-positives and false-negatives, genotype by environment interactions, and different genetic backgrounds of the animals used in the studies. Whereas one polymorphic locus may have an effect on susceptibility to disease in one population/breed in one environment, that same locus may have no effect on susceptibility to the same disease in other populations and environments. Genetic testing companies have recognized this fact; for example, Pfizer Animal Genetics is marketing separate DNA marker panels for Black Angus and Red Angus and for cattle raised in North America and other parts of the world. A "one size fits all" approach may not be appropriate for DNA testing for polygenic, economically important traits.

Another challenge with predicting breeding values for susceptibility to *M. paratuberculosis* infection is the definition of a "resistant" animal. Unlike chickens and to some extent pigs, challenging cattle with a pathogen and measuring their response to the pathogen is often not practical, at least for the number of animals needed for most genetic studies. Bovine geneticists often rely on field data for their research. Cattle that are positive for *M. paratuberculosis* infection by diagnostic testing are clearly susceptible to this infection. However, cattle that are negative for *M. paratuberculosis* infection as determined by available diagnostic tests may not be resistant to this infection. Why could an animal be classified as *M. paratuberculosis* negative?

1. The animal was exposed to *M. paratuberculosis* and either never became infected or eliminated the infection. These animals would be resistant to *M. paratuberculosis* infection.

2. The animal was never exposed to *M. paratuberculosis*. If exposure never occurred, then even susceptible animals would be classified as *M. paratuberculosis* negative.

3. The animal was infected with *M. paratuberculosis*, but the diagnostic test did not detect the infection. A false-negative result has occurred.

The problem is that we cannot know which animals were exposed to *M. paratuberculosis* from field data. We also do not know which test results were false-negatives. Without the benefit of challenge studies, cattle geneticists need to control for exposure levels as much as possible. Only *M. paratuberculosis* test-negative animals from the same herd and at the same age as an *M. paratuberculosis* test-positive animal will be used for genetics studies. Further, only diagnostic tests with low false-positive and false-negative rates should be used to minimize technical variation.

Bovine Respiratory Disease (BRD) complex

The BRD complex in cattle is a disease that can be caused by a large number of pathogens, including bacterial (e.g., *Mannheimia hemolytica*, *Pasteurella multocida*, and mycoplasma) and viral (e.g., Bovine viral diarrhea (BVD) virus, Bovine respiratory syncytial virus, and parainfluenza-3) species. The BVD virus is particularly troubling because cattle can become persistently infected (PI) with BVD and constantly spread the virus to other animals in the herd. The BRD complex is the most economically important infectious disease in beef cattle and causes morbidity and mortality in feedlots. The disease is most common during times of stress, including weaning and entry into the feedlot. Vaccines are available for the various pathogens that cause BRD, but they are not completely effective at controlling the disease. Because of the economic cost of BRD, animal geneticists are investigating the potential for genetic selection for BRD resistance as a complementary tool for controlling the disease.

Like *Mycobacterium paratuberculosis* susceptibility, the heritability of susceptibility to BRD was estimated to be about 0.10. Susceptibility to BRD was also favorably genetically correlated with average daily gain and marbling score in beef cattle. Restated, increased BRD resistance was genetically correlated with higher average daily gains and higher marbling scores. Thus, genetic selection for BRD resistance should result in favorable genetic improvement in average daily gains and marbling scores, with little to no unfavorable genetic changes in other traits. Only one research group to date has published QTL affecting BRD resistance. These QTL were located on bovine chromosomes 2 and 26.

Also like *M. paratuberculosis* susceptibility, the definition of an animal that is truly resistant to the BRD complex is problematic. Complicating the definition of resistance is that numerous pathogens (e.g., bacteria, viruses) can cause BRD and an animal that is resistant to one BRD pathogen may be susceptible to other pathogens that cause BRD. Susceptibility to BRD has primarily been defined in one of two ways: (1) identification of animals that express clinical signs of BRD infection and (2) number of treatments received for BRD. Thus, a "resistant" animal either did not express clinical signs of BRD infection or was not treated for BRD. Lack of clinical signs or treatment for BRD may result because (1) the animal was truly resistant to BRD infection, (2) the animal was not exposed to a BRD-causing pathogen, or (3) clinical signs of BRD infection were not detected by farm personnel. Bovine geneticists often choose BRD-negative animals from the same herds where BRD-positive animals were found. Because BRD-negative animals were selected from the same herds as BRD-positive animals, the BRD-negative animals were more likely to be exposed to the same pathogen loads as the BRD-positive animals.

An alternative approach for defining BRD-resistance and –susceptibility is to focus on presence or absence of a specific pathogen that can cause BRD. For example, studies have focused specifically on cattle that are persistently infected with BVDV instead of the entire BRD complex as a whole. Because only a single pathogen is examined, the definition of our phenotype is more precise, but results may not be extrapolated to all of the pathogens that can cause BRD.

Porcine Reproductive and Respiratory Syndrome (PRRS) virus

The PRRS virus is the most economically important disease in the swine industry. As the name implies, this virus causes reproductive failure and respiratory problems. Additionally, the virus reduces the growth rate of infected pigs. Attempts to develop a vaccine for the PRRS virus have been unsuccessful. Because of the lack of efficacious vaccines and the economic importance of the disease, animal breeders have begun to investigate whether selective breeding could reduce the incidence of PRRS virus infections in swine herds.

Early evidence suggested that clinical signs resulting from PRRS infections were more severe in Hampshire and Hampshire crossbred pigs than in purebred Duroc, Meishan, and crossbred Landrace/Large White pigs. Although the contribution of specific breeds to severity of clinical signs following PRRS infection has not been established, these early results suggested that the severity of clinical signs following PRRS infection was heritable.

Unlike cattle, geneticists studying genetic resistance to PRRS have employed challenge studies. As a result, the number of animals used to study PRRS genetic disease resistance tends to be smaller than for the BRD complex and Johne's disease studies discussed previously. The one major study investigating PRRS genetic resistance used about 600 pigs, compared to (usually) more than 1500 cattle sampled when studying BRD and Johne's genetic disease resistance. One advantage to using challenge studies instead of field data is that we can be more certain that each animal sampled was exposed to the pathogen. More uniform exposure to the pathogen decreases environmental variance, leading to higher heritabilities and higher power for detecting QTL. Thus, all else being equal, smaller sample sizes can be used for challenge studies relative to field data to achieve the same level of statistical power.

For PRRS, a commonly used measure of disease susceptibility/resistance is the development of viremia after challenge with the PRRS virus. Viremia is the presence of virus in the bloodstream. Often a real-time PCR assay is used to measure the amount of PRRS virus in blood, providing geneticists with a quantitative trait that can be used to assess the ability of the pig to control the virus. Average daily gain after challenge with PRRS virus is another phenotype commonly used; pigs that are most susceptible to PRRS virus should have lower average daily gains.

Using these measures of PRRS disease resistance/susceptibility, geneticists discovered a QTL with large effect on PRRS resistance. This QTL was found on chromosome 4 and explained about 16% of the genetic variation for viral load as measured by real-time RT-PCR. Further, the frequency of the favorable allele in this study was only 0.16, suggesting that many animals do not carry the favorable allele for this trait. Taken together, selection on only a single locus may have a significant impact on PRRS resistance in swine herds, which is surprising because loci with major effects have not been found for many other polygenic traits studied to date. This finding still needs to be confirmed in an independent population of pigs, but early results for genetic selection for PRRS resistance are promising.

Challenges faced when studying the genetics of disease resistance in livestock

1. Disease prevalence may be low. This challenge is relevant when geneticists use field data for their research. If disease prevalence is low, many herds will not harbor the pathogen that causes the disease. Animals in these herds cannot be used for this research because many animals that are truly susceptible to the disease will appear resistant, simply because these animals have not been exposed to the pathogen. Advances in genetic resistance to disease will likely be most successful for diseases with moderate to high prevalences in a species, especially for species where challenge studies are infeasible.

2. How do you define disease resistance and susceptibility? As we learned in our discussion on Johne's disease, defining which animals are resistant to disease can be a challenge. Animals can be classified as uninfected or disease free because they are truly resistant to the pathogen, were not exposed to high enough pathogen loads for an infection to occur, or because of a false-positive diagnostic test. The definition of a disease susceptible animal, however, is not always straightforward, too. Upon infection, if one animal becomes sick and recovers and a second animal becomes sick and die, are both animals susceptible to the disease? Both animals are clearly susceptible to infection with the pathogen. However, only one animal dies as a result of the infection. It is critical to use a well-defined disease phenotype when studying the genetics of disease resistance/susceptibility.

3. Challenge methods are sometimes not feasible or ethical. Challenge studies are defined as exposure of a group of animals to a specific amount of a pathogen and measuring their response to the pathogen. Challenge studies are commonly employed when studying vaccine efficacy and other veterinary science questions. These studies often utilize less than 50 individuals. Genetics studies, on the other hand, often require hundreds, if not thousands, of individual phenotypes. For some species such as cattle, it is difficult to justify exposing large numbers of animals to a pathogen. Further, exposure of hundreds to thousands of animals to a pathogen is arguably unethical, especially for diseases that are painful to the animal.

4. Heritability of disease resistance/susceptibility is often low. The heritability of disease resistance/susceptibility is often low, which is expected given the large number of environmental factors that can affect whether an animal develops a disease. Environmental factors that can affect disease incidence include pathogen load, age, season of exposure, stress level of the animal, nutrition, and management. Traits with low heritabilities require geneticists to collect larger numbers of animals to accurately map genetic loci affecting resistance to the disease. Larger numbers of progeny records are also required to estimate genetic merit for disease resistance as well, relative to traits with higher heritabilities.

Should we select for increased disease resistance?

At first glance, this question seems silly. Why wouldn't we want to select for animals that are better protected from disease? Many good reasons for selecting for healthier animals exist, including increased economic productivity and food supplies, reduced poverty in developing countries, and improved animal welfare. We should be cautious about the potential genetic consequences from selection for increased disease resistance, however. Genetic selection can lead to unintended consequences, both in the short and long-term.

Unfavorable **genetic correlations** between resistance to a specific disease and other traits are one of the larger concerns geneticists have when selecting for disease resistance. Starting in the 1970s, mice were divergently selected for high and low antibody responses to sheep and pigeon red blood cells. The mouse line selected for high antibody response was more resistant to extracellular bacteria and produced greater amounts of antibody in response to antigen. Interestingly, the mouse line selected for lower antibody response was more resistant to intracellular bacteria. Further, the high antibody response mouse line was more susceptible to a large number of autoimmune disorders. Focusing on genetic selection for resistance to one pathogen may inadvertently cause decreased resistance to other pathogens and disorders.

Selection for increased disease resistance may also cause a loss in genetic diversity. The maintenance of genetic diversity is an important preventive measure used by populations to defend a species from potential pathogens. The Major MHC Loci are a good example of the importance of maintenance of genetic diversity (discussed in detail in Chapter 25). Most mammalian species harbor a large number of MHC alleles. When a pathogen invades a population, some individuals may harbor alleles that can protect against pathogen infection while other individuals may not possess protective alleles. However, the same allele that is ineffective at protecting an animal from one pathogen may be highly effective at protecting the same animal from a different pathogen. This allelic diversity increases the chance that at least some of the individuals in a population will be protected from any given disease. Strong genetic selection for disease resistance may decrease genetic diversity at loci important for an effective immune response. In the short-term, this genetic selection may decrease disease incidence but we must continuously guard against harmful loss of genetic diversity that could decrease fitness of our livestock populations in the long-term.

Finally, disease resistance could be unfavorably genetically correlated with other traits important to livestock production, for example, fertility, production, survivability, and end-product traits. We need to ensure that the benefits accrued to livestock and companion animals through genetic selection for disease resistance are not outweighed by reductions in other traits.

Summary

- Diseases can be placed into one of four categories: simply-inherited genetic defects, infectious disease, metabolic disease, and autoimmune disease. This chapter was primarily concerned with infectious diseases.
- Geneticists study infectious diseases in livestock for three primary reasons. First, if we can estimate genetic merit for resistance/susceptibility to disease, then we can select for animals that are genetically more resistant to disease. Second, livestock and companion animals often are good biomedical models for understanding the genetics of disease resistance/susceptibility in humans. Third, increased understanding of the genetics of disease resistance/susceptibility may lead to the development of better vaccines and pharmaceuticals.
- Infectious disease resistance/susceptibility is almost always a polygenic trait, meaning that a large number of loci and environmental factors determine the individual's response to pathogen exposure.
- Genetic tools for selecting healthier animals are scarce. The dairy industry estimates genetic merit for somatic cell score, which is an indicator trait for mastitis susceptibility. DNA tests are available for susceptibility to enterotoxigenic *Escherichia coli* infections in swine and scrapie susceptibility in sheep. The poultry industry has been selecting for decreased incidence of Marek's disease for decades.
- When studying the genetics of disease resistance, how "resistance" and "susceptibility" is defined is very important. An uninfected animal is not necessarily more resistant to disease; the animal could be uninfected simply because of lack of pathogen exposure. Similarly, animals that become infected but quickly recover may be considered resistant to the disease despite being susceptible to initial infection.
- Studying the genetics of disease resistance/susceptibility in livestock and companion animals usually requires sampling hundreds to thousands of animals. DNA markers associated with disease resistance cannot be discovered with only a few animals.
- Challenge studies insure that all animals in a study have been exposed to similar pathogen loads, but are difficult to execute in some livestock species.
- When selecting for resistance to a disease, we must be careful not to inadvertently select for susceptibility to other diseases and decreased fertility, production, and end-product traits. We also need to ensure that the species doesn't lose genetic diversity by selecting intensely on a single trait.

Key terms

Autoimmune disease – A disease that causes the immune system to recognize self-antigens as foreign to the body, for example, systemic lupus erythematosus.

Genetic correlation – An association between loci that affect two traits. Usually caused when a locus affects both traits (e.g., pleiotropy) but can also be caused by linkage between loci that each affect one trait.

Indicator trait – A trait that by itself is not important to breeders but is genetically correlated with a trait that breeders are interested in improving. For example, somatic cell score isn't important to dairy cattle breeders, but this trait is genetically correlated with mastitis susceptibility. Thus, somatic cell score is an indicator trait.

Infectious disease – A disease caused by a pathogen, for example, bacteria, virus, parasite, and so on.

Metabolic disease – A disease caused by improper balance between nutritional intake and production demands, for example, milk fever and ketosis.

Polygenic trait – Trait expression is influenced by a large number of loci and environmental factors.

Polymorphism – Variation in DNA sequence in a population.

Quantitative trait locus – A locus that harbors polymorphism(s) affecting a polygenic trait.

Simply-inherited genetic defect – A single locus harboring a polymorphism that causes a developmental defect during gestation or shortly after parturition. Often, but not always, fatal.

Further reading

Amor, S., P.A. Smith, B. Hart, and D. Baker. 2005. Biozzi mice: Of mice and human neurological diseases. *J Neuroimmunol* 165:1–10.

Boddicker, N., E.H. Waide, R.R.R. Rowland, J.K. Lunney, D.J. Garrick, J.M. Reecy, and J.C.M. Dekkers. 2012. Evidence for a major QTL associated with host response to Porcine Reproductive and Respiratory Syndrome Virus challenge. *J Anim Sci* 90:1733–1746.

Falconer, D.S., and T.F.C. Mackay. 1996. *Introduction to Quantitative Genetics*. 4th edn. Longman Group Limited, Essex, UK.

Friars, G.W., J.R. Chambers, A. Kennedy, and A.D. Smith. 1972. Selection for resistance to Marek's disease in conjunction with other economic traits in chickens. *Avian Dis* 16:2–10.

Imran, M., and S. Mahmood. 2011. An overview of animal prion diseases. *Virol J* 8:493.

Jacobsen, M., S.S. Kracht, G. Esteso, S. Cirera, I. Edfors, A.L. Archibald, C. Bendixen et al. 2009. Refined candidate region specified by haplotype sharing for *Escherichia coli* F4ab/F4ac susceptibility alleles in pigs. *Anim Genet* 41:21–5.

Minozzi, G., J.L. Williams, A. Stella, F. Strozzi, M. Luini, M.L. Settles, J.F. Taylor et al. 2012. Meta-analysis of two genome-wide association studies of bovine paratuberculosis. *PLoS One* 7:e32578.

Rupp, R., and D. Boichard. 2003. Genetics of resistance to mastitis in dairy cattle. *Vet Res* 34:671–88.

Snowder, G.D., L.D. Van Vleck, L.V. Cundiff, and G.L. Bennett. 2006. Bovine respiratory disease in feedlot cattle: Environmental, genetic, and economic factors. *J Anim Sci* 84:1999–2008.

Review questions

1. Most infectious disease resistance/susceptibility is polygenic, yet many of the available DNA tests for this trait are single gene tests (e.g., *E. coli* susceptibility in pigs, scrapie in sheep). Why do you think the available DNA tests for disease resistance/susceptibility have focused on simply-inherited traits?

2. Why do USA dairy breed associations select for somatic cell score instead of mastitis susceptibility?

3. A cow-calf producer is having problems with calf scours in his/her herd. The herd consists of about 100 cow-calf pairs. This spring, 13 calves were treated for scours. The producer approaches you with the idea of using his/her herd to develop a DNA test for scours susceptibility in cattle. Using only this herd, is development of a DNA test for scours feasible? Why or why not?

4. What is disease susceptibility? What is disease resistance?

5. Explain why genetic selection for resistance to a disease may have deleterious consequences on other traits.

27 Animal Genetics and Welfare

Amin A. Fadl and Mark E. Cook
Animal Sciences Department, University of Wisconsin–Madison, WI, USA

Introduction

Animal genetics have led to a remarkable increase in animal productivity. Increased production of milk, meat, eggs, and fiber has been critically needed for a growing world. Genetic engineering has provided excellent tools to develop animal models to study and find possible treatment for devastating diseases such as cancer, diabetes, cardiovascular diseases, and Alzheimer's disease. However, **extensive genetic selection based on single production traits can compromise the welfare of animals**. Such welfare-related problems include reduced reproductive efficiency and increased disease susceptibility in dairy cattle, skeletal disorders and behavioral change in poultry, and cardiac arrest and lameness in pigs. To decrease welfare issues in animals, **breeding strategies have had to combine selection for productivity along with welfare to assure sustainable farm animal production**. Establishment of a trait assessing system using physiological, phenotypic, and behavioral indicators will improve genetic selection and speed detection of welfare and production related problems, problems that can be addressed quickly. Breeding and genetic selection is fundamentally based on economics. Increasingly welfare fits into the economics of any breeding program. With the development of new genetic and non-genetic tools such as genomic selection and genetic markers, animal productivity, health, and welfare will continue to improve and advance production goals.

A continued need for genetic improvements and knowledge

Use of modern animal breeding techniques has dramatically improved animal productivity (e.g., growth, feed efficiency, reproduction, and disease resistance). For example, in the last 40 years milk production per cow has nearly doubled due to genetic selection and management strategies to realize genetic potential (Oltenacu and Broom, 2010). Broiler carcass weight increased nearly five-fold during a similar 40-year period (Havenstein et al., 2003). Such advances have assured an affordable supply of animal products for human consumption and efficient use of feedstocks to produce these products. According to the United Nations Food and Agricultural Organization (FAO) food production will need to continue to expand (approximately 1% per year just to meet population growth) for the next few decades in order to feed the world's population; a population expected to reach 9 billion by 2070 (FAO, 2009). As global economies grow, the demand for animal products will also expand (a demand on meat production that exceeds population growth). Global meat consumption has doubled since 1950 and continued expansion is predicted (Brown, 2001). In the past, increased human consumption of animal products has been complemented with increased animal productivity (Christensen, 1998). In the future, expanded demand for increased animal products, particularly in emerging economies and no additional landmass for cultivation, will require accelerated genetic improvements of livestock and poultry growth, feed efficiency, and reproduction using both traditional breeding and genetic engineering.

Genetic advances in agricultural animal productivity will require new methods to manage increased genetic potential. Added productivity without improved management strategies can adversely compromise animal welfare. For example, dairy nutritionists found that the energy demands for high milk producing cows had to be managed through the use of improved feeding practices. Without improved management practices, the genetic potential was either not realized or the animal's welfare was compromised. However, all genetic manipulations do not have management fixes and some may result in chronic welfare concerns. In situations where animal welfare cannot be maintained by the caretaker, scientists may have to step forward and set limits on genetic changes that result in a chronic animal welfare concerns that are not acceptable. Failure to set scientifically based limits could result in public concern and ultimately laws that further regulate the use of animals both in agriculture and in science.

Increasingly, genetics for improved animal productivity are in the hands of fewer and fewer commercial genetics companies. The companies that supply most of the genetics for swine, laying

Molecular and Quantitative Animal Genetics, First Edition. Edited by Hasan Khatib.

hens, and broiler chickens are controlled by a handful of primary breeders. Competition to remain as a primary source of genetics in these animal agricultural industries is fierce. While there have been breeding programs that have adversely affected animal welfare at times, genetics companies cannot remain profitable if the overall welfare of the animal is compromised to a point where animal productivity is adversely affected. Hence, the welfare of the animal is a primary concern to all commercial genetics companies. Just as animal genetics and breeding have greatly contributed towards efficient use of land and agricultural resources leading to increase in farm production and improved food quality and security, genetic selection and engineering will be critical to continued improvement in productivity, food quality, and the welfare of the animal.

In other ways, genetically engineered animals have served as critical models for the advancement of both human and animal medicine. For example, genetically engineered mice have served as key models for the development of therapies for a number of cancers (Hansen and Khanna, 2004). Advancements in the creation of genetically "defective" animals have demanded the development of new methods to manage animal welfare. Mice that were severely immunocompromised could be managed in environmental conditions that minimized exposure to potentially lethal pathogens. As with farmed animals, scientist may find they have to either develop specialize management of genetically engineered models or in some cases decide that welfare is compromised to such a great extent, that the animal model should be abandoned.

Welfare

Welfare defined

Geneticists must have a solid understanding of animal welfare issues when making genetic decisions involving animals. Animal welfare or "faring well" is the state of an animal's condition. Since the means by which an animal is "cared" for by humans can vary greatly, the welfare of the animal is also the result of "the ability of an animal to cope with the environment in which it lives" (as described by United States Department of Agriculture, or USDA, 2012, and the American Veterinary Medical Association, AVMA). Both the USDA and AVMA recognized that "animals may use a variety of behavioral and physiological methods when trying to cope with a perceived stressor within their environment." Hence, the definition of animal welfare is less about the methods of care or the environment that the animal lives in, and more about "the state of the animal" under said care and environment (Swanson, 1995). Swanson (1995) said, "the treatment that an animal receives is covered by other terms such as animal care, animal husbandry and humane treatment." Gonyou (1993) distinguishes the scientific (or "technical") use of the term "animal welfare" from its legal and public usage.

The terms "welfare" and "well-being" are often used as synonyms. Some studies have suggested that animal **welfare** should be used to describe long-term implications for the animal's well-being; and that animal **well-being** be used to describe the current state of the animal (Gonyou, 1993). Other studies have considered animal well-being as an integral part of the welfare. According to Hurnik (1990), animal well-being has been defined as "a state or condition of physical and psychological harmony between the animal and the environment with the absence of any condition that adversely affects health and productivity of the animal." The well-being is assessed based on the physiological and psychological evaluations. Physical well-being has been related to the absence of disease and injury. On the other hand, psychological well-being is more difficult to evaluate in animals, but is often done by assessing animal species-typical behaviors, lack of distress, and ability to effectively cope with the environment (Dawkins, 1990; Duncan and Petherick, 1991; Hetts, 1991; Mench, 1993; Petherick and Duncan, 1989).

Legal definitions are made by legislators, or in some cases public referendum, and serve as law to govern society's use of animals. Laws governing the use of animals do not always follow sound scientific animal welfare guidelines. Public definitions are often influenced by perceived animal need and compassion, and in some cases is an anthropomorphism of animal use. Public perception is a powerful force in the development of laws involving definitions of animal welfare. The following review will focus only on the technical or scientific definition of animal welfare.

Assessment of animal welfare

Several criteria have been used to measure animal welfare; the most widely recognized criteria are referred to as the "five freedoms" (Webster, 1997): (1) freedom from thirst and hunger which is assessed by availability of fresh water and feed; (2) freedom from discomfort; is determined by assessing the animal environment including shelter and a comfortable resting area; (3) freedom from pain, injury, and disease, and availability of veterinary care including preventative care and proper treatment; (4) freedom to express normal behavior, which is judged by the presence of sufficient space, proper facilities and company of the animal's own kind; and (5) freedom from fear and distress, this is achieved by avoiding stress suffering. Other welfare assessment parameters include physiological and visible indicators (Broom, 1993) such as life expectancy, reproductive performance, weight change, disease susceptibility, morphological, and behavioral changes. For example, weight changes, poor weight gain, susceptibility to diseases, anatomical defects, and behavioral change are indicative of poor welfare and most likely to be associated with housing and management as well as environmental factors. Regardless of the method of assessment, for animals under the care of humans, the responsibility of animal welfare and well-being falls upon the human caretaker and those who supervise their development and use.

Genetic advancement and animal welfare

Genetic selection and improved farmed animal welfare

Domestication was selective breeding's greatest contribution to the welfare of agriculture animals. Animal that could be managed by the early domesticator and had the temperament for human control and confinement became established species that were cultivated. Animals capable of adapting to the domesticated environment realized considerable "freedom." The caretaker assumed the responsibility to provide food and water, to shelter, to prevent the fear of suffering and to treat the animal when injured or sick. Genetic selection for behavior is an ongoing field of research and has the opportunity to continue to improve animal welfare and productivity (Newman, 1994).

Numerous examples of modern breeding programs demonstrate the importance of genetic selection and animal welfare.

One breeding strategy that can improve or harm the welfare of farmed animals involves breeding for maternal behavior (Dwyer, 2008). Studies involving sheep have clearly demonstrated bred differences with regards to mothering ability and the subsequent survival of their offspring. Ewes bred for "calm" temperament were better at grooming their lambs than "nervous" ewes. Craig and Muir (1996) showed that group selection of laying hens to prevent cannibalism dramatically reduced mortality in laying hens with intact beaks. This and related works demonstrate that breeding programs can be created to eliminate modern agricultural practices (i.e., beak trimming), which are considered by some to be substitute for good animal welfare practices. Genetic selection for endotoxin resistance, whether inadvertently or intentionally, can dramatically improve animal welfare by reducing inflammatory responses that cause damage to the host (Cheng et al., 2004). Genetic selection for polled cattle, sheep, and goats can also improve animal welfare by reducing injury from animal to animal contact. Genetic selection and breeding have significantly improved productivity and welfare in pigs. For example, when pigs are homozygous for the PSS (Porcine Stress Syndrome) gene, excitement through transport or handling can induce cardiac arrest and death. Pigs heterozygous for the PSS gene do not suffer lethal effects from stress. However, it should be pointed out that the presence of the PSS gene in the genetic pool of swine was also a byproduct of modern breeding techniques for increased productivity. Offspring of PSS homozygous positive boars bred with the PSS homozygous negative sows had a higher percentage of lean muscle and larger loin eyes (Aalhus et al., 1991). Therefore, genetic selection for improved muscling introduced the PSS gene, but breeding practices with knowledge of the PSS gene allowed for improved animal productivity without an adverse effect on animal welfare in the progeny. Genetic selection also has the potential to influence the environment. Williams et al. (2006) showed that genetic selection for improved feed efficiency in broilers has significantly reduced their carbon footprint.

Use of genetic engineering to improve animal welfare

Genetic engineering techniques are well developed for mice; however genetic engineering use in farm animals has not been adopted in animals producing food. Techniques to genetically manipulate farm animals are available, but the main barriers are regulatory issues and in some societies, public resistance to consuming engineered animals. The first genetically modified animal to receive FDA approval was the salmon and it is discussed in another chapter. The use of genetic engineering to modify, insert or remove genes that have the potential to not only improve animal productivity, but also to improve animal welfare are inevitable. Using the previous example, once the PSS gene responsible for pig death due to hyperexcitability was found, the gene could have been silenced, removed, or modified without the use of generations of animal breeding to prevent the gene's adverse effects. Genetic engineering could be used to develop new milk and egg products that are more nutritious or of unique value for certain medicinal uses. Dairy herds could be engineered to create specialized products. For example, whey protein glycomacropeptide is low in phenylalanine and serves as an excellent protein for children born with phenylketonuria. The yield of this protein in milk is low. However, through genetic engineering, specialized cows could produce the protein at high concentrations, making this a readily available protein to the more than 30,000 children born each year with the disease (van Calcar and Ney, 2012). While this may not directly affect the welfare of the cow, indirectly the added value to each animal will likely improve overall welfare standards to meet FDA medicinal requirements. Milk could also be modified through engineering means to provide specific needs for suckling offspring. For example, genetic engineering could provide beneficial supplements to milk such as natural growth factors, improving neonatal health and growth. In this case the welfare benefits would be direct. Antibodies produced in genetically engineered animals may help prevent mastitis and other diseases in animals, improving the well-being of these animals. Genetic modification could have a great effect on feed efficiency and positively impact the environment. Genetic introduction of enzymes such as phytase into swine and poultry have been shown to increase the availability of phosphorus from phytic acid in corn and soy products by increasing digestion of the dietary phytate phosphorus. Increasing phytate digestion reduces phosphorus output in the animal excreta and lessens the environmental impact of phosphorus pollution from animal excreta.

While the concepts have not received commercial application, farm animals have been genetically engineered. Genetic engineering has significantly contributed to the welfare and well-being of animals in several ways. Introduction of certain genes have significantly improved disease resistance in these animals. For example, transgenic cows that produce antimicrobial peptide lysostaphin were protected against *Staphylococcus aureus* mastitis, an infectious disease of mammary glands of high economic importance in dairy animals (Van Hekken et al., 2009). It has also been shown that deletion of genes encoding the intestinal receptor for the k88 antigen on *E. coli* provided resistance against certain strains of the bacteria (Gibbons et al. 1977). Similar approaches could potentially be used to develop transgenic animals resistant to other diseases such as bovine spongiform encephalopathy (BSE) and scrapie in sheep. Most importantly, several genes have been identified that affect fertility and reproductive performance in farm animals. Estrogen receptor (ESR) and the Boroola fecundity (FECB) genes were shown to increase litter size in pigs, compared to pigs that do not contain ESR gene (Davis et al., 1982; Rothschild et al., 1994). The FECB gene was found to be associated with an increase in ovulation rate and fertility in Merino sheep. Improvement in quality, length and color of wool and fiber from sheep and goats are also an area of focus for genetic engineering in livestock. While some of these latter examples do not necessarily lead to improved animal welfare, they do provide examples of how genetic engineering can have broad impacts on both animal productivity and welfare.

Biomedical research that improves human and animal welfare

In addition to improved animal production, genetics have been used to create laboratory animal models to study serious human and animal diseases. Cloned and genetically modified animals used in biomedical research helped to make advances in several fields possible. Mouse models have been genetically engineered to study cancer (e.g. cervical cancer, Hodgkin's lymphoma, mammary tumor, and ovarian cancer), metabolic disorders such as diabetes, and infectious diseases including influenza, scrapie, and gastric ulcers. Studies using these animal models lead to the development of detection methods and new therapies for treatment (Hansen and Khanna, 2004). Moreover, genetic engineering and biotechnology are already helping to preserve endangered and extinct animal species. Another potential use of genetic engineering is to create animals capable of producing biological compounds of therapeutic or nutritional importance that are naturally absent in animals. In these types of animals, a gene of human origin is introduced in the animal genome so that they are able to produce a specific protein, often in the milk, that can then be used in the production of a particular medicine. It is important that the animal's welfare is considered when these laboratory animal models are developed.

Genetic selection that adversely affects farmed animal welfare

While it is clear that traditional animal breeding and modern animal genetic engineering have greatly improved animal production and the economic value of animals, genetic manipulation can also adversely affect animal health, behavior, and welfare. As mentioned, genetic selection and breeding have doubled the production of milk production of each cow in the last 40 years. Extensive selection for milk yield has been associated with a decline in reproductive efficiency, increased the incidence of health problems and decreased longevity in modern dairy cows (Oltenacu and Broom, 2010). Studies have indicated an unfavorable genetic correlation between milk yield and incidence of mastitis and lameness (Dunklee et al., 1994). An increase in the incidence of lameness in dairy cattle has doubled in the last three decades. In turn, lameness is a significant cause of decreased milk production in the dairy cow (Oltenacu and Broom, 2010). An increase in disease susceptibility associated with genetic selection has been cited for reduced ability to breed and consequently decreased longevity of dairy animals. In beef cattle, breeding for increased muscle growth has resulted in heavily muscled cattle; however, the increased muscling has also been related to an increase in the incidence of calving by cesarean section (Broom, 1993).

In pigs, genetic selection for meat production traits has been associated with decreased reproductive ability and increased lameness (Dickerson, 1973). As mentioned, pigs that contained a PSS gene from both the sire and the dam (homozygous positive) were shown to be prone to cardiac arrest when stressed (Aalhus et al., 1991). Selection for leanness in pigs appears to increase their susceptibility to transport death when compared to their less lean counterparts. Genetically engineered pigs carrying human growth hormone showed increased growth rate, however, these genetically altered pigs had increased lameness, uncoordinated gaits, lethargy, mammary development in males, disruption of estrous cycles, and skin and eye problems.

Genetic selection for increased egg and meat production in poultry has also been shown to adversely affect bird welfare. In layers, selection for increased egg production was associated with an increased in skeletal defects. Selection for rapid muscle growth of poultry was shown to cause morphological, physiological, and behavioral changes. In broiler chickens, more skeletal disorders were observed when the birds were reared in cages (Hester, 1994). Breeding has therefore largely confined meat type birds to floor rearing. Physiologically, an increase in muscle mass in broiler chickens without an increased growth of internal organs results in birds with reduced cardiopulmonary capacity, relative to their muscle mass, affecting their physical fitness (Julian, 1993; Julian et al., 1986; Konarzewski et al., 2000). Increased ascites and sudden death syndrome were resulting welfare concerns related to breeding practices. Extensive selection for muscle growth also was related to reduced reproductive efficiency and natural mating ability in turkeys. Today, essentially all commercial turkey breeders are artificially mated. In turkeys, selection for increased muscle growth and large breasts resulted in leg problems and footpad problems, which affected the welfare because of the restriction of natural behavior (Dinnington et al., 1990).

The welfare of pure-bred dogs is also susceptible to adverse events due to breeding. While one might think that welfare is the top concern in the genetic attributes of dogs, selection traits for dogs can have profound effects on the morphology and physiology of a dog breed (Collins et al., 2011). For example, eye problems in collies and structural and back problems in dachshunds have been linked to genetic selection (Ott, 1996). To address breeding problems that can adversely affect the welfare of dogs, standards and databases for dog breeding have been developed by the American Kennel Club to assure that a mating meets health and welfare criteria (Collins et al., 2011).

The public will not turn a blind eye to breeding or genetic engineering practices that adversely affect animal health and welfare. The public has two powerful means to reject practices that they see as cruel. They can simply stop consuming the product (boycott the product) or they can push for the passage of laws that regulate the industry. Fortunately, the animal industry can set standards with regard to the genetics of their animals. The centralization of genetic ownership by a small group of genetics companies has resulted in powerful methods to assure that a breeding program or genetic engineering program is meeting the animal welfare standards. When the animal's welfare is compromised, the animal's productivity often declines and the value of the genetic material is considered inferior. In other words, genetics that compromise animal welfare will no longer be acceptable to the public or the animal producer. Increasingly, modern animal agriculture practices have evolved its own internal system of checks and balances to assure animal welfare.

Welfare concerns in animals used in medical research

The breeding of animal models that spontaneously or were engineered to express genetic defects has had a huge impact in the advancement of medical and pharmacological sciences. An animal model of a disease is almost always required to understand and develop strategies to treat disease or medical complications. Therapies developed in animals to treat disease provide the preclinical data necessary to obtain permission from an Institutional Review Board to conduct a study in humans. Genetically altered animal models have been engineered to study human diseases like diabetes, cancer or Alzheimer's disease. Animal models for breast cancer have been widely used to study various aspects of breast cancer biology. Obesity is a common disorder, and related diseases such as diabetes and cardiovascular diseases are a major cause of mortality and morbidity. Murine models of obesity have been very useful tools to study mechanisms of adipose tissue development and the effects of hormonal therapy. A mouse model of obsessive-compulsive disorder (OCD) in humans was created by deletion of a gene involved in the regulation of serotonin. The OCD transgenic mouse showed an inability to find its way through a maze, a simple problem for the mouse without the genetic defect (Joel, 2006). A mouse model of Lesch–Nyan syndrome in children has also been created. Children afflicted with Lesch–Nyan syndrome self-mutilate and bite themselves. With these two behavioral animal models, we have tools to develop treatments for both of these human behavior disorders (Jinnah et al., 1994).

Unlike unintended adverse consequences associated with farm animal breeding and engineering, the creation of animal models for biomedical research often are associated with intended adverse health events. Often, the welfare of the animal is compromised. Use of genetically altered animal models requires special care and animal management (e.g., strict disease, temperature, and humidity control). Federal, local, and institutional statues and policies are required to house and maintain genetically altered animals. Each research institution has an established Institutional Animal Care and Use Committee (IACUC), a committee enacted by the Animal Welfare Act, which closely monitors research and management involving these animal models. In a Gallup Poll of Public Opinion (Kiefer, 2004), 62% of Americans polled found that the use of animals for medical testing was morally acceptable versus 32% who felt it was morally wrong. As long as the public continues to support the use of animals in medical research, the United States can continue to expect to be a world leader in biomedical research. However, the strong internal rules governing the use of animals in medical experimentation are likely an important factor in assuring continued public support. Researchers and IACUCs must be forever vigilant that the standards and limits they set maintain the highest level of welfare that can be obtained in the quest for new advances in medical sciences.

Welfare: A cornerstone in the genetic advancement of farmed animals

Unintended consequences associated with animal breeding and genetic engineering is unavoidable. However, to maintain a breeding/genetic engineering program that knowingly results in adverse effects on the welfare of the animal is penny-wise and pound-foolish. Short-term gains might be realized, but in the long-term the animal producers who purchase the genetics and the public who purchase the final product will not tolerate compromised animal welfare. Genetic selection is a powerful tool that allows for the correction of mistakes that have resulted from a specific selection program. Some unintended consequences of genetic selection can be mitigated through altered management practices. However, to hold the animal manager and caretaker totally responsible for correcting welfare issues that are a product of breeding is problematic. Breeders and genetic engineers must be as cognizant of their breeding program as the caretaker is of the animal's need. Increasingly, genetics companies will find value and marketability of an animal's ability to faring well in the production setting; a trait that will be valued by the customer in addition to the standard productivity indices. Evidence of a negative correlation between animal productivity and health should send an alarm that productivity and welfare are not balanced.

The public is watching. In response to public concerns, many animal production companies are now taking serious steps towards setting welfare standards and monitoring welfare indicators alongside production traits. Although selection for productivity initially started with few traits such as growth, egg number, and milk quantity, today animal industry select for more than 50 traits, most of which are based on health and welfare (Avendano and Emmerson, 2009; Katanbaf and Hardiman, 2010). As a result, published works now indicate a significant decrease in health and welfare-related problems in the animals. For example, in one study when selection was based on milk production alone and no weight was given to resistance to mastitis, milk yield increased by 1179 kg and mastitis treatments increased to 12.9%. However, when resistance to mastitis was given more weight than milk production trait, the milk production increased by 964 kg and mastitis treatments decreased by 5.5% (Dunklee et al., 1994). Since treatment for mastitis results in a net loss of milk, production of milk was not affected and animal welfare was improved. This example shows that selection for production and welfare-related traits can yield improvements in both productivity and animal welfare.

Summary

Genetic advancement of farmed animals, through breeding and bioengineering, must include animal welfare as a strategic practice in improving the productivity of herds and flocks. While some unintended consequences associated with genetic traits can be addressed by altering the animal's environment or management, geneticists can no longer sit idle assuming the problems with be fixed by someone else. The success or failure of modern genetic companies that provide genetic material to animal producers will increasingly depend on the ability of the animal to fare well in an animal production environment. The public will not tolerate altered genetics that negatively impact farm animal well-being. The public may avoid the consumption of animal products they feel originate from animals whose welfare is compromised, or they may force the creation of new laws that govern animal agriculture.

Genetic engineering and the breeding of genetically altered animals for the purpose of biomedical research, where there is little question that the animal's welfare is compromised, is still acceptable to the general public. Genetically engineered animals with gene defects will be tolerated as long as the use of these animals is scientifically justified and their use has the potential to provide cures for diseases that afflict humans or other animals. The public has and will continue to demand that animals used for research are treated with the highest standard of care available and that their use is continually peer vetted by other scientists and veterinarians.

References

Aalhus, J.L., S.D.M. Jones, W.M. Robertson, A.K.W. Tong, and A.P. Sather. 1991. Growth characteristics and carcass composition of pigs with known genotypes for stress susceptibility over a weight range of 70 to 120 kg. *An. Prod.* 52:347–353.

Avendano, S., and D.A. Emmerson. 2009. Animal welfare and the future of poultry genetics. *Poult. Sci.* 88 (Suppl. 1):2 Abst.

AVMA. What is animal welfare? www.avma.org/kb/resources/reference/animalwelfare/pages/default.aspx (accessed May 14, 2014).

Broom, D.M. 1993. Assessing the welfare of modified or treated animals. *Livestock Prod. Sci.* 36:39–54.

Brown L.R. 2001. *World on the Edge. How to Prevent Environmental and Economic Collapse.* International Publishers, Taylor & Francis Group, Abingdon.

Cheng, H.W., R. Freire, and E.A. Pajor. 2004. Endotoxin stress responses in chickens from different genetic lines. 1. Sickness, behavioral, and physical responses. *Poult. Sci.* 83:707–715.

Christensen, L.G. 1998. Possibilities for genetic improvement of disease resistance, functional traits and animal welfare. *Agriculturae Scand. Sec A., Anim. Sci.* 29:77–89.

Collins, L.M., L. Asher, J. Summers, and P. McGreevy. 2011. Getting priorities straight: Risk assessment and decision-making in the improvement of inherited disorders in pedigree dogs. *Vet J.* 189:147–154.

Craig, J.V., and W.M. Muir. 1996. Group selection for adaptation to multiple-hen cages: beak-related mortality, feathering, and body weight responses. *Poult. Sci.* 75:294–302.

Davis, G.H., G.W. Montgomery, A.J. Allison, R.W. Kelly, and A.R. Bray. 1982. Segregation of a major gene influencing fecundity progeny of Booroola sheep. *New Zealand J. Ag. Res.* 25:525–529.

Dawkins, M.S. 1990. From an animal's point of view: Motivation, fitness, and animal welfare. *Behav. Brain Sci.* 13:1–9.

Dickerson, G. E. 1973. Inbreeding and heterosis in animals. In: *Animal Breeding and Genetics Symposium in Honor of Dr. Jay L. Lush*, pp. 54–77.

Dinnington, E.A., P.B. Siegel, and N.B. Anthony. 1990. Reproduction fitness in selected lines of chickens and their crosses. *J. Hered.* 81:217–218.

Duncan, I.J.H., and J.C. Petherick. 1991. The implications of cognitive processes for animal welfare. *J. Anim. Sci.* 69:5017–5022.

Dunklee, J.S., A.E. Freeman, and D.H. Kelley. 1994. Comparison of Holsteins selected for high and average milk production. 2. Health and reproductive response to selection for milk. *J. Dairy Sci.* 77:3683–3690.

Dwyer, C.M. 2008. Genetic and physiological determinants of maternal behavior and lamb survival: implications for low-input sheep management. *J. Anim. Sci.* 86 (E. Suppl.):E246–E258.

FAO. 2009. State of food insecurity in the world 2009-FTP FAO. ftp://ftp.fao.org/docrep/fao/012/i0876e/i0876e.pdf (accessed May 14, 2014).

Gibbons, R.A., R. Sellwood, M. Burrows, and P.A. Hunter. 1977. Inheritance of resistance to neonatal *E coli* diarrhoea in the pig: examination of the genetic system. *Theor. Appl. Gene.* 51:65–70.

Gonyou, H.W. 1993. Animal Welfare: Definitions and assessment. *J. Agric.* Ethics 6 (Suppl. 2):37–43.

Hansen, K., and C. Khanna. 2004. Spontaneous and genetically engineered animal models: Use in preclinical cancer drug development. *Eu. J. Cancer.* 40:858–880.

Havenstein, G.B., P.R. Ferket, and M.A. Qureshi. 2003. Carcass composition and yield of 1957 versus 2001 broilers when fed representative 1957 and 2001 broiler diets. *Poult. Sci.* 92:1509–1518.

Hester, P.Y. 1994. The role of environment and management on leg abnormalities in meat-type fowl. *Poult. Sci.* 73:904–915.

Hetts, S. 1991. Psychological well-being: conceptual issues, behavioral measures, and implications for dogs. *Vet. Clin. North Am. Small Anim. Pract.* 21:369–387.

Hurnik, J.F. 1990. Animal welfare: ethical aspects and practical considerations. *Poult. Sci.* 69:1827–1834.

Jinnah, H.A. B.E. Wojcik, M. Hunt, N. Narang, K.Y. Lee, M. Goldstein, et al. 1994. Dopamine deficiency in genetic mouse model of Lesch–Nyhan disease. *J. Neurosci.* 14:1164–1175.

Joel, D. 2006. Current animal models of obsessive compulsive disorder: A critical review. Prog Neuro-Pshchopharma. *Biol. Psych.* 30:374–388.

Julian, R.J. 1993. Ascites in poultry. *Av. Path.* 22:419–454.

Julian, R.J., J. Summers, and J.B. Wilson. 1986. Right ventricular failure and ascites in broiler chickens caused by phosphorus-deficient diets. *Avian Dis.* 30:453–459.

Katanbaf, M.N., and J.W. Hardiman. 2010. Primary broiler breeding-striking a balance between economic and well-being traits. *Poult. Sci.* 89:822–824.

Kiefer, H.M. 2004 (May 25). *Americans unruffled by animal testing.* The Gallup Poll of Public Opinion.

Konarzewski, M., A. Gavin, R. McDevitt, and I.R. Wallis. 2000. Metabolic and organ mass responses to selection for high growth rates in the domestic chicken (*Gallus domesticus*). *Physiolo. Biochem. Zool.* 73:237–248.

Mench, J. A. 1993. Problems associated with broiler breeder management. In: *Proceedings of the Fourth European Symposium on Poultry Welfare.* Universities Federation for Animal Welfare, Potters Bar. pp. 195–207.

Newman, S. 1994. Quantitative-and molecular-genetic effects on animal well-being: adaptive mechanisms. *J. Anim. Sci.* 72:1641–1653.

Oltenacu, P.A., and D.M. Broom. 2010. The impact of genetic selection for increased milk yield on the welfare of dairy cows. *Animal Welfare.* 19:39–49.

Ott, R.S. 1996. Animal selection and breeding techniques that create diseased populations and compromised welfare. *J. Am. Vet. Med. Assoc.* 15:1969–1974.

Petherick, J.C., and I.J.H. Duncan. 1989. Behavior of young domestic-fowl directed towards different substrates. *Brit. Poult. Sci.* 30:229–238.

Rothschild, M.F., D.A. Vaske, C.K. Tuggle D.G. McLaren, T.H. Short, G.R. Eckhardt, A.J. et al. 1994. A major gene for litter size in pigs. In: *Proceedings of the Fifth World Congress Genet. Applied to Livestock Production*, 21, 225–228, Guelph, Canada.

Swanson, J.C. 1995. Farm animal well-being and intensive production systems. *J. Anim. Sci.* 73:2744–2751.

USDA. 2012. The United States Comments on the Definition of Animal Welfare. www.aphis.usda.gov/animal_health/vet_accreditation/nvap_modules/AWIC/Assets/NVAP_Mod22_Animal_Welfare_Introduction_Sept2012.pdf (accessed May 15, 2014).

Van Calcar, S.C., and D.M. Ney. 2012. Food products made with glycomacropeptide, a low-phenylalanine whey protein, provide a new alternative to amino acid-based medical foods for nutrition management of phenylketonuria. *J. Acad. Nutr. Diet.* 112:1201–1210.

Van Hekken, D.L., R.J. Wall, G.A. Somkuti, M.A. Powell, M.H. Tunick, and P.M. Tomasula. 2009. Fate of lysostaphin in milk from individual cows through pasteurization and cheesemaking. *J. Dairy Sci.* 92:444–457.

Webster, J. 1997. Applied ethology: what use is it to animal welfare? *Advances in Ethology 32 Supplements to Ethology.* 10.

Williams, A.G., E. Audsley, and D.L. Sandars. 2006. *Determining the environmental burdens and resource use in the production of agricultural and horticultural commodities.* Main Report. Defra Research Project IS0205. www.silsoe.cranfield.ac.uk.

Review questions

1. What are accepted definitions of "welfare" and "well-being"? What criteria are used to measure welfare and well-being (consider the five freedoms)?

2. Indicate how animal genetics and breeding can adversely affect animal welfare? How can animal genetics and breeding improve animal welfare? Be prepared to provide specific examples of each.

3. How can the use of genetic engineering positively contribute to animal welfare and well-being?

4. Even though genetic manipulation of animals can alter animal welfare, how might these manipulations have long term benefit to animal and human health?

5. The final consumer of animal products can have a significant impact on animal welfare. How can the consumer influence genetic and breeding practices in animal industries?

28 Animal Biotechnology: Scientific, Regulatory and Public Acceptance Issues Associated with Cloned and Genetically Engineered Animals

Alison L. Van Eenennaam
Department of Animal Science, University of California, Davis, CA, USA

What is animal biotechnology?

Biotechnology is defined as the application of science and engineering to living organisms. From this definition, it is obvious that animal breeders have been practicing biotechnology for many years. For example, traditional selection techniques involve using measurements on the physical attributes and biological characteristics of the animal (i.e., applying science) to select the parents of the next generation. One only needs to look at the amazing variety of dog breeds (see www.akc.org/breeds/breeds_a.cfm) to realize the influence that animal breeders can have on the appearance and characteristics of animals from a single species. Selection based on appearance is sometimes associated with unwanted deleterious effects on other traits such as fitness. In dogs it has been noted that each of the top 50 breeds has one aspect of breed type that predisposes the breed to a genetic disorder (Asher et al., 2009). For example Bulldogs are prone to airway obstruction syndrome, and Cavlier King Charles Spaniels are affected by a reduced-size malformation of the skull related to strong selection on snout shape and for skull conformations that are steep caudally, respectively.

Although the term biotechnology is often associated with the relatively modern biotechnologies of cloning and genetic engineering (**GE**), which are the foci of this chapter, it is important to realize that other technologies such as progeny recording schemes to objectively measure performance and the application of statistical methods to calculate the genetic merit of an animal have enabled rapid genetic progress in domestic livestock populations. Genetic improvement through selective breeding (i.e., carefully choosing which animals will become parents of the next generation based on their estimated breeding value or genetic superiority) has been an important contributor to the dramatic improvements in animal production that have been achieved over the past 50 years. Perhaps this is nowhere more evident than in poultry breeding.

The body weight of broiler (meat) chickens at 8 weeks of age increased from 0.81 to 3.14 kg between 1957 and 2001, and approximately 80% of this four-fold increase was due to genetic selection (Figure 28.1).

Animals that can be grown to market weight at a younger age use proportionally less of their total feed intake on maintenance energy. In 1960, the average time needed to produce a broiler chicken in the United States was 72 days. By 1995, this was reduced to 48 days, even as the average slaughter weight increased by 0.4 kg. Concurrently, the feed conversion ratio (kg feed/kg gain) was reduced by 15% (Table 28.1).

These remarkable improvements in production efficiency have resulted in a dramatic reduction in the inputs required to produce a kilogram of chicken. From an environmental perspective, this genetic improvement has also resulted in reductions in greenhouse gas emissions and global warming potential per unit of animal product (e.g., dozen eggs or kg of chicken). However, some have argued that productivity improvements were achieved without adequately considering the effects on associated animal well-being and the welfare implications of these genetic improvements.

The global number of livestock animals used in agricultural production has been estimated to be 1.8 billion large ruminants, 2.4 million small ruminants (sheep and goats), 20 billion poultry and nearly one billion pigs (Niemann et al., 2011). Since the early 1960s, livestock production has grown rapidly with a worldwide four-fold increase in the number of chickens, two-fold increase in the number of pigs, and 40–50% increases in the numbers of cattle, sheep, and goats. This so-called "livestock revolution" is being driven by the sharp rise in demand for animal food products in many developing countries, resulting in a pronounced reorientation of agricultural production systems (Delgado, 2003). The United Nations Food and Agriculture Organization predict the global population will rise to approximately 8 billion people by 2030, and will exceed 9 billion people by 2050. Accordingly the demand for animal protein is also expected to grow as consumers in developing countries become more affluent. Although some may yearn for low input, pastoral livestock production systems, the increasing demand for animal protein is likely to require a sustainable intensification of livestock production

Molecular and Quantitative Animal Genetics, First Edition. Edited by Hasan Khatib.

Figure 28.1 Contemporary comparison of 1957 control and 2001 selected broiler carcasses fed the same diet and slaughtered at different ages (from left; 43, 57, 71, and 85 days). Modified from Hill and Kirkpatrick (2010), original photo by G.A. Havenstein. Reprinted with permission from the *Annual Reviews of Animal Biosciences*, Volume 1 © 2013 by Annual Reviews, www.annualreviews.org.

Table 28.1 Typical broiler performance in the USA from (a) Havenstein et al. (2003) and (b) Gordon (1974).

Year	Weeks of age when sold	Live weight (kg)	Feed efficiency (kg feed/kg gain)	Mortality (%)
1923[a]	16.0	1.00	4.7	18.0
1933[a]	14.0	1.23	4.4	14.0
1943[a]	12.0	1.36	4.0	10.0
1953[a]	10.5	1.45	3.0	7.3
1963[a]	9.5	1.59	2.4	5.7
1973[a]	8.5	1.77	2.0	2.7
1957[b]	12.0	1.43	3.84	4.7
2000[b]	6.0	2.67	1.63	3.6

systems. Visit the link, www.youtube.com/watch?v=6B-CH-NCdiY, to view a 5-minute music video contemplating the impact that genetic improvement has had on increasing the productivity of livestock over the past 50 years.

During the past century, several biotechnologies have been incorporated into programs aimed at accelerating the rate of the genetic improvement of livestock. One such technology is artificial insemination (**AI**), which is the deliberate introduction of semen into the reproductive tract of a female for the purpose of fertilization. AI allows the extensive use of well-proven, genetically superior sires and plays a major role in design of breeding programs and dissemination of advanced genetics. AI technology was introduced into the dairy industry and commercialized in the United States during the late 1930s to early 1940s. Today, approximately seventy per cent of all dairy cows in the US are bred using AI, as are virtually all turkeys and chickens. It provides an economical means for livestock breeders to improve their herds utilizing genetically superior males.

Although AI is now used routinely in animal breeding and human medicine, it was initially viewed with skepticism. There was a fear that AI would lead to abnormalities, and influential cattle breeders were originally opposed to the concept as they

Figure 28.2 The Holstein breeding bull, Elevation, lived in Plain City in the 1970s. Roughly half of the Holstein dairy cows in the United States today are believed to descend from Elevation.

believed it would destroy their bull market (Foote, 2002). When independent, university research demonstrated that the technology could be used to provide superior bulls, control venereal disease, and produce healthy calves, subsequent industry adoption was swift. To put the extensive use of AI in the US dairy industry in perspective, a single US bull named Elevation (Figure 28.2), born in 1965, had over 80,000 daughters, 2.3 million granddaughters, and 6.5 million great-granddaughters!

Such extensive use of this single exceptional bull clearly accelerated the rate of genetic gain, but also has the potential to reduce the genetic diversity of the dairy cattle population.

Cloning

Similar concerns regarding abnormal outcomes and reduced genetic diversity have been expressed about the use of animal cloning. A clone is an organism that is descended from, and has the same nuclear genomic DNA as, a single common ancestor. We routinely eat plant clones as many common fruits (e.g., bananas) and vegetables (e.g., potatoes) are clonally propagated. A variety of animals have also been intentionally cloned by animal breeders and researchers. There is a report of a cloned newt being produced as long ago as 1953! There are two basic methods that can be used to produce cloned animals: mechanical embryo splitting and nuclear transfer.

Embryo splitting involves bisecting a multi-cellular embryo at an early stage of development to generate clones or "twins." This type of cloning occurs naturally (e.g., human identical twins result from a spontaneous version of this process), and it can also be performed in a laboratory (Willadsen,1979) where it has been successfully used to produce clones from a number of different animal species. This technique was first used in agriculture to replicate valuable dairy breeding animals in the 1980s. The Holstein Association USA registered their first embryo split clone in 1982, and more than 2300 had been registered by October 2002 (Norman and Walsh, 2004). This method has a practical limitation in that only a small number clones, typically two, can be produced from each embryo and the genetic merit of the embryo is unknown (i.e., there are no individual or progeny performance records available on an embryo to know whether it is a genetically-superior individual).

Cloning can also be performed using a technique called somatic cell nuclear transfer (**SCNT**). Nuclear transfer involves transferring the nucleus from a somatic cell (containing a full diploid set of paired chromosomes) to an unfertilized oocyte that has been "enucleated" by removal of its own haploid set of chromosomes. Oocytes at the metaphase II stage of meiosis are the most appropriate recipient for the production of viable cloned mammalian embryos. In order to begin the development process, the donor nucleus must be fused with the egg through the administration of a brief electrical pulse, and then the egg is activated though exposure to short electrical pulses or a chemical fusion process, after which the embryo starts to divide as if it had been fertilized. These "reconstructed" embryos are typically cultured in petri dishes for 5–7 days until they reach the blastocyst stage. In the case of mammals, the embryo is then placed into the oviducts or uterus of a surrogate or "recipient" dam where it will develop until birth.

The first mammals were cloned via somatic cell nuclear transfer in the early 1980s, almost 30 years after the initial successful experiments with frogs. Numerous mammalian clones followed, including mice, rats, rabbits, pigs, goats, sheep, cattle, and even two rhesus monkeys named Neti (Neti stands for "nuclear embryo transfer infant") and Detto in 1997. The Holstein Association USA registered their first embryo nuclear transfer clone in 1989, and approximately 1200–1500 cows and bulls were produced by embryonic cell nuclear transfer in North America in the 1980s and 1990s. However, all of these clones were produced from the transfer of nuclei derived from early (8–32 cell) embryos. This was based on the assumption that cells from mammalian embryos lose totipotency (ability of a single cell to divide and produce all the differentiated cells in an organism) after the fifth cleavage division, and therefore a theoretical maximum of only 32 clones could be produced from each individual embryo.

This assumption was shattered by the birth of Dolly the sheep on July 5, 1996 (Wilmut et al., 1997). She was the first animal to be cloned via SCNT from a differentiated somatic cell derived from an adult. This result opened up the possibility that clones could be produced from a potentially unlimited number of cells from an adult animal. From an animal breeding perspective, the importance of being able to clone from differentiated cells is that this opened up the possibility of cloning adult animals with known attributes and highly accurate estimated breeding values based on pedigree, progeny, and their own performance records.

Successful cloning from differentiated cells requires a remarkable epigenetic "nuclear reprogramming" to occur in the donor nucleus. This reprogramming involves a series of events where interactions between the donor nucleus and the oocyte cytoplasm induce change in the DNA structure towards a pluripotent (i.e., capable of giving rise to several different cell types) form that is more appropriate for embryonic development. To do this the nucleus must shut down the gene expression profile that was appropriate for its original somatic cell role (e.g., a skin fibroblast), and begin expression of the genes appropriate for embryogenesis. This requires down-regulating the expression of approximately 8000–10,000 somatic cell genes, and initiating expression of an equivalent number of embryonic genes. Currently this reprogramming process is not well understood, and several studies have shown that there appears to be an increased rate of pregnancy, early postnatal loss, and other abnormalities in SCNT clones relative to offspring conceived in the traditional way. However, these problems are not seen in all SCNT clones, and many apparently healthy clones have been born, grown to maturity, and have gone on to conceive and have healthy offspring (Couldrey et al., 2011). Because the abnormalities seen in clones are largely epigenetic, meaning they are not based on changes in the underlying DNA sequence, they are corrected during gametogenesis and analogous problems have not been observed in the sexually-derived offspring of clones.

Significant improvements in the protocols for SCNT cloning have occurred over the past 15 years, and bovine cloning is now achieving efficiencies of 20–25% live cloned offspring per oocyte transferred (Panarace et al., 2007). Most embryonic loses occur in the first 2 weeks after transfer of the reconstructed embryo into the uterus of the recipient cow. This is the time when natural embryonic mortality in pigs and cattle is also high (35–50%). Porcine cloning can produce pregnancy rates as high as 80%, although the average litter size tends to be reduced compared to conventional breeding figures (~6 piglets as compared to 9–10 piglets).

The performance and behavior of cloned offspring that successfully survive the neonatal period are not different from age matched controls. An early study on Dolly suggested that clones might be susceptible to premature aging, due to shortened telomeres in their cells (Shiels et al., 1999). Telomeres are repetitive nucleotide sequences at each end of a chromosome, which protect the end of the chromosome from deterioration and prevent them from fusing with neighboring chromosomes. It was speculated that because the somatic cell nucleus that became Dolly was taken from a 6-year-old sheep, Dolly would have shortened telomeres in all her cells because she was genetically six years old at birth. This is because telomere length is reduced after each cell division and hence telomeres become shorter as an organism ages. Subsequent studies on other SCNT clones have not repeated this finding of shortened telomeres (Betts et al., 2001; Miyashita et al., 2011; Tian et al., 2000), and have shown that SCNT animals have telomeres of normal length. Dolly eventually died from a progressive lung disease in 2003. Roslin scientists stated that they did not consider that her death was the result of being a clone as other sheep on the farm had similar ailments. Such lung diseases are especially a danger for sheep kept indoors, as Dolly had to be for security reasons. Because longevity is a population statistic (i.e., it is the average age at death in a given population and thus cannot be determined based on a single observation), and SCNT cloning from adult cells has only been in general use since 1997, it is too early to assess the effects of cloning on lifespan and senescence (Niemann and Lucas-Hahn, 2012). Although some studies have reported that clones may experience a higher than normal annual mortality rate (Wells, 2005), others indicate no obvious problems with second generations of cloned cattle (Konishi et al., 2011) and mice that have been reiteratively cloned for six generations reveal no aberrant pathology (Wakayama et al., 2000).

A diverse range of 16 animal species have now been successfully cloned from adult tissues using SCNT including mice, rats, zebrafish, rabbits, ferrets, goats, horses, pigs, cattle, deer, camel, dogs, cats, and a range of endangered species including wild cats, muflon, gaur, wolf, and ibex. Although clones carry exactly the same genetic information in their DNA, they may still differ from each other, in much the same way as identical twins do not look or behave in exactly the same way. Clones do not share the same

Box 28.1

How animals are cloned and why problems sometimes occur

Cloning by nuclear transfer is a two-part process. First, scientists remove the nucleus from an egg, and then they fuse it with a somatic cell containing the nucleus and genetic material from another cell by the application of an electrical charge. The fused egg is then placed in a laboratory dish with the appropriate nutrients. Eventually the resulting embryo, which is a genetic copy of the animal that produced the somatic cell and not the egg, is transplanted into a surrogate mother.

The successful production of normal clones from differentiated somatic cells suggests that adult nuclear DNA retains the ability to direct the correct pattern of gene expression for embryogenesis. The process of resetting adult nuclear DNA to the embryonic pattern of gene expression is known as **reprogramming** and likely involves switching off certain genes and turning on others. Errors in reprogramming may lead to abnormalities in gene expression in cloned animals and affect the health and longevity of the animal.

Reprogramming involves changes at the epigenetic level. Epigenetic changes refer to alterations in gene expression resulting from modifications of the genome that do not include changes in the base sequence of DNA. Two key areas of epigenetic control are **chromatin remodeling** and **DNA methylation**. Epigenetic changes may also include the switching off of maternal or paternal copies of certain genes in a process called **imprinting**.

In the case of clones it appears that the reprogramming of somatic cell modifications is sometimes incomplete leading to inappropriate patterns of DNA methylation, chromatin modification, and X-chromosome inactivation in the developing clone. This can result in aberrant gene expression patterns and correspondingly high rates of pregnancy loss, congenital abnormalities, and postnatal mortality.

cytoplasmic inheritance of mitochondria from the donor egg, nor the same maternal environment as they are often calved and raised by different animals (see Box 28.1). It is also important to remember that most traits of economic importance are greatly influenced by environmental factors, and so even identical twins may perform differently under varying environmental conditions.

Applications of cloning

Cloned animals can provide a "genetic insurance" policy in the case of extremely valuable stud animals like Elevation, or produce several identical bulls in production environments where AI is not a feasible option. This so-called "reproductive cloning" could conceptually be used to reproduce a genotype that is particularly well-suited to a given environment. The advantage of this approach is that a genotype that is proven to do especially well in a particular location could be maintained indefinitely, without the genetic shuffle that normally occurs every generation with conventional reproduction and meiosis. However, the disadvantage of this approach is that it freezes genetic progress at one point in time. As there is no genetic variability in a population of clones, within-herd selection no longer offers an opportunity for genetic improvement. Additionally, the lack of genetic variability could render the herd vulnerable to a catastrophic disease outbreak, or singularly ill-suited to changes that may occur in the environment.

Cloning offers an approach to reproduce otherwise sterile animals (e.g., mules or neutered animals). Cloning may also have some utility as one approach contributing towards the preservation of rare and endangered species. It should be noted in this regard, that oocytes can only reprogram and support the development to term where the donor nucleus species is closely related to the species of oocyte origin as was the case when a muflon was cloned in a sheep oocyte, and the gaur with a cow oocyte. Although embryonic development can begin in the case where species are not closely related, such as a cow and a pig, embryonic genome activation does not occur and development is arrested at the early cleavage stages of embryogenesis.

Although cloning does not alter the genetic makeup of the animal, there is a logical partnership between cloning and the process of using recombinant DNA technology to make transgenic or genetically engineered (**GE**) animals. As will be discussed later, cloning can be used to efficiently generate transgenic animals from cultured somatic cells that have undergone precise, characterized modifications of the genome. The first GE mammalian clones were sheep born in 1997 carrying the coding sequences for human clotting factor IX (Schnieke et al., 1997), which is an important therapeutic for hemophiliacs. Cloning has also been used to generate GE cows that produce human polyclonal antibodies (Kuroiwa et al., 2002). It is envisioned that these unique cows will make it possible to create an efficient, safe, and steady supply of human polyclonal antibodies for the treatment of a variety of infectious human diseases and other ailments including organ transplant rejection, cancer and various autoimmune diseases, such as rheumatoid arthritis. Genetically engineered proteins have been made and secreted in milk, blood, urine, and semen of livestock, although to date most commercial systems favor the mammary gland. Cloning also offers the unique opportunity to produce animals from cells that have undergone a targeted "knock out" (see http://learn.genetics.utah.edu/content/science/transgenic/) or deletion of an endogenous gene such as those that encode the allergenic proteins that cause the rejection of animal organs when used in human xenotransplantation surgeries (www.revivicor.com).

Studies examining the composition of food products derived from clones have found that they have the same composition as milk or meat from conventionally-produced animals (Yang et al., 2007). In 2001 the Center for Veterinary Medicine at the US Food and Drug Administration (FDA) undertook a comprehensive risk assessment to identify hazards and characterize food consumption risks that may result from the introduction of SCNT animal clones, their progeny, or their food products (e.g., milk or meat) into the human or animal food supply. As there is no fundamental reason to suspect that clones will produce novel toxins or allergens, the main underlying food safety concern was whether the SCNT cloning process results in subtle changes in the composition of animal food products.

In 2008 the FDA published its final 968-page risk assessment on animal cloning (available at www.fda.gov/AnimalVeterinary/SafetyHealth/AnimalCloning/UCM055489), which examined all existing data relevant to (1) the health of clones and their progeny, and (2) food consumption risks resulting from their edible products, and found that no unique food safety risks were identified in cloned animals. This report, which summarized all available data on clones and their progeny, concluded that meat and milk products from cloned cattle, swine and goats, and the offspring of clones from any species traditionally consumed as food, are as safe to eat as food from conventionally bred animals. The FDA also has made available three public education fact sheets "Myths about Cloning," "Animal Cloning and Food Safety," and "A Primer on Cloning and Its Use in Livestock Operations," on their website (available from www.fda.gov/AnimalVeterinary/SafetyHealth/AnimalCloning/default.htm). Subsequent rodent feeding studies have revealed no obvious food safety concerns related to the consumption of cloned-cattle meat (Yang et al., 2011).

Genetic engineering

Genetic engineering (**GE**) is a process in which scientists use recombinant DNA (rDNA) technology to introduce desirable traits into an organism. DNA is the chemical inside the nucleus of a cell that carries the genetic instructions for making living organisms. Because the genetic code for all organisms is made up of the same four nucleotide building blocks, this means that a gene encodes the same protein whether it is made in an animal, a plant or a microbe. Recombinant DNA refers to DNA fragments from two or more different sources that have been joined together in a laboratory. The resultant rDNA "construct" is usually designed to express a protein(s) that is encoded by the gene(s) included in the construct. Genetic engineering involves producing and introducing the rDNA construct into an organism so new or changed traits can be given to that organism. A GE animal is an animal that carries a known sequence of rDNA in its cells, and which passes that DNA onto its offspring. Genetically engineered animals are sometimes referred to as genetically modified organism (GMO), living modified organism, transgenic, or bioengineered animals. Genetically engineered animals were first produced in the late 1970s. Forty years later GE animals have been produced in many different species, including those traditionally consumed as food although most have not moved from the laboratory to commercialization.

Techniques

The first method to produce GE animals was microinjection of rDNA into blastocysts to produce transgenic mice in 1974 (Jaenisch and Mintz, 1974). However, these mice were mosaic, meaning they did not carry the transgene in all of the cells of their body and most importantly their germ cells (egg and sperm), and so were not able to pass the transgene on to their offspring. Germline transmission (i.e., the rDNA construct is present in gametes produced by the GE animal) of the rDNA was achieved using a technique called pronuclear microinjection.

This technique involves injecting many copies of the recombinant gene into one of the two pronuclei of a newly fertilized single-cell embryo. Transgene integration happens randomly in the genome at sites of DNA double-strand breakage, and typically multiple copies of the transgene integrate into a single chromosomal locus in the embryo. If integration takes place prior to the first nuclear division, then all cells will carry the transgene. In many cases integration happens after the embryo has undergone cell division, which results in a mosaic animal in which some cells contain the gene construct while others do not. After microinjection, eggs are typically surgically transferred into the oviducts of synchronized surrogate females. The offspring resulting from injected eggs may or may not carry the transgene in their somatic and/or germ (sperm and egg) cells. Typically only 1–5% of the implanted embryos will test positive for the transgene. Animals that do have the GE construct integrated into their genome are called founders, and only those that are germline founders reliably transmit the transgene to their offspring.

Only a small fraction of GE founder animals produced using pronuclear microinjection show the expected phenotype. This is mainly due to the random nature of the integration site. The transgene can be affected by the surrounding DNA and result in animals that fail to express the transgene, a phenomenon called "gene silencing." Screening transgene expression levels in founder animals is currently the only way to identify animals with suitable expression patterns. Several alternatives to pronuclear microinjection have been developed to improve the efficiency and reduce the costs associated with generating transgenic animals. These include the use of targeted gene modifications in cell culture followed by SCNT of the modified cell, injection, or infection of oocytes and/or embryos by retro- and lentiviral vectors, cytoplasmic injection of circular plasmids (CPI), sperm mediated gene transfer (SMGT), and intracytoplasmic injection (ICSI) of sperm heads carrying foreign DNA (Figure 28.3).

As discussed previously SCNT offers an approach to clone cells that have been genetically modified in culture and thereby produce GE clones. Unfortunately the success rate of SCNT is also low, and although no mosaic animals are produced when a genetically engineered cell is cloned, reconstructed embryos have a low survival rate and typically only 1–10% of reconstructed embryos result in live births. Cattle seem to be an exception to this rule as levels of 15–20% can be reached (Kues and Niemann, 2004). Cloning also offers the possibility of producing animals from cultured cells that have had selected genes removed, a technique called gene targeting. The first "targeted gene knockout" technique that resulted in the selective inactivation of specific genes was developed in 1987 (Thomas and Capecchi, 1987) and gene targeting was the subject of the 2007 Nobel Prize in medicine (www.nobelprize.org/nobel_prizes/medicine/laureates/2007/advanced.html?print=1#.U3M-kvlSbiQ). This original gene targeting work was carried out in pluripotent embryonic stem cells (ESC) derived from mice. These cells have the ability to participate in organ and germ cell development following injection into the blastocysts. Despite extensive research, stem cells that are able to contribute to the germline are currently only available for rodents and not food animal species. However, gene targeting in somatic cells followed by SCNT offers an approach to allow additional species to employ high efficiency "targeted gene knockout" techniques. Somatic cell gene targeting directly recombines homologous genes in somatic cells and then GE animals can be produced through SCNT. This approach has been successfully used to produce cattle from cells lacking the gene for the prion protein responsible for mad cow disease (Richt et al., 2007), and pigs have been produced that lack the allergenic proteins that are responsible for the rejection of pig organs when used for transfer into human organ-transplantation patients (Whyte and Prather, 2011).

The disadvantage of this approach is that somatic cells have a limited lifespan *in vitro* and aged somatic cells result in a high number of abnormalities in cloned embryos. Recently, gene targeting technologies based on designer nucleases (e.g., zinc finger

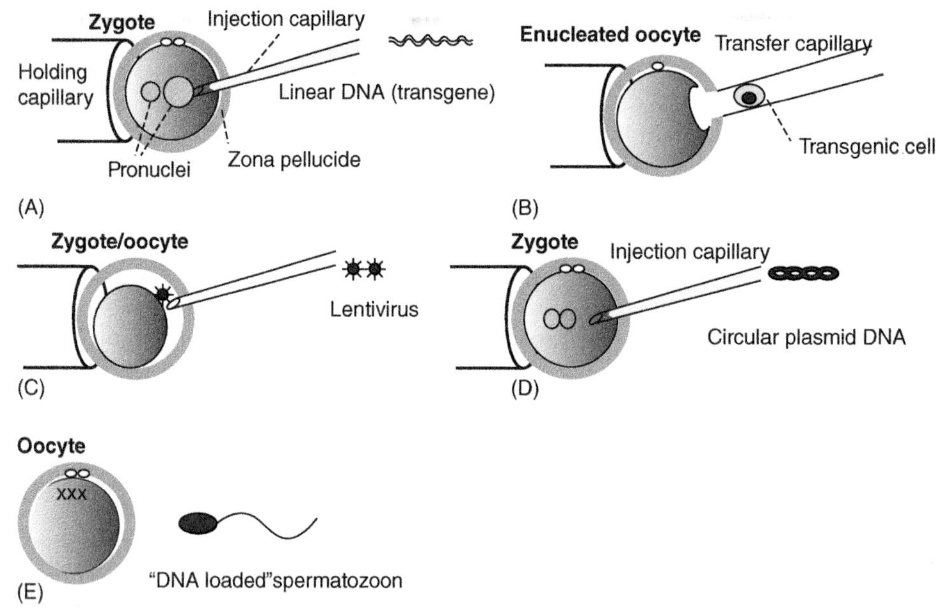

Figure 28.3 Methods for transgenesis in large mammals. Modified from Garrels et al. (2012). (**A**) Pronuclear injection (PNI): With a fine glass capillary linearized DNA molecules are injected into one pronucleus of a zygote. Requires highly skilled experimentalist. Random integration into the genome. High rates of transgene mosaic animals and unwanted concatemeric integrations. Approximately, 1–5% of treated zygotes develop to transgenic offspring. (**B**) Somatic cell nuclear transfer (SCNT): Requires highly skilled experimentalist for enucleation of oocytes and transfer of transgenic somatic cell. Integration into the genome of somatic cells is random in most cases, but can be targeted by homologous recombination. Genetic modification of donor cells with viruses, zinc finger nucleases and transposons, and subsequent use in SCNT has been shown. All offspring should be transgenic, but due to low developmental capacity only 1–5% of reconstructed embryos develop to vital offspring. (**C**) Lentivirus transfection: Requires advanced virus production facility and S2 safety laboratories. Replication-deficient lentiviruses are injected into the perivitelline space. Typically 50–90% of the offspring are transgenic, however, a high mosaicism rate and animals carrying multiple integrations are found. (**D**) Cytoplasmic plasmid injection (CPI): Circular expression plasmids are injected into the cytoplasm by employing transposon systems, active enzyme-catalyzed transgene integration of monomeric units can be achieved. Monomeric insertions into transcriptionally accessible regions are favored. Typically 40–60% of the offspring are transgenic, correlating to 6–10% of treated zygotes. (**E**) Sperm-mediated gene transfer (SMGT) and intracytoplasmic sperm injection (ICSI): For SMGT sperm cells are incubated with DNA, and are subsequently used for artificial insemination, thus avoiding any micromanipulation. However, the transgenesis rates are unpredictable and highly variable between laboratories. A more reliable extension of SMGT is the combination with intracytoplasmic sperm injection (ICSI). In this method sperm cell membranes are damaged (freezing, NaOH or drying) before incubation with DNA, then immobile (dead) spermatozoa are used for ICSI, followed by embryo transfer. However, the ICSI procedure is laborious and requires a highly skillful experimentalist, smoothing out the simplicity of SMGT. Reproduced with permission from Laible, G. Enhancing livestock through genetic engineering – Recent advances and future prospects. *Comparative Immunology, Microbiology and Infectious Diseases* 32, 123–137 (2009).

nucleases, transcription activator-like effector nucleases (TALENS), meganucleases) that target specific sequences in the genome have also been developed. These nucleases are like "molecular scissors" that introduce a double-strand break at a single predetermined location in the genome. They can be used for targeted gene modification including endogenous gene knockouts, targeted gene addition and/or replacement through homologous recombination, and chromosomal rearrangements. Gene knock-out plants, *Drosophila*, zebrafish, rats, pigs, and cattle have been successfully produced by zinc-finger nucleases (Miyashita et al., 2011). Recent progress in reprogramming somatic cells to become pluripotent stem cells that can divide indefinitely will likely further improve the efficiency of targeted gene modifications in the future.

Applications of genetically-engineered animals

Genetically engineered animals can be divided into six broad classes based on the intended purpose of the genetic modification: (1) to develop animal models for research purposes (e.g., pigs as models for cardiovascular diseases); (2) to produce products intended for human therapeutic use (e.g., pharmaceutical products); (3) to enrich or enhance the animals' interactions with humans (e.g., new color varieties of pet fish); (4) to produce industrial or consumer products (e.g., fibers for multiple uses); (5) to enhance production attributes or food quality traits (e.g., faster growth); and (6) to improve animal health (e.g., disease resistance). Some of the most notable genetically engineered animals have been developed for a variety of reasons ranging from biomedical research to food production. All GE animals must receive regulatory approval before the products they produce can be commercialized for pharmaceutical or food purposes.

Regulation of genetically-engineered animals

The FDA is the lead agency responsible for the regulation of GE food animals in the United States. In 2009, the FDA outlined its science-based regulatory process (www.fda.gov/downloads/AnimalVeterinary/GuidanceComplianceEnforcement/GuidanceforIndustry/UCM113903.pdf) to assess GE animals and their edible products. To evaluate a GE animal the FDA requires the company interested in commercializing the GE animal to provide data to enable analyses of the following seven points:

1. Product definition: what does the GE animal do? For example, grow faster, disease resistant;

2. Molecular characterization of the construct: a description of the rDNA construct and how it was assembled;

3. Molecular characterization of the GE animal lineage: how was the rDNA construct introduced into the animal and whether it is stably maintained over time;
4. Phenotypic characterization of the GE animal: comprehensive data on the characteristics of the GE animal and its health;
5. Durability plan: plan to show that GE modification is stable over time, and will continue to have the same effect;
6. Environmental and food/feed safety: assessment of any environmental impacts, and for GE animals intended for food, that food from those GE animals is safe to eat for humans and/ or animals;
7. Claim validation: does it do what it is meant to do?

In the United States, any animal containing an rDNA construct is subject to regulation by the FDA prior to commercialization. However, based on risk, there are some GE animals for which the FDA exercises something called "enforcement discretion," meaning they do not require an approval prior to commercialization. In general, this includes transgenic laboratory rodents such as mice and rats that have become increasingly important for biological and biomedical research and are sold to researchers around the world. GE livestock are also being developed specifically as biomedical research models. Several groups have created GE pigs with alterations in key genes in disease pathways to provide models for human disease (Whyte and Prather, 2011). Pigs are anatomically and physiologically similar to humans and these models will help to improve our understanding of the causes and potential therapies for human disease. The emerging technologies for gene targeting will likely mean more GE animals from a variety of species will be produced as valuable models to study human disease and therapies in the future. The FDA does not plan on exercising enforcement discretion for any GE animal of a species traditionally consumed as food. On a case-by-case basis, the FDA may consider exercising enforcement discretion for GE animals of very low risk, such as it did for Glofish (see www.youtube.com/watch?v=SA9PEBPnhWU), a GE aquarium fish that glows in the dark (www.glofish.com).

Pharming is a term used to describe the production of pharmaceutical proteins or drugs in GE animals following the introduction of a gene construct that directs the production of that drug. The mammary gland of dairy animals is a logical place to produce therapeutic proteins as it has the ability to produce large amounts of protein, and milk is easily harvested from the animal. In 2009, the first GE animal producing a pharmaceutical product, a GE goat (http://www.fda.gov/downloads/ForConsumers/ConsumerUpdates/UCM144055.pdf) synthesizing recombinant human antithrombin III in its milk (http://www.atryn.com), was approved by the FDA. This drug is an anticoagulant for the treatment of individuals with hereditary antithrombin deficiency, a blood-clotting disorder. Subsequently, a human recombinant C1 plasma protease inhibitor produced in transgenic rabbit milk was approved in Europe for treatment of patients with hereditary angioedema (www.pharming.com). Transchromosomal cattle carrying a human artificial chromosome harboring the entire sequence of the human major histocompatability complex have been made and these animals are able to make human polyclonal antibodies (Kuroiwa et al., 2002; www.hematech.com).

Agricultural applications of genetic engineering include making animals with improved food products, animal welfare (e.g., disease-resistant animals), and animals with a reduced environmental footprint per unit of food (e.g., egg or serving of milk and meat) (Figure 28.4). There is a much higher economic incentive associated with the production of GE animals for human medicine applications, than for agricultural applications.

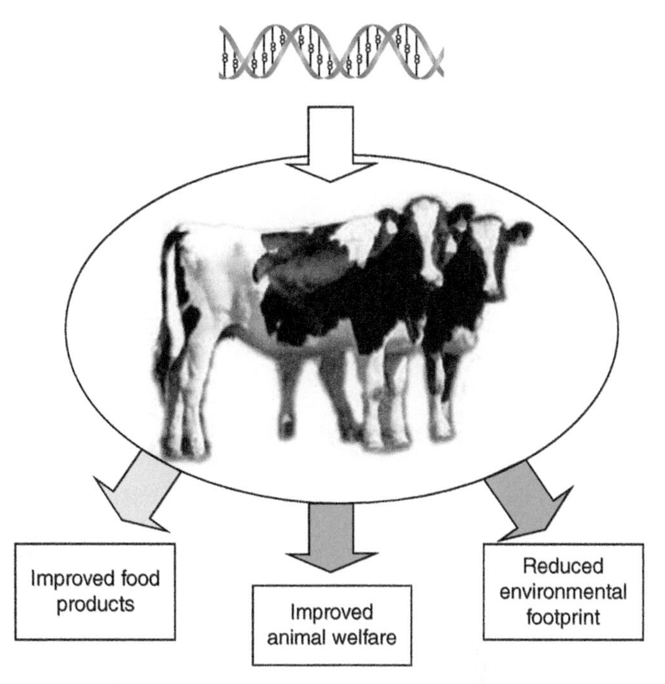

Figure 28.4 Main objectives of agricultural applications for transgenic livestock technology. Image from Laible (2009).

The use of GE animals for agricultural applications tends to generate greater public scrutiny than the biomedical and pharmaceutical applications previously discussed. This may be partly due to the fact that GE animals for agricultural applications will enter the food supply. The advantage of GE for animal breeding is that unlike traditional selection approaches (Figure 28.4), the technology is not restricted by the species barrier and so entirely novel and unique characteristics can be introduced using genes derived from unrelated species. Traditional selection schemes make relatively slow genetic progress and are imprecise, meaning that selection for one characteristic is often accompanied by undesired changes in associated traits (e.g., production and fertility). Some of the GE animals that have been developed for agricultural applications are listed in Table 28.2, although none have yet received regulatory approval for commercialization and entry into the food supply.

A company called AquaBounty requested regulatory approval for a GE line of growth June 2014 enhanced Atlantic salmon intended for food. The AquAdvantage Atlantic salmon reaches market size twice as fast as wild-type salmon. Consisting of an "all fish" construct, the transgenic salmon contain an ocean pout antifreeze promoter driving a Chinook salmon growth hormone gene that allows the fish to grow up to six times larger than non-transgenic salmon of the same age (Du et al., 1992). The company has completed all of the major studies required to gain regulatory approval for the transgenic salmon to be consumed in the US. The data package included regulatory studies to address food safety, allergenicity, nutrient content, and genetic stability through inheritance, and evaluation was completed in September 2010. As of June 2014, the FDA had not made a decision as to whether this fish will be the first GE animal approved to enter the food supply.

Table 28.2 Some examples of traits targeted for improvement in GE animals for agricultural applications. Modified from Kues and Niemann (2004).

Transgenic trait	Key molecule	Gene transfer method	Species	Ref.
Increased growth rate	Growth hormone (GH)	Microinjection	Pig	Nottle et al., 1999
Increased growth rate	Insulin-like growth factor-1 (IGF-1)	Microinjection	Pig	Pursel, 1999
Increased muscle mass	Slaon–Kettering virus	Microinjection	Pig	Pursel et al., 1992
Resistant to heat stress	Heat-shock protein	Microinjection	Pig	Chen et al., 2005
Increased ovulation rate	B-cell Leukemia 2	Microinjection	Pig	Guthrie et al., 2005
Increased muscle mass	Myostatin pro-domain	Microinjection	Pig	Mitchell and Wall 2008
Increased level of polyunsaturated fatty acids in pork	Desaturase (from spinach) Desaturase (from *C. elegans*)	Microinjection Somatic cloning	Pig Pig	Saeki, 2004 Lai et al., 2006
Phosphate metabolism	Phytase	Microinjection	Pig	Golovan et al., 2001
Milk composition	α-Lactalbumin	Microinjection	Pig	Wheeler et al., 2001
Influenza resistance	Mx protein (Myxovirus resistance 1, interferon-inducible protein)	Microinjection	Pig	Muller et al., 1992
Enhanced disease resistance	Immunoglobulin (IgA)	Microinjection	Pig, sheep	Lo et al., 1991
Wool growth	Insulin-like growth factor-1 (IGF-1)	Microinjection	Sheep	Damak et al., 1996
Visna virus resistance	Visna virus envelope	Microinjection	Sheep	Clements et al., 1994
Bovine spongiform encephalopathy (BSE) resistance	Prion protein gene	Somatic cloning	Sheep	Denning et al., 2001
Milk fat composition	Stearoyl desaturase	Microinjection	Goat	Reh et al., 2004
Milk composition (increase of whey proteins)	β-Casein κ-Casein	Somatic cloning	Cattle	Brophy et al., 2003
Milk composition (increase of lactoferrin)	Human lactoferrin	Microinjection	Goat	Maga et al., 2006
Mastitis resistance	Lysostaphin	Somatic cloning	Cattle	Wall et al., 2005
Bovine spongiform encephalopathy (BSE) resistance	Prion protein gene	Somatic cloning	Cattle	Richt et al., 2007
Influenza resistance	Short hairpin RNA	Lentiviral transduction	Chicken	Lyall et al., 2011
Increased growth rate	Growth hormone (GH)	Microinjection	Salmon	Du et al., 1992

The extensive regulatory process to document the food and environmental safety of GE animals bred for agricultural applications is unique to GE technology. For example, genetic modifications that result from using traditional animal breeding approaches to select for faster growing salmon undergo no analogous regulatory scrutiny. While there may be some risks that are uniquely associated with some GE animals (e.g., potential introduction of an allergenic protein from a different species), there are other risks where there is no difference between those associated with GE animals and risks associated with conventionally-bred animals. For example, environmental risks associated with fast growing GE salmon would be similar to those associated with fast growing strains of farmed salmon developed using traditional selection for faster growth (Schiermeier, 2003). Subjecting conventionally-bred and GE animals to discordant regulatory requirements despite similar risks is inconsistent from a scientific perspective, and places a disproportionate regulatory burden on the development of GE technology. Commercialization of agricultural applications of GE animals in the US is currently being delayed by concerns about the cost and timelines associated with the regulatory process.

Yonathan Zohar, a professor at The University of Maryland, wrote an opinion piece entitled "Genetically modified salmon can feed the world" on the GE salmon. Read his opinion piece at http://edition.cnn.com/2010/OPINION/09/22/zohar.genetically.engineered.salmon/ and then consider the question in Box 28.2.

The University of Guelph in Canada was also interested in obtaining regulatory approval for its Enviropig – a GE pig that produces the enzyme phytase in its saliva (Golovan et al., 2001; www.uoguelph.ca/enviropig). This bacterial enzyme enables the Enviropig to process indigestible phosphorus in the form of phytate and better absorb the phosphate in its diet, thereby eliminating the need to supplement the diet with readily-available forms of phosphate supplement. As a consequence the phosphorus content of Enviropig's manure is reduced by as much as 60%. This pig is discussed on a CNN report entitled "Enviropig: the next transgenic food?" Watch the video at http://eatocracy.cnn.com/2010/09/25/enviropig-the-next-transgenic-food/ and then consider question in Box 28.3.

In May, 2012 the University of Guelph closed down its Enviropig project after failing to find an industry partner to

Box 28.2

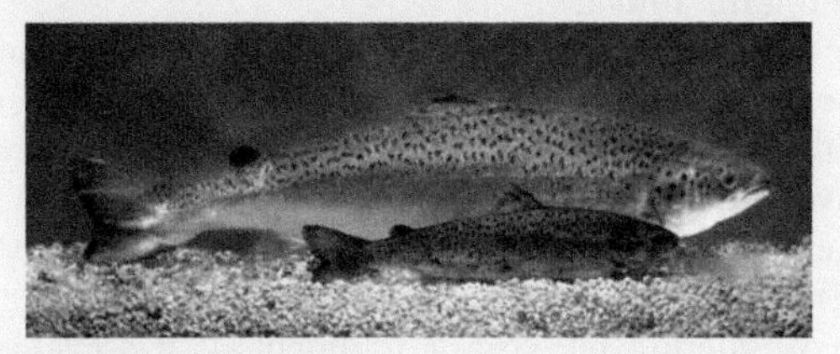

The AquAdvantage™ salmon

Since the mid-1980s, the yield of food fish from wild capture fisheries has been static at about 60 mMT per year. The growth of the fish supply since that time has largely come from aquaculture. It has been calculated that an extra 52 mMT of aquaculture production will be needed by 2025 if the current rate of fish consumption is to be maintained. Atlantic salmon remain the most important farmed food fish in global trade. The AquAdvantage™ salmon is an Atlantic salmon carrying a Chinook salmon growth hormone gene controlled by an antifreeze protein promoter from a third species, the ocean pout. The mature weight of these fish remains the same as other farmed salmon, but their growth rate is increased, with a concomitant 25% decrease in feed input, decreased waste per unit of product, and decreased time to market. The application to market this fish for food purposes has been going through the FDA regulatory approval process for over a decade. Do you think this fish should be approved for commercialization? Give three reasons to support your answer.

Photo Courtesy of AquaBounty Technology.

Box 28.3

The "Enviropig"

Given the large increase that is expected in both pig and poultry production in the developing world over the next 20 years as a result of population growth and increased income, decreasing the phosphorus levels in the manure of these monogastric species would likely have a huge worldwide environmental benefit. However, using GE to reduce the levels of this important pollutant in swine manure has been subject to the criticism that this kind of approach encourages "non-sustainable, unecological approaches to livestock management." Critics argue that if farmers really want to be environmentally friendly, they should let pigs graze on pasture instead of feeding them grain. However, outdoor pig farming can itself exacerbate nutrient leaching into the soil and groundwater. What is your opinion of the Enviropig – can it help reduce phosphorous pollution in environment; or do you think this GE animal is a bad idea? Give three reasons to support your answer.

Photo by Cecil Forsberg.

continue to fund the project that began in 1999. Prior to that time the pig producer industry association "Ontario Pork" had financially supported the research. The University's applications for food approval of the Enviropig with Health Canada and the Food and Drug Administration in the United States will remain active until a regulatory decision is made, or until such time that the University no longer desires to obtain a final decision from the regulatory evaluators. The shelving of this GE animal project example emphasizes the fact that although the potential of transgenic livestock is tremendous, there are still significant scientific, regulatory and public acceptance issues that need to be resolved before this technology is widely adopted on farms.

Ethical, moral, and animal welfare concerns

The use of animals for any purpose is associated with ethical and moral concerns. Many people who are opposed to GE and cloning of animals tend to oppose all research using animals. The following discussion emphasizes the key moral and ethical issues specifically associated with GE and cloning technologies. An excellent resource explaining why animals and their treatment raise ethical and moral questions is a booklet entitled "Ethics, Morality, and Animal Biotechnology" prepared by the Biotechnology and Biological Sciences Research Council (BBSRC) in the United Kingdom (see www.bbsrc.ac.uk/web/files/policies/animal_biotechnology.pdf). That booklet discusses some of the key considerations and schools of thought when considering morality and ethics as it relates to animals. One important point that is stressed in that booklet is the difference between moral and ethical concerns. The distinction between morals and ethics is explained by the BBSRC as follows:

> Everybody (except perhaps the psychopath) can be said to have moral views, beliefs and concerns, to the effect that certain things are right or wrong and that certain actions ought or ought not to be performed. What issues arouse most moral concern will of course vary enormously between different individuals, cultures and periods of history... Such moral concerns may result from a lot of deliberation and reflection, or from very little; they may be firmly grounded in a consistent set of carefully considered principles, or they may not. We all probably hold some moral views almost unthinkingly, perhaps as a result of our upbringing. We may just "feel" that certain things are right or wrong; we have a "gut reaction" about them; and that may be the sum total of some people's "morality."
>
> Ethics is a narrower concept than morality, and it can be used in several different, though related, senses. The most general of these: " ... suggests a set of standards by which a particular group or community decides to regulate its behavior – to distinguish what is legitimate or acceptable in pursuit of their aims from what is not." Hence we talk of "business ethics" or "medical ethics." More technically, ethics can also refer to a particular branch of philosophy which tries to analyze and clarify the arguments that are used when moral questions are discussed and to probe the justifications that are offered for moral claims. So ethics in this sense puts our moral beliefs under the spotlight for scrutiny.

Genetic engineering and cloning may be considered by some to be intrinsically wrong, meaning they are morally wrong under any circumstances, regardless of their consequences and intentions. If someone considers a practice to be intrinsically wrong, then no further discussion can reverse their belief of that intrinsic wrongness. One of the intrinsic arguments that is often heard when scrutinizing animal biotechnology is that GE and cloned animals are "unnatural" and therefore they should not be allowed. However, there are a number of ethical questions that are raised when considering that moral argument. The first is that the techniques used to produce GE and cloned animals employ natural processes such as DNA repair mechanisms. And so this raises the question of what exactly is natural, and does the fact that something is natural make it right? Vaccinations are not natural and yet we routinely employ them to protect ourselves from disease. It might be more natural to let people die from exposure to naturally-occurring viruses like smallpox, but is that the ethically correct choice? The notion of "nature" and "natural" tends to be an interpretation drawn from the observer's perspective. In the context of ethical judgments the notion of nature is more a conclusion than an argument. There is a presupposition that because something is natural it is ethically correct. Asserting that GE animals are unnatural does not allow a conclusion about whether they should not be allowed. From a description of what is, there is no logical way to prescribe what ought to be. George Edward Moore first observed this fact in 1903 and called the conclusion of an "ought" from an "is" the "naturalistic fallacy."

Some may hold religious views that GE and cloning are intrinsically blasphemous and that humans are intruding into areas that are the realm of God. This argument obviously is one that will only be persuasive for people that believe in a Creator. However, not all religions share this perspective and there is no unanimous condemnation of cloning or GE among religious groups *per se*. Creation is defined as "bringing something out of nothing," and some may argue that GE animals and clones are produced from something (i.e., living cells) and hence this does not meet the definition of creation. There is also some support for the idea that God gave humans a position of "dominion" over Nature. Some may even see biotechnology as an opportunity for humans to work with God as "co-creators." Others may argue that GE is intrinsically wrong because it moves genes from one species to another. However, this occurs routinely in nature, although some GE animals like the phytase pig where a rDNA bacterial phytase gene driven by a porcine promoter was integrated into a pig chromosome could only exist as the result of human intervention. Many religions do not hold that the boundaries between species are sacred and immutable, nor indeed that they are so regarded by God. From an animal breeding perspective, it is exactly this ability to bring entirely new traits into an animal from a different species (e.g., disease resistance genes) that makes the potential benefits of animal GE so compelling.

Another intrinsic argument is that making GE animals interferes with the integrity or "telos" of the animal. Telos is defined as:

> ... the set of needs and interests which are genetically based, and environmentally expressed, and which collectively constitute or define the form of life or way of living exhibited by that animal, and whose fulfillment or thwarting matter to that animal.
>
> *(Holland and Johnson, 1998).*

However, as discussed at the beginning of this chapter, all domesticated animals show characteristics that have been produced by selective breeding and that represent changes to their telos – for example, reduced aggression. Those who oppose GE because it alters an animal's telos must consider whether they would also raise ethical objections to the selective breeding methods that have produced all domestic pets (e.g., breeds of dogs) and farm animals.

Public opinion polls have repeatedly shown that the public acceptance is influenced by the utility or reason that a genetically engineered animal is being created. Medical applications are viewed more favorably than food applications and food applications with consumer benefits are viewed more favorably than those with producer benefits (e.g., increased growth rate). Here, we move into the realm of considering the extrinsic arguments, that is, evaluating the consequences and intentions associated with the production of GE and cloned animals and determining whether the benefits outweigh the risks. A patient awaiting an organ transplant from a GE pig may have a different view on the appropriateness of using GE to produce transplantation-friendly pigs than someone who is not facing a similar life-threatening situation.

Critics of GE contend that the risks involved are so great that any use of GE is irresponsible; that it is the particular and potentially dire risks associated with these techniques that make them ethically unjustifiable. Others, including the National Academy of Sciences, argue that there are no unique risks associated with GE and cloning that do not also arise from other genetic improvement techniques including conventional breeding. The risks that are associated with each unique rDNA/animal combination, will vary from case to case making generalizations about the "safety" of GE animals virtually impossible. However, excessive caution does not necessarily reduce risk. Abandoning research and development in all forms of GE animals might prevent the development of a technique or product that could allow animals to better adapt to climate change, help feed the world's growing population, or prove invaluable in the treatment of serious diseases in 50 years' time.

Methodologies to produce cloned and GE animals themselves sometimes create animal welfare concerns, not the least of which is the current inefficiencies of the techniques that result in the use of many more experimental animals than would be needed if success rates were higher. However, efficiencies have been increasing as researchers improve experimental protocols. Additionally the use of SCNT in conjunction with GE cells results in 100% GE clones thereby avoiding the inefficiencies associated with pronuclear microinjection where only a small fraction of microinjected eggs result in a GE animal, and even fewer of these turn out to be germline transgenic.

Some of the reproductive manipulations (e.g., embryo transfer, superovulation) that are required for the production of clones may cause pain or discomfort to the animal, but again these are not new or unique concerns to cloning as these techniques are commonly employed by commercial livestock breeders, and have been for many years. A problem that is often seen with bovine embryos cultured using *in vitro* embryo culture techniques (e.g., SCNT clones) is that the resultant calves tend to have high birth weights and long gestational periods. This phenomenon, known as large offspring syndrome, can result in calving difficulties and an increased rate of caesarian section for the dam. These abnormalities have predominately been observed in ruminants (sheep and cattle), and mice (Niemann and Lucas-Hahn, 2012). Other naturally-occurring breeds of cattle have analogous calving difficulties. For example double-muscled cows of the Belgian Blue breed (see Box 28.4), routinely require a caesarean section to safely deliver their muscular calves.

An animal welfare concern that is more specifically associated with GE animals is poorly controlled expression of the introduced

Box 28.4

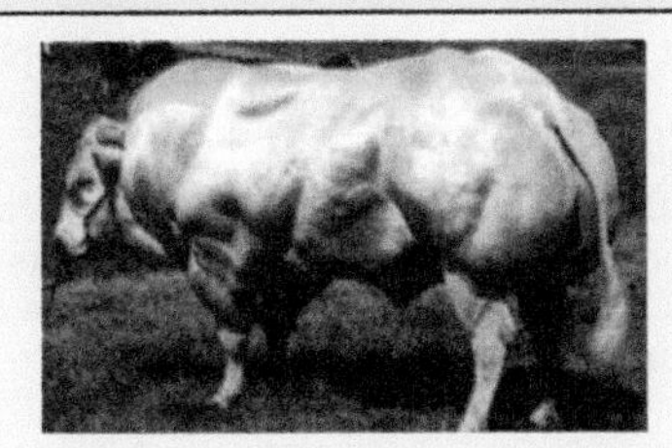

Myostatin GE cattle

Natural mutations in the myostatin gene result in the "double muscled" appearance of some beef breeds (e.g., Belgian Blue), and a 20% increase in muscle mass, the source of beef. This natural mutation is associated with major calving difficulties as the increased muscle mass in affected calves makes it difficult for the cow to deliver her calf. Additionally, cows who are themselves double-muscled have calving difficulties even when carrying an unaffected calf because of their narrower birth canal. In this example, there is a potential conflict between animal productivity and animal welfare arising from a naturally-occurring mutation. This mutation could be introduced into other breeds of cattle using traditional crossbreeding and marker-assisted selection to maintain the natural mutation in the new breed – a process called introgression. New gene editing approaches also offer an approach to introduce the same mutation into other breeds of cattle using GE. Are either of these approaches to introduce this mutation into new breeds ethically acceptable? Do concerns arise based on the process used to introduce the gene (i.e., selective breeding versus GE), the attributes of the animal (i.e., double muscling), or the combination of the two?

Reprinted by permission from Macmillan Publishers Ltd: *Nature Genetics*, vol. 17 issue 1. Luc Grobet, Luis José Royo Martin, Dominique Poncelet, Dimitri Pirottin, Benoit Brouwers, Juliette Riquet, Andreina Schoeberlein, Susana Dunner, François Ménissier, Julio Massabanda, Ruedi Fries, Roger Hanset, Michel Georges. A deletion in the bovine myostatin gene causes the double-muscled phenotype in cattle.

gene. Various growth abnormalities have been noted in GE animals that are expressing a growth hormone transgene (Pursel et al., 1989). Many of the problems that were encountered in these early experiments have been minimized by the use of tissue-specific promoters that result in more targeted expression of the transgene. As technologies to make transgenic animals improve through the use of more sophisticated targeted gene modification approaches, it is likely these unintended effects will become increasingly rare. Of course GE animals that are produced as models for human disease are intentionally modified to have a disease phenotype, and they raise a distinct set of welfare issues. Reduction in animal welfare is intrinsic to the objective of this research and is therefore inevitable while for other applications animal suffering, when it occurs, might be seen as incidental. The decision as to whether the benefits derived from creating diseased animals outweigh the adverse animal welfare effects falls into the realm of ethics.

Public acceptance of agricultural applications of GE has generally been lower than that associated with medical applications of this technology (e.g., recombinant insulin is used routinely by people with diabetes), and public acceptance may be even more of an issue when considering animal agricultural applications of this technology. In a 2012 survey commissioned by the International Food Information Council (see www.foodinsight.org/Content/5438/FINAL%20Executive%20Summary%205-8-12.pdf), about one-third (33%) of the US respondents were somewhat or very favorable towards animal biotechnology and slightly more than one-quarter (26%) were somewhat or very unfavorable. The primary reasons consumers give for being "not favorable" (i.e., somewhat or very unfavorable or neutral) toward animal biotechnology relate to lack of information and not understanding the benefits of animal biotechnology: More than half (55%) of not favorable consumers chose "I don't have enough information" about animal biotechnology as their primary reason, while 42% cited "I don't understand the benefits of using biotechnology with animals." Ironically, the development of GE animals with direct consumer benefits is unlikely to occur if developers are concerned about public acceptance – somewhat of a "Catch-22" situation.

Paradoxically it often seems that the arguments for and against GE animals overlap. Groups opposed to the technology argue that the risks GE animals pose to food safety, animal health, and the environment are too great to allow the technology to move forward. Proponents of the technology see the potential benefits for GE animals to produce safer food, improve animal health, and reduced environmental impact as too great to forgo the use of this technology in animal agriculture production systems. As with many complex issues there is no right or wrong answer. Polarizing the issue of GE and cloned animals into "all is permitted" or "nothing is permitted" prevents rational social progress on the issue. There are both benefits and risks associated with all technologies. Effective and responsible communication among scientific, community, industry and government stakeholders is essential to reach a societal consensus regarding the appropriate use of these technologies.

Summary

Animal biotechnology is a general term that encompasses older, well-accepted technologies for the genetic improvement of animals such as selective breeding and artificial insemination, and also the more recent "modern" biotechnologies of cloning and genetic engineering. Cloning entails making a genetically identical copy of an individual, whereas genetic engineering involves the use of rDNA to intentionally make changes in the genetic makeup of an individual. There are a number of different techniques that can be used to make cloned and genetically engineered animals, and the optimal approach will vary depending upon the desired outcome. The coordinated use of the two techniques simultaneously can greatly improve the efficiency of producing genetically engineered animals. The genetic modification of animals using any technique is associated with animal welfare, ethical, and moral concerns. Opinions about the appropriate use of cloning and genetic engineering vary greatly. Many countries are currently evaluating both the benefits and risks associated with these technologies, and wrestling to come to a societal consensus as to the appropriate use of these technologies when it comes to genetically modifying animals.

Further reading

National Academies' National Research Council. *Animal Biotechnology: Science Based Concerns.* The National Academy Press, Washington, DC (2002).

Straughen, R. *Ethics,* Morality and Animal Biotechnology The Biotechnology and Biological Science Research Council (BBSRC), UK (2000).

Van Eenennaam, A. L. *Animal Biotechnology.* 30 minute Peer-reviewed video. Animal biotechnology encompasses a broad range of techniques for the genetic improvement of domesticated animal species including selective breeding, artificial insemination, cloning, and genetic engineering. Learn about both biomedical and agricultural applications of animal biotechnology and some of the science-based and ethical concerns that are engendered by certain applications: www.youtube.com/watch?v=qClvAuwaf-o (2008).

Van Eenennaam, A. L., Hallerman, E. M., and Muir, W.M., *The Science and Regulation of Food from Genetically Engineered Animals.* Council for Agricultural Science and Technology (CAST) Commentary QTA2011-2. CAST, Ames, Iowa (2011).

References

Asher, L., Diesel, G., Summers, J.F., McGreevy, P.D., and Collins, L.M. Inherited defects in pedigree dogs. Part 1: Disorders related to breed standards. *Veterinary Journal* 182, 402–411 (2009).

Betts, D.H., Bordignon, V., and King, W.A. Reprogramming of telomerase activity and rebuilding of telomere length in cloned cattle. *Proceedings of the National Academy of Sciences of the United States of America* 98, 1077–1082 (2001).

Brophy, B., Smolenski, G., Wheeler, T., et al. Cloned transgenic cattle produce milk with higher levels of beta-casein and kappa-casein. *Nature Biotechnology* 21, 157–62 (2003).

Chen, M.Y., Tu, C.F., Huang, S.Y., et al. Augmentation of thermotolerance in primary skin fibroblasts from a transgenic pig overexpressing the porcine HSP70.2. *Asian-Australasian Journal of Animal Sciences* 18, 107–112 (2005).

Clements, J.E., Wall, R.J., Narayan, O., et al. Development of Transgenic Sheep That Express the Visna Virus Envelope Gene. *Virology* 200, 370–380 (1994).

Couldrey, C., Wells, D.N., and Lee, R.S.F. DNA Methylation Patterns Are Appropriately Established in the Sperm of Bulls Generated by Somatic Cell Nuclear Transfer. *Cellular Reprogramming* 13, 171–177 (2011).

Damak, S., Su, H.Y., Jay, N.P., and Bullock, D.W. Improved wool production in transgenic sheep expressing insulin-like growth factor 1. *Bio-Technology* 14, 185–188 (1996).

Delgado, C.L. Rising consumption of meat and milk in developing countries has created a new food revolution. *Journal of Nutrition* 133, 3907s–3910s (2003).

Denning, C., Burl, S., Ainslie, A., et al. Deletion of the alpha(1,3)galactosyl transferase (GGTA1) gene and the prion protein (PrP) gene in sheep. *Nature Biotechnology* 19, 559–562 (2001).

Du, S.J., Gong, Z.Y., Fletcher, G.L., et al. Growth Enhancement in Transgenic Atlantic Salmon by the Use of an All Fish Chimeric Growth-Hormone Gene Construct. *Bio-Technology* 10, 176–181 (1992).

Foote, R.H. The history of artificial insemination: Selected notes and notables. *Journal of Animal Science* 80, 1–10 (2002).

Garrels, W., Ivics, Z., and Kues, W.A. Precision genetic engineering in large mammals. *Trends in Biotechnology* 30, 386–93 (2012).

Golovan, S.P., Meidinger, R.G., Ajakaiye, A., et al. Pigs expressing salivary phytase produce low-phosphorus manure. *Nature Biotechnology* 19, 741–745 (2001).

Gordon, J.F. Broilers fifty year-old meat industry presents outstanding picture of specialization. in *American Poultry History 1923–1973,* Hanke, O.A., Skinner, J.L., and Florea, J.H. (eds). American Printing and Publishing Inc., Madison, WI (1974).

Guthrie, H.D., Wall, R.J., Pursel, V.G., et al. Follicular expression of a human beta-cell leukaemia/lymphoma-2 (Bcl-2) transgene does not decrease atresia or increase ovulation rate in swine. *Reprod Fertil Dev* 17, 457–66 (2005).

Havenstein, G., Ferket, P., and Qureshi, M. Growth, livability, and feed conversion of 1957 versus 2001 broilers when fed representative 1957 and 2001 broiler diets. *Poultry Science* 82, 1500–1508 (2003).

Hill, W.G. and Kirkpatrick, M. What animal breeding has taught us about evolution. *Annual Review of Ecology, Evolution, and Systematics* 41, 1–19 (2010).

Jaenisch, R. and Mintz, B. Simian virus 40 DNA sequences in DNA of healthy adult mice derived from preimplantation blastocysts injected with viral DNA. *Proceedings of the National Academy of Sciences of the United States of America* 71, 1250–1254 (1974).

Konishi, K., Yonai, M., Kaneyama, K., et al. Relationships of Survival Time, Productivity and Cause of Death with Telomere Lengths of Cows Produced by Somatic Cell Nuclear Transfer. *Journal of Reproduction and Development* 57, 572–578 (2011).

Kues, W.A. and Niemann, H. The contribution of farm animals to human health. *Trends in Biotechnology* 22, 286–294 (2004).

Kuroiwa, Y., Kasinathan, P., Choi, Y.J., et al. Cloned transchromosomic calves producing human immunoglobulin. *Nature Biotechnology* 20, 889–894 (2002).

Lai, L.X., Jang, J.X., Li, R., et al. Generation of cloned transgenic pigs rich in omega-3 fatty acids. *Nature Biotechnology* 24, 435–436 (2006).

Laible, G. Enhancing livestock through genetic engineering – Recent advances and future prospects. *Comparative Immunology, Microbiology and Infectious Diseases* 32, 123–137 (2009).

Lo, D., Pursel, V., Linton, P.J., et al. Expression of mouse Iga by transgenic mice, pigs and sheep. *European Journal of Immunology* 21, 1001–1006 (1991).

Lyall, J., Irvine, R.M., Sherman, A., et al. Suppression of Avian Influenza Transmission in Genetically Modified Chickens. *Science* 331, 223–226 (2011).

Maga, E.A., Shoemaker, C.F., Rowe, J.D., et al. Production and processing of milk from transgenic goats expressing human lysozyme in the mammary gland. *Journal of Dairy Science* 89, 518–524 (2006).

Mitchell, A.D. and Wall, R.J. Body composition of transgenic pigs expressing the myostatin pro domain. *Journal of Animal Science* 86 (2008).

Miyashita, N., Kubo, Y., Yonai, M., et al. Cloned Cows with Short Telomeres Deliver Healthy Offspring with Normal-length Telomeres. *Journal of Reproduction and Development* 57, 636–642 (2011).

Muller, M., Brenig, B., Winnacker, E.L. and Brem, G. Transgenic pigs carrying CDNA copies encoding the murine Mx1 protein which confers resistance to influenza-virus infection. *Gene* 121, 263–270 (1992).

Niemann, H. and Lucas-Hahn, A. Somatic Cell Nuclear Transfer Cloning: Practical Applications and Current Legislation. *Reproduction in Domestic Animals* 47, 2–10 (2012).

Niemann, H., Kuhla, B. and Flachowsky, G. Perspectives for feed-efficient animal production. *Journal of Animal Science* 89, 4344–4363 (2011).

Norman, H.D. and Walsh, M.K. Performance of dairy cattle clones and evaluation of their milk composition. *Cloning and Stem Cells* 6, 157–164 (2004).

Nottle, M.B., Nagashima, H., Verma, P.J., et al. Production and analysis of transgenic pigs containing a metallothionein porcine growth hormone gene construct. in *Transgenic Animals in Agriculture,* Murray, J.D., Oberbauer, A.M., and McGloughlin, M.M. (eds) pp. 145–156. CABI Publ., New York, USA (1999).

Panarace, M., Agüero, J.I., Garrote, M., et al. How healthy are clones and their progeny: 5 years of field experience. *Theriogenology* 67, 142–151 (2007).

Pursel, V.G. Expression of insulin-like growth factor-I in skeletal muscle of transgenic pigs, in *Transgenic Animals in Agriculture*. Murray, J.D., Oberbauer, A.M. and McGloughlin, M.M. (eds) 131–144. CABI Publ., New York, USA (1999).

Pursel, V.G., Pinkert, C.A., Miller, K.F., et al. Genetic engineering of livestock. *Science* 244, 1281–8 (1989).

Pursel, V.G., Sutrave, P., Wall, R.J., Kelly, A.M., and Hughes, S.H. Transfer of c-SKI gene into swine to enhance muscle development. *Theriogenology* 37, 278 (1992).

Reh, W.A., Maga, E.A., Collette, N.M.B., et al. Hot topic: Using a stearoyl-CoA desaturase transgene to alter milk fatty acid composition. *Journal of Dairy Science* 87, 3510–3514 (2004).

Richt, J.A., Kasinathan, P., Hamir, A.N., et al. Production of cattle lacking prion protein. *Nature Biotechnology* 25, 132–138 (2007).

Saeki, K. Functional expression of a Delta12 fatty acid desaturase gene from spinach in transgenic pigs. *Proc Natl Acad Sci U S A* 101, 6361–6 (2004).

Schiermeier, Q. Fish farms' threat to salmon stocks exposed. *Nature* 425, 753 (2003).

Schnieke, A.E., Kind, A.J., Maycock, K., et al. Human factor IX transgenic sheep produced by transfer of nuclei from transfected fetal fibroblasts. *Science* 278, 2130–2133 (1997).

Shiels, P.G., Kind, A.J., Campbell, K.H., et al. Analysis of telomere lengths in cloned sheep. *Nature* 399, 316–317 (1999).

Thomas, K.R. and Capecchi, M.R. Site-directed mutagenesis by gene targeting in mouse embryo-derived stem cells. *Cell* 51, 503–512 (1987).

Tian, X.C., Xu, J. and Yang, X.Z. Normal telomere lengths found in cloned cattle. *Nature Genetics* 26, 272–273 (2000).

Wakayama, T., Shinkai, Y., Tamashiro, K.L., et al. Cloning of mice to six generations. *Nature* 407, 318–319 (2000).

Wall, R.J., Powell, A.M., Paape, M.J., et al. Genetically enhanced cows resist intramammary Staphylococcus aureus infection. *Nature Biotechnology* 23, 445–451 (2005).

Wells, D.N. Animal cloning: problems and prospects. *Revue Scientifique Et Technique-Office International Des Epizooties* 24, 251–264 (2005).

Wheeler, M.B., Bleck, G.T. and Donovan, S.M. Transgenic alteration of sow milk to improve piglet growth and health. *Reprod Suppl* 58, 313–24 (2001).

Whyte, J.J. and Prather, R.S. Genetic Modifications of pigs for medicine and agriculture. *Molecular Reproduction and Development* 78, 879–891 (2011).

Willadsen, S.M. A method for culture of micromanipulated sheep embryos and its use to produce monozygotic twins. *Nature* 277, 298–300 (1979).

Wilmut, I., Schnieke, A.E., McWhir, J., Kind, A.J. and Campbell, K.H.S. Viable offspring derived from fetal and adult mammalian cells. *Nature* 385, 810–813 (1997).

Yang, B.-C., Lee, N.-J., Im, G.-S., et al. A diet of somatic cell nuclear transfer cloned-cattle meat produced no toxic effects on behavioral or reproductive characteristics of F1 rats derived from dams fed on cloned-cattle meat. *Birth Defects Research Part B: Developmental and Reproductive Toxicology* 92, 224–230 (2011).

Yang, X., Tian, X.C., Kubota, C., et al. Risk assessment of meat and milk from cloned animals. *Nature Biotechnology* 25, 77–83 (2007).

Review questions

1. Breeds of dogs are all derived from a common wolf ancestor by selective breeding – are dogs genetically modified? If so, should their breeding be regulated by the FDA? Why or why not?

2. What is the difference between cloning and GE?

3. How can SCNT cloning be used to help improve the efficiency of GE?

4. Can you describe a use/application of GE animals that you consider to be ethically acceptable? How would you discuss your idea with someone who is morally opposed to GE engineering?

5. In 1918, an avian influenza epidemic killed more than 20 million people. If GE could be used develop influenza-resistant poultry, do you think chicken and eggs derived from these birds would be accepted by consumers?

6. The science-based regulatory review process undertaken by the FDA is designed to provide a predictable science-based framework that will ensure the safety and safe use of GE animals. Moral, ethical and broader social issues are not included in its review process. How should these issues be addressed in deciding which applications of animal biotechnology are acceptable?

7. How should society go about making decisions on technologies that are considered to be intrinsically or morally wrong by some members of society, and highly beneficial by other members? Can you think of other technologies that have faced this predicament, and how did society address this dilemma?

29 Intellectual Property Rights and Animal Genetic Resources

Jennifer Long[1] and Max F. Rothschild[2]
[1]Bureau for Food Security, USAID, Washington, DC, USA
[2]Department of Animal Science Iowa State University, IA, USA

Introduction

The rapid pace of developments in the fields of animal genetics and genomics, and the recent and emerging technologies related to animal breeding, make the study of intellectual property (IP) issues an increasingly important area for students of animal genetics. This chapter will describe what intellectual property is, explore case studies to illustrate these concepts in action, and summarize the broader international questions related to the movement and use of animal genetic resources. A number of books have been written on intellectual property and are referenced in the literature citations at the end of this chapter for further study.

Old McDonald's Farm meets Dolly (and her lawyer)

Since the domestication of livestock and the development of breeds, genetic pedigrees and breeding societies have enabled breeders to maintain some element of control over elite bloodlines and to ensure they recoup their investments (Rothschild and Newman, 2002). As the methods to carry out animal breeding have become more sophisticated, as with the advent of biotechnology, property protection norms have been co-evolving. To this end, intellectual property protections are an increasingly important feature in the animal breeding landscape and are a requisite area of knowledge for students in the field. To explore these issues, this chapter will begin with an explanation of what intellectual property is and how it has been applied, particularly since 1980 when the first patent on a living organism was authorized in the US by the US Supreme Court, to innovations in animal genetics and genomics research. The chapter will conclude with two case studies.

What is intellectual property?

Intellectual property relates to the legal rights resulting from intellectual activity in the industrial, scientific, literary, and artistic fields (WIPO, 2004). The first major international agreement dealing with intellectual property rights was the Paris Convention for the Protection of Industrial Property, signed in Paris on March 2, 1883. The Convention has been revised seven times since it went into force, and it is still in force today. The 1967 Act of the Convention established the World Intellectual Property Organization (WIPO) and set forth the parameters for what can be defined as intellectual property. These include scientific works, inventions in all fields of human endeavor, and scientific discoveries. As outlined in the Convention, intellectual property can take different forms, such as copyright protections, patents, and trademarks.

Forms of intellectual property

Different forms of intellectual property protection exist, emerging from the need to protect various types of expression, discoveries, and innovation (Table 29.1).

The following are key areas relevant to animal agriculture. **Copyright** laws involve the protection of works of creative expression, but can include written product descriptions or product label directions (Smiler and Erbisch, 2004). Copyright protections relate to the form in which information is communicated, not the ideas conveyed in such communications (WIPO, 2004). Most students are aware of copyrights on textbooks, articles, designs and even recorded music. **Patents** are used in the United States to protect genes and DNA markers and their uses in selection and breeding, whole animals (such as the "Harvard mouse") or a transgenic livestock, as well as processes associated with animal breeding, genetics, and genomics such as diagnostic assays. **Trademarks** are the signs, symbols, phrases or other marks that "individualize the goods of a given enterprise and distinguishes them from the goods of its competitors" (WIPO, 2004). **Trade secrets** are private information that can be anything from a recipe (e.g., the formula for Coca Cola®) to a list of customers. As long as the owner of the trade secret actively works to maintain the secrecy of the information, there are certain legal protections afforded to the owner. Trade secrets are important in animal agriculture, particularly in poultry and swine. For example, access to elite pure parent lines of poultry is carefully

Molecular and Quantitative Animal Genetics, First Edition. Edited by Hasan Khatib.

Table 29.1 Forms of intellectual property in the United States most relevant to animal genetics research.

Form of Intellectual Property	Definition	Example
Copyright	Protections accorded literary, artistic, and scientific works	Software, web page advertisements
Patent	Protection that excludes others from making, using, offering for sale, selling the invention or importing for limited duration in exchange for the invention's public disclosure when the patent is granted	Use in selection of DNA sequence for gene with known function
Trademark	Protection for words, names, symbols, etc. that individualize products and distinguish them from competitors	Brand names with design logos, e.g., Newsham Choice Genetics®
Trade secrets	Information that companies keep secret to give them an advantage over their competitors	Inbred poultry lines, breeding value formulas

Source: US Patent & Trademark Office, www.uspto.gov

controlled, and only the hybrid progeny are sold, thus, the parent lines are considered trade secrets (Narrod and Fuglie, 2000). Given the significance of patents in animal agriculture, additional information on patents is provided next.

Patents

A patent is a document issued by a government, or a regional office acting for several countries, which describes an invention and provides the legal basis for ownership of the invention.

> While the owner is not given a statutory right to practice his invention, he is given a statutory right to prevent others from ... exploiting his invention, which is frequently referred to as a right to exclude others from making, using or selling the invention.
>
> *(WIPO, 2004).*

In return for this exclusive right, which is time limited (usually 20 years from the patent application filing date), the inventor discloses the invention so that, the broader public can access and benefit from the invention. While many examples of animal genetic patents exist, they are too numerous to list here. However, some obvious examples include patents for gene markers and their use to select animals for a specific trait, the use of specialized statistical formulas to select or genetically evaluate animals, and transgenic animals developed for specific purposes, such as for research models to study human diseases.

To qualify for patent protection in the US, inventions must be considered **patentable subject matter**, and also be novel, useful, and non-obvious. Similar criteria apply in all countries that issue patents, but the scope of what is patentable subject matter varies. The issue of whether animals represent patentable subject matter is a question countries have engaged both domestically and in various international fora. In the United States, the door to patenting living organisms was opened with the landmark case, *Diamond vs Chakrabarty* in 1980. In this case, a bacterium with applications for bioremediation, to breakdown components of crude oil, and the US Supreme Court majority opinion deemed that the bacterium was "not a hitherto unknown natural phenomenon, but to a nonnaturally occurring manufacture or composition of matter – a product of human ingenuity." (*Diamond vs Chakrabarty*, 1980). Subsequently, in 1987, the Commissioner of Patents and Trademarks at the US Patent and Trademark Office issued a notice that the USPTO "would now consider nonnaturally occurring, nonhuman multicellular living organisms, including animals, to be patentable subject matter within the scope of 35 U.S.C. 101" (USPTOa, 1987). Guidelines on the question of the patentability of living organisms have been developed and clarify that, while animals and other multi-celled organisms are acceptable patentable subject matter, "human organisms are not patent-eligible subject matter" (USPTOb, 2012). While the Trade-Related Aspects of Intellectual Property Rights (TRIPS) agreement under the World Trade Organization's Uruguay Round Agreements requires parties to allow patenting of microorganisms, patenting of animals is not required and can be determined at the national level. For a review of various countries positions on patenting of animals, please see Blattman et al. (2002).

Utility, novelty, and non-obviousness

There are three primary criteria for determining whether appropriate subject matter can be patented. An invention must be useful (**utility**), **novel**, and **non-obvious**.

Usefulness, or utility, requirements are significant considerations in genetics research. As noted in Lesser (2002), early patent submissions for expressed sequence tags identified from the Human Genome Project were rejected on the basis that there was no known use for the gene.

An invention is novel only if the information is not publicly disclosed before the submission of the patent application. The stringency of this requirement varies among countries, but most require absolutely no prior public disclosure. Once a patent is filed in one country that is a party to the Paris Convention, applicants generally have one year to file patent applications in other Convention countries, and the original patent application is not considered a public disclosure that would preclude patentability. This is known as the **right of priority**. However, the patent submissions in all countries must be exactly the same content as the original submission. This prevents an inventor from continuously refining the invention and getting a longer period of protection through subsequent patent submissions in other countries (WIPO, 2004).

Non-obviousness requires that the "invention is not expected by a person with ordinary skill in the art." This requirement has some latitude, but clearly the invention should not be easily anticipated based on existing technology

Here a patent, there (not) a patent

Intellectual property is only protected in the jurisdictions where protections have been sought and granted to the owner of the invention. That is to say, a patent obtained in the United States does not confer protections beyond the jurisdiction of the United States. To ensure effective protection, inventors must seek intellectual property protections in all relevant countries (that is, those countries in which the invention might be used and is economically important). If the additional countries where protections are sought are members of the Paris Convention, then foreign applicants have a right to obtain the same type of protection as that available to a national of the country where protection is being sought. This is known as the **right to national treatment**. A number of Treaties and Conventions have been developed since the late nineteenth century to refine the scope for intellectual property protections and to clarify methods and processes for seeking protections in multiple countries. These legal agreements (Table 29.2), from the Paris Convention to the Patent Cooperation Treaty, have helped to address the diversity of laws that were developed to protect intellectual property and to facilitate the development of uniform legislation among countries party to the agreements.

One problem that results from the requirement that patents be obtained in all countries where manufacture or use might occur is that the costs for such patent protection are extremely high. Hence, inventors often choose to protect their inventions only in those countries that have the most opportunity to yield significant economic returns for their inventions.

Though countries have agreed to the use of patent protections through the legal agreements noted previously, certain exceptions are allowed based on the three-step test per WTO TRIPS rules (Misati and Adachi, 2010). Exceptions are allowed if they are (1) limited, (2) do not unreasonably conflict with normal exploitation, and (3) do not unreasonably prejudice the legitimate interests of the patent holder (Misati and Adachi, 2010). One important exception is the "research exception" and interpretation varies greatly among countries. The US allows limited non-commercial use of patented innovations, while other countries, such as Brazil, provide for broader exceptions for research use of patented innovations. More on different countries' approaches to implementing the research exception can be found in Misati and Adachi (2010).

Forms of payment or remuneration

Once intellectual property protections have been obtained, the owner can choose to license the innovation to one or more third parties for manufacture or use. It is this right, particularly for those intellectual property owners who are not well positioned to make use of their innovation directly, that serves to incentivize inventors. This would enable the owner of the invention to be remunerated for the innovation without having to manufacture or apply the innovation by him/herself. Forms of direct monetary compensation can include lump sum payments, royalties, or fees (WIPO, 2004), or any form of payment agreed upon by both the patent owner and the licensee. Lump sum payments are calculated in advance and can be paid all at once or in installments, while royalties are recurring payments that can be calculated based on economic results of use of the intellectual property (e.g., production units, product sales) (WIPO, 2004). There are risks to either form of payment to the intellectual property owner – as a lump sum could underestimate the actual commercial value of the intellectual property (an instance where a royalty based remuneration structure would have yielded greater returns) while a royalty-based remuneration structure could be a disappointment if sales are low (e.g., assuming sales are the basis of the royalty payment). To the licensee, the record keeping burdens of the need to document use or sales of licensed products may result in a preference for a lump sum payment. Non-monetary and indirect forms of payment also exist, such as dividends, in-kind exchange (e.g., market data for access to the innovation), and cost sharing. More detailed description of these forms of compensation can be found in the WIPO Intellectual Property Handbook (WIPO, 2004).

Table 29.2 Summary of international conventions and treaties related to intellectual property and animal genetics.

Convention/Treaty	Entry into force	Key provisions/objectives
The Paris Convention for the Protection of Industrial Property	1884	Right to national treatment, Right of priority, Common rules for establishing rights and responsibilities, Administrative framework
Berne Convention for the Protection of Literary and Artistic Works	1886	Right to national treatment, No formal registration process, Independence of protection (foreign protection possible even if not protected in country of origin)
Madrid Agreement Concerning the International Registration of Marks	1891	International registration affords protection in all countries party to the agreement, Each country has right to refuse recognizing registration
Patent Cooperation Treaty	1978	Facilitates international cooperation for filing, searching, and patent application examination
Budapest Treaty on the International Recognition of the Deposit of Microorganisms for the Purposes of Patent Procedure	1980	Recognizes patented microorganisms regardless of country in which they are deposited (e.g., culture collection)
Agreement on Trade-Related Aspects of Intellectual Property Rights ("TRIPS")	1995	Strengthens other IP conventions and links to trade obligations, Includes dispute settlement provisions
Patent Law Treaty	2005	Streamlines national and regional patent applications and patents; Major outcome, Reduces inadvertent loss of rights due to procedural factors
Nagoya Protocol, Convention on Biological Diversity	2010	Defines Access & Benefit Sharing (ABS) obligations and procedures

Source: WIPO Intellectual Property Handbook. 2008 (reprint). World Intellectual Property Organization Publication No. 489 (E). Available at: www.wipo.int/about-ip/en/iprm/

Case studies

Despite the seemingly straightforward nature of requirements for filing and being awarded a patent, a number of patent cases within the field of animal genetics and breeding merit study to understand how various patent offices apply these criteria in patent examination. Next are two case studies which illustrate a number of issues related to patenting in the animal breeding and genomics fields. These cases raise different issues and have very different outcomes that ultimately deal with the challenges of the current patent system given the highly technical and fast-pace of change in the field of breeding and genetics.

Test day models in dairy cattle

Enhancing the genetic potential of dairy herds through a systematic analysis of data on factors including average lactation yields and pedigree has been practiced widely in the US in dairy cattle genetics since the early part of the twentieth century. One such method was patented and has been the subject of intense debate in North America and Europe. The Method of Bovine Herd Management was invented by a Cornell University faculty member, Robert Everett, and the patent assigned to the Cornell Research Foundation. While the patent still stands in Canada and the United States (where it expired in 2013), it was revoked by the European Patent Office in 2007 on the primary grounds that there was no inventive step (EPO, 2007). A detailed analysis of this case can be found in Schaeffer (2002), and is distilled here from that source. We will explore the outstanding questions next.

In the US, dairy cow output is often measured once a month and then the amount for the entire month is estimated from the yield on the single test day knowing the age of the cow and the stage of lactation. This routine process is at the heart of the US patent. The US patent describes a process for improving the milk productivity at the herd level, through a process for collecting and collating routine test day data, a mathematical model that models milk productivity at the individual cow level, and the use of the data by dairy producers to inform management and breeding decisions. With regards to the novelty of collecting milk productivity data, this has been a norm among US dairy producers since the early part of the twentieth century. Test days involve a formalized process for collecting milk samples from individual cows on a routine basis and measuring various parameters (such as total milk yield, and yield of fat and protein content) and recording additional information related to each cow (e.g., age, stage of lactation, etc.) Geneticists used these test day data to develop 305-day lactation yields of cows to predict the genetic merit of daughters of a given bull and hence to inform sire selection. Hence, the novelty of gathering test day data at an individual cow level related to milk productivity is seriously questioned by many in the US, Canadian, and European dairy industries.

Computational power and statistical methodology advanced significantly from 1935 through the 1980s. As described in Schaeffer (2002), other models were developed over this time using test day data to estimate milk yields and to determine the genetic variation to inform sire and dam selection. However, the Everett patent asserts that its mathematical model is distinct in that it accounts for 40% of the environmental variation in the milk yield. As models are tested, revised, and updated with new types of information, such as genomic data, it is possible to develop a model with greater ability to account for non-genetic effects or residual effects in the model. As such, it seems that this method does not satisfy the basic requirement of being novel, as it is simply a variation on a standard theme used previously in the industry.

Though the lack of novelty of collecting test day data and developing a mathematical model could have been possible grounds for revoking the patent, it was the issue of how the information was used that led to the European Patent Office's action to revoke the patent in 2007. The Boards of Appeal of the European Patent Office indicated that the basis of their review and decision related to the question of whether the mathematical model was a technical solution to a technical problem. They explained that the model's information would allow for modifications to herd management (e.g., culling less productive cows, mating high performing sires with more cows within the herd, etc.), "but the method does not in itself result in a change in the amount of milk" produced (EPO, 2007). Thus, they deemed that the method ultimately only provided information "for a decision making process in operations" (EPO, 2007), and was therefore not a technical solution *per se*.

The consequences of this decision have not affected the standing of the patent in the US or Canada, and as a result, the US dairy industry has largely avoided use of test day data-driven methods to inform herd management due to the logistical challenges associated with coordinating royalty payments (Averdunk and Gotz, 2004).

Genetic marker for marbling in beef cattle

This case study examines the patenting of a genetic marker for marbling in beef cattle. In this example, distilled from Barendse (2002), the researchers identified a polymorphism in the thyroglobulin gene (TG5) that is used to predict marbling in beef. Marbling, or intramuscular fat, is a characteristic with significant commercial value, but is difficult to breed for as it is a quantitative trait and because more definitive forms of assessment are performed on the carcasses of progeny, making testing expensive and time consuming. Developing a series of tests for genetic markers associated with the loci that contribute to improved marbling would expedite breeding progress in this trait. To that end, the researchers in this example evaluated a number of polymorphisms and their associations with marbling scores. Ultimately, breeding for improved marbling will emerge when tests are available for a number of loci and there is greater understanding of how each of these loci, and their interactions with environmental factors (e.g., feeding regimen) contribute to marbling. However, incremental progress can be made when using even only a few markers, called marker assisted selection, given the costs associated with alternative approaches to breeding for this trait. The researchers identified *TG5*, a single nucleotide polymorphism (SNP) in the thyroglobulin gene (*TG*). As thyroid hormones are associated with fat cell growth and differentiation (Barendse, 2002), the researchers asserted the association made sense. Interestingly, it must be distinguished that the TG5 test is used to predict marbling (a physical characteristic of an animal), and not to predict other polymorphisms (a causative mutation perhaps) in the *TG* gene itself. If it were the latter, it is possible that the discovery would infringe on a patent already issued (US Patent 5,612,179) that "has a claim to the use of any DNA test based on an intronic sequence ... to predict genotype at a 'non-intronic' sequence" (Barendse, 2002). However, because the polymorphism predicts the physical characteristics of the animal, and not a DNA sequence, it can be considered to be a different invention. But this raises a series of questions about how information can be used as we learn more about genetics and genomics. Every year, more data are being generated, and those data are more efficiently analyzed, enabling us to test for associations of genetic markers with phenotypes, markers associated with known genes and gene discovery, and ultimately these analyses help us to untangle complex genome by environment interactions. As prior art is examined to establish whether something can be patented, as each field matures and greater understanding underpins subsequent innovation and research, overly broad claims will not be allowed by the US Patent and Trademark Office. Thus, as greater understanding is achieved, by default, the US patent system itself naturally shifts the threshold for defining something as novel, useful, and non-obvious. However, patents granted in previous eras of genetic/genomics research may limit the latitude to conduct research that builds on the wealth of information constantly emerging in the field of animal genetics and genomics, not to mention building on research in other complementary fields.

This case study demonstrates that as the field of animal genetics evolves, new information is more readily available and the costs of collecting and analyzing that information decrease, such as the case with new high density single nucleotide polymorphism (SNP) chips now used. Previous assertions of intellectual property rights within the field might not only hamper subsequent studies in ways that are predictable, but because of the rapid pace of change in the field, we might also be impeding future progress in ways we can't even conceive at present. The scope of patent rights granted is becoming more narrow as the body of prior art in this field increases. For example, a patent for a specific polymorphism demonstrated in one animal species, but claimed in multiple species, was once allowed but is no longer patentable. However, the legal landscape is complicated because of the number of broad patents issued in earlier periods, the increase in applications to patent DNA sequences in recent years, and the patenting of processes related to genetic selection, such as the DNA tests in the marbling case study. Given these trends, it is important for legal practitioners to have access to technical experts in the field of animal genetics to facilitate interpretation of study findings, patent examinations and patent searches. Professional opportunities for students interested in animal genetics are much broader than a career in breeding alone, and career opportunities such as being a technical expert assisting in this area of law is relevant to those interested in careers in both the public and private sectors.

E-I-E-I-O: The alphabet soup of domestic and international issues

Intellectual property issues affect those conducting research in both the public and private sectors, domestically and internationally. Specific changes to US domestic legislation, described in this section, enabled public sector researchers to assert intellectual property protections on their innovations. The evolving legal landscape on intellectual property, especially given the increase in public-private sector collaborations, has led to new developments in the international public sector agricultural research community. These issues, along with the emergence of international agreements that affect the animal genetics research community through rules on access to genetic resources, are increasingly affecting how research is conducted and will be described in this section.

Public sector research and IP – domestic and international

Intellectual property issues have become an increasingly important feature in the agricultural research landscape in recent years, for both public and private sector researchers. During the 1960s and 1970s, such as during development of Green Revolution technologies, intellectual property issues were not considered by public sector researchers. However, with the Supreme Court decision to allow the patenting of living organisms, followed by the decision of the US Patent and Trademark Office (PTO) to allow patents on living animals, intellectual property issues have become increasingly relevant to research on farm animals. Further, the US Bayh-Dole Act, passed by Congress in 1980 (35 U.S.C. 200–212), provided consistent policy guidance across all federal government agencies to enable institutions that developed innovations resulting from federally funded research, such as US universities, to pursue intellectual property protections over these innovations. To that end, US Universities set up systems to help faculty and staff patent their innovations, facilitate market access and collect remuneration through the development of technology transfer offices. Federal research agencies, such as the US Department of Agriculture's Agricultural Research Service, pursue intellectual property protection for research carried out by its scientists funded with public funds. Thus, intellectual property issues will remain important regardless of whether one is working in the public or private sectors.

Internationally, the publicly funded international agricultural research centers that make up the Consultative Group on International Agricultural Research (CGIAR) can also assert intellectual property protections over their innovations. These centers were established in the 1960s and 1970s, and individual CGIAR Centers had, over time, established their own policies and procedures for managing intellectual assets and engaging with the private sector. However, during a recent reform process with the system, the issue of IP resulting from public sector funding emerged as an area where uniform rules and procedures should be established for the Centers. As there are many international donor governments to the system, many different perspectives on the role of intellectual property and publicly funded research had to be negotiated to define principles to govern the Centers' intellectual property management practices. After a year of negotiation, donor governments to the CGIAR established the CGIAR *Principles on the Management of Intellectual Assets*, which can be found at www.cgiarfund.org.

Access to animal genetic resources

Though this chapter has focused on the downstream issues associated with animal breeding and genetics and the issues around asserting intellectual property protections, upstream issues are critically important as well. Agricultural research depends on access to information and biological materials to advance the field. Thus, international agreements that may encumber access to these materials can significantly affect the future of the field, and how new discoveries will be made.

With the arrival of the World Trade Organization TRIPS agreement, there was a growing chorus of dissent around the world about how intellectual property protections were provided for innovations made using biological materials obtained from countries that did not share in the financial benefits derived from the innovation (Safrin, 2004). This has been, on occasion, called biopiracy. The Convention on Biological Diversity (CBD), which entered into force in 1993, took up this issue, known as Access and Benefit Sharing, resulting in the Nagoya Protocol (NP) which was adopted in 2010.[1] The objective of the Nagoya Protocol is to provide a "transparent legal framework for the effective implementation of ... fair and equitable sharing of benefits arising out of the utilization of genetic resources." (CBD, 2010). The contracting Parties to the Nagoya Protocol are obligated to "take measures in relation to access to genetic resources, benefit-sharing and compliance." (CBD, 2010) As Nagoya Protocol implementation is effected at the national level, and this process is only just beginning, it is unclear how implementation will affect international research collaborations. Some have voiced concern that the Nagoya Protocol could stifle academic research through the limitations on sharing of genetic resources (Jinnah and Jungcurt, 2009). Thus, given the collaborative nature of agricultural research, and the increasingly international scope of animal breeding and genetics/genomics research, researchers and practitioners need to be aware of any obligations under the Nagoya Protocol. Given the national level of implementation, systems and procedures will likely vary by country.

Another important international body related to animal breeding and genetic resources is the Commission on Genetic Resources for Food and Agriculture (CGRFA), based at the United Nations Food and Agriculture Organizations. The Commission, composed of government officials from member countries, has been exploring how to address the Access and Benefit Sharing discussion for all areas of agricultural genetic resources. The Commission has an International Technical Working Group on Animal Genetic Resources that addresses the full range of issues associated with the movement, use and conservation of agricultural animal genetic resources. A 2007 technical conference in Interlaken, Switzerland (FAO, 2007) was the starting point for subsequent discussion within the Commission on whether and how a tailored approach to access and benefit sharing for agricultural animals could be devised in harmony with the Nagoya Protocol. Discussions continue on this point in the various fora mentioned above and action is likely to be incremental.

Given the array of international institutions where issues relating to animal genetic resources, animal breeding and genetics research, and intellectual property have been engaged, public and private sector researchers in the field of animal genetics would be well advised to stay abreast of emerging debates and issues. The challenges of doing collaborative international research in the emerging environment are not inconsequential. The "interactive spiral of increased enclosure" (Safrin, 2004) from patenting and the national assertions of sovereignty over genetic resources defined in the Convention on Biological Diversity, referred to as hyperownership (Safrin, 2004), does present challenges for agricultural researchers who are accustomed to broad, open international collaborations in their research pursuits. As these issues could affect how research in animal breeding and genetics may be conducted in the near future, there is significant value in having more researchers in the field weighing in on these international, multidisciplinary debates.

[1] All countries, except the USA, Andorra, South Sudan, and the Holy See, have ratified the Convention on Biological Diversity (CBD), though the USA has signed the Convention. Countries that have signed and/or ratified the Nagoya Protocol can be found at www.cbd.int/abs/nagoya-protocol/signatories/

Summary

Patents and trade secrets are a way of life in the current animal genetics and genomics research landscape. For students, a career in the public or private sectors will require understanding of how and when these issues are relevant. As international institutions are established for animal genetic resources, such as the emerging access and benefit sharing framework, practitioners will need to stay well informed about these changes as they are implemented through national legislation, policies and procedures. Further, as the fields of animal genetics and genomics rapidly evolve and new information is made available, the number and scope of relevant patents will likely change, and conducting research and doing business within the continually evolving legal landscape could become increasingly complex. Thus, understanding intellectual property concepts is critical for the animal breeding and genetics student. To strengthen understanding of these concepts, we have prepared questions for further study next.

Further reading

Blakeney, M. 2009. *Intellectual Property Rights and Food Security*. CABI. Cambridge University Press.

Van Doormaal, B. 2004. *Current Situation of the Cornell Patent in Canada*. Available at: www.cattlenetwork.net/docs/paris_march_2004/Van_Doormaal_Cornell_Patent%20.pdf (accessed May 14, 2014).

References

Averdunk, G. and K. Gotz. 2004. *Possible Consequences of The Cornell Test-Day Model Patent in Europe 7th Conferenza Mondiale Allevatori Razza Bruna, Verona, 3–7 Marzo 2004*. Available at: www.anarb.it/Bruna2004/inglese/Presentazioni/Presentazioni_IN.htm. (accessed May 14, 2014).

Barendse, W. 2002. Development and Commercialization of a Genetic Marker for Marbling of Beef in Cattle: a Case Study. In: *Intellectual Property Rights in Animal Breeding and Genetics*, Rothschild, M. and Newman, S. (eds). CAB International, New York, pp. 197–212.

Blattman, A. J. McCann, C. Bodkin, and J. Naumoska. 2002. Global Intellectual Property: International Developments in Animal Patents. In: *Intellectual Property Rights in Animal Breeding and Genetics*, Rothschild, M. and Newman, S. (eds). CAB International, New York, pp. 63–84.

CBD. www.CBD.INT available at: www.cbd.int/abs/about/#objective (Accessed May 14, 2014).

Diamond vs Chakraborty 1980. 447 U.S. 303, section III.

EPO. European Patent Office website. T 0365/05 (Bovine herd management /Cornell Research Foundation of 19.6.2007). Available at: www.epo.org/law-practice/case-law-appeals/recent/t050365eu1.html (Accessed May 14, 2014).

FAO. 2007. *Animal Genetic Resources Information.* Special Issue: Interlaken International Conference.

Jinnah, S. and Jungcurt, S. 2009. Global biological resources: Could access requirements stifle your research? *Science* 323:464–465.

Lesser, W. 2002. Patents, trade secrets, and other forms of intellectual property rights. In: *Intellectual Property Rights in Animal Breeding and Genetics* Rothschild, M. and Newman, S. (eds). CAB International, New York, pp. 1–15.

Misati, E. and Adachi, K. 2010. UNCTAD- ICTSD Project on IPRs and Sustainable Development. *Policy Brief* 7, March.

Narrod, C. and Fuglie, K. 2000. Private Investment in Livestock Breeding with Implications for Public Research Policy. *Agribus iness,* 16(4):457–470.

Rothschild, M. and S. Newman. 2002. *Intellectual Property Rights in Animal Breeding and Genetics.* CAB International, New York.

Safrin, S. 2004. Hyperownership in a time of biotechnological promise: the international conflict to control the building blocks of life. *American Journal of International Law*, Vol. 98, October 2004; Available at SSRN: http://ssrn.com/abstract=658421.

Schaeffer, L. 2002. Dairy cattle test day models: A case study. In: *Intellectual Property Rights in Animal Breeding and Genetics*, Rothschild, M. and Newman, S. (eds). CAB International, New York, pp. 233–246.

Smiler, B. and F.H. Erbisch. 2004. *Introduction to Intellectual Property.* In: *Intellectual Property Rights in Agricultural Biotechnology* Erbisch, F.H. and Maredia, K.M. (eds). 2nd edn. CAB International. Cambridge University Press. pp. 1–21.

USPTOa. *Animals – Patentability*, 1077 O.G. 24, April 21, 1987.

USPTOb. *Guidelines on the patentability of living subject matter.* 2105 Patentable Subject Matter – Living Subject Matter [R-9]. Available at www.uspto.gov/web/offices/pac/mpep/s2105.html (accessed May 14, 2014).

WIPO 2004. *WIPO Intellectual Property Handbook.* 2nd edn. No. 489 E. Available at: www.wipo.int/about-ip/en/iprm/ (accessed May 14, 2014).

Review questions

1. Go to the US patent database. Find two gene marker patents in pigs. Describe these animal breeding inventions and how they are used.

2. Go to the US patent database and find a patent related to cloned or transgenic farm animals.

3. If you were a company and made a discovery, explain reasons either to patent the work or to consider keeping it as a trade secret.

4. If a company uses an invention in a country not covered by existing patents, have they infringed the patents? Explain.

5. Use the Internet to find three examples of trademarks used in animal breeding and genetics.

6. Explain the difference between exclusive and nonexclusive licenses. Provide reasons why inventors might prefer or be required to do one or the other of these.

7. What does "someone skilled in the art" mean for practitioners in the field of animal genetics and genomics, and how is this important to obtaining a patent?

Index

Printed and bound by CPI Group (UK) Ltd, Croydon, CR0 4YY